Practical Hydrogeology: Concepts and Field Applications

Practical Hydrogeology: Concepts and Field Applications

Editor: William Sobol

STATES
ACADEMIC PRESS
www.statesacademicpress.com

States Academic Press,
109 South 5th Street,
Brooklyn, NY 11249, USA

Visit us on the World Wide Web at:
www.statesacademicpress.com

ISBN: 978-1-63989-432-1 (Hardback)

Cataloging-in-Publication Data

Practical hydrogeology : concepts and field applications / edited by William Sobol.
p. cm.
Includes bibliographical references and index.
ISBN 978-1-63989-432-1
1. Hydrogeology. 2. Groundwater. 3. Hydrology. I. Sobol, William.
GB1003.2 .A56 2022
551.49--dc23

Table of Contents

Preface

Hydrogeology is the study of occurrence, distribution and movement of groundwater. Groundwater contamination, conservation of supplies and water quality are the key concerns under hydrogeology. It deals with the flow of water through aquifers and other shallow porous media. The prediction of the behavior of an aquatic system with the help of an analysis of past and present observations is the primary goal of hydrogeology. It also studies the hydraulic conductivity, porosity and transmissivity of aquifers. The advancements in topographical mapping are also playing a crucial role in enhancing groundwater sustainability. From theories to research to practical applications, case studies related to all contemporary topics of relevance to hydrogeology have been included in this book. The topics included herein are of utmost significance and bound to provide incredible insights to readers. Coherent flow of topics, student-friendly language and extensive use of examples make this book an invaluable source of knowledge.

The information contained in this book is the result of intensive hard work done by researchers in this field. All due efforts have been made to make this book serve as a complete guiding source for students and researchers. The topics in this book have been comprehensively explained to help readers understand the growing trends in the field.

I would like to thank the entire group of writers who made sincere efforts in this book and my family who supported me in my efforts of working on this book. I take this opportunity to thank all those who have been a guiding force throughout my life.

Editor

1

Groundwater dynamics under water-saving irrigation and implications for sustainable water management in an oasis: Tarim River basin of western China

Z. Zhang[1], **H. Hu**[1], **F. Tian**[1,*], **X. Yao**[2], **and M. Sivapalan**[3]

[1]State Key Laboratory of Hydroscience and Engineering, Department of Hydraulic Engineering, Tsinghua University, Beijing 100084, China
[2]Xinjiang Tarim River Basin Management Bureau, Bayangol Prefecture, Xinjiang 841000, China
[3]Department of Civil and Environmental Engineering, Department of Geography and Geographic Information Science, University of Illinois at Urbana-Champaign, Urbana, IL 61801, USA
[*]now at: State Key Laboratory of Hydroscience and Engineering, Department of Hydraulic Engineering, Tsinghua University, Beijing, China

Correspondence to: F. Tian (tianfq@tsinghua.edu.cn)

Abstract. Water is essential for life. Specifically in the oases of inland arid basins, water is a critically limited resource, essential for the development of the socio-economy and the sustainability of eco-environmental systems. Due to the unique hydrological regime present in arid oases, a moderate groundwater table is the goal of sustainable water management. A shallow water table induces serious secondary salinization and collapse of agriculture, while a deep water table causes deterioration of natural vegetation. From the hydrological perspective, the exchange flux between the unsaturated vadose zone and groundwater reservoir is a critical link to understanding regional water table dynamics. This flux is substantially influenced by anthropogenic activities. In the Tarim River basin of western China, where agriculture consumes over 90 % of available water resources, the exchange flux between the unsaturated vadose zone and groundwater reservoir is influenced strongly by irrigation. Recently, mulched drip irrigation, a sophisticated water-saving irrigation method, was widely applied in the Tarim River basin, which greatly impacted the exchange flux and thus the regional groundwater dynamics. Capitalizing on recent progress in evaporation measurement techniques, we can now close the water balance and directly quantify the exchange flux at the field scale, thus gaining a better understanding of regional groundwater dynamics. In this study, comprehensive observations of water balance components in an irrigated cropland were implemented in 2012 and 2013 in a typical oasis within the Tarim River basin. The water balance analysis showed that the exchange flux and groundwater dynamics were significantly altered by the application of water-saving irrigation. The exchange flux at the groundwater table is mostly downward (310.5 mm $year^{-1}$), especially during drip irrigation period and spring flush period, while the upward flux is trivial (16.1 mm $year^{-1}$) due to the moderate groundwater table depth (annual average depth 2.9 m). Traditional secondary salinization caused by intense phreatic evaporation (fed by upward exchange flux) is alleviated. However, a new form of secondary salinization may be introduced unwittingly if there is lack of water for periodic flushing, especially when brackish water is used in the irrigation. Furthermore, the water saved via drip irrigation has been used in further growth of irrigated lands instead of supporting the ecological system. This could lead to an increased risk of eco-environmental degradation and calls for improved governance schemes. The insights gained from this study can be potentially applied to other arid inland areas (e.g., central Asia) which face similar water shortages and human development problems.

1 Introduction

According to the FAO (http://www.fao.org/docrep/t0122e/t0122e03.htm), one-third of the world's total land area is classified as arid. In these arid regions, the limits to water availability make water a critical factor in socio-economic development (Hanasaki et al., 2013). Agriculture in arid regions greatly relies on irrigation and in addition the delicate natural ecosystem is also vulnerable due to the exacerbation of water scarcity by human extraction (Shen and Chen, 2010). Recently, water scarcity has become more acute all over the world due to the effects of both climate change and population growth, further aggravating the water problems, especially in arid regions (Xu et al., 2010). Water-saving irrigation, as one approach to mediate water conflicts between humans and nature, has been popularized in many arid regions (Christen et al., 2007; Ibragimov et al., 2007; Scanlon et al., 2010). However, it is still not clear how water-saving irrigation alters overall water balance dynamics and, in the long term, how it affects the evolution of human–water systems. There is also concern about whether sustainable development can be maintained by the application of water-saving irrigation, and what additional steps we should take to implement better water management in the future. The Tarim River basin in western China provides an excellent case study to generate insights into all of the above questions.

The Tarim River basin (TRB) is the largest inland basin of China. It is also a typical hyper-arid basin in central Asia, itself one of the biggest arid zones of the world. Within the TRB, water resources are quite scarce due to its hyper-arid climate, characterized by limited precipitation ($<50\,mm\,year^{-1}$) and extremely high potential evaporation ($>2000\,mm\,year^{-1}$) (Xu et al., 2010). As the home to an ancient human civilization dating back thousands of years, water has always played a key role in the growth and evolution of human settlements within the TRB (Xu et al., 2004). Nowadays, about 10 million people live in the oases located along the river banks and in the alluvial plains downstream of the TRB (Zhang et al., 2010).

The TRB remains an agricultural society, with agriculture accounting for more than 90 % of all water consumption (Shen and Chen, 2010). However, oases continue to play an important role in the societal and economic development within the TRB. Since the 1950s, the hydrologic system has been significantly disturbed by the expansion of irrigation in this region (Xu et al., 2007; Zhang et al., 2012). An ambitious plan for wasteland reclamation within the TRB was conducted by the Chinese central government, and as a result the areal extent of the oases began to sharply increase (Bruelheide et al., 2003). Water consumption also rapidly increased, associated with the extensive wasteland reclamation and increased human population. More water had been extracted from the river for flood irrigation, a traditional form of irrigation, which resulted in negative long-term consequences. A continuous drying-up of the main stream of the Tarim River has been taking place since 1972 and natural vegetation has degraded due to the drawdown of groundwater table within the riparian areas (Shen and Chen, 2010). Soil salinization induced by the intense phreatic evaporation under the shallow groundwater tables within the irrigated districts greatly hindered further agricultural development (Tang et al., 2007a). Large-scale irrigation has also significantly altered the groundwater table dynamics within this region.

Recently, water-saving irrigation has been popularized within the TRB to enhance the irrigation efficiency and mitigate soil salinization in the irrigated farmlands where the groundwater table is quite shallow (Dou et al., 2011; Ma et al., 2010; Wang et al., 2011; Zhang et al., 2014a). Mulched drip irrigation (MDI), a new micro-irrigation approach incorporating surface drip irrigation and the film mulching technique, has been widely adopted by local farmers. It is therefore crucial to understand the impacts of irrigation, including the transformation of irrigation methods from flood irrigation to water-saving drip irrigation, on the regional groundwater table dynamics, which is bound to further influence agricultural development and ecological system health within TRB.

The key control on regional water table dynamics is the exchange flux (EF), which is defined as the water flux into or out of the lower boundary of the soil control volume representing the vadose zone (Tang et al., 2007a; Timms et al., 2012). Since EF is substantially influenced by irrigation and the conversion from flood irrigation to drip irrigation, any study of groundwater table dynamics in this region requires that EF be appropriately quantified under these different circumstances.

Water balance analysis is a widely adopted approach to quantify components of the water balance equation that cannot otherwise be directly measured (Evett et al., 2012; Tang et al., 2007b; Tennakoon and Milroy, 2003; Wilson et al., 2001). The method lends itself to the estimation of EF between the vadose zone and the groundwater table. In general, EF can be directly quantified by water balance analysis, provided that all other components in the water balance equation are known. In the past it has been difficult to obtain the rates of all components to close the water balance equation, especially evapotranspiration (Allen et al., 2011). However, due to the recent rapid progress in measurement of evaporation, EF between the vadose zone and groundwater table can now be quantified directly by means of water balance analysis at the field scale, which can then be used to gain improved understanding of regional groundwater dynamics.

In this paper, we present the results of comprehensive measurements and water balance analyses carried out within the irrigated cropland area associated with the Kaidu–Kongqi River basin, which is one of the four headwater basins of the Tarim River (Chen et al., 2010). EF at the seasonal and annual timescales under water-saving irrigation are estimated based on detailed water balance analysis, and the resulting groundwater table dynamics are quantified. The results of these analyses are used to draw inferences regarding the impacts of anthropogenic activities, especially the conversion from flood irrigation to drip irrigation, on regional groundwater dynamics, and for eventual achievement of water sustainability in a highly dynamic region of China.

This paper is organized as follows: Sect. 2 provides a detailed description of the Kaidu–Kongqi River basin. Section 3 presents the methods and instruments used as part of the comprehensive measurement system. The methodology for analysis of water balance and groundwater dynamics is then proposed. The results of the study are presented in Sect. 4, including seasonal and annual water balances and groundwater dynamics within the irrigated cropland. Based on the results, in Sect. 5 we discuss the human–water interactions within the oases of inland arid basins and the anthropogenic effects on groundwater, with the conclusions being presented in Sect. 6.

2 Description of Tarim River and Kaidu–Kongqi River basins

The main stream of Tarim River is 1321 km long, flowing west to east along the northern edge of Taklimakan Desert, the largest desert in China. At present, only four headwater streams, the Aksu River, Yarkand River, Hetian River, and Kaidu–Kongqi River, feed the main stream of the Tarim River (Fig. 1). Except for the Kaidu–Kongqi River, the other three headwater streams directly flow into the main stream at the upper reaches of the Tarim River, with the annual surface runoff of $4.6 \times 10^9\,m^3$ at the confluence.

The Kaidu–Kongqi River naturally fed into the main stream of the Tarim River at the lower reaches before the 1950s. However, due to excessive water extraction and stream regulation, Kaidu–Kongqi River lost its natural surface hydraulic connection with the main stream, aggravating the ecosystem degradation in the lower reaches of the Tarim River (Xu et al., 2007). Recently, in an attempt to restore the degraded ecosystem downstream, the Chinese government initiated an "eco-water diversion" project, aimed at conveying water from the Kaidu–Kongqi River to the Tarim River by artificial canals (Kuta Canal). Since 2000 this project has been activated 14 times, and to date approximately $4.2 \times 10^9\,m^3$ of water in total has been released into the Tarim River (http://tahe.gov.cn/zhuanti/shushui/shouye.html). As a result of the water conveyance project, the groundwater table along the Tarim River bank has risen from 6–8 m to 2–4 m, and the natural vegetation in the riparian areas has begun to recover (Chen et al., 2010).

Hydrologically, the Kaidu–Kongqi River basin (KKRB) is one of the typical inland basins, with a drainage area of $8.9 \times 10^4\,km^2$ (Gao et al., 2008; Xia et al., 2003) (Fig. 2). Glacier and snowmelt of southern Tianshan Mountains are concentrated in the months of April and May (Zhang et al., 2007). Rainfall concentrated in the months of May to August, contributing to 57–70 % of the annual rainfall amount (Li et al., 2012). The precipitation decreases when the elevation decreases (Wang et al., 2012). The Kaidu River flows through the middle and lower mountainous areas and alluvial plains, and finally reaches Bosten Lake. Annual runoff of the Kaidu River is $3.41 \times 10^9\,m^3$ (Wang et al., 2013). The contributions of rainfall and snow-/glacier melt to the total runoff are 85 and 15 %, respectively (Gao and Yao, 2005).

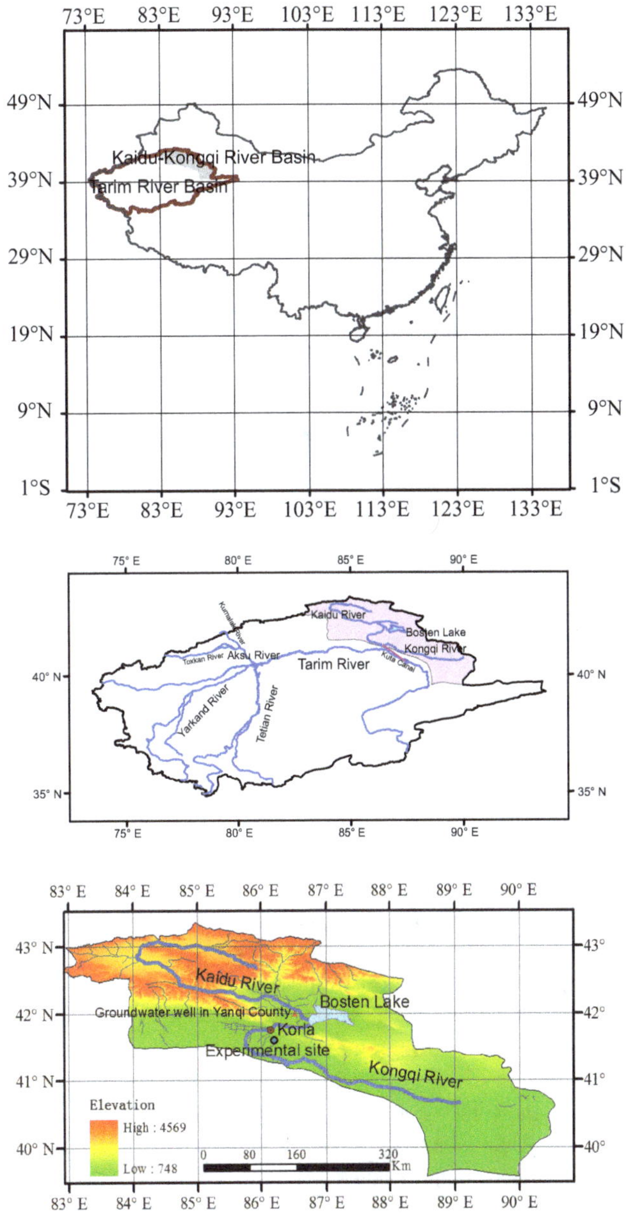

Figure 1. Geographic location of Tarim River basin and Kaidu–Kongqi River basin.

Bosten Lake is the largest inland freshwater lake in China, with a total volume of $8.8 \times 10^9\,m^3$ (Wan et al., 2006). It is the final destination of the Kaidu River and also the source of the Kongqi River. The annual runoff of Kongqi River is $1.19 \times 10^9\,m^3$, 80 % of which is extracted for irrigation

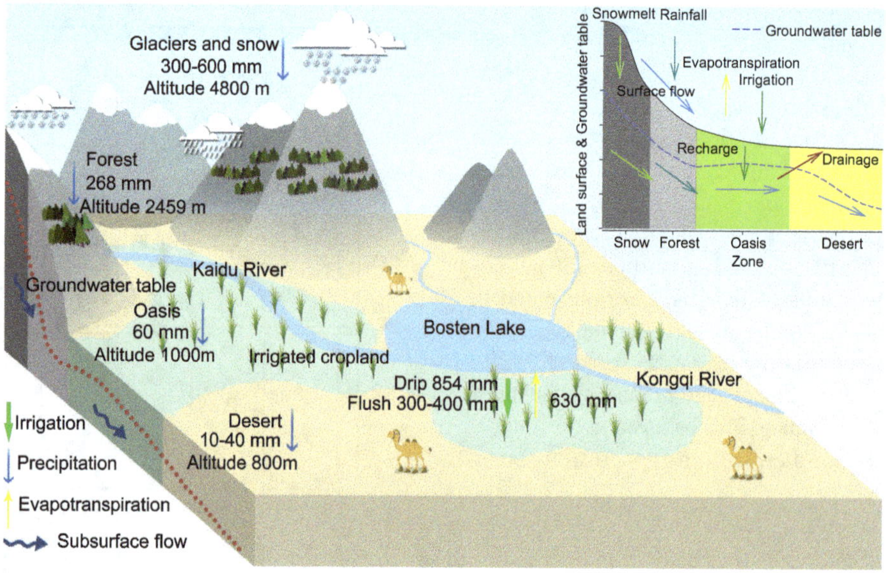

Figure 2. Sketch of hydrologic cycle in Kaidu–Kongqi River basin.

(Zhou et al., 2001). Kongqi River used to flow to the Lop Nor before the 1950s, but the downstream dried up after large-scale reclamation of wasteland within the KKRB. Oases are clustered along the Kaidu and Kongqi Rivers in the alluvial plain and the annual precipitation is approximately 60 mm only. Humans settled down in these oases and extracted water from the river for agricultural, industrial and domestic use (Feng et al., 2000). In the desert areas precipitation is rarer and ranged from 10 to 40 mm only.

Within the KKRB, agriculture greatly depends on irrigation, which accounts for 96.2 % of all water consumption (Zhang, 2012). Cotton is the most important economic crop grown within the KKRB and cotton fields occupy more than 50 % of the total agricultural area (the statistical Yearbook of Tarim River basin, Yearbook hereafter). In 2010, the fraction of water-saving irrigated area to the total irrigated area is more than 54 % within the KKRB, with mulched drip irrigation being the most extensive (Yearbook). With the wide application of water-saving irrigation, the irrigation quota during the growth period was 854 mm in 2011 (Yearbook). The fraction of area salinized to the total agricultural area is approximately 60 % (Xia et al., 2003; Zhang, 2012). To mitigate soil salinization, a non-growth season flushing approach in the amount of 300–400 mm has been adopted for decades to leach soil salt from the surface into the groundwater. Annual evapotranspiration within the irrigated cotton fields is about 630 mm based on the measurements carried out as part of this study. The environmental problems such as vegetation degradation and soil salinization are severe due to the large-scale irrigation in the KKRB, which also applies to the whole Tarim River basin (Zhang, 2012).

3 Methods and materials

3.1 Field experimental site and cotton planting

The field experimental site (86°12′ E, 41°36′ N; 886 m a.s.l.; see Fig. 1) is located in one of the oases scattered in the alluvial plain of the Kaidu–Kongqi River basin. The experimental field has an area of 3.48 ha, and cotton (*Gossypium hirsutum* L.) is planted under mulched drip irrigation. The surrounding fields have the same cotton planting and irrigation conditions as the experimental field, which ensures the measurements can represent typical irrigated cropland conditions in this region. In the experimental site, the mean annual precipitation is approximately 60 mm, whereas the mean annual potential evaporation measured by a Φ20 evaporation pan (pan diameter 20 cm) is 2788 mm. The annual mean temperature is 11.48 °C, and the annual total sunshine duration is 3036 h, which is favorable for cotton growth. The major soil type in the experimental field is silt loam, and the contents of sand, silt and clay are 32.8, 62.4, and 4.8 %, respectively. The soil porosity is 0.42 which was directly determined in the laboratory using the known volume of undisturbed soil columns collected in the experimental field.

Within the KKRB, cotton is planted in April and harvested from September to November. The style of cotton planting and drip pipe arrangement is referred to as the "one pipe, one film, and four rows of cotton arrangement" (Hu et al., 2011), which means that one drip pipe beneath the film is located in the middle of four rows of cotton. The width of film is 1100 mm, and the inter-film zone is 400 mm. The cotton seeds are sown at 0.1 m intervals in each row to yield an anticipated population of 260 000 plants ha^{-1}. But the emergence rates are usually about 60 % due to sandstorm and freezing damage, resulting in an actual plant density of only about 160 000 plants ha^{-1}. The maximum root depth of cotton is approximately 500 mm under mulched drip irrigation.

3.2 Data and measurements

The field experiment was conducted from April 2012 to September 2013. A stationary tower was erected in the middle of the experimental field to mount an eddy covariance (EC) system, which is known as a reliable method for obtaining direct field ET measurements (Baldocchi et al., 2001). At this site,the EC system consists of a fast-response open-path infrared gas (H_2O and CO_2) analyzer (model EC150, Campbell Scientific Inc., Logan, UT, USA), a fast-response 3-D sonic anemometer (model CSAT3, Campbell Scientific Inc., Logan, UT, USA), an air temperature/humidity sensor (model HMP155A, Vaisala Inc., Woburn, MA, USA), and a micro-logger (model CR3000, Campbell Scientific Inc., Logan, UT, USA). Since the maximum crop height is 0.7 m, the EC system was installed 2.25 m above the ground level to maintain the appropriate footprint. The CSAT3 sensor was oriented towards the predominant wind direction with an azimuth angle of 50° from true north.

Energy balance closure is used to evaluate the reliability of eddy covariance (EC) measurements (Foken, 2008). Net radiation (R_n) was measured at a height of 2.25 m (model LITE2, Kipp and Zonen, Delft, the Netherlands). Meanwhile, two soil heat flux plates (model HFP01SC, Hukseflux, the Netherlands) were placed 0.05 m below the ground surface in the film-mulched zone and inter-film zone, respectively, to obtain the soil heat flux (G). A rain gauge, a groundwater well, and two profiles for soil water content (SWC) measurements were set up around the tower. Precipitation was measured by a tipping bucket rain gauge (model TE525MM, Campbell Scientific Inc., Logan, UT, USA) which was mounted at 0.7 m above the ground. Groundwater table level was measured using an automatic water depth sensor (model HOBO U20 Titanium Water Level Data Logger, Onset Computer Corporation, Inc., Pocasset, MA, USA) installed in this groundwater well. More than 30 soil sensors (three models, i.e., Hydra Probe, Stevens Water Monitoring System, Inc., Beaverton, OR, USA; Digital TDT, Acclima Inc., Meridian, ID, USA; CS616, Campbell Scientific Inc., Logan, UT, USA) were placed in the film-mulched zone and inter-film zone in two profiles, respectively. The SWC were measured at 50, 100, 150, 200, 300, 400, 500, 600 and 900 mm depth below the ground by these sensors and stored every 1 h. In order to measure SWC in the zones deeper than 900 mm, soil samples were collected once a month with 10 replications using an auger (model Auger Edelman combination 50 mm, Eijkelkamp Agrisearch Equipment, Giesbeek, the Netherlands) from the boreholes. The samples were obtained both from the film-mulched and inter-film zones, respectively at depths of 1000, 1200, and 1500 mm. The SWC for the soil sample was determined using the gravimetric method.

The drip irrigation amount was measured by water meters installed on the branch pipes of the drip irrigation system, whereas the periodic flush amounts were obtained by measurements of depth and velocity of flow in the channel. The irrigation schedule adopted in 2012 and 2013 has been summarized in Table 1. The annual irrigation amount was approximately 940 mm in this site, including the flushing during the non-growth season and drip irrigation during the growth season.

3.3 Methodology

3.3.1 Water balance analysis

In this study, since the deepest soil sensor was installed at a depth of 900 mm, the depth of control volume (a cylindrical soil column of unit surface area) was extended to the 900 mm soil depth. The water balance of the control volume can be written as follows (Evett et al., 2012):

$$I + P + \mathrm{EF} = \mathrm{ET} + R + F_L + \Delta S, \tag{1}$$

where I is irrigation; P is precipitation; EF is exchange flux (EF) at the lower boundary of the control volume (900 mm soil depth); ET is evapotranspiration; ΔS is the change of soil water storage within 900 mm soil depth and is calculated from the differences in soil water content at the beginning and end of the period under consideration; R is the overland flow for the control volume; and F_L is the lateral flow in the control volume. I, P, ΔS, and ET are measured at this site as described in Sect. 3.2. During most of the year, the precipitation and amount of drip irrigation were limited, and the control volume was above the groundwater table. Therefore, R and F_L were trivial and can be ignored during the calculation. During the period of spring flush, the groundwater table exceeded the bottom boundary of control volume. Under this situation, R and F_L also can be ignored due to the homogeneous irrigation condition and short duration of water ponding and shallow groundwater table. Therefore, the Eq. (1) also can be applied to the period of spring flush.

The seasonal and annual trends of EF at 900 mm soil depth were obtained based on the daily water balance analysis using Eq. (1) during the experimental period (Fig. 3). This method is expected to provide reliable estimates of EF since the other four items of the water balance equation were relatively well gauged at this site, and this measurement site well represents the surrounding vegetated environment and water availability.

3.3.2 Groundwater table dynamics

Groundwater is replenished by both local recharge and regional (lateral) flow (Tang et al., 2007b). In the irrigated cropland, the local recharge is the EF. The lateral flow refers

Table 1. Irrigation schedule.

Cotton growth stage	Spring flush	Squaring stage				Flower stage			Bolls stage					Total drip irrigation	Total
2012 irrigation date	3-25 to 3-29	6-10/11 and 6-14/15	6-21	6-28	7-6/7	7-15/16	7-26	8-4/5	8-8	8-12/13	8-17	8-22/23	8-27/28		
Amount (mm)	375	65.17	34.35	35.32	36.77	33.26	44.1	40	59.28	46.73	42.19	50.84	52.22	540.23	915.23
2013 irrigation date	3-22 to 3-26	6-12	6-18/19 and 6-26/27	7-1/2	7-7/8	7-14/15 and 7-20	7-26	8-1/2	8-7/8	8-11/12	8-16/17	8-21/22			
Amount (mm)	375	48.45	62.93	39.16	76.02	46.55	38.97	53.09	63.23	51.83	52.10	58.77		591.09	966.09

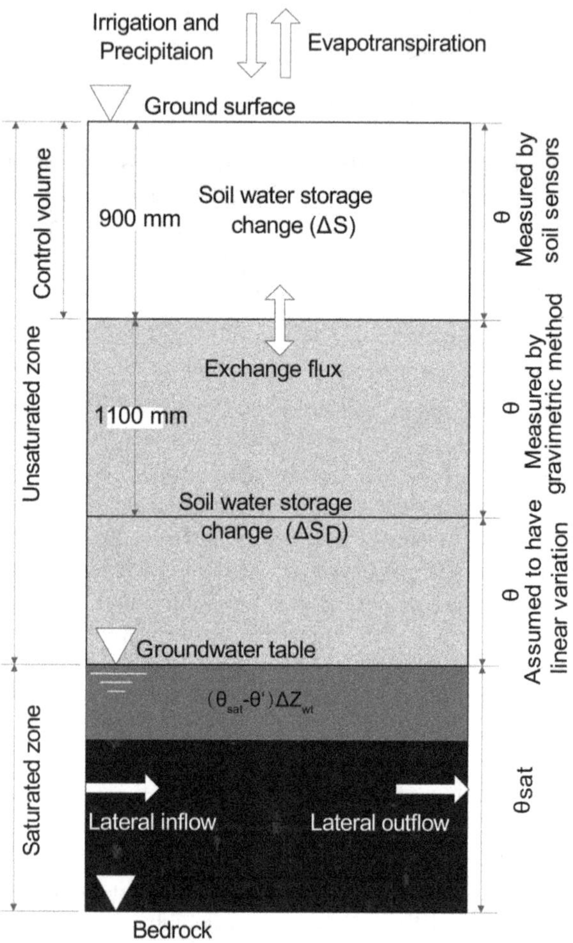

Figure 3. Analysis zone for water balance and groundwater dynamics.

to the subsurface inflow from mountainous areas and the subsurface outflow to the drainage canals or the desert areas. Because the groundwater always contains a high level of salinity in this region, there is only small-scale groundwater withdrawal for human activities (Gao and Yao, 2005; Zhang and Hu, 2008).

In this study, the groundwater dynamics can be written as follows (Tang et al., 2007b):

$$\mathrm{LF} = \mathrm{EF} + (\theta_{\mathrm{sat}} - \theta^{'})\Delta z_{\mathrm{wt}} + \Delta S_{\mathrm{D}}, \tag{2}$$

where EF is the exchange flux at 900 mm soil depth (local recharge), LF is the net lateral flow (regional flow), θ_{sat} is the SWC at saturation, θ is the SWC within the range of groundwater table changes, Δz_{wt} is the change of groundwater table depth, $(\theta_{\mathrm{sat}} - \theta^{'})\Delta z_{\mathrm{wt}}$ is the soil water storage change associated with a falling or rising groundwater table, and ΔS_{D} is the soil water storage change in the zone between 900 mm soil depth (the bottom of control volume) and upper boundary of groundwater table variation.

Using Eq. (2), the groundwater dynamics were analyzed at this experimental site (Fig. 3). As described in Sect. 3.3.1, the EF at 900 mm soil depth was obtained based on the measurement data. The SWC at the depth of 1000, 1200 and 1500 mm were measured once a month, and assumed to apply to soil levels at 900–1100 mm, 1100–1300 mm, and 1300–2000 mm, respectively. In addition, θ was assigned to the layer between the 2000 mm depth and the groundwater table, and assumed to have a linear variation in this zone. θ_{sat} was determined from the porosity with the value of 0.42.

4 Results

In consideration of irrigation, cotton growth and climate variability, we divided the whole year into five analysis periods, i.e., period 1 is the period after cotton is sown and before drip irrigation, period 2 is the period of drip irrigation, period 3 is the period after drip irrigation and before soil freezing, period 4 is the period of winter, and period 5 is the period of spring flush. The seasonal water balance and groundwater dynamics were analyzed during each of these five analysis periods, after which the annual characteristics were also explored.

4.1 Energy balance closure of eddy covariance and seasonal ET variation

The evapotranspiration (ET) was measured by eddy covariance (EC) in this study. Energy balance closure is used to evaluate the reliability of EC measurements. The variations

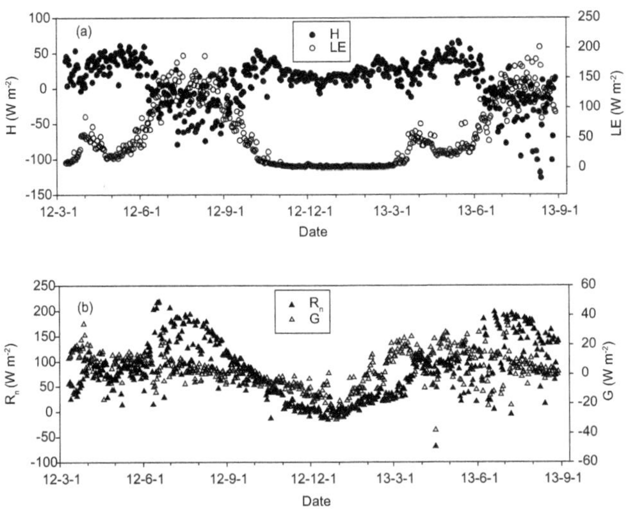

Figure 4. The daily variations in averaged sensible heat (H), latent heat (LE), net radiation (R_n) and soil heat flux (G) during 2012–2013.

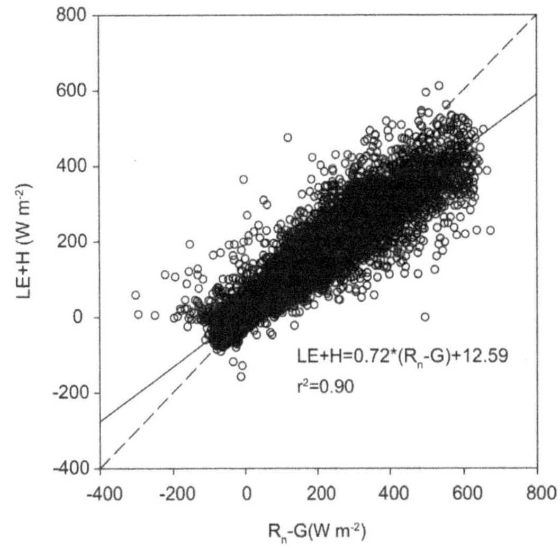

Figure 5. Slope, intercept and r^2 of the half-hourly energy balance closure.

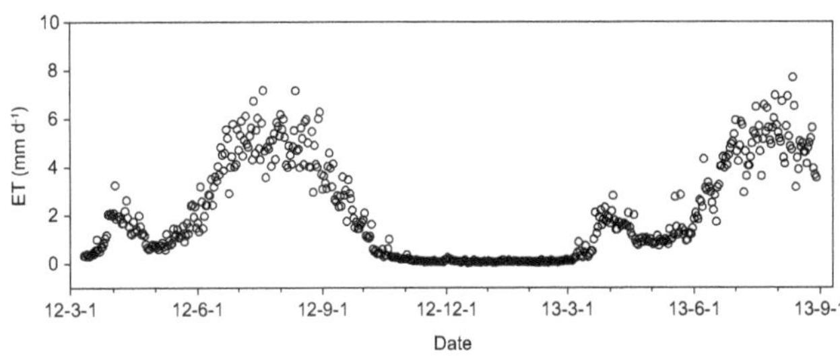

Figure 6. Seasonal evapotranspiration variation during 2012–2013.

in daily averaged sensible heat (H), latent heat (LE), net radiation (R_n) and soil heat flux (G) during 2012–2013 are shown in Fig. 4. Using all valid half-hourly data in 2012 and 2013 (data points, $n = 21\,886$), the slope between the available energy flux ($R_n - G$) and the sum of sensible (H) and latent heat fluxes (LE) for this site was 0.72, the intercept was 12.59 W m^{-2}, and the coefficient of determination (r^2) was 0.90 (Fig. 5). Based on the previously obtained slope values of 0.53–0.99 for the energy closure (Wilson et al., 2002), these performance results are deemed reasonable and provide us confidence in the eddy covariance measurements at this site (Zhang et al., 2014b).

The seasonal ET variation is presented in Fig. 6. There were two ET peaks within the whole year due to the flush in spring and drip irrigation in summer. The maximum ET rate was almost 8 mm day^{-1} and happened in July and August. The total evapotranspiration in cotton growth period (29 April 2012 to 31 October 2012) was approximately 560 mm, which was consistent with the result obtained in the cotton field under mulched drip irrigation in northern Xinjiang (538 mm, Zhou et al., 2011). However, ETs in the cotton fields under flood irrigation were much higher (670 mm in USA, Chavez et al., 2009; 735 mm in Australia, Tennakoon and Milroy, 2003), suggesting that water-saving irrigation significantly reduced the ET in the farmland.

4.2 Seasonal water balance analysis and exchange flux

4.2.1 Period after cotton is sown and before drip irrigation

Cotton is usually planted in late April and drip irrigation starts in early June. This period after cotton is sown and before drip irrigation (period 1) is the seedling stage of cotton growth. In 2012 and 2013, upward EFs at 900 mm soil depth derived from water balance analysis during period 1 were 53.5 and 36.5 mm in total (Table 2), respectively, suggesting that the soil moisture went upward from deep soil zone and the groundwater table to the land surface.

During period 1, the potential evaporation was high due to higher temperature and net radiation. However, there was no irrigation during this period, and soil water storage and precipitation cannot completely satisfy the evapotranspiration water demand. On the other hand, the spring flush implemented about 40 days earlier led to the rise of groundwater table and high soil water content in the deep soil zone, thus leading to the upward EF that was estimated. Most of the EF was consumed by evapotranspiration, while the soil storage contributed a small part to ET. It is worth noting that precipitation impacted the EF significantly. The precipitation was higher in 2013 than in 2012 during period 1, with the result that the EF was lower in 2013. The sum of precipitation and EF was almost the same for these 2 years (2012: $3.5 + 53.5 = 57$ mm; 2013: $23.2 + 36.5 = 59.7$ mm), indicating the consistent water demand for evapotranspiration during period 1.

Table 2. The water balance in the control volume (unit: mm).

Period	Start and end date	Exchange flux at 90 cm depth[a]	$\frac{\sigma_{EF}}{EF}$	Evapotranspiration	Irrigation	Precipitation	Soil water storage change ΔS[b]
Period 1	29-04-2012 10-06-2012	+53.5	0.02	61.6	0.0	3.5	−4.6
Period 2	11-06-2012 04-09-2012	−133.4	0.21	409.3	540.3	38.3	+35.9
Period 3	05-09-2012 22-11-2012	+46.6	0.18	90.7	0.0	1.4	−42.7
Period 5	22-03-2013 09-04-2013	−296.4	0.07	34.2	375.0	0.0	+44.4
Period 1	30-04-2013 11-06-2013	+36.5	0.04	66.8	0.0	23.2	−7.1
Period 2	12-06-2013 29-08-2013	−252.5	0.12	375.2	591.1	36.9	+0.3

[a] Positive represents upward fluxes; [b] Positive represents increased storage; σ_{EF} are the standard errors for EF.

The daily EF is presented in Fig. 7a and b. Most days during period 1 experienced upward fluxes, while there were some exceptions during the rainy days because of infiltration. EF gradually became larger as time went by, indicating that more water was required to satisfy the intense evapotranspiration. The groundwater table continuously declined during this period. The relationship between EF and groundwater dynamics will be analyzed in more detail in Sect. 4.3. The results above were consistent during both years.

4.2.2 Period of drip irrigation

The drip irrigation schedules during the study period are shown in Table 1. The period of drip irrigation (period 2) covered the flower, square, and bolling stages of cotton growth. The total EFs were downward with the value of 133.4 and 252.5 mm in 2012 and 2013 during period 2, respectively. The downward EFs were induced by drip irrigation conducted during this period and more irrigation amount in 2013 led to more leakage.

On the one hand, the irrigation water was sufficient to cause the leakage; on the other hand, some upward EF still occurred just before the next irrigation event when the interval between irrigation events was long (Fig. 7c and d). The irrigation frequency was lower in 2012 than in 2013, resulting in more upward EFs to satisfy the water requirement of plant transpiration and soil evaporation. The results indicate that more scientific irrigation schedules can further improve irrigation efficiency. In fact, the evapotranspiration values were 375–400 mm during period 2 (Table 2), far less than the sum of irrigation and precipitation which was approximately 600 mm. There is therefore still great potential for water saving under current irrigation conditions.

However, compared with flood irrigation, the downward EF was remarkably reduced due to its limited irrigation amount each time. In fact, the irrigation amount each time under flood irrigation is about 300 mm in this region, leading to notable leakage and groundwater table rise, whereas the irrigation amount is only 30–80 mm under drip irrigation, and the groundwater table is relatively stable during the irrigation periods (Fig. 7c and d).

4.2.3 Period after drip irrigation and before soil freezing

The period after drip irrigation and before soil freezing (period 3) is the open boll stage for cotton growth. This period is characterized by no irrigation, decreased potential evaporation and declining groundwater table. The upward EF during period 3 was 46.6 mm in 2012 (Table 2) and contributed to more than 50 % of the evapotranspiration during this period. Meanwhile, the soil water content rapidly decreased and provided the remainder of the water demand for evapotranspiration.

Since the evaporative demand rapidly decreased during this period due to the climatic condition, the upward EFs were significant in September and became smaller in October and November, as shown in Fig. 7e. The sharply declined groundwater table was another reason for decreased upward EFs. In fact, the soil water storage decrease within the 900 mm depth was also concentrated in September with a total value of 45.3 mm. In October and November the soil water storage even increased by 2.6 mm.

4.2.4 Period of winter and spring flush

The hydrologic processes slowed down in winter (period 4). No irrigation and only rare precipitation (less than 5 mm) occurred during this period. The soil was frozen up to 1000 mm deep and soil moisture was immovable in the frozen zone. The groundwater table declined to below 4.5 m.

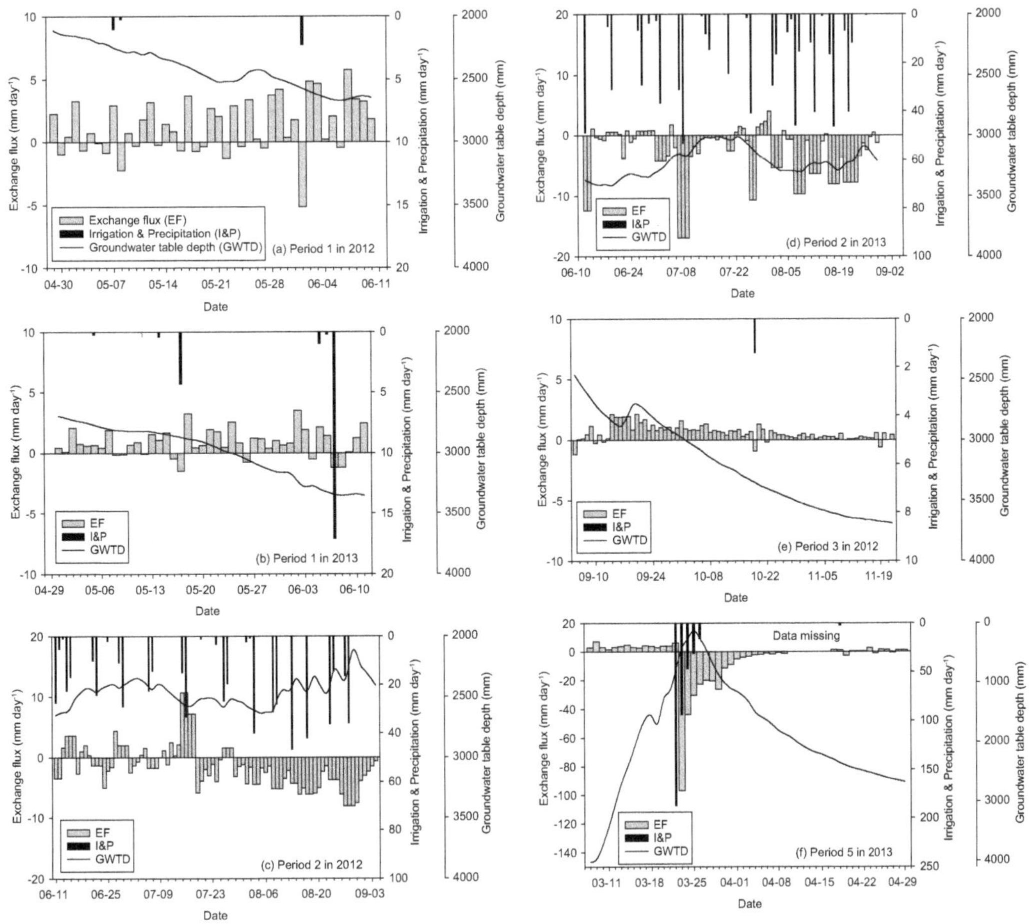

Figure 7. Exchange flux at 900 mm soil depth and groundwater table variation. Positive EFs represent upward fluxes and negative EFs represent downward fluxes.

The evapotranspiration was very low and the total value was only 9.4 mm for 105 days from 23 November 2012 to 7 March 2013. Since the soil sensors cannot provide an accurate measurement in frozen soil, we cannot obtain soil water storage change to close the water balance. Therefore, the detailed analysis was not conducted in period 4.

The spring flush is usually conducted in March for the purpose of leaching soil salt and enhancing soil moisture (period 5). In 2013, the spring flush was conducted in the period 22 to 26 March at the experimental site. The groundwater table increased before 22 March due to the spring flush in the surrounding fields (Fig. 7f). Data was missing during 10 to 16 April because of sowing. The total downward EF during period 5 (22 March to 9 April) was 296.4 mm, accounting for 79 % of the amount of water flushed (Table 2). Meanwhile, the soil water storage increased by 44.4 mm. The groundwater table continually declined after spring flush during this period.

4.3 Seasonal groundwater table dynamics

The seasonal groundwater dynamics are analyzed in this section using Eq. (2). Since the available soil moisture data is quite limited in the deep unsaturated zone, only the seasonal trends were analyzed rather than the daily trends, and the results are shown in Table 3.

The EF above the groundwater table was downward in all periods except period 1 in 2012, indicating that salinization caused by phreatic evaporation had been controlled under the current irrigation condition. The groundwater table rose during period 2 and period 5, while it always declined during the other periods. The rate of decline was highest during period 5 due to the groundwater plateau formed by spring flush. Groundwater plateau was also formed during period 2 because of drip irrigation. The rate of decline was as high as 36.3 mm day^{-1} during the first 10 days of period 3 just after the drip irrigation period in 2012. During period 1, since the plateau effects caused by spring flush were attenuated, the rates of decline were not very high, with a value of 12.8 to 15.8 mm day^{-1}.

The net lateral flow was obtained by groundwater balance analysis using Eq. (2) and the flow direction was outward during all the analysis periods, indicating that groundwater plateau caused by irrigation existed within the irrigated croplands. To present the regional groundwater balance clearly, representative sketches of seasonal groundwater dynamics are presented in Fig. 8. When irrigation was implemented during period 2 and period 5, the groundwater table rose significantly. Then the groundwater table declined after irrigation periods and the rates gradually decreased when the water table became lower, which has been shown in Fig. 7e and f. The lateral flow out of the analysis zone during periods 2 and 5 was expected to be high due to the recharge caused by irrigation and the high groundwater table. However, the outflow rate was only 1.0 and 3.0 mm day^{-1} during period 2, indicating that the lateral flow into this zone was also significant. In fact, snowmelt happened during spring and summer, and the precipitation was also concentrated in the summer period in the mountainous areas. They resulted in significant subsurface flow into this zone during period 2 (Fig. 8). Similarly, snowmelt in spring led to the subsurface flow into this zone, resulting in the fact that the net lateral flow was larger during period 3 than during period 1.

4.4 Annual water balance and groundwater dynamics

Annual water balance and groundwater dynamics were analyzed based on the data collected from 29 April 2012 to 28 April 2013. The evapotranspiration was 637.5 mm for the whole year and precipitation was 48.9 mm. The total irrigation amount was 915.3 mm, which was composed of drip irrigation (540.3 mm) and spring flush (375 mm). Soil water storage increased by 16.2 mm in the unsaturated zone during this year. Therefore, the total downward EF above the groundwater table was 310.5 mm. The downward flux recharged the groundwater, which eventually flowed into the desert or drained to the Bosten Lake.

The groundwater table declined by 540.1 mm from 29 April 2012 to 28 April 2013, resulting in 45.1 mm groundwater storage decrease. Therefore, after taking the downward EF of 310.5 mm into account, the annual net lateral flow at the experimental site was 355.6 mm, flowing out of this zone to the desert or the lake. The groundwater table slightly declined during this year mainly due to the limited irrigation amount.

4.5 Error analysis

The exchange flux (EF) is calculated in Sect. 4.2 by the following equation:

$$\mathrm{EF} = \mathrm{ET} + \Delta S - P - I. \tag{3}$$

Barry (1978) indicated that when a final result is calculated from direct measurements, its precision is a function of the variability in the direct measurements. The exchange flux is computed from direct measurements including evapotranspiration, soil water content, precipitation and irrigation. Therefore, the standard error (SE) of EF can be expressed by SE of the direct measurements:

$${\sigma_{\mathrm{EF}}}^2 = {\sigma_{\mathrm{ET}}}^2 + {\sigma_{\Delta S}}^2 + \sigma_P^2 + {\sigma_I}^2, \tag{4}$$

where $\sigma_{\mathrm{EF}}, \sigma_{\mathrm{ET}}, \sigma_{\Delta S}, \sigma_P$ and σ_I are the standard errors for EF, ET, ΔS, P and I, respectively. The variabilities of EF, ET, ΔS, P and I are assumed to be normally distributed and independent since the ET, ΔS, P and I are separately measured. We can rewrite Eq. (4) and express the variabilities of all parameters relative to their respective means:

$$\frac{\sigma_{\mathrm{EF}}}{\mathrm{EF}} = \left\{\left(\frac{\mathrm{ET}}{\mathrm{EF}}\right)^2\left(\frac{\sigma_{\mathrm{ET}}}{\mathrm{ET}}\right)^2 + \left(\frac{\Delta S}{\mathrm{EF}}\right)^2\left(\frac{\sigma_{\Delta S}}{\Delta S}\right)^2 + \left(\frac{P}{\mathrm{EF}}\right)^2\left(\frac{\sigma_P}{P}\right)^2 + \left(\frac{I}{\mathrm{EF}}\right)^2\left(\frac{\sigma_I}{I}\right)^2\right\}^{\frac{1}{2}}. \tag{5}$$

This analysis shows that the variabilities of all components of water balance affect the accuracy of the EF estimate. The effect of one specific component will be more significant when its value becomes larger. In this study, ET and I are relatively larger compared with ΔS and P. Therefore, $\frac{\sigma_{\mathrm{ET}}}{\mathrm{ET}}$ and $\frac{\sigma_I}{I}$ play an important role in the variability of EF. Since ET measured by eddy covariance is relatively stable, we can suppose that $\frac{\sigma_{\mathrm{ET}}}{\mathrm{ET}}$ is quite small with the value of 0.001 (Zhang et al., 2014b). Water is delivered precisely to the cotton plants under drip irrigation, thus the water availability in the field is homogeneous and $\frac{\sigma_I}{I}$ is assigned a value of 0.05. The variability of P is supposed to be small because of the small area of the experimental field, and $\frac{\sigma_P}{P}$ is assigned a value of 0.02. Soil moisture is highly spatially variable. In this study, we only considered the change of soil water storage (ΔS) during water balance analysis. Therefore, although there will be huge spatial heterogeneity in SWC, the SWC change for each measured location is relatively consistent and reliable. $\frac{\sigma_{\Delta S}}{\Delta S}$ is assigned a value of 0.2.

The $\frac{\sigma_{\mathrm{EF}}}{\mathrm{EF}}$ results are shown in Table 2. $\frac{\sigma_{\mathrm{EF}}}{\mathrm{EF}}$ is large during periods 2 and 3, and small during period 1. The results indicate that the exchange flux is difficult to evaluate under the following situations: (1) ΔS is the large component of EF. Soil water storage is highly variable and difficult to measure. When it is significant in the EF estimate, the variability of EF will increase. (2) EF is relatively small when the other components of water balance are large. That is to say, it is difficult to evaluate the small EF at the bottom when the surface fluxes are significant. However, it is worth noting that the analysis above is based on the standard error, and only represents variation relative to the mean, and is not an indication of accuracy.

Table 3. Evaluation of groundwater and lateral flow (unit: mm).

Period	Start and end date	Average ground-water depth	Exchange flux at 90 cm depth[a]	Soil water storage change ΔS_d^b	Exchange flux above groundwater table[a]	Groundwater table change[c,e]	Groundwater storage change[b]	Net lateral flow[d,e]
Period 1	29-04-2012 10-06-2012	2416.0	+53.5	−37.4	+16.1	−551.7(−12.8)	−40.1	−24.0(−0.6)
Period 2	11-06-2012 04-09-2012	2457.2	−133.4	+39.9	−93.5	+239.1 (+2.8)	+7.2	−86.3(−1.0)
Period 3	05-09-2012 22-11-2012	3210.6	+46.6	−109.3	−62.7	−1284.1(−16.3)	−104.1	−166.9(−2.1)
Period 5	20-03-2213 09-04-2013	1038.9	−296.4	+87.2	−209.2	−1056.4(−55.6)	−63.1	−272.3(−14.3)
Period 1	30-04-2013 11-06-2013	2996.6	+36.5	−59.8	−23.3	−677.3(−15.8)	−29.5	−52.8(−1.2)
Period 2	12-06-2013 29-08-2013	3218.4	−252.5	+11.3	−241.2	+143.8(+1.8)	+1.3	−239.8(−3.0)

[a] Positive represents upward fluxes;
[b] Positive represents increased storage;
[c] Positive represents rising water table;
[d] Positive represents flow into this region;
[e] The value per day in the brackets.

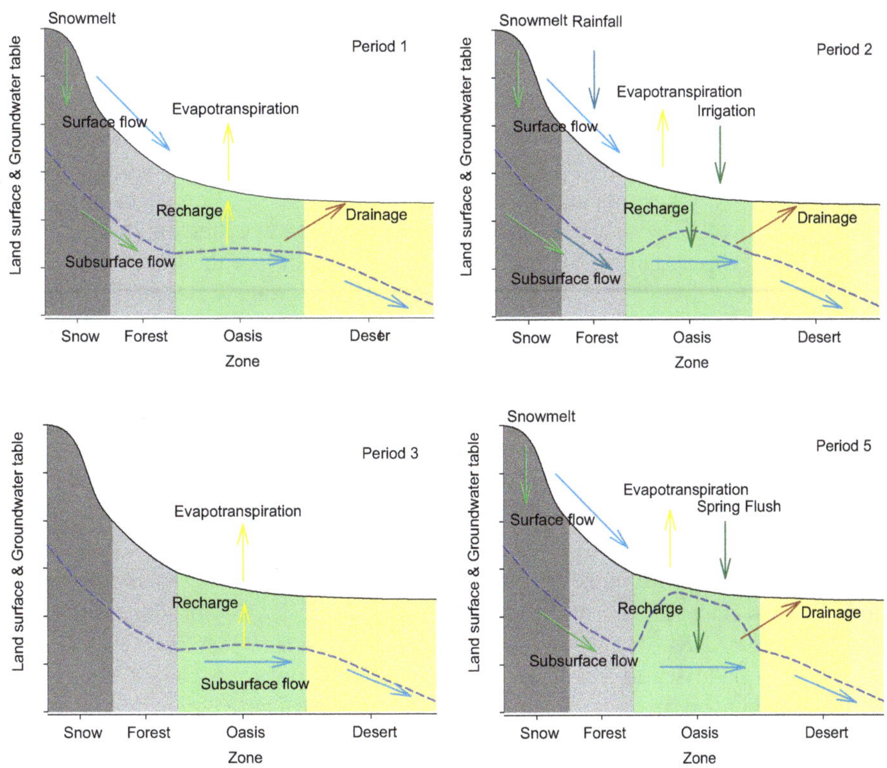

Figure 8. Sketch of seasonal groundwater dynamics (typical qualitative patterns in this region).

5 Discussion

In this study, water balance and groundwater table dynamics under water-saving irrigation in an oasis within the Tarim River basin were analyzed based on comprehensive measurements that helped to close the water balance. The study region represents the typical social and hydrologic system of oases in inland basins. Therefore, the conclusions drawn in this study can potentially be applied to other arid inland areas (e.g., in central Asia) which face similar water shortage and development problems (Christen et al., 2007).

5.1 Three stages of human–water system development

Due to the unique hydrological regime in oases, a moderate groundwater table is the desired goal of sustainable water management (El Bastawesy et al., 2013; Zhu et al., 2004). The groundwater table is significantly affected by anthropogenic activities, especially irrigation (Tang et al., 2007a). To discuss the irrigation effects on groundwater table, we can divide the evolution of the human–water system into these inland basins into three stages in terms of water use, i.e., the natural stage before the 20th century; the exploitation stage in the 20th century; and the balanced development stage in 21st century. As shown in this study, Tarim River basin presents us with a good case study to gain insights into the effects of irrigation on groundwater dynamics, and associated human development.

5.1.1 Natural stage: limited anthropogenic impacts on groundwater

During the natural stage, the anthropogenic activities had limited impacts on water resources. The groundwater table did not change much and the water table was relatively deep in the irrigated districts (Fig. 9). Oases were restricted along the river and isolated from each other. Natural factors dominated the hydrologic cycle and socio-economic development (Liu et al., 2014). Rivers had sufficient water to support the natural vegetation in the riparian areas. The relationship between surface water and groundwater was quite weak during this stage.

5.1.2 Exploitation stage: salinization and natural vegetation degradation

During the exploitation stage during the 20th century, water balance and groundwater dynamics were significantly altered by anthropogenic activities, especially by the rapid development of agriculture (Cui and Shao, 2005; El Bastawesy et al., 2013). Groundwater tables significantly rose in the irrigated districts due to flood irrigation and canal leakage, resulting in severe soil salinization. Within the KKRB, the groundwater table in most irrigated croplands had risen to less than 1 m below the surface due to long-term flood irrigation in the late 1990s (Cui and Shao, 2005). More than 60 % of the irrigated croplands suffered severe salinization due to intense phreatic evaporation induced by high groundwater tables. As the area of irrigated cropland greatly expanded, more water drained into Bosten Lake from the irrigated district. During the period 1958 to 2002, $1.9 \times 10^8\ m^3\ year^{-1}$ saline water from the irrigated district directly drained into the Bosten Lake, contributing to 9 % of the total influx of this lake (Gao and Yao, 2005). The salinity of the lake increased from $0.38\ g\ L^{-1}$ in 1956 to $1.32\ g\ L^{-1}$ in 2005 (Xie et al., 2011) as a result. The water quality in the lake changed from fresh water into brackish water as a consequence. At the same time, the ecosystem of the lake suffered severe degradation. The area covered by reeds within the lake decreased from $558\ km^2$ in 1958 to $300\ km^2$ in 1998 (Wan et al., 2006).

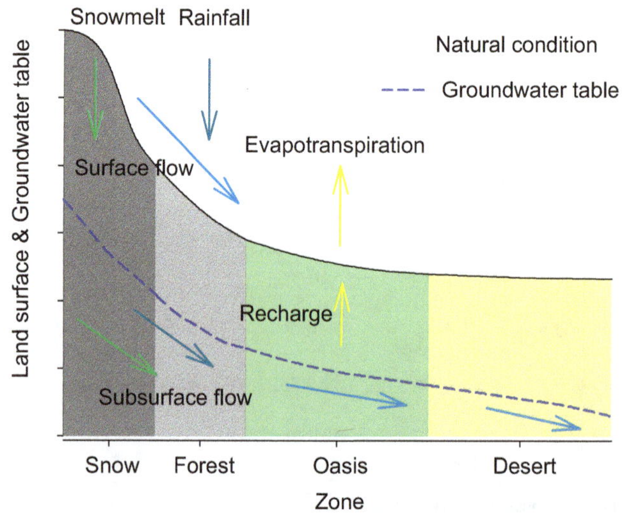

Figure 9. Groundwater dynamics in the natural stage (typical qualitative patterns in this region).

During the exploitation stage, when the groundwater table sharply rose in the irrigated districts, the water table in riparian areas in the lower reaches obviously declined. The large amount of water extraction from the river for irrigation led to the interruption of the lower reaches, resulting in the drawdown of the groundwater table (Chen et al., 2010). Within the KKRB, the area of irrigated croplands increased rapidly from 39.9×10^3 ha to 101.2×10^3 ha from 1950 to 1979, while the water use for agriculture increased faster from $0.6 \times 10^9\ m^3$ to $1.83 \times 10^9\ m^3$ (Table 4). The irrigation quota reached the peak of 1811.3 mm in 1979. During the period 1979 to 1999, irrigation quota decreased to 1037.7 mm benefiting from the improved canal systems and reduced leakage. However, the cropland area still rapidly expanded and the annual water withdrawal did not decrease too much and remained at the value of $1.78 \times 10^9\ m^3$ (Gao and Yao, 2005; Wang et al., 2002; Xia et al., 2003; Zhou et al., 2001). The natural ecosystem in the lower reaches suffered severe degradation due to the drawdown of the groundwater table. Thus, irrigation dramatically altered the natural hydraulic river regimes and regional hydrologic cycle in inland basins during the exploration stage.

5.1.3 Balanced development stage: drawdown of groundwater table and efficiency paradox

In order to mitigate soil salinization within the irrigated croplands, the groundwater table should be lowered to an appropriate depth. In order to leave more water in the river for ecologic use, water extraction from the river should be reduced. Since agriculture consumes more than 90 % of water in most

Table 4. Development of irrigation in Kaidu–Kongqi River basin.

Year	1950	1979	1989	1999	2005	2011
Agricultural water use (10^9 m^3)	0.60	1.83	1.66	1.78	1.81	1.76
Cropland area (10^3 ha)	39.9	101.2	134.4	171.1	193.9	206.3
Irrigation quota (mm)	1503.8	1811.3	1236.2	1037.7	935.4	853.7

inland basins, water-saving irrigation is the best choice to realize the two goals above. In fact, water-saving irrigation has been popularized in many inland basins in recent years of the 21st century (Ibragimov et al., 2007; Rajak et al., 2006; Zhang et al., 2012). After the 2000s, the government began to promote the development of water-saving irrigation through the employment of subsidies in TRB (Ma et al., 2010). Water-saving irrigation not only improved water-use efficiency, but also significantly enhanced crop yield, and thus it was quickly and enthusiastically adopted by local farmers (Liu et al., 2014; Wang et al., 2011).

Groundwater tables significantly declined with the implementation of water-saving irrigation during the balanced development stage. Within the KKRB, as we showed in this study, the mulched drip irrigation amount is approximately 625 mm in the cotton field. The total irrigation amount for the whole year is about 925–1025 mm including the non-growth season flush. The annual recharge to groundwater was only 310.5 mm due to the limited irrigation amount, accounting for 32 % of the total amount of irrigation and precipitation. It should also be noted that the daily downward fluxes were small during the drip irrigation period, avoiding a significant rise of groundwater table. The long-term series data observed in Yanqi County within the KKRB showed that the groundwater table in this region continually declined in the past decade and has been managed to the favorable depth needed for the plant growth, which is in the range of 2–4 m (Fig. 10). The relatively stable groundwater table under mulched drip irrigation helped to reduce the intense phreatic evaporation, which was common after previous flood irrigation events (El Bastawesy et al., 2013; Tang et al., 2007a). The seasonal analysis in this study showed that the upward EFs from the groundwater only happened during spring with a trivial amount. Since the groundwater is 4–6 times more saline than the irrigated water from the canal, the reduced upward EFs are a significant benefit towards salinization control as well. Thus, the purpose of soil salinization mitigation is realized to a large extent by the implementation of water-saving irrigation.

The yearbook shows that the irrigation quota dropped from 1037.7 mm in 1999 to 853.7 mm in 2011 following the adoption of the new irrigation methods within KKRB (Table 4). If the water savings gained by the replacement of flood irrigation with water-saving irrigation remained in the river for ecological use, the recovery of ecosystem and environment could be easily realized (Scheierling et al., 2006). How-

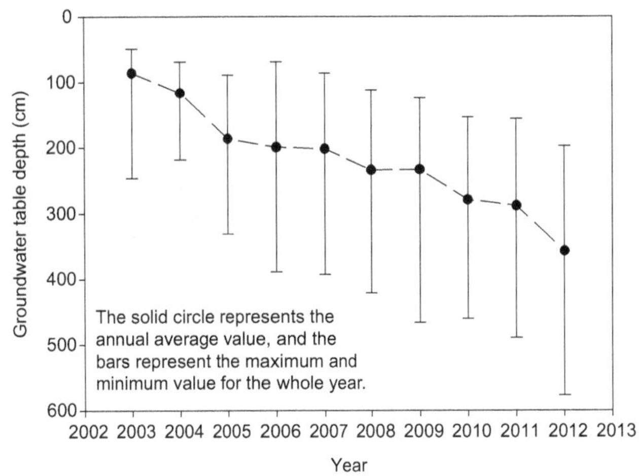

Figure 10. Long-term groundwater table variation in Yanqi County in the KKRB.

ever, the water "saved" can also lead to increased use of the water through irrigation area expansion (Ward and Pulido-Velazquez, 2008). This phenomenon, which is known as "efficiency paradox", has also happened in many other regions, such as Spain, Chile and the United states (Scott et al., 2014). Within the KKRB, subsidies provided by the government not only popularized water-saving irrigation, but also stimulated farmers to reclaim more wastelands where they could reuse the saved water. The consequence was that more water is extracted to irrigate more fields with the smaller irrigation quota. In 2011, the area of the irrigated cropland amounted to 206.3×10^3 ha, and the water withdrawal also was up to 1.76×10^9 m^3 (Table 4). As predicted by the "efficiency paradox", water-saving irrigation indeed resulted in more (not less) water usage in this basin.

In fact, as shown in this study, high irrigation efficiency under water-saving irrigation means high water consumption by evapotranspiration and reduced recharge to the groundwater (Perry, 2007). Then the groundwater table will continuously decline in the whole region. Drawdown of the groundwater table is good for salinization control in irrigated districts. However, it is also the main reason for the disappearance of natural vegetation and desert expansion. Therefore, the water savings gained through advanced drip irrigation should remain within the river basin to recharge the riparian groundwater and support the natural vegetation in future.

Meanwhile, soil salinization is problematic under the water-saving irrigation and may threaten sustainable development (Zhang et al., 2014a). The mechanism of salinization is different from that caused by flood irrigation. Intense phreatic evaporation induced by a shallow groundwater table is the main reason for salinization under flood irrigation. However, under water-saving irrigation, the salinization is mainly caused by deficient leaching of water. For instance, under mulched drip irrigation, the dissolved salts accumulate at the periphery of the wetted soil mass and increase with distance from the emitters in the soil profile (Palacios-Diaz et al., 2009; Wang et al., 2011). This kind of salinization is serious, especially when brackish water is used for irrigation (Phocaides, 2007). Since precipitation is rare in inland basins, irrigation plays an important role flushing and draining the salt from the surface soil to the groundwater. Therefore, a more comprehensive treatment of water and salt balances is essential in these regions.

5.2 Future: water-saving irrigation with restriction of reclamation

To avoid further deterioration of the environment and maintain sustainable development in oases in inland basins, water management, which takes ecosystem and environment into account, should be implemented across the whole region (Ward and Pulido-Velazquez, 2008). On the one hand, the groundwater table should remain deep in the irrigated districts by conversion from flood irrigation to water-saving irrigation; on the other hand, the agriculture field expansion should be strictly controlled. Most of the water savings which derive from increasing irrigation efficiency should remain in the river to recharge the groundwater in the lower reaches. The application of water-saving irrigation and restriction of cultivated land area will dramatically reduce water extraction from the river. The groundwater table in irrigated districts will then continually decline and salinization will be mitigated. The water shortage of the ecosystem in the lower reaches can be addressed and the riparian vegetation can gradually recover from degradation. The lake can also be fed by more fresh water from the river rather than by saline water from the irrigated croplands, and the salinity level of the lake water will thus decrease. The non-growth season flush is highly recommended for salt leaching in inland basins. However, the frequency and amount of flushing still should be optimized, which requires more study.

In this study, the exchange fluxes between the unsaturated zone and the groundwater table were obtained by water balance analysis based on comprehensive field measurements. The wide application of eddy covariance provided the possibility to perform this kind of analysis to quantify the exchange fluxes. The exchange flux as the linkage of groundwater dynamics between field scale and oasis scale is important for water management (Tang et al., 2007a; Timms et al., 2012). Therefore, the methodology of this study can be applied to evaluate the exchange fluxes and predict the salinization trend in arid regions.

Anthropogenic activities significantly affect water balance and groundwater table dynamics in the oases located in inland basins. The study showed that the water saving not only needs advanced irrigation technology, but also enlightened scientific management. It is important to understand the dynamics of human–water systems to maintain sustainable development. This study can be regarded as the first step towards quantitative studies of process socio-hydrology, which can generate much insight into human–water system dynamics and water management in inland basins (Sivapalan et al., 2012). The anthropogenic effects on human–water systems should be clarified further by both field observations and monitoring and through the use of an integrated model. This study could potentially contribute towards the development of such an integrated model.

6 Conclusions

In this study the seasonal water balance and groundwater dynamics under water-saving irrigation have been quantified based on comprehensive measurements conducted in an oasis within the Tarim River basin. The effects of conversion of irrigation methods on human–water systems were discussed to maintain sustainable water management in the future. In general, anthropogenic activities have significantly affected the water balance and groundwater dynamics in this arid area. After the application of water-saving irrigation, the downward exchange flux is greatly reduced during irrigation periods, while the upward flux is trivial during non-irrigation periods due to the moderate groundwater table depth. Some problems such as salinization induced by shallow groundwater table are mitigated, while new challenges also emerge. The water savings resulting from water-saving irrigation have not remained in the river to recharge the groundwater for ecologic use. In fact, they have instead been reused towards the expansion of irrigation croplands, resulting in even more water consumption. The continued decline of the groundwater table will lead to the destruction of the natural vegetation that relies on the groundwater and this will further result in the degradation of the whole ecosystem. In addition, the salinization risk still exists due to the insufficient leaching water under water-saving irrigation.

To be noted, the water balance terms in the agricultural oasis have been comprehensively measured in our study. However, the information about the snow, forest and desert areas (see Fig. 2) is still limited and their groundwater dynamics can only be qualitatively analyzed. In further studies, the detailed measurement should be implemented in the whole region. In addition, a more comprehensive treatment of water and salt balances should be introduced into the water management practice. The frequency and amount of the non-growth season flushing should be optimized to avoid the potential risk of secondary salinization, and groundwater dynamics under other crops should be explored.

To address the water challenges in the arid area, improved irrigation efficiency is, of course, important. However, improved understanding of human–water systems is also significant. With the adoption of a socio-hydrologic framework, there is potential for the dynamics of interactions between people and water to be further clarified, and for sustainable development to be achieved.

Acknowledgements. This research was funded by the National Science Foundation of China (NSFC 51190092, 51109110, and 51222901), SRF for ROCS, SEM, and the Foundation of the State Key Laboratory of Hydroscience and Engineering of Tsinghua University (2012-KY-03). We also thank the staff at Tsinghua University–Korla Oasis Eco-hydrology Experimental Research Station for their excellent work.

Edited by: T. J. Troy

References

Allen, R. G., Pereira, L. S., Howell, T. A., and Jensen, M. E.: Evapotranspiration information reporting: I. Factors governing measurement accuracy, Agric. Water Manage., 98, 899–920, 2011.

Baldocchi, D., Falge, E., Gu, L. H., Olson, R., Hollinger, D., Running, S., Anthoni, P., Bernhofer, C., Davis, K., Evans, R., Fuentes, J., Goldstein, A., Katul, G., Law, B., Lee, X. H., Malhi, Y., Meyers, T., Munger, W., Oechel, W., Paw, K. T., Pilegaard, K., Schmid, H. P., Valentini, R., Verma, S., Vesala, T., Wilson, K., and Wofsy, S.: FLUXNET: a new tool to study the temporal and spatial variability of ecosystem-scale carbon dioxide, water vapor, and energy flux densities, B. Am. Meteorol. Soc., 82, 2415–2434, 2001.

Barry, B. A.: Errors in practical measurement in science, engineering and technology, Wiley, New York, 1978.

Bruelheide, H., Jandt, U., Gries, D., Thomas, F. M., Foetzki, A., Buerkert, A., Gang, W., Zhang, X., and Runge, M.: Vegetation changes in a river oasis on the southern rim of the Taklamakan Desert in China between 1956 and 2000, Phytocoenologia, 33, 801–818, 2003.

Chavez J. L., Howell T. A., and Copeland K. S.: Evaluating eddy covariance cotton ET measurements in an advective environment with large weighing lysimeters, Irrig. Sci., 28, 35–50, 2009.

Chen, Y. N., Chen, Y. P., Xu, C., Ye, Z., Li, Z., Zhu, C., and Ma, X.: Effects of ecological water conveyance on groundwater dynamics and riparian vegetation in the lower reaches of Tarim River, China, Hydrol. Process, 24, 170–177, 2010.

Christen, E., DeLange, S., Patti, T., and Hornbuckle, J.: Soil salinity in drip irrigated vineyards of the MIA, Irrigation Research & Extension Committee Farmers' Newsletter, 176, 54–57, 2007.

Cui, Y. and Shao, J.: The role of ground water in arid/semiarid ecosystems, Northwest China, Ground Water, 43, 471–477, 2005.

Dou, C., Kang, Y., Wan, S., and Hu, W.: Soil salinity changes under cropping with Lycium Barbarum L. and irrigation with saline-sodic water, Pedosphere, 21, 539–548, 2011.

El Bastawesy, M., Ramadan Ali, R., Faid, A., and El Osta, M.: Assessment of waterlogging in agricultural megaprojects in the closed drainage basins of the Western Desert of Egypt, Hydrol. Earth Syst. Sci., 17, 1493–1501, doi:10.5194/hess-17-1493-2013, 2013.

Evett, S. R., Kustas, W. P., Gowda, P. H., Anderson, M. C., Prueger, J. H., and Howell, T. A.: Overview of the bushland evapotranspiration and agricultural remote sensing experiment 2008 (BEAREX08): a field experiment evaluating methods for quantifying ET at multiple scales, Adv. Water Resour., 50, 4–19, doi:10.1016/j.advwatres.2012.03.010, 2012.

Feng, Q., Cheng, G. D., and Masao, M. K.: Trends of water resource development and utilization in arid north-west China, Environ. Geol., 39, 831–838, 2000.

Foken T.: The energy balance closure problem: An overview, Ecol. Appl., 18, 1351–1367, 2008.

Gao, H. and Yao, Y.: Quantitative effect of human activities on water level change of Bosten Lake in recent 50 years, Scientia Geographica Sinica, 25, 3305–3309, 2005 (in Chinese with English abstract).

Gao, Q., Wang, R., and Giese, E.: Impact of climate change on surface runoff of Tarim River originating from the south slopes of the Tianshan Mountains, Journal of Glaciology and Geocryology, 30, 1–11, 2008 (in Chinese with English abstract).

Hanasaki, N., Fujimori, S., Yamamoto, T., Yoshikawa, S., Masaki, Y., Hijioka, Y., Kainuma, M., Kanamori, Y., Masui, T., Takahashi, K., and Kanae, S.: A global water scarcity assessment under Shared Socio-economic Pathways – Part 2: Water availability and scarcity, Hydrol. Earth Syst. Sci., 17, 2393–2413, doi:10.5194/hess-17-2393-2013, 2013.

Hu, H., Tian, F., and Hu, H.: Soil particle size distribution and its relationship with soil water and salt under mulched drip irrigation in Xinjiang Province of China, Sci. China Tech. Sci., 54, 1–7, 2011.

Ibragimov, N., Evett, S. R., Esanbekov, Y., Kamilov, B. S., Mirzaev, L., and Lamers, J. P. A.: Water use efficiency of irrigated cotton in Uzbekistan under drip and furrow irrigation, Agric. Water Manage., 90, 112–120, 2007.

Li, Q., Li, L., and Bao, A.: Snow cover change and its impact on streamflow in Kaidu River Basin, Resources Science, 34, 91–97, 2012 (in Chinese with English abstract).

Liu, Y., Tian, F., Hu, H., and Sivapalan, M.: Socio-hydrologic perspectives of the co-evolution of humans and water in the Tarim River basin, Western China: the Taiji-Tire model, Hydrol. Earth Syst. Sci., 18, 1289–1303, doi:10.5194/hess-18-1289-2014, 2014.

Ma, Y., He, J., Hong, M., and Zhao, J.: Analysis of development process and tendency of drip irrigation under film technology in Xinjiang, Water saving irrigation, 12, 87–89, 2010 (in Chinese with English abstract).

Palacios-Diaz, M. P., Mendoza-Grimon, V., Fernandez-Vera, J. R., Rodriguez-Rodriguez, F., Tejedor-Junco, M. T., and Hernandez-Moreno, J. M.: Subsurface drip irrigation and reclaimed water quality effects on phosphorus and salinity distribution and forage production, Agric. Water Manage., 96, 1659–1666, 2009.

Perry, C.: Efficient irrigation; Inefficient communication; Flawed recommendations, Irrig. Drain., 56, 367–378, 2007.

Phocaides, A.: Water quality for irrigation, in: Handbook on pressurized irrigation techniques, 2nd edition, FAO, Rome, Italy, Chapter 7, 2007.

Rajak, D., Manjunatha, M. V., Rajkumar, G. R., Hebbara, M., and Minhas, P. S.: Comparative effects of drip and furrow irrigation on the yield and water productivity of cotton (Gossypium hirsutum L.) in a saline and waterlogged vertisol, Agric. Water Manage., 83, 30–36, 2006.

Scanlon, B. R., Gates, J. B., Reedy, R. C., Jackson, W. A., and Bordovsky, J. P.: Effects of irrigated agroecosystems: 2. Quanlity of soil water and groundwater in the southern High Plain, Texas, Water Resour. Res., 46, W09538, doi:10.1029/2009WR008428, 2010.

Scheierling, S. M., Young, R. A., and Cardon, G. E.: Public subsidies for water-conserving irrigation investments: Hydrologic, agronomic, and economic assessment, Water Resour. Res., 42, W03428, doi:10.1029/2004WR003809, 2006.

Scott, C. A., Vicuña, S., Blanco-Gutiérrez, I., Meza, F., and Varela-Ortega, C.: Irrigation efficiency and water-policy implications for river basin resilience, Hydrol. Earth Syst. Sci., 18, 1339–1348, doi:10.5194/hess-18-1339-2014, 2014.

Shen, Y. and Chen, Y.: Global perspective on hydrology, water balance, and water resources management in arid basins, Hydrol. Process, 24, 129–135, 2010.

Sivapalan, M., Savenije, H. H. G., and Blöschl, G.: Socio-hydrology: A new science of people and water, Hydrol. Process, 26, 1270–1276, 2012.

Tang, Q., Hu, H., and Oki, T.: Groundwater recharge and discharge in a hyperarid alluvial plain (Akesu, Taklimakan Desert, China), Hydrol. Process, 21, 1345–1353, 2007a.

Tang, Q., Hu, H., Oki, T., and Tian, F.: Water balance within intensively cultivated alluvial plain in an arid environment, Water Resour. Manage., 21, 1703–1715, 2007b.

Tennakoon, S. B. and Milroy, S. P.: Crop water use and water use efficiency on irrigated cotton farms in Australia, Agric., Water Manage., 61, 179–194, 2003.

Timms, W. A., Young, R. R., and Huth, N.: Implications of deep drainage through saline clay for groundwater recharge and sustainable cropping in a semi-arid catchment, Australia, Hydrol. Earth Syst. Sci., 16, 1203–1219, doi:10.5194/hess-16-1203-2012, 2012.

Wan, H., Sun, Z., and Wang, R.: Study on the evaluation of ecological frangibility of the wetlands in the Bosten Lake Region, Arid Land Geography, 29, 248–254, 2006 (in Chinese with English abstract).

Wang, R., Kang, Y., Wan, S., Hu, W., Liu, S., and Liu, S.: Salt distribution and the growth of cotton under different drip irrigation regimes in a saline area, Agric. Water Manage., 100, 58–69, 2011.

Wang, T., Xie, H., and Jia, B.: A preliminary study on regulation mode of oasis ecological system in Peacock River Valley of Xinjiang, Journal Arid Land Resources and Environment, 16, 7–11, 2002 (in Chinese with English abstract).

Wang, W., Wang, X., Jiang, F., and Peng, D.: Temperature and precipitation along the Kaidu River over the past 50 years, Arid Land Geography, 35, 746–753, 2012 (in Chinese with English abstract).

Wang, W., Wang, X., Jiang, F., and Peng, D.: Response of runoff volume to climate change in the Kaidu River Basin in recent 30 years, Arid Zone research, 30, 743–748, 2013 (in Chinese with English abstract).

Ward, F. A. and Pulido-Velazquez, M.: Water conservation in irrigation can increase water use, P. Natl. Acad. Sci., 105, 18215–18220, 2008.

Wilson, K. B., Hanson, P. J., Mulholland, P. J., Baldocchi, D. D., and Wullschleger, S. D.: A comparison of methods for determining forest evapotranspiration and its components: sap-flow, soil water budget, eddy covariance and catchment water balance, Agric. For. Meteorol., 106, 153–168, 2001.

Wilson, K., Goldstein, A., Falge, E., Aubinet, M., Baldocchi, D., Berbigier, P., Bernhofer, C., Ceulemans, R., Dolmanh, H., Field, C., Grelle, A., Ibrom, A., Lawl, B. E., Kowalski, A., Meyers, T., Moncrieffm, J., Monsonn, R., Oechel, W., Tenhunen, J., Valentini, R., and Verma, S.: Energy balance closure at FLUXNET sites, Agric. For. Meteorol., 113, 223–243, 2002.

Xia, J., Zuo, Q., and Shao, M.: Sustainable utilization of water resources in Bosten Lake: Theory, method and practice, Science Press, Beijing, China, 2003 (in Chinese).

Xie, G., Zhang, J., Tang, X., Cai, Y., and Gao, G.: Spatio-temporal heterogeneity of water quality (2010–2011) and succession patterns in Lake Bosten during the past 50 years, Journal of lake sciences, 837–846, 2011 (in Chinese with English abstract).

Xu, H., Ye, M., Song, Y., and Chen, Y.: The natural vegetation responses to the groundwater change resulting from ecological water conveyances to the Lower Tarim River, Environ. Monit. Assess., 131, 37–48, 2007.

Xu, Z., Chen, Y., and Li, J.: Impact of climate change on water resources in the Tarim River basin, Water Resour. Manage., 18, 439–458, 2004.

Xu, Z., Liu, Z., Fu, G., and Chen, Y.: Trends of major hydroclimatic variables in the Tarim River basin during the past 50 years, J. Arid Environ., 74, 256–267, 2010.

Zhang, J.: Study of optimized allocation of water resources in Kaidu River- Kongqi River Basin, M. S. thesis, Xinjiang Agricultural University, China, 60 pp., 2012 (in Chinese with English abstract).

Zhang, J. and Hu, S.: Salinity Analysis of Bosten Lake, Arid Environmental Monitoring, 22, 19–23, 2008 (in Chinese with English abstract).

Zhang, Q., Xu, C., Tao, H., Jiang, T., and Chen, Y.: Climate changes and their impacts on water resources in the arid regions: a case study of the Tarim River basin, China, Stoch. Environ. Res. Risk Assess., 24, 349–358, 2010.

Zhang, Q., Xu, H., Li, Y., Fan, Z., Zhang, P., Yu, P., and Ling, H.: Oasis evolution and water resource utilization of a typical area in the inland river basin of an arid area: a case study of the Manas River valley, Environ. Earth Sci., 66, 683–692, 2012.

Zhang, Y., Li, B., Bao, A., Zhou, C., Chen, X., and Zhang, X.: Study on snowmelt runoff simulation in the Kaidu River basin, Sci. China Earth Sci., 50, 26–35, 2007 (in Chinese with English abstract).

Zhang, Z., Hu, H. C., Tian F., Hu, H. P., Yao, X., and Zhong, R.: Soil salt distribution under mulched drip irrigation in an arid area of northwestern China, J. Arid. Environ., 104, 23–33, 2014a.

Zhang, Z., Tian, F., Hu, H., and Yang, P.: A comparison of methods for determining field evapotranspiration: photosynthesis system, sap flow, and eddy covariance, Hydrol. Earth Syst. Sci., 18, 1053–1072, doi:10.5194/hess-18-1053-2014, 2014b.

Zhou, C., Luo, G., Li, C., Tang, Q., Li, H., Wang, Q., and Fukui, H.: Environmental change in Bosten Lake and its relation with the oasis reclamation in Yanqi Basin, Geogr. Res., 20, 14–23, 2001 (in Chinese with English abstract).

Zhou, S., Wang, J., Liu, J., Yang, J., Xu, Y., and Li, J.: Evapotranspiration of a drip-irrigated, film-mulched cotton field in northern Xinjiang, China, Hydrol. Processes, 26, 1169–1178, doi:10.1002/hyp.8208, 2011.

Zhu, Y., Wu, Y., and Drake, S.: A survey: obstacles and strategies for the development of ground-water resources in arid inland river basins of Western China, J. Arid Environ., 59, 351–367, 2004.

2

A propensity index for surface runoff on a karst plateau

Christian Reszler[1], **Jürgen Komma**[2], **Hermann Stadler**[1], **Elmar Strobl**[3], **and Günter Blöschl**[2]

[1]JR-AquaConSol GmbH, Steyrergasse 21, 8010 Graz, Austria
[2]Vienna University of Technology, Institute for Hydraulic and Water Resources Engineering, and Centre for Water Resource Systems, Karlsplatz 13, 1040 Vienna, Austria
[3]Technisches Büro – Ingenieurbüro Mag. Dr. Elmar Strobl, Grubbergweg 4, 8062 Kumberg, Austria

Correspondence: Christian Reszler (christian.reszler@jr-aquaconsol.at)

Abstract. Karst aquifers are an important water resource, but are particularly vulnerable to pollution due to the typically short residence times. As the rainwater runs off on the surface it may collect contamination from faeces and other sources, before infiltrating the surface. It is therefore important to understand the spatial distribution of the frequency of surface runoff in karst areas. This paper proposes a new field-mapping method for the ability of the landscape to produce and convey surface runoff. The mapping method is based on (i) prior spatial information (e.g. geological map, terrain model), (ii) a visual assessment from a distance at the landscape scale (e.g. traces of surface runoff) and (iii) local data collection in the field (e.g. soil moisture, grain size distribution). The focus on variables that can be assessed from a distance in the landscape makes the method suitable for mapping larger areas than traditional field mapping. The mapping method is developed and tested for the 60 km^2 Hochschwab area in Austria. The field mapping is used to specify a surface runoff propensity index which is tested against the spatial distribution of observed sink holes in the area. The mapping indicates that, in the study region, runoff occurs much more frequently in the poorly karstified dolomitic areas than in the limestone areas that are highly karstified. On dolomites, low permeable soils or debris prevail, often resulting in a permanent surface drainage network. On karstified limestone, sometimes overlaid by debris, surface runoff only occurs through infiltration excess at high rainfall intensities. Overall the analyses suggest that the mapping method is suitable for efficiently and reliably identifying spatial patterns of the ability of the landscape to produce and convey surface runoff in karst areas.

1 Introduction

Karst groundwater is an important source for drinking water in many countries around the world. In Austria, numerous karst springs exist, e.g. in the Northern Calcareous Alps, where the water is used for the drinking water supply of Vienna (e.g. Plan et al., 2010). In karst areas, soils tend to be shallow, and the fast drainage through the karst system make karst springs very vulnerable to pollution in the source areas. Pollution may include faeces from cattle farming and wild animals, tourism, and accidental fuel spills on forest roads. Safeguarding the quality of karst springs therefore involves land management of the source areas in order to minimize the risk of pollution. An understanding of the processes of surface runoff may enhance the efficiency of the land management by quantifying the hazard of contaminant input into the karst system, particularly during high-intensity convective storms. During convective storms, flow processes tend to accelerate (Bonacci, 2001a); they therefore play an important role in material mobilization and transport. The longer contaminants are in contact with the surface water, the more likely is their transport to sink holes and further into the karst system towards the spring. Frequent surface runoff in karst areas can therefore be associated with high contamination risk, all other factors being equal.

At the plot scale, the propensity of areas to produce surface runoff is usually mapped by field surveys (e.g. Peschke et al., 1998; Scherrer and Naef, 2003) based on a detailed mapping of hydrological properties, often including sprinkling experiments (Markart et al., 2004). Such mapping exercises have been used to develop methodologies for identifying hydrological response units (HRUs). HRUs are based on the idea that areas of similar physiographic properties, such as slope aspect, vegetation type and soil type, imply hydrological similarity, i.e. similar runoff response (Flügel, 1995). The way the layers of information are combined can have various degrees of process representation. For example, Flügel (1995) combined the layers by reasoning for a unit to represent "rangeland on gley soil at the valley floor with shallow groundwater over impervious bedrock".

In karstic catchments, traditional mapping methods tend to focus on the geology and the hydrogeology (e.g. Andreo et al., 2006; Goldscheider, 2015), which comprise the lithology, the existence of faults and their permeability, the location of karst forms (e.g. dolines, poljes) and, less often, soil surveys. The focus on geology may be related to the wider availability of geological maps than soil maps and to the deep subsurface flow paths feeding karst springs. Sprinkling experiments in the limestone region of southeastern Spain were used for elucidating the mechanisms of runoff generation and infiltration. For example, Calvo-Cases et al. (2003) found that runoff on bare patches was mainly generated by the Hortonian mechanism. At the slope scale, Hortonian discontinuous runoff occurred on the most degraded slopes or during high-intensity rain events, and mixed runoff generation on less degraded slopes or on previously wet soils. For both types, there was a patchwork of runoff and runon areas. Li et al. (2011) found vegetation cover, litter and rainfall intensity to be the main controls of infiltration rates, and the presence of bare soil and rainfall intensity to be the main controls of the runoff coefficient. Canton et al. (2016) found generally higher clay contents towards the foot of hillslopes which explained the increase of soil moisture downslope at depths of 0.1–0.35 m. Macropores (rock and soil cracks) disconnected the surface water fluxes and reduced runon to the lower part of the hillslope during wet periods. The spatial connectivity of flow processes tends to enhance fast contaminant transport (Western et al., 1998, 2001), but Fiener et al. (2011) point out that the effects of patchiness and the spatial organization of landscape characteristics on flow and transport processes are still poorly understood.

Z. Fu et al. (2016) studied the role of "epikarst" in near-surface hydrological processes in a subtropical karst region of southwestern China based on plot-scale rainfall simulation experiments. They found that more than 70 % of the total rainfall moved vertically through the shallow soil layer and was redistributed as subsurface flow at the soil–epikarst interface or percolated deeper into the rock. T. Fu et al. (2015, 2016) and Chen et al. (2010) identified the spatial variability of the topsoil saturated hydraulic conductivity, surface soil water content and soil moisture dynamics as the main controls of infiltration in a small karst catchment in a subtropical region of China, and Peng and Wang (2012) found that surface runoff from the karstic hillslopes in southwestern China was much less than that from non-karstic areas and that limestone fissures and fractures played important roles in surface runoff generation on karst limestone slopes due to their large infiltration and storage capacities.

At the scales of a couple of square kilometres or more, which are more relevant from a management perspective, karst studies are traditionally based on GIS tools with little field validation. Vulnerability is often mapped by combining various layers of spatial data such as vegetation, land use, soil texture and geology (e.g. Flügel, 1995; Cost Action 620, 2003). Walter et al. (2000) presented a GIS-based conceptual approach to assessing water quality risk in agricultural catchments based on the variable source area concept, following Frankenberger et al. (1999). They used the term hydrologically sensitive area (HSA) to refer to areas that are especially prone to generating runoff and therefore transporting contaminants to surface water bodies. HSAs are similar to HRUs, but differ from them in terms of their stronger emphasis on flow connectivity and transport of contaminants. The method was further developed by Thomas et al. (2016) by the use of lidar elevation models and by validation against rainfall and quick-flow measurements. Similar to the HSA concept, Plan et al. (2009) identified "ultra-vulnerable areas" using a mapping procedure that involves the automatic delineation of local catchment boundaries of individual karst forms.

The purpose of this paper is to propose a method that allows the mapping of surface runoff propensity in karstic environments at the catchment scale. The main innovation of the method is that it allows the more rapid mapping of larger areas than the traditional geological mapping paradigm which consists of traversing the entire region of interest on foot. The study is set in a 63 km^2 karst plateau in Austria which is the recharge area of the main springs of the Vienna drinking water supply.

2 Case study area

The study area is part of the Hochschwab massif, one of the important karst areas in Austria. It consists of Triassic limestones and dolostones with a thickness of up to 2000 m. The karstified rocks sit on shales and sandstones of the Werfen formation (Mandl et al., 2002). The geological set-up has led to intensive karstification including numerous cave systems and a range of surface karst forms such as dolines, karren, polje and ponors.

Due to the tectonic structure (Decker et al., 2006), the main karst spring (LKAS2) is situated at the northern slope of the Hochschwab massif. The catchment (Fig. 1) of the main spring covers an area of about 60 km^2. Elevations range from 650 to 2277 m a.s.l. with a mean elevation of 1720 m a.s.l. (Stadler and Strobl, 2006). The vegetation mainly comprises forests, krummholz areas, alpine pastures on the plateau and badlands in the peak regions. Alpine farming and touristic activities during the summer months are important land use activities and potential sources for contam-

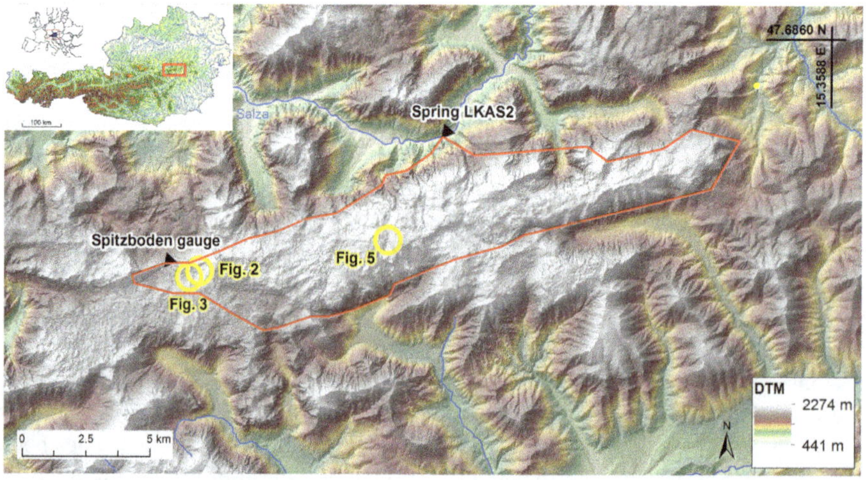

Figure 1. Area subject to mapping (red line) in the catchment of the Hochschwab spring (LKAS2). Yellow circles indicate the locations of Figs. 2, 3 and 5.

ination of the karst aquifer. Additionally, faeces of wildlife may contaminate the aquifer.

A hydrogeological map was compiled based on a detailed geological map (Mandl et al., 2002) and information from additional mapping surveys. The basic concept of compiling the map was the hydrogeological quality rating of the rocks, according to the "rock quality designation" (RQD) concept of exploration geology (Stadler et al., 2016). This hydrogeological quality rating describes fundamental properties of the rocks such as karstification capacity; tectonic stress related to water storage capacity; and permeability, bedding, underline and fall. The objective was to combine different lithological units with units with similar hydrogeological behaviour.

The delineation of the catchment boundaries of the spring is based on two pieces of information: the detailed geological mapping and investigation of geological structures, and a long-term analysis of stable isotopes, ^{18}O and deuterium, which provided robust estimates of the mean altitude of the spring (Stadler and Strobl, 2006). Spring discharge measurements with high temporal resolution (15 min) have been conducted by Vienna Water since 1992. A stream gauge at the Spitzboden site is used to assess whether the local stream is permanent or not. The catchment boundary differs significantly from the topographic boundary of the spring.

3 Method of process-based mapping of surface runoff propensity

The proposed method of mapping runoff characteristics consists of combining (i) prior spatial information (e.g. geological map, digital elevation model), (ii) a visual assessment from a distance at the landscape scale (e.g. traces of surface runoff, spatial extent of hydrologically homogeneous areas) and (iii) local data collection in the field (e.g. soil moisture, grain size distribution). The mapping results in a classification of the entire landscape, which is used for calculating a surface runoff propensity index. The mapping method was developed and tested for the Hochschwab area.

Before starting the field campaign, available spatial information of the study region was analysed. A lidar data set at 1 m resolution was used to obtain a prior understanding of the geomorphology of the area (including karst forms), and for orientation in the field. A hydrogeological map was analysed to obtain an understanding of the bedrock (Stadler and Strobl, 2006; based on geological and tectonic information by Mandl et al., 2002; Decker, 2006; and Bryda et al., 2013). On the basis of both maps, strategic lookout points were identified that were suitable for a landscape-scale assessment. These were mainly vegetation-free hilltops or locations on counter slopes, preferably with the line of sight orthogonal to the land surface, and that were accessible during the campaign.

During the field campaign, the field researcher positioned him- or herself at a lookout point. On the basis of a visual assessment of the landscape, its geomorphology, apparent soil characteristics, surface runoff traces, water logging and vegetation patterns ("reading the landscape") the field researcher then tentatively delineated polygons on the map with approximately homogenous characteristics (HRUs) and specified these characteristics. Next he or she selected representative points in these polygons, still from the lookout point, to test their position and characteristics. Selection and number of points depended on the size of the polygon (more points in larger areas), line of sight from the lookout point (no interruption by vegetation), and the spatial patterns of similarity indicators such as morphology and slope. He or she then walked to these representative points and performed local measurements (soil depth, grain size, soil moisture, etc.) which they used to revise the polygons or characteristics, if necessary. The procedure was repeated starting with the next lookout point. In some cases it was found to be more practical to apply the method in reversed order. Representative points, i.e. positions where a significant change in properties were found while walking on a slope, were mapped locally and the transfer to the area was done afterwards from a lookout point.

The mapping of the polygons, or HRUs, involves four characteristics, and each polygon consists of a combination of these characteristics (Table 1):

a. *Geology (3 classes).* Three geological classes occur in the study region according to the geological map: limestone (KAL), dolomite (DOL) and calcareous sand stone (KSS). The three classes mainly differ in their karstification potential. KAL is much more karstified than DOL and KSS, and thus cracks and fissures predominantly lead to immediate percolation into the karst system. DOL and KSS bedrock often show ponding characteristics, which favour the generation of surface runoff and lateral subsurface flow towards a drainage network. The classification was based on the hydrogeological map and was checked during the field campaign, and was complemented by an assessment of the karstification capacity and tectonic stress related to water storage capacity based on the quality designation concept. Finally, the infiltration capacity of the bedrock was classified into high, medium and low on the basis of the geological classes and the additional hydrogeological assessment above for each polygon.

b. *Soils and debris (8 classes).* The main mapping item was the type of the loose material, debris and soils along with their infiltration characteristics. Grain sizes were visually assessed. One soil sample per site at selected mapping points was sieved to obtain information about water storage capacity and infiltration capability, using a 2 mm sieve to separate sand and gravel (i.e. fraction of material larger than 2 mm in vol %). The 2 mm threshold was chosen as it lends itself to dry sieving and because of its frequent use in soil mapping as a threshold between fine and coarse material (AG Boden, 1994). These measurements are quick to make and require a minimum of equipment. A total of 26 points was selected for detailed analyses during the field campaigns. Soil depths were mapped based on visible soil profiles at slope cuttings. They were classified into four groups (< 0.1, 0.1–0.5, 0.5–1, > 1 m). Deeper soils indicate larger storage capacity and delayed runoff. Table 3 provides a summary of the soil and debris type classification.

   An important item in this study was the mapping of surface runoff traces at hillslopes, such as washed out sediments and bent vegetation. Characteristics related to surface runoff were visually mapped from a distance, particularly observable permanent flow (drainage network and springs), indications for temporary surface flow (such as dry creeks), bent vegetation (post-event), erosion traces on the hillslopes or sediment that accumulated during surface flow. The observations on a particular day reflected all visible traces from the events during a period preceding the mapping date. Some of the surveys were performed immediately after heavy storms (e.g. a storm on 4 August 2013), which allowed mapping of particularly clear traces of surface runoff (post-event surveys; see for example Borga et al., 2008). Intense and large storm events may also leave traces of infiltration excess runoff on the highly permeable debris. Temporary and permanent creeks were visually mapped (also from a distance), and the order of magnitude of the flow width was estimated visually (0, 0–10, 10–100, > 100 m). Water logging occurrences, also found using mapping vegetation wetness indicators (Markart et al., 2004; Rogger et al., 2012), assisted in identifying areas where saturation excess is deemed to occur frequently. Water logging and surface runoff traces are related to the soils and were therefore used as one of the indicators for the soil classification.

   Another spatio-temporal characteristic variable is soil moisture, which was measured during the field campaigns by portable TDR ("ThetaProbe", UMS) equipment down to 6 cm depth. Standard equations recommended by the sensor producer were used to transform the measured dielectric constant to volumetric water content for organic and mineral soils (UMS, 2001). Measurements at characteristic points were performed primarily to confirm (or otherwise) the assessed drainage characteristics of the corresponding polygon. The measurements were conducted to assess the spatial variability of soil moisture (e.g. seven points in the Spitzboden test area) and the temporal variability (e.g. 11 points along an easily accessible transect on three dates).

   Typical karst forms were identified from a distance, and the locations were cross-checked with the hydrogeological map and the 1 m elevation model. Only those karst forms were mapped that were considered significant regarding surface runoff and subsequent direct infiltration. These were generally non-filled sinks with runoff traces indicating a sizeable local catchment. Finally, the infiltration capacity of the soils and debris layer was classified into high, medium and low on the basis of the soils and debris information described above for each polygon, including general morphology (i.e. apparent karst forms), soil moisture measurements, water logging and surface runoff traces.

c. *Vegetation (2 classes).* Vegetation cover was assigned one of two classes, dense vegetation (SVE) and sparse vegetation (GVE). Dense alpine forest does not occur on

Table 1. Process-oriented mapping catalogue involving geology, soils and debris, vegetation, and infiltration capacity. An "x" symbol indicates whether information from existing data, local mapping or from a distance has been used. An asterisk (*) indicates that information has been combined from the previous rows.

Main property	Item	Mapping: existing data	Mapping: from a distance	Mapping: local mapping	Synthesis	Method	Examples	Processes
Geology	Classification	x		x		Geological map, randomly checked in the field (hydrochloric acid test)	Limestone (KAL), dolomite (DOL), calcareous sandstone (KSS), 3 classes	Percolation into karst system vs. lateral subsurface flow towards drainage network
	Assessment	x		x		Rock quality designation	Karstification capacity, tectonic stress related to water storage capacity	Percolation into karst system vs. lateral subsurface flow towards drainage network
	Classification of infiltration capacity of the bedrock				*	Synthesis of geological information	3 classes: high, medium and low infiltration capacity of the bedrock	
Soils and debris	Grain sizes			x		Descriptive	Silt, sand, gravel, etc.	General soil physical properties
(surface material)	Fraction larger than 2 mm			x		In situ sieving at selected mapping points		Reduced storage capacity and increased hydraulic conductivities
	Soil depths			x		Morphological estimates based on visible soil profiles at slope cuttings	Four depth classes	Increased storage capacity, runoff delay
	Surface runoff traces		x	x		Visual examination	Gullies, erosion traces, sediments, bent vegetation	Surface runoff generation propensity and frequency: temporary and permanent creeks
	Indication for waterlogging		x	x		Visual examination	Swamps, ponds	None or very low storage capacity, immediate surface runoff generation

Table 1. Continued.

Main property	Item	Mapping: existing data	Mapping: from a distance	Mapping: local mapping	Synthesis	Method	Examples	Processes
	Soil moisture at time of survey			x		TDR (portable)	Relative differences between mapping points (spatially) and between different dates (temporally)	Distinction between generally wet or dry areas (spatially), wetting and drying dynamics (temporally)
	Karst forms and surface runoff towards these	x	x	x		Visual examination, assisted by 1 m DTM and hydrogeological map	Dolines, sinks	
	Classification of soil type	x	x	x	*	Based on soil and debris descriptors	OLG, BOF, HSG, etc. (8 classes); see Table 3	
	Classification of infiltration capacity of the soils and debris				*	Synthesis of soil and debris descriptors	3 classes: high, medium and low infiltration capacity of soils and debris	
Vegetation	Classification		x	x		Visual examination	2 classes: GVE, SVE	Interception and ET processes (of minor relevance in this work)
	Wetness indicators		x	x		According to Markart et al. (2004)	Hydrophilic plants	Water logging, high water availability for plants (e.g. high local groundwater table), low storage capacity, preferential surface flow paths
Infiltration capacity	Classification				*	Combination of infiltration capacity classes of the bedrock with those of the soils and debris, accounting for vegetation effects	3 classes: high, medium and low overall infiltration capacity (HIN, MIN, GIN)	

the plateau. Areas with dense grassland (SVE) mostly exhibit soils with significant organic matter and high moisture content. Sparse vegetation was assigned to an area not fully covered by vegetation, i.e. included areas with bare soil or bare rock. Since the campaigns were conducted in the summer and the focus was on summer conditions, the vegetation state during the campaigns was considered representative. Wetness indicators based on hydrophilic plants were mapped according to Markart et al. (2004) (also see Rogger et al., 2012, for application of the method). The occurrence of wetness indicators indicates low storage capacity (high local groundwater table), reduced infiltration capacity and the potential for preferential surface flow paths.

d. *Infiltration capacity (3 classes).* As a synthesis of the classifications of the infiltration capacity of the bedrock and that of the soils and debris, and accounting for vegetation effects, the overall infiltration capacity of each polygon was classified in situ by the field researcher (Table 1) into three classes: high (HIN), medium (MIN) and low (GIN) infiltration capacity. Conceptually it was considered to represent the infiltration capacity for a typical summer storm, across a polygon. This classification is an attempt of an overall evaluation of the infiltration capacity of the rock–soil–vegetation continuum in the field. The overall capacity was therefore determined as the minimum of the infiltration capacities of the two layers: soils–debris and bedrock. For example, low infiltration capacity of the soils and debris and high infiltration capacity of the bedrock was considered to result in an overall low infiltration capacity. High infiltration capacity of the soils and debris and medium infiltration capacity of the bedrock was considered to result in an overall medium infiltration capacity. The minimum was chosen as an approximation to the effective hydraulic conductivity of a layered porous medium with flow perpendicular to the layers which is the harmonic mean of the individual conductivities. To account for the effect of vegetation on the infiltration capacity, in case of dense grassland (SVE) the overall infiltration capacity was reduced from high to medium (the other two classes were not changed). This change reflects the uptake of water by the vegetation and enhanced surface runoff on mountain meadows (Markart et al., 2004). In some areas, the classification was clear. For example, on the high alpine, intensely karstified plateau with no significant soils, no vegetation, but clearly visible karst forms, the high infiltration capacity classification was obvious. In other areas, the characteristics within the polygons were heterogeneous, so the classification was more difficult. Typically, units with a scale of more than 100 m were resolved (e.g. a debris area of 200 m length) while smaller units were lumped into the same polygon (e.g. an area with debris at the foot of a slope of 20 m length). This synthesis step also allowed a plausibility check of the polygon delineation in terms of the homogeneity of the polygons and their mapped properties.

Table 2. Hydrological response units of Fig. 2.

No.	Geology	Soils, debris	Vegetation	Infiltration capacity
1	Limestone	Bare bedrock	Sparse	High
2	Limestone	Coarse debris	Sparse	High
4	Limestone	Boulders	Sparse	High
7	Limestone	Loam	Dense	Medium
8	Dolomite	Bare bedrock	Sparse	Low
12	Dolomite	Loam	Dense	Medium
13	Calcareous sandstone	Loam	Dense	Medium
16	Calcareous sandstone	Boulders	Dense	High
17	Calcareous sandstone	Organic soils	Dense	Low

This approach contrasts with the traditional geological mapping paradigm which consists of traversing the entire region on foot. The proposed approach uses variables that can be assessed from a distance in the landscape. While the accuracy may be somewhat lower, it enables the rapid mapping of larger areas. In this study, a total area of 63 km^2 was mapped by one person within a total of 12 days, a 2-day campaign in July 2012 and a campaign from 1 July to 6 September 2013.

4 Examples of mapping results

Figure 2 shows an example of the mapped polygons, with the explanation of the units in Table 2. Polygons no. 1, no. 4 and no. 7 consist of limestone (KAL) and differ in their overlying material and debris: in polygon no. 1 bare rock occurs, no. 4 is a rockfall area and no. 7 is flatter and exhibits a thin soil layer. Soil samples indicated fine soils (BOF, Table 3). All three polygons show sparse vegetation (GVE). No. 13, no. 16 and no. 17 are examples for underlying calcareous sandstone (KSS) with low permeabilities compared to KAL. They differ in their slopes, which has led to different soil sedimentation processes. The flat area in no. 17 shows organic soils with high wetness (BOO). A drainage network originates in this polygon, which forms a permanent creek. In polygon no. 13 much less organic material has been deposited, but there is a high fraction of fine material in the soil (BOF) (identified by detailed mapping). Polygon no. 16 is a rockfall area, but the vegetation cover (SVE) indicates some water storage in the debris fed by runoff from the adjacent bare rock areas and supported by the ponding properties of the underlying KSS. Except for no. 17, surface runoff traces such as gullies are not pronounced in the polygons, which indicates low surface runoff propensity and significant deep percolation.

Figure 3 shows a cross section of the mapping results in the Spitzboden area (see Fig. 1). Bare dolomitic rock at the Plankogel (far southwest, left in Fig. 3) shows low infiltration capacity, because of its small fissures. Precipitation falling

Table 3. Classification of "soils and debris" type into 8 classes.

Type (acronym)	Name	Description
OLG	Bare rock	Bare rock (no soils), properties of geology control runoff generation
HSF	Fine debris	Debris predominantly fine material from dolomitic rocks, low permeability, medium to high storage capacity
HSG	Coarse debris	Debris predominantly coarse material, often at the bottom of steep calcareous slopes
BST	Boulders	Rockfall material: large boulders, infiltration properties of underlying geology (mainly limestone) becomes dominant
FLS	Fluviatile sand	Fine fluviatile sediments (small depth), connected to permanent water flow (creek bed, riparian zone), occurs rarely
BOF	Loam	Soil with high fraction of fine material, mainly Neogene ("red loam", secondary sediments from crystalline), comprise organic material, potential humus, grain size: silt–clay, very low permeability
BOG	Coarse sand	Soil with low/no fraction of fine material, developed in situ mainly over limestone, small depths, occurs also at steeper slopes, very low storage
BOO	Organic soil	Soil with significant organic components (humus), mainly above BOF with shallow impermeable layers, low storage

Figure 2. Example of delineation of response units based on Table 1. For an explanation of units see Table 2. To illustrate the scale, the diameter of polygon no. 16 from left to right is approximately 300 m.

on these areas will mostly flow towards the debris and the drainage network on the surface, due to the steep slopes. Only at high precipitation intensities will water flow through the debris and reach the drainage network as surface runoff; otherwise it is expected to infiltrate into the debris and contribute to the springs. In these ditches connected to the neighbouring rock walls only occasional surface runoff traces were found in the field. The springs form a drainage system and a small creek, which sinks into a karst doline right after the stream gauge location. The bottom of the Spitzboden (centre in Fig. 3) is dominated by low permeable sediments overlain by soils with high organic contents. A permanent drainage system exists, which facilitates surface and subsurface flow generated by the precipitation on the area itself and subsurface flow from the neighbouring dolomitic debris. Due to the available but low storage capacity of the organic soil (BOO), surface runoff is expected to be slightly delayed (saturation overland flow). On the neighbouring debris, temporary sur-

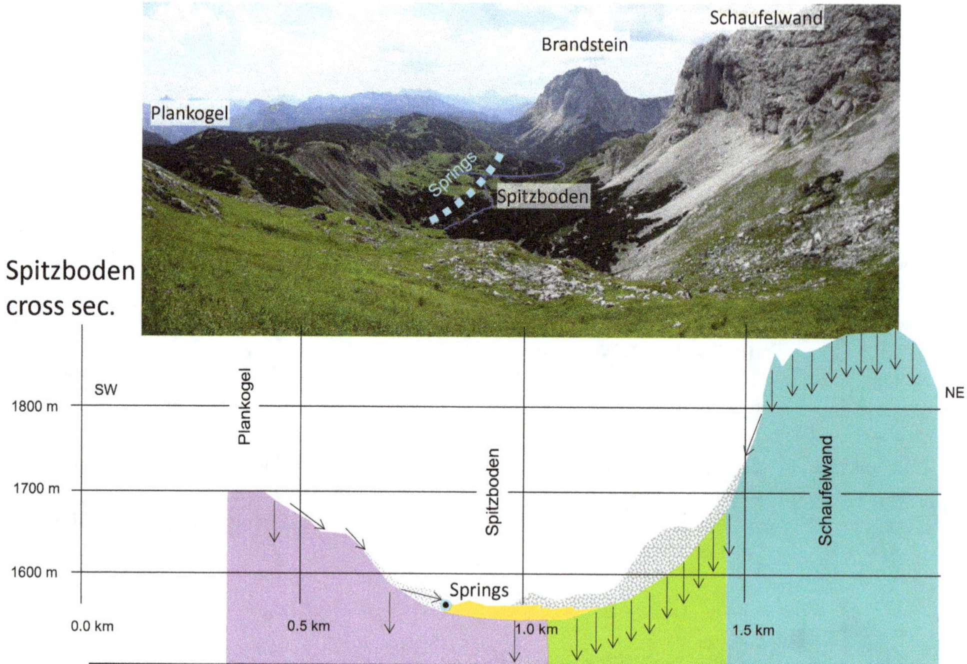

Geology	Dolomite			Calcareous sandstone		Limestone	
Soils and debris	No/minor	Debris fine	Organic soil	Organic soil	Rock fall	Debris coarse	No/minor
Vegetation	Sparse	Sparse	Dense	Dense	Dense	Sparse	Sparse
Infiltration	Low	Medium	Low	Low	High	High	High

Figure 3. Cross section of the Spitzboden area illustrating the mapped properties (geology, soils and debris, vegetation, and infiltration capacity; see Table 1). Yellow and grey shading relates to organic soil layer and coarse debris from the wall in the northeast (NE), respectively. Blue arrows indicate dominant drainage directions during a typical summer storm; a higher density of arrows indicates larger fluxes.

face runoff traces (small ditches) were observed, but springs at the bottom of the slope indicate that, most of the time, precipitation infiltrates and subsurface flow tends to occur at the interface to the relatively low permeability underlying dolomite (interflow, shallow groundwater flow). Permanent spring discharge could be observed, which contributes to the Spitzboden drainage network. However, the small ditches indicate surface runoff towards the lower permeable organic soils, i.e. the network, at high rainfall intensities. Rocks in the northeastern parts (right in Fig. 3) are calcareous sandstones and limestones with high infiltration capacities. From the Spitzboden to the foot of the Schaufelwand the bare rock is covered by coarse debris (gravel) and rockfall material with very high permeability. It is very likely that precipitation falling on these areas contributes very little to the Spitzboden creek. Direct infiltration is dominant with subsequent deep percolation into the karst system. Exceptions are the very steep slopes where surface runoff can occur but will infiltrate immediately into the debris at the slope bottom, as inferred by the short flow lengths of the gullies observed.

5 Definition of the surface runoff propensity index

An index is proposed based on the mapped infiltration capacity to summarize typical patterns of surface runoff. We term this index the "surface runoff propensity index". It is intended to reflect the likelihood of a particular location to exhibit surface runoff during high-intensity summer storms. Since the rainfall mechanisms are rather uniform in the study area, the index focuses on the infiltration capacity of the soil as well as topographic slope. The latter reflects the field observation in the study area that steep slopes tend to exhibit surface runoff more frequently than flatter slopes with the same soil types. Even in intensely karstified areas with coarse debris on steep slopes, surface runoff was observed to occur. Traces (gullies) were observed resulting from runon from bare rock areas above and surface runoff generated on the area itself. For example, Fig. 5 shows traces of a temporary flow path above highly permeable sediment after the extreme event on 4 August 2013. The flow path collected surface runoff from the steep upslope areas (bare rock and debris) associated with the infiltration excess mechanism. Incorporating topographic slope in the surface runoff propensity index is mainly intended to represent the steep slopes of karstified bare rock and debris with high and medium infiltra-

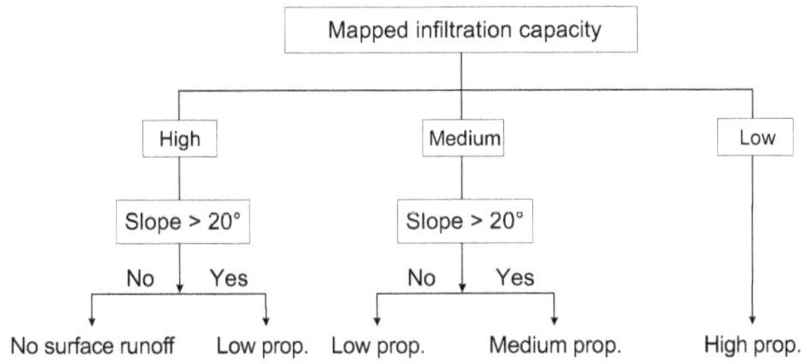

Figure 4. Decision tree for assigning the propensity index to the mapped polygons of infiltration capacity.

tion capacities. For areas with low infiltration capacity, slope was not used in the index.

The index values are estimated by a decision tree from the classified infiltration capacity and topography (Fig. 4). Four classes were defined: no surface runoff and low, medium and high surface propensity. No surface runoff can be expected at locations with coarse sediments and debris and intensely karstified rocks with high infiltration capacity (HIN) in relatively flat areas. The surface roughness of these areas is high, and rainfall as well as rare infiltration excess surface flow will infiltrate immediately. The field mapping and morphologic analyses showed that 20° is the range of threshold slopes beyond which gullies occur much more frequently, so the threshold slope was set to 20°. Low propensity is assigned to areas with intensely karstified rocks with high infiltration (HIN) and steep slopes (slope $> 20°$) as well as areas with medium infiltration (MIN) and flat slopes. Only short gullies can sometimes be observed on steep slopes in these areas. Clear surface runoff traces were not observed, but ephemeral surface runoff with very short flow lengths (< 10 m) and subsequent infiltration in adjacent areas may sometimes occur. Medium propensity is assigned to areas with medium infiltration (MIN) and steep slopes. Traces of temporary surface runoff are clearly evident (gullies with surface flow lengths of up to about 100 m) in these areas. High propensity is assigned to areas with low infiltration (GIN), showing a pronounced drainage network with surface flow lengths generally much larger than 100 m. This class is independent of the slope.

6 Testing the surface runoff propensity index

The spatial pattern of the surface runoff propensity index (four classes, Fig. 4) was tested by comparing it with patterns of sink holes identified from the digital terrain model (DTM). The sink holes were identified by aggregating the 1 m DTM to 3 m and delineating pixels without outflow. The sink hole distribution was not used for estimating the index, so can be used for testing it. The index was processed on a raster with a cell size of 10 m.

Figure 6 shows a detail of the study area including the Spitzboden catchment, where permanent runoff was observed. The location of the sinks (Fig. 6b) corresponds very well with the areas of index no. 1, which indicates no surface runoff. High propensity is assigned to the areas with less karstified bare rock (DOL), sometimes overlaid by low permeability debris and organic soils. In these areas surface runoff can occur very often and, depending on the morphology, a permanent drainage network is likely to develop. Even formation of ponds and small lakes, such as the one in the southeastern part of the test catchment, is possible. In the large karstified areas the distinction between low (no. 2) and medium (no. 3) is mainly a result of differences in the topographic slope (Fig. 4). In these areas only temporary surface runoff occurs with very short flow lengths until the flow sinks into highly permeable debris or dolines.

In Fig. 7 the surface runoff propensity index in the whole study area (Hochschwab region, Fig. 7a) is plotted and compared to the sink holes (Fig. 7b). No surface runoff is calculated in 11 % (7 km^2), low propensity in 65 % (41 km^2), medium propensity in 21 % (13 km^2) and high propensity in 3 % (2 km^2) of the study area.

7 Discussion

7.1 Strengths and limitations of the method

The main advantage of the proposed mapping method over traditional field mapping (e.g. Scherrer and Naef, 2003) is its ability to use variables that can be assessed from a distance in the landscape, so large areas can be mapped efficiently. The main advantage over traditional vulnerability mapping (e.g. Cost Action 620, 2003; Andreo et al., 2006) is the incorporation of hydrological processes in the mapping.

Rogger et al. (2012) used geologic maps, orthophotos and geomorphic information to identify the dominant subsurface processes in Alpine catchments. These were combined with the surface runoff process maps of Markart et al. (2004) that are based on sprinkling experiments using vegetation and soil characteristics as indicators for event runoff coefficients. Sprinkling experiments are logistically challenging in remote alpine areas of the size of the study area. Instead, the mapping method in this study puts more emphasis on "reading the landscape" for delineating HRUs and classifying them according to geology, soils and debris, vegetation, and infiltration capacity. Spotting karst forms and surface runoff traces are an important part of the method. The field mapping is supported by morphological and geological maps, but they are not directly used for HRU delineation as in traditional vulnerability mapping. Additionally, the location of springs provides in situ information on the hydrogeology (see for example Cervi et al., 2017).

The mapping catalogue proposed here is defined in a way that it maximizes the information that can be mapped in

Figure 5. Examples of traces of surface runoff paths from the extreme event on 4 August 2013 (photos taken 4 days after event at location indicated in Fig. 1). Thin blue arrows: surface runoff traces on highly permeable gravel and debris; thick blue arrows: larger surface runoff paths, collecting runoff from larger areas during the event. The main water flow path in **(a)** ends at the edge of the rock wall, and **(b)** shows the situation at the foot of the wall. The red circle indicates the leafless vegetation caused by a temporary waterfall (water and debris). For illustrating the scale: the flow length of the main flow path in **(a)** is approximately 20 m.

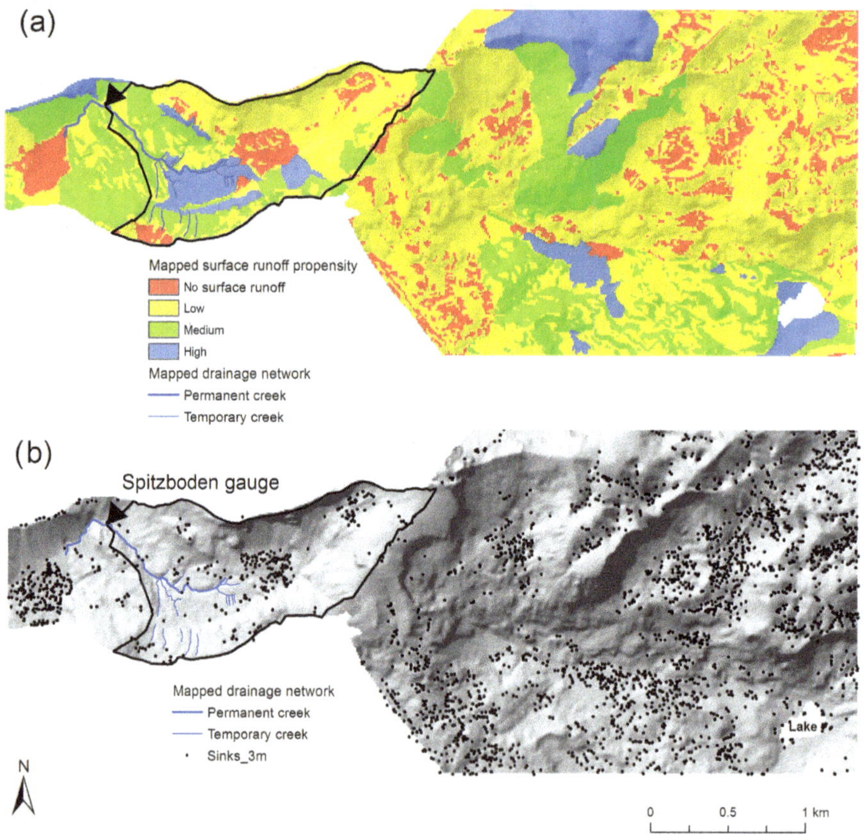

Figure 6. Comparison of the surface runoff propensity index **(a)** with sink holes **(b)**. Sink holes have been automatically generated from the lidar terrain model at 3 m resolution. Shading represents topography from the 1 m terrain model. Black polygon shows the topographic catchment of the Spitzboden stream gauge.

the field with a view of surface runoff propensity, including infiltration and storage characteristics as well as immediate response vs. delayed response. Of course, like in any field-mapping method, some subjectivity is involved but minimized by the use of a catalogue. Used as a "recipe" it will provide similar but not identical polygons if used by differ-

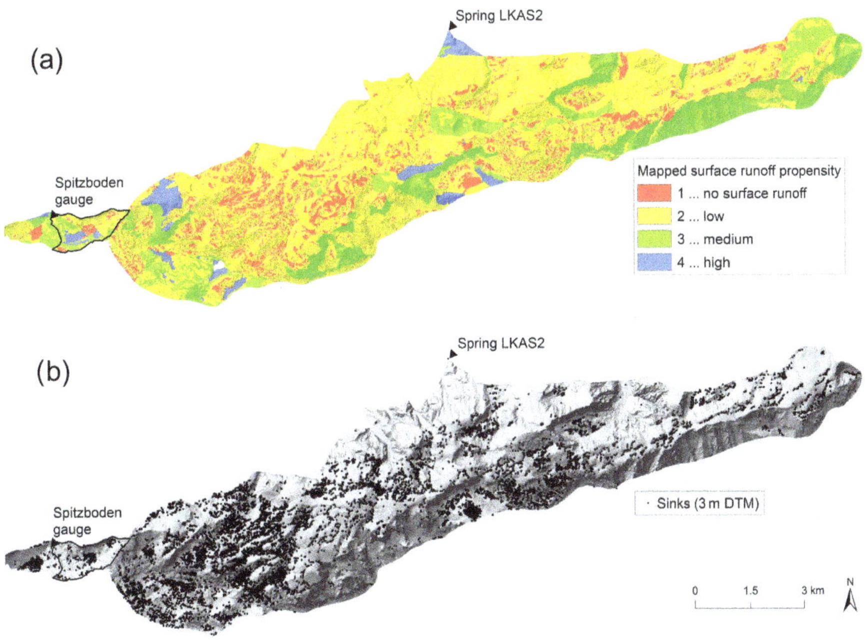

Figure 7. (a) Surface runoff propensity index in the whole study region (Hochschwab region). **(b)** Sink holes from the terrain model. Shading represents topography from the 1 m terrain model.

ent field researchers, due to differences in the path choices of the individuals and selection of representative points to verify the properties. Also, the mapping scale has an element of subjectivity, i.e. which sub-scale variability is allowed. However, a careful application of the method should provide very similar areas with high propensity (e.g. bare dolomitic rock with low infiltration capacity of the small fissures and permanently wet areas). The date of the field work may sometimes influence the results as visual traces of surface runoff are related to intense rainfall events in the preceding weeks. We would like to point out that alternative methods usually also involve a certain level of subjectivity in terms of the choice of variables and model structure. In vegetation mapping, for example, due to the typically large diversity it is difficult to assure that all species within an area are resolved. A trade-off always exists between mapping effort and mapping scale (resolution), particularly when mapping large areas, and every method has limitations in terms of accessibility and visibility in remote regions.

We used the mapping results for estimating a surface runoff propensity index at the pixel scale ($10\,\text{m} \times 10\,\text{m}$) that is a measure of the frequency with which surface runoff occurs. Although the pixel scale is 10 m, the aggregation scale is larger as it is mainly controlled by the size of the mapped polygons, which are on average 32 ha in area. The hydrological characteristics do vary within the polygons. This means that the smaller scale spatial variability is not explicitly represented by the index.

The index also points towards possible flow lengths, as areas of more frequent runoff generation in the study region are typically also those with larger flow lengths. The propensity index is derived from the mapped infiltration capacity and topographic slope. It compares well against the sink hole distribution. The index can be used for a rapid regional assessment of contamination risk for water resource purposes. Additionally, the index could be used for supporting the parameter estimation of a rainfall–runoff model in the context of the "dominant processes concept" (e.g. Grayson and Blöschl, 2000; Reszler et al., 2008) and more generally for regionalizing hydrological characteristics related to surface runoff (Blöschl et al., 1995; Blöschl, 2006).

7.2 Surface runoff generation processes

In karstic areas, surface runoff occurrence depends on the infiltration properties of the soil, debris and the underlying geology. In the study area, organic soils have mostly developed above poorly karstified rocks and hence low-permeability lithology such as dolomite. In these areas, a permanent drainage network is prone to occur and the soils have small storage capacities, and water logging, including the formation of ponds, is likely to occur, leading to frequent surface runoff. Both the local drainage network and the ponds drain into the karst. Processes on these areas have substantial similarity to the saturation overland flow mechanism (Dunne, 1983), which is common in humid climates at the foot of hillslopes as they flatten out. Such areas are important to identify, but their extents in karstic catchments are sometimes small. In the Hochschwab study region, 3 % of the area was classified as high surface runoff propensity based on the index. When substantial debris is found above the ponding layer, mostly in dolomitic areas with higher slopes, local springs indicate the dominance of storage and subsurface flow that exfiltrates and feeds the local drainage network. Traces of surface runoff paths above the debris were visible during the surveys, but pronounced continuous flow paths were hardly found. This implies that flow connectivity to the creek is only established at high rainfall intensities.

The mean altitude of the recharge area of the spring is about 1700 m a.s.l. and soils are shallow and sparsely vegetated. In some parts of the area, losses due to evapotranspiration may play a role for the general soil moisture status in summer (e.g. in areas with high local groundwater levels, water-logged areas). These characteristics are captured by the mapping and are classified as a relative high surface runoff propensity. It is possible that, after long dry periods, these areas are drier than predicted by the index. However, given the depth to the groundwater table in most of the area, long-term spring water balances could not be used to validate the index (Bonacci, 2001b).

In the highly karstified areas of the study region, often overlaid by debris and/or a thin humus layer, surface runoff occurs only during high-intensity storms and is therefore very rare. This is consistent with the literature (e.g. Zhang et al., 2011; Peng and Wang, 2012). The storm of 4 August 2013 (return period of about 10 years) allowed the mapping of particularly clear surface runoff traces during a post-event survey. Rainfall intensities of about 100 mm in 75 min were recorded at the nearby stations, which is on the order of the rainfall necessary for generating surface runoff obtained by sprinkling experiments in semi-arid karstic environments (Li et al., 2011; Calvo-Cases et al., 2003). The index results suggest that in about 75 % of the area no surface runoff occurs or the surface runoff propensity is low. However, some high-intensity (local) rainfall events are likely to occur every summer, which do produce surface runoff even in areas that are highly karstified.

7.3 Transferability of the method

The index proposed in this paper could also be adopted for non-karstic regions. The overall methodology of combining prior spatial information, a visual assessment from a distance at the landscape scale and local data collection in the field, as well as the combination of geological, soils and debris, and vegetation information may remain similar. However, the individual geological, soils and vegetation classes will obviously change. Also, some of the field instrumental methods may change, depending on the local hydrology, as other processes may gain more importance, e.g. lateral subsurface flow, shallow groundwater level fluctuations, connectivity of flow paths, runoff routing. Of course, stream gauging would become more important in non-karstic areas. Runoff dynamics may point towards possible runoff components, and water balance considerations relying on accurate catchment boundaries may indicate losses by evapotranspiration that can assist in estimating soil moisture dynamics during the period of field work. Also, surface runoff traces will be much more frequent in non-karstic areas, and capturing events by post-event surveys will become more likely.

In any case, the field researchers play a very important role in assessing the hydrological situation. Even though this implies some subjectivity, we believe the additional information obtained during the field campaigns outweighs any subjectivity. Aerial photos such as from drones could be useful to resolve more spatial detail than the aerial photos used here, but visibility in the forests is an issue.

8 Conclusions

A new method for mapping the propensity for surface runoff in karst areas is proposed. The method puts emphasis on "reading the landscape". It includes information that can be assessed from a distance, such as visible traces of overland flow, so large areas can be mapped efficiently. The outcome of the mapping is a tessellation of the landscape into hydrological response units, based on a classification of the geology, soils and debris, vegetation, and infiltration capacity. The method was applied to the karstic Hochschwab area in Austria. In this region, surface runoff is generally more frequent in less karstified dolomitic areas than in limestone areas. Low-permeability soils or debris are more likely to occur above dolomitic geology, where a permanent drainage network tends to develop. A "surface runoff propensity index" is estimated from the classified infiltration capacity obtained by the mapping. It represents an index of how frequently surface runoff may occur. The index compares well with the spatial distribution of sink holes. The surface propensity index can be used to assist in safeguarding the quality of the water supply from karst aquifers, particularly for optimizing land management and formulating water safety plans in a risk-based procedure by comparing the patterns with potential contamination loads. Since the overall conceptualization of the mapping procedure is generic to karst areas, it may be applied to other karst regions around the world.

Author contributions. The methodology was developed by all co-authors. ES performed the field work. CR computed and processed the results. CR and GB prepared the paper with contributions of all co-authors.

Competing interests. The authors declare that they have no conflict of interest.

Acknowledgements. This paper is dedicated to the memory of our wonderful colleague, Hermann Stadler, who recently passed away. Funding from the Vienna Water Works through two projects of "Mapping and modelling spatial patterns of surface runoff in the karstic catchments Hochschwab and Zeller Staritzen" is gratefully acknowledged. We would like to thank Gerhard Kuschnig and Wolfgang Zerobin for their support of these projects. Funding from the Austrian Science Foundation as part of the Vienna Doctoral Programme on Water Resource Systems (DK Plus W1219-N22) is also acknowledged.

Edited by: Fabrizio Fenicia

References

AG Boden: Bodenkundliche Kartieranleitung – 4. Auflage, Bundesanstalt für Geowissenschaften und Rohstoffe und geologische Landesämter der Bundesrepublik Deutschland, Hannover, 1994.

Andreo, B., Goldscheider, N., Vadillo, I., Vías, J. M., Neukum, C., Sinreich, M., Jiménez, P., Brechenmacher, J., Carrasco, F., Hötzl, H., Perles, M. J., and Zwahlen, F.: Karst groundwater protection: First application of a Pan-European Approach to vulnerability, hazard and risk mapping in the Sierra de Líbar (Southern Spain), Sci. Total Environ., 357, 54–73, 2006.

Blöschl, G.: Hydrologic synthesis – across processes, places and scales, Water Resour. Res., 42, W03S02, https://doi.org/10.1029/2005WR004319, 2006.

Blöschl, G., Grayson, R. B., and Sivapalan, M.: On the representative elementary area (REA) concept and its utility for distributed rainfall-runoff modelling, Hydrol. Process., 9, 313–330, 1995.

Bonacci, O.: Analysis of the maximum discharge of karst springs, Hydrogeol. J., 9, 328–338, 2001a.

Bonacci, O.: Monthly and annual effective infiltration coefficients in Dinaric karst: example of the Gradole karst spring catchment, Hydrol. Sci., 46, 287–299, 2001b.

Borga, M., Gaume, E., Creutin, J. D., and Marchi, L.: Surveying flash floods: gauging the ungauged extremes, Hydrol. Process., 22, 3883–3885, https://doi.org/10.1002/hyp.7111, 2008.

Bryda, G., van Husen, D., Kreuss, O., Koukal, V., Moser, M., Pavlik, W., Schönlaub, H. P., and Wagreich, M.: Erläuterungen zu Blatt 101 Eisenerz, Geologische Karte der Republik Österreich 1 : 50 000, Geologische Bundesanstalt, Wien, 223 pp., 2013.

Calvo-Cases, A., Boix-Fayos, C., and Imeson, A.: Runoff generation, sediment movement and soil water behaviour on calcareous (limestone) slopes of some Mediterranean environments in SE Spain, Geomorphology, 50, 269–291, 2003.

Canton, Y., Rodríguez-Caballero, E., Contreras, S., Villagarcia, L., Li, X. Y., Solé-Benet, A., and Domingo, F.: Vertical and lateral soil moisture patterns on a Mediterranean karst hillslope, J. Hydrol. Hydromech., 64, 209–217, 2016.

Cervi, F., Blöschl, G., Corsini, A., Borgatti, L., and Montanari, A.: Perennial springs provide information to predict low flows in mountain basins, Hydrolog. Sci. J., 62, 1–13, https://doi.org/10.1080/02626667.2017.1393541, 2017.

Chen, H., Zhang, W., Wang, K., and Fu, W.: Soil moisture dynamics under different land uses on karst hillslope in northwest Guangxi, China, Environ. Earth Sci., 61, 1105–1111, 2010.

Cost Action 620: Vulnerability and Risk Mapping for the Protection of Carbonate (Karst) Aquifers: Scope – Goals – Results, edited by: Zwahlen, F., European Commission, Directorate-General XII – Science, Research and Development, Luxembourg, 2003.

Decker, K.: Tektonische Karte des Hochschwabmassivs, Tektonisch/Strukturgeologischer Endbericht, Erläuterungen zur tektonischen Karte, Geo-Zentrum der Universität Wien, Wien, 41 pp., 2006.

Decker, K., Plan, L., and Reiter, F.: Tectonic assessment of deep Groundwater Pathways in fractured and karstified Aquifers, Hochschwab Massif, Austria, in: Proceedings All about Karst and Water, Vienna, 138–142, 2006.

Dunne, T.: Relation of field studies and modelling in the prediction of storm runoff, J. Hydrol., 65, 25–48, 1983.

Fiener, P., Auerswald, K., and Van Oost, K.: Spatio-temporal patterns in land use and management affecting surface runoff response of agricultural catchments – A review, Earth-Sci. Rev., 106, 92–104, 2011.

Flügel, W.-A.: Delineating hydrological response units by geographical information system analyses for regional hydrological modelling using PRMS/MMS in the drainage basin of the River Bröl, Germany, Hydrol. Process., 9, 423–436, 1995.

Frankenberger, J. R., Brooks, E. S., Walter, M. T., Walter, M. F., and Steenhuis, T. S.: A GIS-based variable source area hydrology model, Hydrol. Pprocess., 13, 805–822, 1999.

Fu, T., Chen, H., Zhang, W., Nie, Y., Gao, P., and Wang, K.: Spatial variability of surface soil saturated hydraulic conductivity in a small karst catchment of southwest China, Environ. Earth Sci., 74, 2381–2391, 2015.

Fu, T., Chen, H., Fu, Z., and Wang, K.: Surface soil water content and its controlling factors in a small karst catchment, Environ. Earth Sci., 75, 1406, https://doi.org/10.1007/s12665-016-6222-0, 2016.

Fu, Z., Chen, H., Xu, Q., Jia, J., Wang, S., and Wang, K.: Role of epikarst in near-surface hydrological processes in a soil mantled subtropical dolomite karst slope: implications of field rainfall simulation experiments, Hydrol. Process., 30, 795–811, https://doi.org/10.1002/hyp.10650, 2016.

Goldscheider, N.: Overview of Methods Applied in Karst Hydrogeology, in: Karst Aquifers – Characterization and Engineering, Professional Practice in Earth Sciences, edited by: Stevanović, Z., Springer International Publishing, Switzerland, 127–145, https://doi.org/10.1007/978-3-319-12850-4_4, 2015.

Grayson, R. and Blöschl, G.: Summary of pattern comparison and concluding remarks, chap. 14, in: Spatial Patterns in Catchment Hydrology: Observations and Modelling, edited by: Grayson, R. and Blöschl, G., Cambridge University Press, Cambridge, UK, 355–367, 2000.

Li, X. Y., Contreras, S., Solé-Benet, A., Cantón, Y., Domingo, F., Lázaro, R., Lin, H., Van Wesemael, B., and Puigdefábregas, J.: Controls of infiltration–runoff processes in Mediterranean karst rangelands in SE Spain, Catena, 86, 98–109, 2011.

Mandl, G. W., Bryda, G., Kreuss, O., Moser, M., Pavlik, W., Decker, K., Draxler, I., Krystyn, L., and Piros, O.: Erstellung moderner geologischer Karten als Grundlage für karsthydrogeologische Spezialuntersuchungen im Hochschwabgebiet, Bund-Bundesländer-Kooperation, Endbericht, Geologische Bundesanstalt, Wien, 211 pp., 2002.

Markart, G., Kohl, B., Sotier, B., Schauer, T., Bunza, G., and Stern, R.: Provisorische Geländeanleitung zur Anschätzung des Oberflächenabflussbeiwertes auf alpinen Boden-/Vegetationseinheiten bei konvektiven Starkregen (Version 1.0), Dokumentation Nr. 3, Bundesministerium für Land- und Forstwirtschaft, Umwelt und Wasserwirtschaft, 2004.

Peng, T. and Wang, S. J.: Effects of land use, land cover and rainfall regimes on the surface runoff and soil loss on karst slopes in southwest China, Catena, 90, 53–62, 2012.

Peschke, G., Etzenberg, C., Müller, G., Töpfer, J., and Zimmermann, S.: Das wissensbasierte System FLAB – Ein Instrument zur rechnergestützten Bestimmung von Landschaftseinheiten mit gleicher Abflussbildung, IHI-Schr. 10, Int. Hochschulinstitut Zittau, Zittau, Deutschland, 1998.

Plan, L., Decker, K., Faber, R., Wagreich, M., and Grasemann, B.: Karst morphology and groundwater vulnerability of high alpine karst plateaus, Environ. Geol., 58, 285–297, https://doi.org/10.1007/s00254-008-1605-5, 2009.

Plan, L., Kuschnig, G., and Stadler, H.: Case Study: Kläffer Spring – the major spring of the Vienna water supply (Austria), in: Groundwater Hydrology of Springs – Engineering, Theory, Management, and Sustainability, edited by: Kresic, N. and Stevanovic, Z., 411–427, https://doi.org/10.1016/C2009-0-19145-6, 2010.

Reszler, C., Komma, J., Blöschl, G., and Gutknecht, D.: Dominante Prozesse und Ereignistypen zur Plausibilisierung flächendetaillierter Niederschlag-Abflussmodelle, Hydrol. Wasserbewirt., 52, 120–131, 2008.

Rogger, M., Kohl, B., Pirkl, H., Viglione, A., Komma, J., Kirnbauer, R., Merz, R., and Blöschl, G.: Runoff models and flood frequency statistics for design flood estimation in Austria – Do they tell a consistent story?, J. Hydrol., 456–457, 30–43, https://doi.org/10.1016/j.jhydrol.2012.05.068, 2012.

Scherrer, S. and Naef, F.: A decision scheme to indicate dominant hydrological flow processes on temperate grassland, Hydrol. Process., 17, 391–401, https://doi.org/10.1002/hyp.1131, 2003.

Stadler, H. and Strobl, E.: Hydrogeology Hochschwab. Summary, Technical Report, Joanneum Research Graz, Graz, 76 pp., 2006.

Stadler, P, Häusler, H., Rogger, M., Savio, D., and Stadler, H.: A field work orientated approach for complex karst aquifer characterization, in: Karst without boundaries, CRC Taylor & Francis, London, 179–197, https://doi.org/10.1201/b21380-16, 2016.

Thomas, I. A., Jordan, P., Mellander, P. E., Fenton, O., Shine, O., Ó hUallacháin, D., Creamer, R., McDonald, N. T., Dunlop, P., and Murphy, P. N. C.: Improving the identification of hydrologically sensitive areas using LiDAR DEMs for the delineation and mitigation of critical source areas of diffuse pollution, Sci. Total Environ., 556, 276–290, 2016.

UMS: Benutzerhandbuch zum ThetaProbe Soil Moisture Sensor ML2x, UMS GmbH München, München, 2001.

Walter, M. T., Walter, M. F., Brooks, E. S., Steenhuis, T. S., Boll, J., and Weiler, K.: Hydrologically sensitive areas: variable source area hydrology implications for water quality risk assessment, J. Soil Water Conserv., 55, 277–284, 2000.

Western, A. W., Blöschl, G., and Grayson, R. B.: How well do indicator variograms capture the spatial connectivity of soil moisture?, Hydrol. Process., 12, 1851–1868, 1998.

Scherrer, S. and Naef, F.: A decision scheme to indicate dominant hydrological flow processes on temperate grassland, Hydrol. Process., 17, 391–401, https://doi.org/10.1002/hyp.1131, 2003.

Stadler, H. and Strobl, E.: Hydrogeology Hochschwab. Summary, Technical Report, Joanneum Research Graz, Graz, 76 pp., 2006.

Stadler, P, Häusler, H., Rogger, M., Savio, D., and Stadler, H.: A field work orientated approach for complex karst aquifer characterization, in: Karst without boundaries, CRC Taylor & Francis, London, 179–197, https://doi.org/10.1201/b21380-16, 2016.

Thomas, I. A., Jordan, P., Mellander, P. E., Fenton, O., Shine, O., Ó hUallacháin, D., Creamer, R., McDonald, N. T., Dunlop, P., and Murphy, P. N. C.: Improving the identification of hydrologically sensitive areas using LiDAR DEMs for the delineation and mitigation of critical source areas of diffuse pollution, Sci. Total Environ., 556, 276–290, 2016.

UMS: Benutzerhandbuch zum ThetaProbe Soil Moisture Sensor ML2x, UMS GmbH München, München, 2001.

Walter, M. T., Walter, M. F., Brooks, E. S., Steenhuis, T. S., Boll, J., and Weiler, K.: Hydrologically sensitive areas: variable source area hydrology implications for water quality risk assessment, J. Soil Water Conserv., 55, 277–284, 2000.

Western, A. W., Blöschl, G., and Grayson, R. B.: How well do indicator variograms capture the spatial connectivity of soil moisture?, Hydrol. Process., 12, 1851–1868, 1998.

3

Predicting groundwater recharge for varying land cover and climate conditions

Chinchu Mohan[1], **Andrew W. Western**[1], **Yongping Wei**[2], **and Margarita Saft**[1]

[1]Department of Infrastructure Engineering, University of Melbourne, Melbourne, Victoria, Australia
[2]School of Geography, Planning and Environmental Management, The University of Queensland, Brisbane, Australia

Correspondence: Chinchu Mohan (mohanc@student.unimelb.edu.au)

Abstract. Groundwater recharge is one of the important factors determining the groundwater development potential of an area. Even though recharge plays a key role in controlling groundwater system dynamics, much uncertainty remains regarding the relationships between groundwater recharge and its governing factors at a large scale. Therefore, this study aims to identify the most influential factors of groundwater recharge, and to develop an empirical model to estimate diffuse rainfall recharge at a global scale. Recharge estimates reported in the literature from various parts of the world (715 sites) were compiled and used in model building and testing exercises. Unlike conventional recharge estimates from water balance, this study used a multimodel inference approach and information theory to explain the relationship between groundwater recharge and influential factors, and to predict groundwater recharge at 0.5° resolution. The results show that meteorological factors (precipitation and potential evapotranspiration) and vegetation factors (land use and land cover) had the most predictive power for recharge. According to the model, long-term global average annual recharge (1981–2014) was 134 mm yr^{-1} with a prediction error ranging from -8 to 10 mm yr^{-1} for 97.2 % of cases. The recharge estimates presented in this study are unique and more reliable than the existing global groundwater recharge estimates because of the extensive validation carried out using both independent local estimates collated from the literature and national statistics from the Food and Agriculture Organization (FAO). In a water-scarce future driven by increased anthropogenic development, the results from this study will aid in making informed decisions about groundwater potential at a large scale.

1 Introduction

Human intervention has dramatically transformed the planet's surface by altering land use and land cover and consequently the hydrology associated with it. In the last 100 years the world population has quadrupled, from 1.7 billion (in 1900) to more than 7.3 billion (in 2014), and is expected to continue to grow significantly in the future (Gerland et al., 2014). During the last century, rapid population growth and the associated shift to a greater proportion of irrigated food production led to an increase in water extraction by a factor of ~ 6. This eventually resulted in the overexploitation of both surface and groundwater resources, including the depletion of 21 of the world's 37 major aquifers (Richey et al., 2015). This depletion threatened human lives in many ways, ranging from critical reductions in water availability to natural disasters such as land subsidence (Chaussard et al., 2014; Ortiz-Zamora and Ortega-Guerrero, 2010; Phien-Wej et al., 2006; Sreng et al., 2009). Therefore, there is a need to closely examine approaches for sustainably managing this resource by controlling withdrawal from the system.

Groundwater recharge is one of the most important limiting factors for groundwater withdrawal and determines the groundwater development potential of an area (Döll and Flörke, 2005) Groundwater recharge connects atmospheric, surface and subsurface components of the water balance and is sensitive to both climatic and anthropogenic factors (Gurdak, 2008; Herrera-Pantoja and Hiscock, 2008; Holman et al., 2009; Jyrkama and Sykes, 2007). Various studies have employed different methods to estimate groundwater recharge including tracer methods, water table fluctuation methods, lysimeter methods and simple water balance techniques. Some of these studies input recharge to numerical groundwater models or dynamically link it to hydrological models to estimate variations under different climate and land cover conditions (Aguilera and Murillo, 2009; Ali et al., 2012; Herrera-Pantoja and Hiscock, 2008; Sanford, 2002).

In the last few decades, interest in global-scale recharge analysis has increased for various scientific and political reasons (Tögl, 2010). L'vovich (1979) made the first attempt at a global-scale analysis by creating a global recharge map using baseflow derived from river discharge hydrographs. The next large-scale groundwater recharge estimate was done by Döll (2002) who modelled global groundwater recharge at a spatial resolution of 0.5^0 using the WaterGAP Global Hydrological model (WGHM; Alcamo et al., 2003; Döll, 2002). In this study, the runoff was divided into fast surface runoff, slow subsurface runoff and recharge using a heuristic approach. This approach considered relief, soil texture, hydrogeology and occurrence of permafrost and glaciers for the runoff partitioning. However, WGHM failed to reliably estimate recharge in semi-arid regions (Döll, 2002). Importantly, in that study, there was no consideration of the influence of vegetation which has been reported to be the second most important determinant of recharge by many researchers (Jackson et al., 2001; Kim and Jackson, 2012; Scanlon et al., 2005). In subsequent years, several researchers have attempted to model global groundwater recharge using different global hydrological models and global-scale land surface models (Koirala et al., 2012; Scanlon et al., 2006; Wada et al., 2010).

Although a fair amount of research has been carried out to model groundwater recharge at a global scale, most studies compared results to country-level groundwater information from the FAO (FAO, 2005). FAO statistics were based on estimates compiled from national institutions. The data estimation and reporting capacities of national agencies vary significantly and raise concerns about the accuracy of the data (Kohli and Frenken, 2015). In addition, according to FAO AQUASTAT reports, most national institutions in developing countries prioritise subnational-level statistics over national-level statistics, and in most cases data are not available for all subnational entities. This decreases the accuracy of country-wide averages and raises concerns about the reliability of using them as standard comparison measures. Only a few studies have validated modelled estimates against small-scale recharge measurements. Döll and Fiedler (2008) used 51 recharge observations from arid and semi-arid regions to correct model outputs. This study develops a recharge model and undertakes a more extensive validation of it using 715 local recharge measurements. Moreover, previous research has mostly been restricted to studying meteorological influences on recharge, few studies have systematically explored global-scale factors governing recharge. Much uncertainty still exists about the relationship between groundwater recharge and topographical, lithological and vegetation factors. Without adequate knowledge of these controlling factors, our capacity to sustainably manage groundwater globally will be seriously compromised.

The major objectives of this study are to identify the most influential factors of groundwater recharge and to develop an empirical model to estimate diffuse rainfall recharge. Specifically, to quantify regional effects of meteorological, topographical, lithological and vegetation factors on groundwater recharge using data from 715 globally distributed sites. These relationships are used to build an empirical groundwater recharge model and then the global groundwater recharge is modelled at a spatial resolution of $0.5° \times 0.5°$ for the time period 1981–2014.

2 Methods

2.1 Dataset

This study is based on a compilation of recharge estimates reported in the literature from various parts of the world. This dataset is an expansion of previously collated sets of recharge studies along with the addition of new recharge estimates (Döll and Flörke, 2005; Edmunds et al., 1991; Scanlon et al., 2006; Tögl, 2010; Wang et al., 2010). The literature search was carried out using Google Scholar, Scopus and Web of Science with related keywords "groundwater recharge", "deep percolation", "diffuse recharge" and "vertical groundwater flux". Several criteria were considered in including each study. To ensure that the data reflects all seasons, recharge estimates for time periods less than 1 year were excluded. The sites with significant contribution to groundwater from streams or by any artificial means were also eliminated as the scope of this research was to model naturally occurring recharge. In order to maximise the realistic nature of the dataset, all studies using some kind of recharge modelling were removed from the dataset. After all exclusions, 715 data points spread across the globe remained (Fig. 1) and were used for further analysis. Of these studies, 345 were estimated using the tracer method, 123 using the water balance method and the remaining studies used baseflow method, lysimeter or water table fluctuation method. This diversity in recharge estimation has enabled us to evaluate systematic differences in various measurement techniques. The year of measurement or estimation of recharge estimates in the final dataset differed (provided as supplementary material), and ranged from 1981 to 2014 (Fig. 2a). This inconsistency in the data raised a challenge when choosing the time frame for factors in the modelling exercise, particularly those showing inter-annual variation. Moreover, the compiled dataset does not represent all climate zones well (Fig. 2c), as most of the studies used were done either in arid, semi-arid or temperate zones. Pasture and cropland were the dominant land uses in the dataset (Fig. 2b).

The next step was to identify potential explanatory factors that could influence recharge (henceforth referred to as

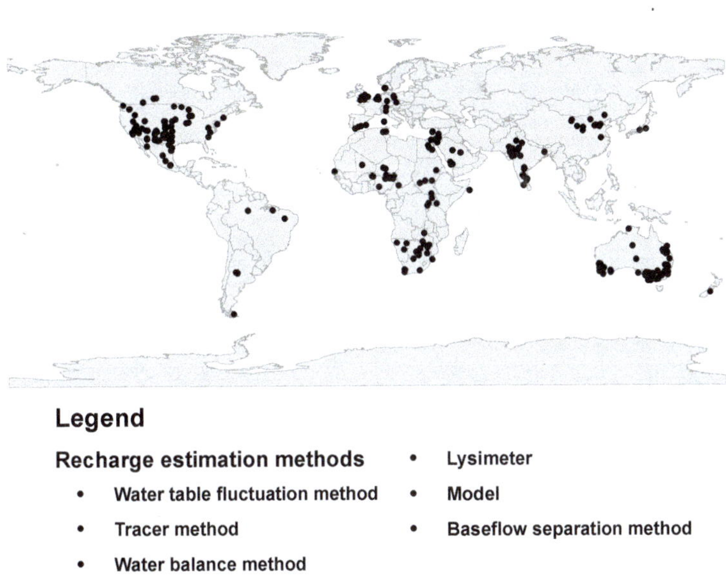

Figure 1. Locations of the 715 selected recharge estimation sites and the corresponding recharge estimation method, used for model building.

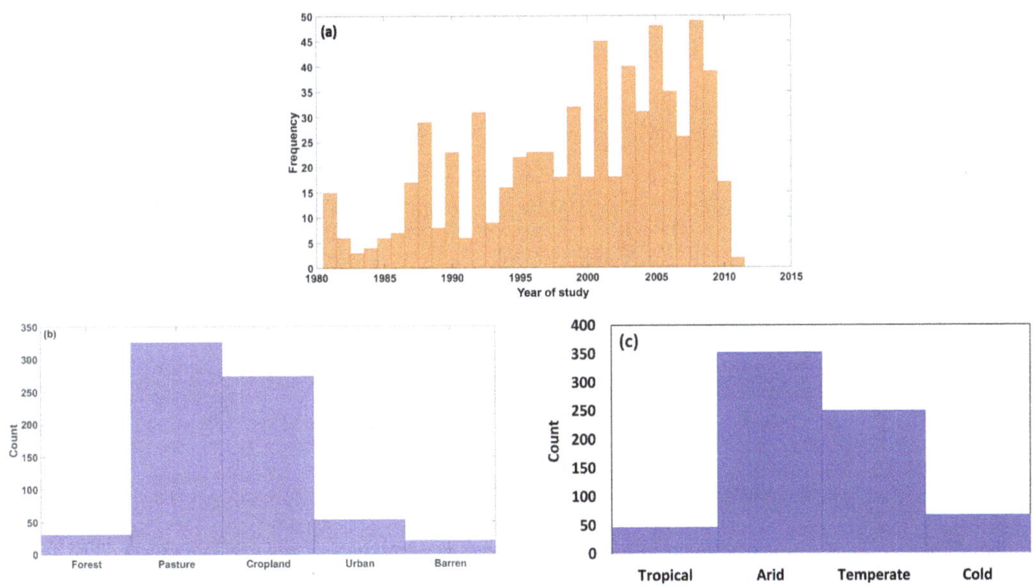

Figure 2. Histograms showing frequency of **(a)** study year, **(b)** land use and **(c)** Köppen–Geiger climate zones for the recharge estimates used.

predictors). Potential predictors that were reported in the literature as having some influence on recharge were identified (Athavale et al., 1980; Bredenkamp, 1988; Edmunds et al., 1991; Kurylyk et al., 2014; Nulsen and Baxter, 1987; O'Connell et al., 1995; Pangle et al., 2014). The choice of predictors was made based on the availability of global gridded datasets and their relative importance in a physical sense, as informed by the literature. According to the literature, the water availability on the surface for infiltration and the potential of the subsurface system to intake water are the two major controls on recharge. Different variables that can potentially represent these two factors were chosen as predictors in this study. The water availability is represented mainly by using meteorological predictors including precipitation, potential evapotranspiration, aridity index and number of days with rainfall and vegetation characteristics (land use and land cover), whereas the intake potential is represented using various quantifiable characteristics of the vadose zone. We employed 12 predictors comprising meteorological factors, soil/vadose zone factors, vegetation factors and topographic factors. However, other factors which could have a sizable influence on recharge were not included in this

Table 1. Description of predictors used for recharge model building.

Predictors	Symbol	Unit	Resolution	Temporal span	Source	Description	Reference
Precipitation	P	mm yr^{-1}	$0.5° \times 0.5°$	1981–2014	Climatic Research Unit, University of East Anglia, United Kingdom	Mean annual precipitation	Harris et al. (2014)
Mean temperature	T	°C	$0.5° \times 0.5°$	1981–2014	Climatic Research Unit, University of East Anglia, United Kingdom	Mean annual temperature	Harris et al. (2014)
Potential evapotranspiration	PET	mm yr^{-1}	$0.5° \times 0.5°$	1981–2014	Climatic Research Unit, University of East Anglia, United Kingdom	Penman–Monteith reference crop evapotranspiration	Harris et al. (2014)
No. of rainy days	R_d		5 arcmin	1981–2014	AQUAMAPS, FAO	Average number of wet days per year defined as having ≥ 0.1 mm of precipitation	New et al. (2002)
Slope	S	fraction	$0.5° \times 0.5°$	–	Earthdata, NASA	Mean surface slope	Verdin (2011)
Saturated hydraulic conductivity	k_{sat}	cm day^{-1}	$1° \times 1°$	–	Earthdata, NASA	Saturated hydraulic conductivity at 0–150 cm depth	Webb et al. (2000)
Soil water storage capacity	SWSC	mm	$1° \times 1°$	–	Earthdata, NASA	Texture-derived soil water storage capacity in soil profile (up to 15 m depth)	Webb et al. (2000)
Excess water (without irrigation)	EW	mm	–	1981–2014	–	$\sum_{i=1}^{12} (P_i - \text{PET}_i)$ where $P_i > \text{PET}_i$	
Aridity index	AI	–	–	1981–2014	–	$\text{AI} = P/\text{PET}$	
Clay content	Clay	%	$1° \times 1°$	–	Earthdata, NASA	0–150 cm profile	DAAC (2016)
Bulk density	ρ_b	gm cm^{-3}	$1° \times 1°$	–	Earthdata, NASA	0–150 cm profile	DAAC (2016)
Land use land cover	LU	–	15 arcsec	–	USGS/Literature	Forest, pasture, cropland, urban/built up, barren	Kim and Jackson (2012), Broxton et al. (2014)

study because of insufficient data. Thus, we did not consider the effects of irrigation on recharge, limiting the scope of the study to rainfall-induced recharge. Subsurface lithology which could be another important recharge factor, was also eliminated from the study, due to a lack of suitable lithological and geological datasets at a larger scale. Better quality information about various predictors would have been desirable to enhance the accuracy of prediction. Details of predictors are given in Table 1.

Data for the chosen predictors corresponding to 715 recharge study sites were extracted from global datasets. Meteorological datasets (P, T and PET) were obtained from the Climatic Research Unit, University of East Anglia, United Kingdom. Even though daily data were available from 1901 to 2014 at a resolution of $0.5° \times 0.5°$, in this study the mean annual average of the latest 34 years (1981 to 2014) was used to reduce the inconsistency in year of recharge measurements in the final dataset. Topographic and soil data were acquired from the NASA Earth observation dataset. Both datasets were of $0.5° \times 0.5°$ spatial resolution. A few of the predictors, including number of rainfall days (R_d) and land use/land cover (LU) data were obtained from AquaMaps (by FAO) and USGS (United States Geological Survey) at a spatial resolution of $0.5° \times 0.5°$ and 15 arcmin respectively. Thus-obtained LU data were compared with land cover reported in literature and corrected for any discrepancies. The spatial resolution of the different data used was diverse. This was dealt with by extracting the values for each recharge site from the original grids using the nearest neighbour interpolation method. As a result, predictor data extracted for each recharge site will differ from the actual value due to scaling and interpolation errors. Out of the 12 predictors LU was not a quantitative predictor and was transformed into a categorical variable in the modelling exercise.

2.2 Recharge model development

With empirical studies, the science world is always sceptical about whether to use a single best-fit model or to infer results from several better-predicting and plausible models. The former option is feasible only if there exists a model which clearly surpasses other models, which is rare in the case of complex systems like groundwater. Usually cross correlation and multiple controlling influences on the system lead to more than one model having similarly good fits to the observations. Thus choosing explanatory variables and model structure is a significant challenge. In the past this challenge was often addressed using various step-wise model construction methods, with the final model being selected based on some model fit criteria that penalise model complexity (Fenicia et al., 2008; Gaganis and Smith, 2001; Jothityangkoon et al., 2001; Sivapalan et al., 2003). These approaches were pragmatic responses to the large computational load involved in trying all possible models. The disadvantage of this method is that the final model will be dependent on the step-wise selection process used (Sivapalan et al., 2003). An alternative approach for addressing this high level of uncertainty in model structure is to adopt a multi-model inference approach that compares many models (Duan et al., 2007; Poeter and Anderson, 2005). It typically results in multiple final models and an assessment of the importance of each explanatory variable. Therefore, this approach was used to develop an understanding of the role of different controlling factors on recharge in a data-limited condition.

Choosing predictors that are capable of representing the system and selecting the right models for prediction are the key steps in the multi-model inference approach. Here, models were chosen by ranking the fitted models based on performance, and comparing this to the best-performing model in the set (Anderson and Burnham, 2004). This model ranking also provided a basis for selecting individual predictors. The analysis progressed through three key stages: exploratory analysis, model building and model testing.

2.2.1 Multi-model analysis

A multi-model selection process aims to explore a wide range of model structures and to assess the predictive power of different models in comparison with others. Essentially, models with all possible combinations of selected predictors are developed and assessed via traditional model performance metrics (discussed later). By conducting such an exhaustive search, multi-model analysis avoids the problems associated with selection methods in step-wise regression approaches (Burnham and Anderson, 2003). Importantly, it reduces the chance of missing combinations of predictors with good predictive performance. However, a disadvantage of this approach is that the number of predictor combinations grows rapidly with the number of factors considered. To make the analysis computationally efficient, we set an upper limit for the number of predictors used. Another problem with this approach is that it can result in overfitting. To address this issue we evaluated model performance with metrics that penalise complexity and tested the model robustness with a cross-validation analysis. The model development procedure using multi-model analysis is described in detail below.

(a) Exploratory analysis

Firstly, all the chosen predictors were individually regressed against the compiled recharge dataset. This was carried out with the main objective of finding the predictors with significant control over recharge and to gain an initial appreciation of how influential each predictor is compared to the others. This understanding will aid in eliminating the least influential predictors from further analysis. Then assumptions involved in regression analysis, such as linearity, low multicollinearity (important for later multivariate fitting) and independent identically distributed residuals were analysed using residual analysis. Following the residual analysis, various data transformations (square root, logarithmic and reciprocal) were carried out to reduce heteroscedasticity and improve linearity of the variables. The square root transformed recharge along with non-transformed predictors gave the most homoscedastic relationships (results not shown). Therefore, these transformed values were used in further model-building exercises. Predictors were selected and eliminated based on statistical indicators such as adjusted coefficient of determination (R^2_{adj}) value and root mean square error (RMSE).

(b) Model building

Multiple linear regression was employed for building the models as the transformed dataset did not exhibit any nonlinearity. Furthermore, the presence of both negative and positive values in the dataset restricted the applicability of other forms of regression analysis like log-linear and exponential (Saft et al., 2016). Linear regression is known for its simple and robust nature in comparison to higher-order analysis. The robustness of linear regression helped to maintain parsimony together with reasonable prediction accuracy. A rigorous model-building approach was adopted in order to capture the interplay between predictors with combined/interactive effects on groundwater recharge. This is an exhaustive search in which all candidate models are fitted and inter-compared using performance criteria. In a way, this modelling exercise used a top-down approach, starting with a simple model which is expanded as shortcomings were identified (Fenicia et al., 2008).

(c) Model testing

The analysis above provided insight into the relative performance of the models. However, it is also important to assess the dependence of the results on the particular sample. Therefore, we conducted a subsample analysis in which the same method was re-applied to subsamples of the data. Finally, predictive uncertainty was estimated through leave-one-out cross validation. In the first case, the whole model development process was redone multiple times using subsamples of the data. To achieve this, the entire dataset was randomly divided into 80 and 20 % subsets and 80 % of the data were used for building the model. The predictive performance of the developed model was tested against the omitted 20 % of data. This was repeated 200 times in order to eliminate random sampling error. The leave-one-out cross validation was applied to the best few individual model structures and provided an estimate of predictive performance for those particular models. It also gave an indication of data quality at each point.

In summary the key steps in the multi-model analysis were

1. selecting predictors;
2. fitting all possible models consisting different combinations of predictors;
3. calculating model performance metrics for each model;
4. calculating the "weight of evidence" for each predictor based on the performance metric of all models containing that predictor;
5. testing the predictive performance of the models.

2.2.2 Ranking models and predictors

This part of the analysis has closely followed the approach developed in Saft et al. (2016). Model performance was evaluated using several information criteria. These information criteria include a goodness-of-fit term and an overfitting penalty based on the number of predictors in the model. In this study we used R^2_{adj}, the consistent Akaike information criterion (AIC_c) and the complete Akaike information criterion (CAIC) as the performance evaluation criteria. These criteria differ in terms of penalising overfitting. R^2_{adj} penalises overfitting the least, AIC_c moderately and CAIC heavily. However, when we are unsure of the true model and whether it overfits or not, there is some advantage in employing several criteria as it gives insight into how the results depend on the criteria used. Suitability of the information criteria also varies with the sample size. CAIC acts as an unbiased estimator for large sample size with relatively small candidate models, but produces large negative bias in other cases. Conversely, AIC_c is well suited for small-sample applications (Cavanaugh and Shumway, 1997; Hurvich and Tsai, 1989). The formulas for the above criteria are as follows:

$$\text{AIC} = -2 \times \text{llf} + 2 \times k \text{ (Akaike, 1974)}, \quad (1)$$

$$\text{AIC}_\text{c} = \text{AIC} + (2 \times (k-1) \times \frac{k+2}{n-k-2}) \text{ (Hurvich and Tsai, 1989)}, \quad (2)$$

$$\text{CAIC} = -2 \times \text{llf} + k \times (\ln(n)+1) \text{ (Bozdogan, 1987)}, \quad (3)$$

$$R^2 = 1 - \left[\frac{n-1}{n-k-1}\right] \times \left[1 - R^2\right] \text{ (Ezekiel, 1929; Wang and Thompson, 2007)}, \quad (4)$$

where "llf" is the log-likelihood function, k is the dimension of the model and n is the number of observations.

When assessing candidate models there are two questions which are of particular interest. (1) Which models are better? (2) How much evidence exists for each predictor in predicting recharge? Analysis of the AIC_c and CAIC was used to answer both these questions. Models were ranked using information criteria, with smaller values indicating better performance. Information criteria are more meaningful when they are used to evaluate the relative performance of the models (Poeter and Anderson, 2005). Models were ranked from best to worst by calculating model delta values (Δ) and model weights (W) as follows:

$$\Delta_i = \text{AIC}_i - \text{AIC}_{\text{min}}, \quad (5)$$

$$W_i = \exp(-0.5 \times \Delta_i) / \Sigma \exp(-0.5 \times \Delta_m,) \quad (6)$$

where AIC_{min} is the information criteria value of the best model. Δ_i and W_i represent the performance of ith model in comparison with the best-performing model in the set of M models.

Evidence ratios were then calculated as the ratio of the ith model weight to the best model weight. They can be used as a measure of the evidence for the ith model compared to the other models. They also provide means to estimate the importance of each predictor. This involves transformation of evidence ratios into a proportion of evidence (PoE) for each predictor. PoE for a predictor is defined as the sum of weights of all the models containing that particular predictor. PoE ranges from 0 to 1. The closer the PoE of a predictor is to 1, the more influential that predictor is.

2.3 Global groundwater recharge estimation

The best model (model 1, Table 3) from the above analysis was used to build a global recharge map at a spatial resolution of 0.5° × 0.5°. Recharge estimation was done annually for a study period of 34 years (1981–2014), and the estimated

Table 2. Summary statistics of potential predictors from the dataset used in this study.

Parameters	Minimum	Maximum	Range	Mean	Standard deviation
P (mm yr^{-1})	1.30	2627.00	2625.70	572.82	305.65
T (°C)	1.60	30.62	29.02	17.73	6.04
PET (mm yr^{-1})	6.60	2600.00	2593.40	1356.17	401.77
R_d (day yr^{-1})	2.00	270.00	268.00	85.89	42.78
S	0.00	10.16	10.15	0.84	1.17
k_{sat} (cm day^{-1})	0.00	265.75	265.75	60.61	59.50
SWSC (mm)	2.00	1121.00	1119.00	517.38	240.81
AI	0.00	68.18	68.18	0.70	3.74
EW (mm yr^{-1})	0.01	1467.87	1467.86	125.41	188.07
ρ_b (gm cm^{-3})	0.15	1.67	1.51	1.44	0.20
Clay (%)	1.87	52.51	50.64	23.77	7.66
LU	1.00	5.00	4.00	2.58	0.81
Recharge (mm yr^{-1})	0.00	1375.00	1375.00	73.22	125.94

groundwater recharge was then averaged over the 34-year period to produce a global map. In addition to this, maps showing percentage of rainfall becoming recharge and the standard deviation of annual recharge over the 34 years were also generated. As recharge data from regions with frozen soil were scarce in the model-building dataset, the model predictions in those regions particularly for regions with Köppen–Geiger classification Dfc, Dfd, ET and EF were not highly reliable. EF regions of Greenland and Antarctica were excluded from the final recharge map due to lack of both recharge and predictor data. However, the modelled recharge for Dfc, Dfd and ET regions were included because of the availability of predictor data. In addition, the modelled recharge values were compared against country-level statistics from FAO (2005) for 153 countries.

3 Results

The results address three important questions. (1) What are the most influential predictors of groundwater recharge? (2) What are better models for predicting recharge? (3) How does groundwater recharge vary over space and time? The first question was answered by carrying out an exploratory data analysis and also by estimating the PoE for each predictor, the second using information criteria and the third by mapping recharge at 0.5° × 0.5° using the best model.

3.1 Exploratory data analysis

Table 2 gives the statistical summary of predictors and groundwater recharge at 715 data sites. It is apparent from the table that predictors varied considerably between sites, consistent with inter-site variability in regional physical characteristics. This variability provided an opportunity to explore recharge mechanisms in a range of different physical environments. As we used linear regression to study the one-to-one relationship of recharge with each of the predictors, RMSE and bias of fitting were used to identify the predictors with the most explanatory power. In this case, RMSE values ranged between 23.2 mm yr^{-1} for P and 30.21 mm yr^{-1} for S. Predictive potential of meteorological predictors was greater than for other classes of predictor (Fig. 3). P, AI, EW and ρ_b had a negative bias, whereas all other predictors had a positive bias.

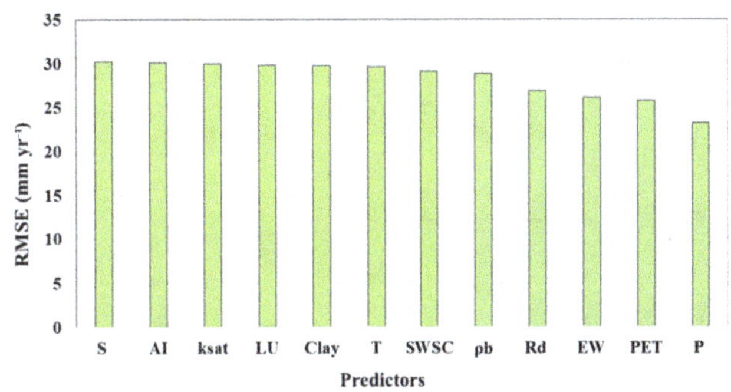

Figure 3. Model fit performance criteria for single predictor regressions.

3.2 Multi-model analysis

3.2.1 Proportion of evidence (PoE) for individual predictors

Figure 4 shows the PoE of the 12 predictors used in this study. According to this analysis, 3 of the 12 predictors stood out as having the greatest explanatory power (Fig. 4). Precipitation (P), potential evapotranspiration (PET) and land use and land cover (LU) had the highest proportions of evidence (~ 1). Subsurface percentage of clay (clay) and saturated hydraulic conductivity (k_{sat}) also had an important influence on recharge with PoE ~ 0.4. Aridity index (AI),

Table 3. Coefficient of predictors used in the top 10 models, ranked based on CAIC.

P	T	PET	R_d	S	k_{sat}	SWSC	AI	EW	ρ_b	Clay	LU	Constant	R^2_{adj}
0.0081		−0.0043									0.9567	5.3543	0.35
0.0086		−0.0044								−0.0606	1.0335	6.3781	0.25
0.0078		−0.0041							−1.9083		0.9667	7.8822	0.25
0.0076		−0.0055	−0.0247		0.0089			0.0040	−2.5857		1.0131	11.8652	0.34
0.0084		−0.0053	−0.0195					0.0036		−0.0758	1.0189	9.4112	0.33
0.0092		−0.0052	−0.0128							−0.0631	1.0409	8.2317	0.33
0.0075		−0.0050	−0.0194					0.0034	−2.3410		0.9370	11.2147	0.35
0.0084		−0.0049	−0.0130						−2.0104		0.9716	9.8549	0.35
0.0086		−0.0050	−0.0122								0.9607	7.0692	0.33
0.0086		−0.0053	−0.0166		0.0075				−2.1688		1.0402	10.2082	0.33

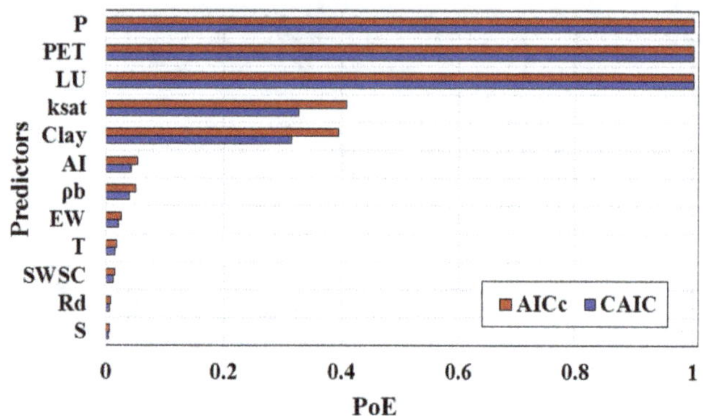

Figure 4. Proportion of evidence according to AIC_c and CAIC for 12 predictors (sorted in descending order of PoE).

rainfall days (R_d), mean temperature (T), bulk density (ρ_d), slope (S), excess water (EW) and soil water storage capacity at root zone (SWSC) were in the lower PoE range (< 0.1 according to both the criteria). There was some variation in the PoE value of the predictors depending on the performance metric, due to the diversity in overfitting penalty. However, ranking of the variables was identical irrespective of the performance metric used. The "best" and "worst" predictors ranked according to R^2_{adj} were also in agreement with the PoE analysis (not shown). In addition, results of the subsample analysis gave similar results (not shown).

3.2.2 Better-performing models

According to information criteria, the performance of models can only be evaluated relative to the best-performing model in the set. In this study, as per the model weights, no model exhibited apparent dominance. The evidence ratio (ratio between the weights of the best model and nth model) suggested that the best model according to CAIC was only 1.04 times better than the second best model. However, the evidence ratio increased exponentially with increase in model rank and there was a clear distinction between better models and worse models. Similar results were reported by Saft et al. (2016) in her work for modelling rainfall–runoff relationship shift. The choice of better models was made by considering the PoE of individual predictors (refer Sect. 3.2.1) and the number of predictors in the model (V). Figure 5 shows the performance criteria for the top three models for different V values. The model performance increased with V up to 6 to 7 depending on the different criteria. After that, AIC_c, CAIC, RMSE and R^2_{adj} values remained almost constant, indicating that further addition of predictors did not improve the model performance. In particular CAIC shows reaches a minimum at $V = 7$ and it penalises model complexity more rigorously. Table 3 illustrates the predictors in the top 10 models selected based on CAIC. All the top 10 models had $V <= 7$. P, PET and LU repeatedly appeared in the predictor list of the top 10 models substantiating their high predictive capacity, and the top ranked model includes these three predictors only. In this particular case, top-performing models according to both information criteria were the same; therefore results from only one criteria (CAIC) will be discussed.

3.2.3 Model testing

Models ranking from 1 to 10 according to CAIC (Table 3) were tested using both model testing techniques discussed in Sect. 2.2.1c. Figure 6 depicts model fit and model prediction RMSE values of 200 subsample tests. It is clear from the box plots that the difference between the RMSE of the 1st and the 10th model during both model fitting and prediction is less than 1 mm yr^{-1}. In subsample tests, R^2_{adj} of the best model ranged from 0.42 to 0.56 implying 42 to 56 % of the variance was explained (please see Sect. 3.2.3 for details on sub-sample testing). The model errors at each data point ranged from −8 to 28 mm yr^{-1}. However, 97.2 % of the points had errors between −8 and 10 mm yr^{-1}. Figure 7 shows the relationship between precipitation and model errors and it is evident from this scatter plot that model predictions were not greatly influenced by low or high precipitation. In other words, the model was unbiased by precipitation trends. Similar checking was done for all other predictors

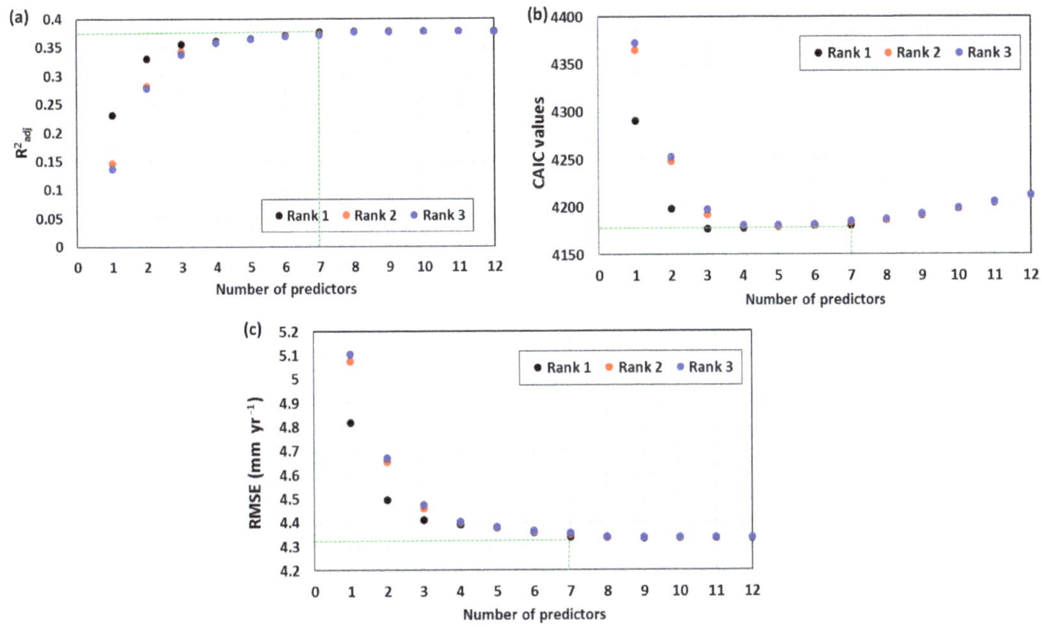

Figure 5. (a) R^2_{adj}, **(b)** CAIC and **(c)** RMSE for the top three models with 1 to 12 predictors and the green dotted lines representing the number of predictors for the best performance criteria value.

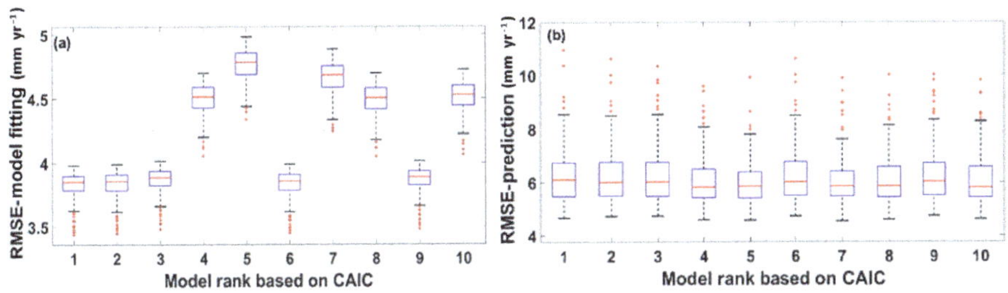

Figure 6. RMSE of sub-sample **(a)** model fitting and **(b)** model prediction of top 10 models according to CAIC.

(not shown) which all showed a similar pattern to precipitation. The dataset was classified based on recharge estimation techniques and model performance was tested with results showing no systematic difference (not shown).

3.3 Global groundwater recharge

The global long term (1981–2014) mean annual groundwater recharge map at a spatial resolution of 0.5° was made by the model developed in Sect. 3.2 (Fig. 8). In this study, the best model as defined by CAIC (model 1 in Table 3) was used to generate the recharge map. However, due to the similarity in structure of the top 10 models (Table 3), all models were equally good at predicting groundwater recharge and gave similar results (not shown). Grid-scale recharge ranged from 0.02 to 996.55 mm yr^{-1} with an average of 133.76 mm yr^{-1}. The highest recharge was associated with very high rainfall (> 4000 mm yr^{-1}). Humid regions such as Indonesia, Philippines, Malaysia, Papua New Guinea, Amazon, western Africa, Chile, Japan and Norway had very high recharge (> 450 mm yr^{-1}), whereas arid regions of Australia, the Middle East and Sahara had very low recharge (< 0.1 mm yr^{-1}). In humid areas, percentage of rainfall becoming groundwater recharge (> 40 %) was found to be very high in comparison to other parts of the world. However, the mean percentage of rainfall becoming recharge is only 22.06 % across the globe. Among all the continents, Australia had the lowest annual groundwater recharge rate.

Over the 34 years, global annual mean recharge followed the same pattern as that of global annual mean precipitation (Fig. 9). Least recharge was predicted in the year 1987 (groundwater recharge = 95 mm yr^{-1}), where the annual average rainfall was < 180 mm yr^{-1}. Variation in recharge over the years was maximal in arid regions of Australia and North Africa (Fig. 10a). However, the standard deviation of recharge was higher in humid areas than in arid regions (Fig. 10b). This indicates that standard deviation did not clearly represent year-to-year variations in recharge. Potentially, the advantage of using coefficient of variation over standard deviation is that it can capture variations even when

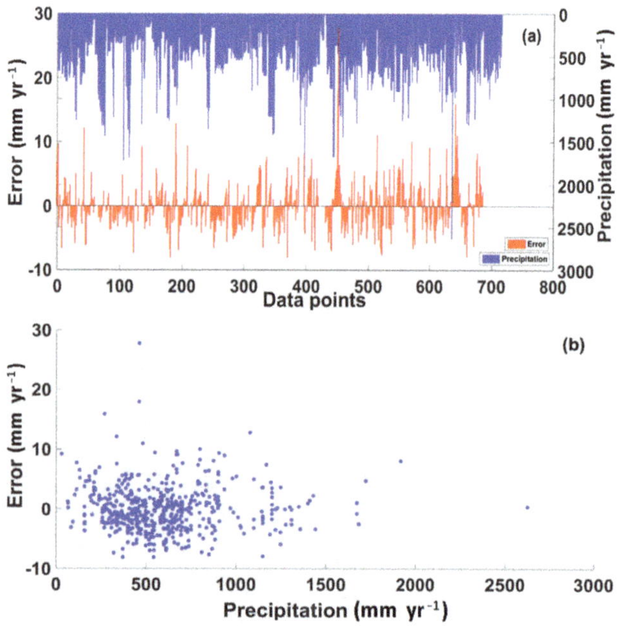

Figure 7. (a) Error at each data point along with the corresponding rainfall obtained using the leave-one-out model testing procedure and **(b)** Scatter plot between error at each data point and corresponding precipitation.

mean values are very small. In this case precipitation and potential evapotranspiration were the two major predictors of recharge. Globally, variability in evapotranspiration is much less than variability in rainfall (Peel et al., 2001; Trenberth and Guillemot, 1995). Therefore, variability of groundwater recharge both temporally and spatially is due to variability in precipitation, which implies that arid regions are more susceptible to inter-annual variation in groundwater recharge. A comparison of predicted recharge against country-level recharge estimates from FAO (2005) shows that the model tends to overpredict recharge, particularly for low recharge areas. However, due to inaccuracies in the FAO estimates this cannot be considered as a reliable comparison (Fig. 11a). Recharge estimates from the best models in the present study were compared to recharge estimates from the complex hydrological model (WaterGAP; Fig. 11b). Even though the model in this study overestimates recharge for countries with fewer data points, the scatter shows a smaller spread compared to the FAO estimates. Figure 12 shows the country-wide distribution of errors in model prediction in comparison with FAO statistics. Very high errors were found in countries with fewer model-building data points. The model considerably overestimated recharge for Russia, Canada, Brazil, Indonesia, Malaysia and Madagascar.

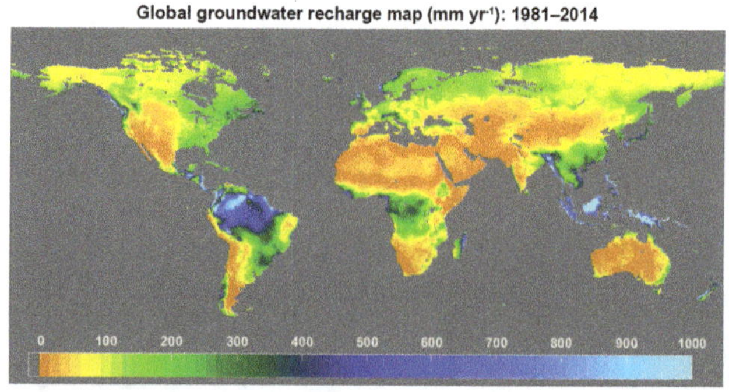

Figure 8. Long-term (1981–2014) average annual groundwater recharge estimated using the developed model.

4 Discussion

The aims of this study were to identify the factors with the most influence on groundwater recharge, and to develop a global model for predicting groundwater recharge under limited data conditions, without extensive water balancing. In this study, an empirical model-building exercise employing linear regression analysis, multimodel inference techniques and information criteria was used to identify the most influential predictors of groundwater recharge and use them to build predictive models. Finally, a global groundwater recharge map was created using the developed model. The key findings from this study and their implications for future research and practice with respect to global groundwater recharge are discussed below.

One of the findings to emerge is that, out of numerous models developed in this study there was no single best model for groundwater recharge. Instead, there were clear sets of better and worse models. However, there were predictors which stood out as having greater explanatory power. Of the 12 predictors chosen for the analysis, meteorological (P, PET) and vegetation predictors (LU) had the most explanatory information followed by saturated hydraulic conductivity and clay content. Thus models using these predictors ranked higher according to information criteria. It is reasonable that meteorological factors had the most explanatory information. In most cases, especially dry regions, groundwater recharge is controlled by the availability of water at the surface, which is mainly controlled by precipitation, evapotranspiration and geomorphic features (Scanlon et al., 2002). Numerous studies agree with this finding. For example, in south-western USA, 80 % of recharge variation is explained by mean annual precipitation (Keese et al., 2005). However, the influence of meteorological factors on groundwater recharge is highly site specific (Döll and Flörke, 2005). The effect of meteorological factors can also depend on whether the season or year is wet or dry, type of aquifer and irrigation intensity (Adegoke et al., 2003; Moore and Rojstaczer, 2002; Niu et al., 2007).

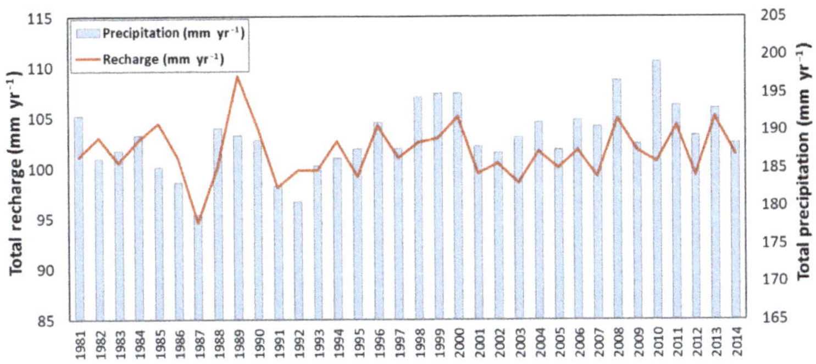

Figure 9. Temporal distribution of total global recharge along with total global precipitation of corresponding years for a period of 1981 to 2014.

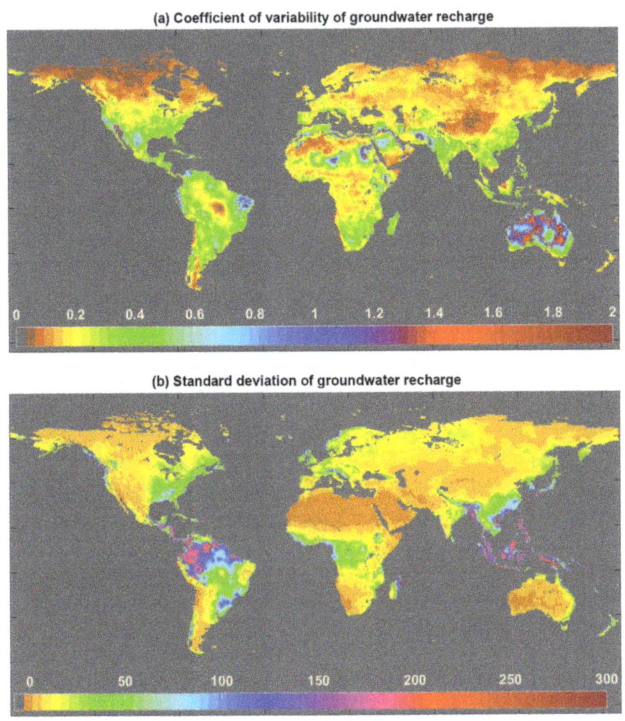

Figure 10. Map showing **(a)** coefficient of variability and **(b)** standard deviation of annual groundwater recharge from 1981 to 2014.

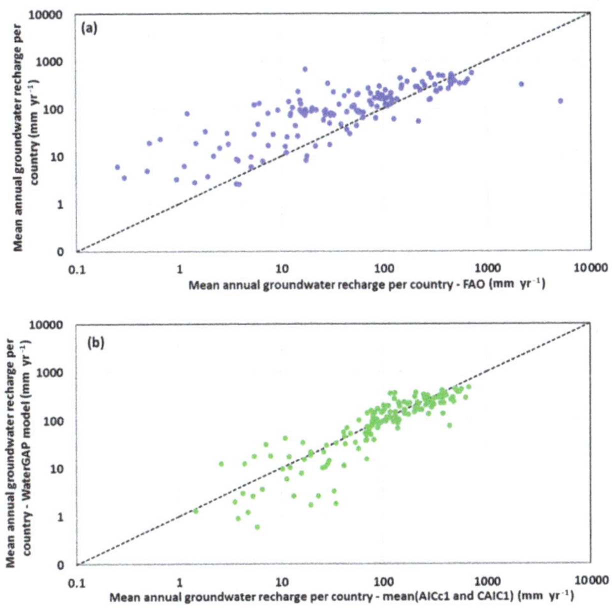

Figure 11. Comparison of predicted recharge against country-level estimates from **(a)** FAO and **(b)** WaterGAP model.

Many studies have reported vegetation-related parameters as the second most influential predictor of groundwater recharge. Vegetation has a high correlation with other physical variables such as soil moisture, runoff capacity and porosity, which adds to its recharge explanatory power (Kim and Jackson, 2012; Scanlon et al., 2005). In this study land use (LU) was used as a proxy for vegetation. According to the results, LU was found to be one of the predictors having the highest proportion of evidence (PoE; Fig. 4). In addition, all the better-performing models included LU as one of the predictors, which clearly indicates that vegetation is one of the most influential factors for groundwater recharge. Results indicates that recharge rates were high where runoff water had greater retention time on the surface. This was mainly observed in areas with shallow-rooted vegetation like grasslands. In deep-rooted forest areas recharge was reduced because of increased evapotranspiration (Kim and Jackson, 2012). However, not all studies reported are in agreement with vegetation as an important predictor of recharge. For example, Tögl (2010) failed to find a correlation between vegetation/land cover and recharge. This may be the result of some peculiarity in the study dataset. Apart from the predictors discussed above, depth to groundwater and surface drainage density were also identified as potential predictors of recharge from literature (Döll and Flörke, 2005; Jankiewicz et al., 2005). Despite this they were excluded from this study because of the lack of appropriate resolution global datasets.

The total recharge estimated in this study is strongly consistent with results from complex global hydrological mod-

Table 4. Global estimates of groundwater recharge.

Model used	Spatial resolution	Temporal range	Total global recharge ($km^3\,yr^{-1}$)	Reference
Empirical model	0.5°	1981–2014	13 600	Current study
WaterGAP 2	0.5°	1961–1990	14 000	Döll (2002)
WaterGAP	0.5°	1961–1990	12 666	Döll and Flörke (2005)
PCR GlobWB	0.5°	1958–2001	15 200	Wada et al. (2010)
PCR GlobWB	0.5°	1960–2010	17 000	Wada et al. (2012)
MATSIRO	1°	1985–1999	29 900	Koirala et al. (2012)
FAO Statistics	Country	1982–2014	10 613	FAO (2016)

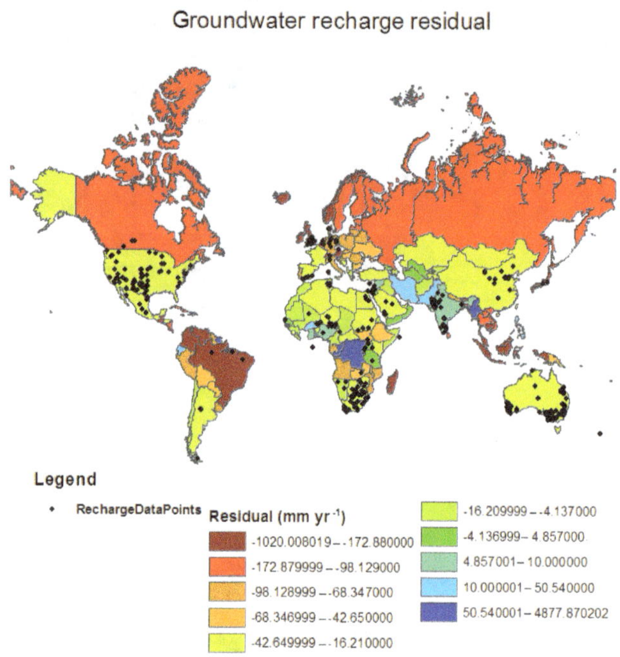

Figure 12. Spatial distribution of groundwater recharge residual (FAO estimates less model estimates) along with recharge sites selected for model building.

els. Long-term average annual recharge was found to be $134\,mm\,yr^{-1}$. The total recharge estimated in this study ($13\,600\,km^3\,yr^{-1}$) was very close to existing estimates of complex hydrological models except those using MATSIRO, which overestimates recharge in humid regions (Koirala et al., 2012). The results shown in Table 4 indicate that, compared to existing techniques, the model developed in this study can make recharge assessments with the same reliability but with fewer computational requirements. Moreover, the error in recharge prediction in this study was low, ranging from only -8 to $10\,mm\,yr^{-1}$ for 97.2 % of cases.

The global recharge map developed showed a similar pattern to recharge maps produced using complex global hydrological models. The results of this study indicate that recharge across the globe varied considerably as a function of spatial region, and was analogous to global distribution of climate zones (Scanlon et al., 2002). Humid regions had very high recharge compared to arid (semi-arid) regions, which is obviously due to the higher availability of water for recharge. Recharge was also affected by climate variability and climate extremes at a regional level (Scanlon et al., 2006; Wada et al., 2012). However, an effect of climate variability on inter-annual recharge at a global scale was not pronounced in our results. The potential reason for this is that the El Niño Southern Oscillation (ENSO), the primary factor determining climate variability globally, has converse effects in different parts of the world. The effects of increased precipitation in some parts of the world would have been counteracted by reductions in precipitation in other areas resulting in relatively small effects on inter-annual variation in global recharge.

5 Conclusion

This study presents a new method for identifying the major factors influencing groundwater recharge and using them to model large-scale groundwater recharge. The model was developed using a dataset compiled from the literature and containing groundwater recharge data from 715 sites. In contrast to conventional water balance recharge estimation, a multi-model analysis technique was used to build the model. The model developed in this study is purely empirical and has fewer computational requirements than existing large-scale recharge modelling methods. The 0.5° global recharge estimates presented here are unique and more reliable because of the extensive validation done at different scales. Moreover, inclusion of a range of meteorological, topographical, lithological and vegetation factors adds to the predictive power of the model. The results of this investigation show that meteorological and vegetation factors had the most predictive power for recharge. The high dependency of recharge on meteorological predictors make it more vulnerable to climate change. Although it is a computationally efficient modelling method, the approach used in this study has some limitations. Firstly it does not include direct anthropogenic effects on the groundwater system or recharge by natural or artificial means, suggesting scope for further future development. Secondly, the recharge dataset used in this study did not include data points from frozen regions. Therefore, Greenland and Antarctica were excluded from the final recharge map. However, the model developed in this study and the recharge maps produced will aid policy makers in predicting future scenarios with respect to global groundwater availability.

Competing interests. The authors declare that they have no conflict of interest.

Acknowledgements. This project was partially supported by the Australian Research Council through project FT130100274. The authors would like to acknowledge the University of Melbourne for providing computational and other technical facilities for this research, and also the international agencies that provided the data required for this study.

Edited by: Nandita Basu

References

Adegoke, J. O., Pielke Sr., R. A., Eastman, J., Mahmood, R., and Hubbard, K. G.: Impact of irrigation on midsummer surface fluxes and temperature under dry synoptic conditions: A Fregional atmospheric model study of the US High Plains, Mon. Weather Rev., 131, 556–564, 2003.

Aguilera, H. and Murillo, J.: The effect of possible climate change on natural groundwater recharge based on a simple model: a study of four karstic aquifers in SE Spain, Environ. Geol., 57, 963–974, 2009.

Akaike, H.: A new look at the statistical model identification, IEEE T. Automat. Control, 19, 716–723, 1974.

Alcamo, J., Döll, P., Henrichs, T., Kaspar, F., Lehner, B., Rösch, T., and Siebert, S.: Development and testing of the WaterGAP 2 global model of water use and availability, Hydrolog. Sci. J., 48, 317–337, 2003.

Ali, R., McFarlane, D., Varma, S., Dawes, W., Emelyanova, I., and Hodgson, G.: Potential climate change impacts on the water balance of regional unconfined aquifer systems in southwestern Australia, Hydrol. Earth Syst. Sci., 16, 4581–4601, https://doi.org/10.5194/hess-16-4581-2012, 2012.

Anderson, D. and Burnham, K.: Model selection and multi-model inference, in: 2nd Edn., Springer-Verlag, New York, 2004.

Athavale, R., Murti, C., and Chand, R.: Estimation of recharge to the phreatic aquifers of the Lower Maner Basin, India, by using the titrium injection method, J. Hydrol., 45, 185–202, 1980.

Bozdogan, H.: Model selection and Akaike's information criterion (AIC): The general theory and its analytical extensions, Psychometrika, 52, 345–370, 1987.

Bredenkamp, D.: Quantitative estimation of ground-water recharge in dolomite, in: Estimation of Natural Groundwater Recharge, Springer, Dordrecht, 449–460, 1988.

Broxton, P. D., Zeng, X., Sulla-Menashe, D., and Troch, P. A.: A Global Land Cover Climatology Using MODIS Data, J. Appl. Meteorol. Clim., 53, 1593–1605, https://doi.org/10.1175/jamc-d-13-0270.1, 2014.

Burnham, K. P. and Anderson, D. R.: Model selection and multimodel inference: a practical information-theoretic approach, Springer Science & Business Media, New York, 2003.

Cavanaugh, J. E. and Shumway, R. H.: A bootstrap variant of AIC for state-space model selection, Statist. Sin., 7, 473–496, 1997.

Chaussard, E., Wdowinski, S., Cabral-Cano, E., and Amelung, F.: Land subsidence in central Mexico detected by ALOS InSAR time-series, Remote Sens. Environ., 140, 94–106, 2014.

DAAC: Spatial Data Access Tool (SDAT), in: ORNL Distributed Active Archive Center, https://daac.ornl.gov/get_data/, last access: 23 September 2016.

Döll, P.: Impact of climate change and variability on irrigation requirements: a global perspective, Climatic Change, 54, 269–293, 2002.

Döll, P. and Fiedler, K.: Global-scale modeling of groundwater recharge, Hydrol. Earth Syst. Sci., 12, 863–885, https://doi.org/10.5194/hess-12-863-2008, 2008.

Döll, P. and Flörke, M.: Global-Scale estimation of diffuse groundwater recharge: model tuning to local data for semi-arid and arid regions and assessment of climate change impact, Frankfurt Hydrology Paper 03, Institute of Physical Geography, Frankfurt University, Frankfurt, Germany, http://www.geo.uni-frankfurtde/ipg/ag/dl/fpublikationen/2005/FHP03DoellFloerke2005.pdf (last access: May 2018), 2005.

Duan, Q., Ajami, N. K., Gao, X., and Sorooshian, S.: Multi-model ensemble hydrologic prediction using Bayesian model averaging, Adv. Water Resour., 30, 1371–1386, 2007.

Edmunds, W., Gaye, C., and Fontes, J.-C.: A record of climatic and environmental change contained in interstitial waters from the unsaturated zone of northern Senegal, in: International Symposium on Isotope Techniques in Water Resources Development, Vienna, 533–549, 1991.

Ezekiel, M.: The application of the theory of error to multiple and curvilinear correlation, J. Am. Stat. Assoc., 24, 99–104, 1929.

FAO – Food and Agriculture Organization of the United Nations: Aspects of FAO's policies, programmes, budget and activities aimed at contributing to sustainable development, Rome, Italy, 2005.

FAO: AQUASTAT Main Database, http://www.fao.org/nr/water/aquastat/sets/index.stm, last access: 20 March 2016.

Fenicia, F., Savenije, H. H., Matgen, P., and Pfister, L.: Understanding catchment behavior through stepwise model concept improvement, Water Resour. Res., 44, W01402, https://doi.org/10.1029/2006WR005563, 2008.

Gaganis, P. and Smith, L.: A Bayesian approach to the quantification of the effect of model error on the predictions of groundwater models, Water Resour. Res., 37, 2309–2322, 2001.

Gerland, P., Raftery, A. E., Ševčíková, H., Li, N., Gu, D., Spoorenberg, T., Alkema, L., Fosdick, B. K., Chunn, J., and Lalic, N.: World population stabilization unlikely this century, Science, 346, 234–237, 2014.

Gurdak, J. J.: Ground-water vulnerability: nonpoint-source contamination, climate variability, and the high plains aquifer, VDM Verlag Publishing, Saarbrücken, Germany, p. 223, 2008.

Harris, I., Jones, P., Osborn, T., and Lister, D.: Updated high-resolution grids of monthly climatic observations – the CRU TS3.10 Dataset, Int. J. Climatol., 34, 623–642, 2014.

Herrera-Pantoja, M. and Hiscock, K.: The effects of climate change on potential groundwater recharge in Great Britain, Hydrol. Process., 22, 73–86, 2008.

Holman, I., Tascone, D., and Hess, T.: A comparison of stochastic and deterministic downscaling methods for modelling potential groundwater recharge under climate change in East Anglia, UK: implications for groundwater resource management, Hydrogeol. J., 17, 1629–1641, 2009.

Hurvich, C. M. and Tsai, C.-L.: Regression and time series model selection in small samples, Biometrika, 76, 297–307, 1989.

Jackson, R. B., Carpenter, S. R., Dahm, C. N., McKnight, D. M., Naiman, R. J., Postel, S. L., and Running, S. W.: Water in a changing world, Ecol. Appl., 11, 1027–1045, 2001.

Jankiewicz, P., Neumann, J., Duijnisveld, W., Wessolek, G., Wycisk, P., and Hennings, V.: Abflusshöhe, Sickerwasserrate, Grundwasserneubildung – Drei Themen im Hydrologischen Atlas von Deutschland, Hydrol. Wasserbewirt., 49, 2–13, 2005.

Jothityangkoon, C., Sivapalan, M., and Farmer, D.: Process controls of water balance variability in a large semi-arid catchment: downward approach to hydrological model development, J. Hydrol., 254, 174–198, 2001.

Jyrkama, M. I. and Sykes, J. F.: The impact of climate change on spatially varying groundwater recharge in the grand river watershed (Ontario), J. Hydrol., 338, 237–250, 2007.

Keese, K., Scanlon, B., and Reedy, R.: Assessing controls on diffuse groundwater recharge using unsaturated flow modeling, Water Resour. Res., 41, W06010, https://doi.org/10.1029/2004WR003841, 2005.

Kim, J. H. and Jackson, R. B.: A global analysis of groundwater recharge for vegetation, climate, and soils, Vadose Zone J., 11, https://doi.org/10.2136/vzj2011.0021RA, 2012.

Kohli, A. and Frenken, K.: Renewable Water Resources Assessment – 2015 AQUASTAT methodology review, Food and Agricultural Organisation of the United Nations, Rome, Italy, 1–6, 2015.

Koirala, S., Yamada, H., Yeh, P., Oki, T., Hirabayashi, Y., and Kanae, S.: Global simulation of groundwater recharge, water table depth, and low flow using a land surface model with groundwater representation, J. Jpn. Soc. Civ. Eng., 68, I_211–I_216, 2012.

Kurylyk, B. L., MacQuarrie, K. T., and Voss, C. I.: Climate change impacts on the temperature and magnitude of groundwater discharge from shallow, unconfined aquifers, Water Resour. Res., 50, 3253–3274, 2014.

L'vovich, M. I.: World water resources and their future, American Geophysical Union, Washington, D.C., USA, 1979.

Moore, N. and Rojstaczer, S.: Irrigation's influence on precipitation: Texas High Plains, USA, Geophys. Res. Lett., 29, 1755, https://doi.org/10.1029/2002GL014940, 2002.

New, M., Lister, D., Hulme, M., and Makin, I.: A high-resolution data set of surface climate over global land areas, Clim. Res., 21, 1–25, 2002.

Niu, G. Y., Yang, Z. L., Dickinson, R. E., Gulden, L. E., and Su, H.: Development of a simple groundwater model for use in climate models and evaluation with Gravity Recovery and Climate Experiment data, J. Geophys. Res.-Atmos., 112, D07103, https://doi.org/:10.1029/2006JD007522, 2007.

Nulsen, R. and Baxter, I.: Water use by some crops and pastures in the southern agricultural areas of Western Australia, Report 32, Department of Agriculture and Food, Western Australia, 1987.

O'Connell, M. G., O'Leary, G. J., and Incerti, M.: Potential groundwater recharge from fallowing in north-west Victoria, Australia, Agr. Water Manage., 29, 37–52, 1995.

Ortiz-Zamora, D. and Ortega-Guerrero, A.: Evolution of long-term land subsidence near Mexico City: Review, field investigations, and predictive simulations, Water Resour. Res., 46, W01513, https://doi.org/10.1029/2008WR007398, 2010.

Pangle, L. A., Gregg, J. W., and McDonnell, J. J.: Rainfall seasonality and an ecohydrological feedback offset the potential impact of climate warming on evapotranspiration and groundwater recharge, Water Resour. Res., 50, 1308–1321, 2014.

Peel, M. C., McMahon, T. A., Finlayson, B. L., and Watson, F. G.: Identification and explanation of continental differences in the variability of annual runoff, J. Hydrol., 250, 224–240, 2001.

Phien-Wej, N., Giao, P., and Nutalaya, P.: Land subsidence in Bangkok, Thailand, Eng. Geol., 82, 187–201, 2006.

Poeter, E. and Anderson, D.: Multimodel ranking and inference in ground water modeling, Groundwater, 43, 597–605, 2005.

Richey, A. S., Thomas, B. F., Lo, M. H., Reager, J. T., Famiglietti, J. S., Voss, K., Swenson, S., and Rodell, M.: Quantifying renewable groundwater stress with GRACE, Water Resour. Res., 51, 5217–5238, 2015.

Saft, M., Peel, M. C., Western, A. W., and Zhang, L.: Predicting shifts in rainfall–runoff partitioning during multiyear drought: Roles of dry period and catchment characteristics, Water Resour. Res., 52, 9290–9305, 2016.

Sanford, W.: Recharge and groundwater models: an overview, Hydrogeol. J., 10, 110–120, 2002.

Scanlon, B. R., Healy, R. W., and Cook, P. G.: Choosing appropriate techniques for quantifying groundwater recharge, Hydrogeol. J., 10, 18–39, 2002.

Scanlon, B. R., Reedy, R. C., Stonestrom, D. A., Prudic, D. E., and Dennehy, K. F.: Impact of land use and land cover change on groundwater recharge and quality in the southwestern US, Global Change Biol., 11, 1577–1593, 2005.

Scanlon, B. R., Keese, K. E., Flint, A. L., Flint, L. E., Gaye, C. B., Edmunds, W. M., and Simmers, I.: Global synthesis of groundwater recharge in semiarid and arid regions, Hydrol. Process., 20, 3335–3370, 2006.

Sivapalan, M., Blöschl, G., Zhang, L., and Vertessy, R.: Downward approach to hydrological prediction, Hydrol. Process., 17, 2101–2111, 2003.

Sreng, S., Li, L., Sugiyama, H., Kusaka, T., and Saitoh, M.: Upheaval phenomenon in clay ground induced by rising groundwater level, Proceedings of the fourth biot conference on poromechanics, Columbia University, New York, 106–203, 2009.

Tögl, A. C.: Global groundwater recharge: Evaluation of modeled results on the basis of independent estimates, Institut für Physische Geographie Fachbereich Geowissenschaften/ Geographie, Goethe-Universität, Frankfurt am Main, 2010.

Trenberth, K. E. and Guillemot, C. J.: Evaluation of the global atmospheric moisture budget as seen from analyses, J. Climate, 8, 2255–2272, 1995.

Verdin, K.: ISLSCP II HYDRO1k elevation-derived products, in: ISLSCP Initiative II Collection, edited by: Hall, F.

G., Collatz, G., Meeson, B., Los, S., Brown de Colstoun, E., and Landis, D., Oak Ridge National Laboratory Distributed Active Archive Center, Oak Ridge, Tennessee, USA, https://doi.org/10.3334/ORNLDAAC/1007, 2011.

Wada, Y., van Beek, L. P., van Kempen, C. M., Reckman, J. W., Vasak, S., and Bierkens, M. F.: Global depletion of groundwater resources, Geophys. Res. Lett., 37, L20402, https://doi.org/10.1029/2010GL044571, 2010.

Wada, Y., Beek, L. P., Sperna Weiland, F. C., Chao, B. F., Wu, Y. H., and Bierkens, M. F.: Past and future contribution of global groundwater depletion to sea-level rise, Geophys. Res. Lett., 39, L09402, https://doi.org/10.1029/2012GL051230, 2012.

Wang, L., Dochartaigh, B., and Macdonald, D.: A literature review of recharge estimation and groundwater resource assessment in Africa, Natural Environment Research Council (NERC), Nottingham, England, 2010.

Wang, Z. and Thompson, B.: Is the Pearson r^2 biased, and if so, what is the best correction formula?, J. Exp. Educ., 75, 109–125, 2007.

Webb, R. W., Rosenzweig, C. E., and Levine, E. R.: Global soil texture and derived water-holding capacities (Webb et al.), Data set, Oak Ridge National Laboratory Distributed Active Archive Center, Oak Ridge, Tennessee, USA, available on-line at: http://www.daac.ornl.gov (last access: 15 July 2016), 2000.

4

A risk assessment methodology to evaluate the risk failure of managed aquifer recharge in the Mediterranean Basin

Paula Rodríguez-Escales[1,2,*], Arnau Canelles[1,2,*], Xavier Sanchez-Vila[1,2], Albert Folch[1,2], Daniel Kurtzman[3], Rudy Rossetto[4], Enrique Fernández-Escalante[5], João-Paulo Lobo-Ferreira[6], Manuel Sapiano[7], Jon San-Sebastián[5], and Christoph Schüth[8]

[1]Dept. of Civil and Environmental Engineering. Universitat Politècnica de Catalunya, Barcelona, Spain
[2]Associated Unit: Hydrogeology Group (UPC-CSIC)
[3]Institute of Soil, Water and Environmental Sciences, The Volcani Center, Agricultural Research Organization, Rishon LeZion, Israel
[4]Institute of Life Sciences, Scuola Superiore Sant'Anna, Pisa, Italy
[5]Empresa de Transformación Agraria (TRAGSA), R&D department, Madrid, Spain
[6]Laboratorio Nacional de Engenharia Civil, Lisbon, Portugal
[7]Energy and Water Agency, Luqa, Malta
[8]Institute of Applied Geosciences, Technische Universität Darmstadt, Darmstadt, Germany
[*]These authors contributed equally to this work.

Correspondence: Paula Rodríguez-Escales (paula.rodriguez.escales@upc.edu)

Abstract. Managed aquifer recharge (MAR) can be affected by many risks. Those risks are related to different technical and non-technical aspects of recharge, like water availability, water quality, legislation, social issues, etc. Many other works have acknowledged risks of this nature theoretically; however, their quantification and definition has not been developed. In this study, the risk definition and quantification has been performed by means of "fault trees" and probabilistic risk assessment (PRA). We defined a fault tree with 65 basic events applicable to the operation phase. After that, we have applied this methodology to six different managed aquifer recharge sites located in the Mediterranean Basin (Portugal, Spain, Italy, Malta, and Israel). The probabilities of the basic events were defined by expert criteria, based on the knowledge of the different managers of the facilities. From that, we conclude that in all sites, the perception of the expert criteria of the non-technical aspects were as much or even more important than the technical aspects. Regarding the risk results, we observe that the total risk in three of the six sites was equal to or above 0.90. That would mean that the MAR facilities have a risk of failure equal to or higher than 90 % in the period of 2–6 years. The other three sites presented lower risks (75, 29, and 18 % for Malta, Menashe, and Serchio, respectively).

1 Introduction

Water scarcity, the chronic lack of sufficient quality water to supply a specific area is one of the major global challenges. In the Mediterranean Basin, due to low overall precipitation and a pronounced irregularity of rainfall events, it has direct impacts on economic sectors that depend on water, such as agriculture, tourism, and related industries (Fader et al., 2016; Maliva and Missimer, 2012; Navarro-Ortega et al., 2012; Stanhill et al., 2015). Besides this, the population in the Mediterranean area increased from 81 million in 1960 to 145 million in 2011 (European Environment Agency, 2015), placing additional stress upon existing water resources. Moreover, the Mediterranean Basin is one of the most sensitive regions of the world to climate changes resulting from human activities; according to the latest IPCC projections, average precipitation could decrease by more than

10 %, with a larger decrease in summer and in the southern areas (Pachauri et al., 2014).

At the same time, large water quantities are lost to the Mediterranean Sea as surface runoff and discharges from rivers, treated and untreated wastewater, or excess water from various sources during periods of low demand. These alternative water sources can potentially help to increase water availability, both in general terms and in periods of high demand, therefore improving water security. The main factors hindering the effective use of such waters are related to concerns about water quality and the lack of sufficient low-cost intermediate storage options. In principal, large storage capacity is available in shallow aquifers, mostly in thick unsaturated zones or in already depleted overexploited aquifers. Managed aquifer recharge (MAR) takes advantage of this available storage.

MAR is defined as the intentional infiltration of water into aquifers with the purpose of either later recovering that water for different uses (agricultural, industrial, or urban) or obtaining an environmental benefit (Dillon et al., 2009). MAR includes a range of recharge options (surface or subsurface) and water sources (natural, reclaimed, or desalinated) (Bouwer, 2002; Dillon, 2005; Dillon et al., 2009; Maliva and Missimer, 2012; Sprenger et al., 2017). Furthermore, MAR can involve different engineering solutions, among them are infiltration ponds, surface spreading, bank filtration, and wells infiltrating into either the unsaturated or the saturated zones. In addition, water quality can be improved through MAR due to the combination of chemical and biological reactions during transport of the infiltrated water. Water can either be recovered at the point of infiltration (ASR – aquifer storage and recovery) or some distant downgradient (ASTR – aquifer storage, transport, and recovery). The infiltrated water can enable hydraulic control of an aquifer, e.g., to prevent seawater intrusion, aid aquifer quality recovery (amelioration of the groundwater quality), or sustain groundwater-dependent ecosystems such as wetlands or marshes. Altogether, MAR links water reclamation, water reuse, and water resources management.

Due to these beneficial effects, MAR is now widely regarded as a useful tool to ensure a safe and good quality water source for the increasing demand. However, to guarantee the success of any MAR project, some essential elements need to be considered (Dillon et al., 2009): (i) an adequate source of water for recharge, (ii) a suitable aquifer to store and recover water, (iii) available land to construct the facilities, (iv) a sufficient demand for the recovered water, and (v) the capability to efficiently manage such a project. If any of these elements fail, a MAR project is usually not viable. However, the listed factors seem to involve infrastructural and management aspects only, and ignore legal, social, economic, and political constraints that can significantly entangle MAR application, eventually leading to failure of the project as a whole. In addition, the analysis of the potential success of a MAR project should account for both the initial set up of installation and also for the potential issues that will arise during its operation (European Community, 2000).

The most common identified technical risks for MAR facilities (e.g., Asano and Cotruvo, 2004; Gale et al., 2006; Leviston et al., 2006; Maliva and Missimer, 2012) are those related to the following: (i) the operation of the facility (low recovery rates, clogging, mechanical or structural damage, low storage efficiency, high energy consumption); (ii) water quality, either recharged or extracted; (iii) hydraulic engineering impacts, such as rock fracturing, subsidence, or host porous media dissolution; and (iv) environmental impacts, including reduced water outflow to springs and rivers, proliferation of pests and odors, and the impact on aquifer-dependent ecosystems.

MAR facilities can also be affected by legal, social, economic, and political issues, which increase the risk of failure, meaning that the facility would not begin or continue operation. Therefore, a full and complete risk assessment must encapsulate all relevant constraints and their confidence levels at a given time and projected into the future. In addition, risk evaluations might include the implementation of measures to control risk by either diminishing the probability of occurrence of a given hazard or by reducing and correcting its effects if they eventually occur.

The implementation of a MAR facility is therefore subject to a relatively high degree of uncertainty (Bouwer, 2003; Dillon et al., 2009). Uncertainty can be managed using probabilistic risk assessment (PRA), a concept used in various fields of science and engineering. Risk is defined here as the probability of an undesired outcome to happen (evaluated in terms of percentage of occurrence) that causes a damage (directly, i.e., breakage of pipes, or indirectly, i.e., perception of effectiveness) to the recharge system and therefore causes a failure during operation or design of the MAR facility. Different definitions for risk in MAR are available in the literature; Maliva and Missimer (2012) defined it as the feasibility (technical and economic) to meet regulatory requirements for aquifer recharge.

Several methods are available for risk evaluation. One such method is the development of "fault trees", already used in engineered systems (Bedford and Cooke, 2003; Vesely et al., 1981). Since MAR systems comprise a mixture of natural and engineered components, this approach has received some attention in the hydrological community (e.g., Bolster et al., 2009). The basic idea of PRA based on fault trees (PRA-FT) is to take a complex system, difficult to be handled as a whole, and to divide it into a series of quasi-independent simpler events that are manageable individually (i.e., basic events). Once probabilities of occurrence of basic events are computed, they are recombined in a systematic manner to provide the overall risk assessment of the system as a whole. Examples of applications of PRA-FT in hydrogeology include De Barros et al. (2011, 2013) and Jurado et al. (2012).

Although some approaches to evaluate the risk of a MAR system have been developed (Assmuth et al., 2016;

Ayuso-Gabella et al., 2011; Dillon et al., 2016; Ji and Lee, 2016, 2017; Juntunen et al., 2017; Page et al., 2010; Toze et al., 2010), comprehensive studies that integrate both non-technical and technical factors are absent (Nandha et al., 2015). In this study, we (i) precisely present an integrated PRA-FT that is applicable for a general MAR facility and (ii) apply it to six different MAR facilities that were part of the EU FP7 project MARSOL, located in five different Mediterranean countries: Portugal (1), Spain (2), Italy (1), Malta (1), and Israel (1). To achieve these goals, basic events that can lead to MAR failure were first compiled based on a literature review of 51 MAR facilities worldwide, and also on data from the MARSOL project. The next step was the development of six individual fault trees for the tested sites, and assigning probabilities of occurrence for these events. Finally, we used the six sites to compare the different realities and to test the relative relevance of technical versus non-technical events.

2 Literature review – events involved in MAR failure

MAR failure is defined as the need to stop operation of the facility. This failure can be either complete or partial. Partial failure means that it is possible to mitigate the problem in a short period of time so that the facility can be put back to operation, where complete failure implies that the facility needs to undertake significant changes and reparations in order to work again (or even not working ever again after the failure). In this paper, the failures exposed are considered as partial failures due to the fact that none of the MAR facilities reviewed in the literature permanently stopped working after those failures occurred.

Basic events that can lead to MAR failure were compiled based on a literature review of the problems encountered by different facilities around the world (Aiken and Kuniansky, 2002; Alazard et al., 2016; Assmuth et al., 2016; Bhusari et al., 2016; Chaoka et al., 2006; Flint and Ellett, 2005; Masetti et al., 2016; Murray and Ravenscroft, 2010; Petersen and Glotzbach, 2005; Schneider et al., 1987; Izbicki et al., 2006; Sultana and Ahmed, 2016; Tredoux et al., 2009; Tredoux and Cain, 2010; Tripathi, 2016). We revised 51 MAR facilities at 47 sites (some sites involved more than one facility) located in different countries and climatic conditions worldwide: Australia, Belgium, Botswana, China, Finland, France, Germany, India, Israel, Italy, Jordan, Namibia, South Africa, Spain, Tunisia, and USA. We classified the facilities according to infiltration typology: deep wells (24), surface infiltration (22), and vadose infiltration (5). A summary of the facilities and details can be found in Sect. S1 of the Supplement.

We then sorted the main causes of MAR failure in terms of frequency of appearance for deep wells and infiltration basins (Fig. 1). Furthermore, we classified these problems into technical and non-technical problems and sub-classified them into different categories. The technical problems with the most occurrence were clogging and the presence of nutrients; they were present in 40–50 % of the reviewed facilities (Fig. 1) and in all types of MAR facilities. Three types of clogging were reported: being in order of decreasing importance, biological, physical, and chemical. On the other hand, the nutrient issues were mainly related to the presence of nitrogen and phosphorus in the recharge water, mostly associated with the use of insufficiently treated reclaimed or surface water, with high nutrients levels, for recharge.

In general terms, quality and infiltration issues were the main aspects that limited the viability of MAR facilities. In fact, the six first technical reasons of MAR failure were the same in all facilities: clogging, nutrients, metals, droughts, low infiltration rate, and salinity–sodicity. This can be explained because MAR facilities are often in semi-arid countries were droughts (a main problem from the quantitative point of view) are common. Quantity issues were seldom relevant, and only in infiltration basins. Civil work failures and natural hazards were rarely reported as problems.

Regarding the non-technical aspects, they were classified into four groups: legal constraints, economic constraints, social unacceptance, and governance-related problems. The actual issues identified in the MAR facilities revised were thus related to cost (maintenance and installation of the MAR facility), legal aspects (mainly sanitary issues for the infiltrated or the reclaimed waters), and local constraints (land permissions and urban planning issues).

3 Methodology – development of the fault trees and risk evaluation.

The methodology used consisted of four steps, modifying the scheme followed by Bedford and Cooke (2003): (1) the definition of the concept of system failure and the identification of the basic events that would potentially result in such failure; (2) construction of the fault tree depicting the combination of events, seeking the combination of all possible events that may contribute to system failure (where all events should be as independent from each other as possible); (3) developing a probabilistic representation of the fault tree using Boolean algebra; and (4) computing the individual probabilities of event occurrence using conservative approaches and individual event probabilities and upscaling to calculate the global risk of the facility.

3.1 Failure definition and identification of basic events

The first step was the identification of the basic events that can produce a failure in a general MAR facility by reviewing the literature (Sect. 2) and, as a second step, by an extension based on the knowledge and the experience of the facility managers. Failure was based on the operation stage, which

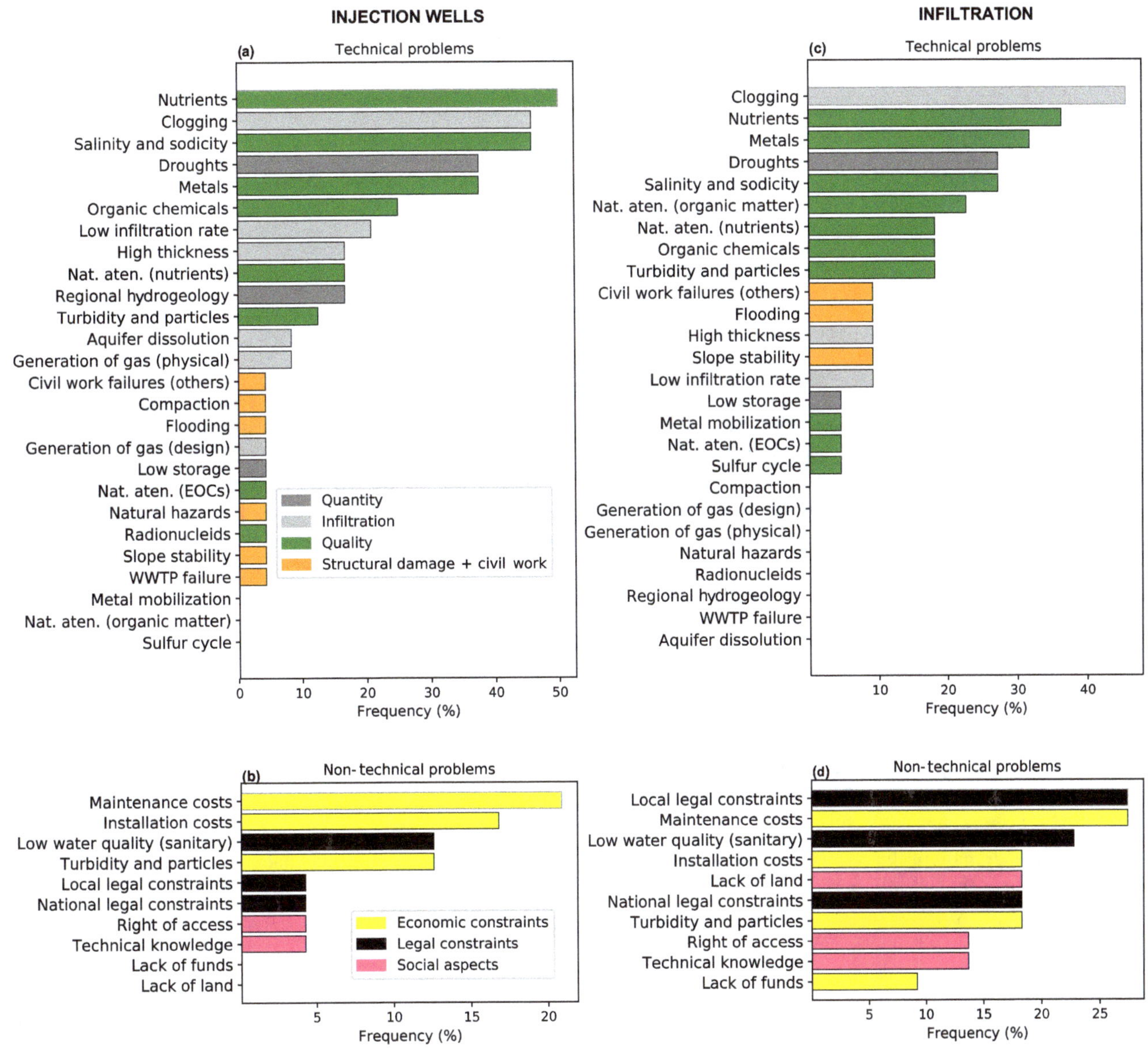

Figure 1. Sorted list in terms of frequency of appearance of the main problems observed in reviewed facilities of deep well injection and infiltration basin. Problems are classified into categories (four for technical, three for non-technical) that are visualized as colors.

implies the improper functioning of the MAR facility, or the cease of its operation for a prolonged time.

3.2 Fault tree construction

The eight categories defined before (technical and non-technical) are described by a few basic events, giving a total of 65 events (44 technical and 21 non-technical; see Fig. 2). A short definition of all the events can be found in Fig. 2 and in Sect. S2 of the Supplement.

3.3 Probabilistic representation of the fault tree

The probabilistic analysis is based on two steps: (1) defining a specific probability for each key event to occur and (2) combining the different events' probabilities, using Boolean algebra, to assign probabilities to the boxes (events) into the one placed immediately above. In this section we illustrate this approach for simplicity and completeness. Additional details about this methodology can be found in Tartakovsky (2007).

For each event, we specify a number of sub-events following two models: (1) if any sub-event occurs, then the event will also occur, thus equivalent to an "OR" operator in Boolean logic; (2) all sub-events must occur for the event to

Table 1. Events of the fault tree divided by categories for non-technical and technical issues.

Non-technical	Legal constraints (LEG): health, urban, environment, construction permits
	Economic constraints (ECO): lack of funds, maintenance or installation costs, macro and microeconomical problems, etc.
	Social unacceptance (SO): health perception, cost perception, effectiveness perception, etc.
	Governance (GOV): coordination between governmental agencies and technical knowledge about the MAR issues
Technical	Structural damage (SD): damage to the MAR infrastructure due to natural hazards, civil works failure, etc.
	Not enough water or quantity (QUAT): low water quality (physical, chemical, and biological), water scarcity (climate, river regulation, waste water treatment plant (WWTP) failure, quantity recharged does not reach some target value that makes it economically feasible) and clogging (physical, biological, and chemical) water available does not reach the quality standards needed to allow it to be used in the recharge facility.
	Unacceptable water quality (QUAL): problems with natural attenuation (nutrients, organic matter, and emerging organic compounds), metabolites (nitrogen cycle, other nutrients like H_2S, etc.), and mobilization of metals. The water finally resulting in the aquifer does not meet some quality standards once it reaches some sensitive location (river, supply well, wetland, etc.).
	Specific targets (ST): failure to achieve targets related to seawater barriers, protected water bodies, and water levels. Seawater intrusion is not sufficiently contained, a protected water body is reached by polluted water or water levels at the target surface water bodies (river, spring, wetland) are not reached.

take place, characteristic of the "AND" operator. Therefore, denoting E as the event and $e_i, i = 1, \ldots, n$ as the sub-events, the "OR" operators involves

$$E(\text{or}) = U_{1,\ldots,n} e_i, \quad (1)$$

while the "AND" operator results in

$$E(\text{and}) = \cap_{1,\ldots,n} e_i. \quad (2)$$

We illustrate this with a simple example (Fig. 3), considering that an issue with either social or political implications increase the probability of having a non-technical MAR failure (which in turn would increase the chance of a general MAR failure). According to the methodology described, we can obtain the probability of the main event (NT), $P[\text{NT}]$, as a function of those of the events SO (social unacceptance) and PO (lack of political implication).

$$P(\text{NT}) = P(\text{SO} \cup \text{PO}) = P(\text{SO}) + P(\text{PO}) - P(\text{SO} \cap \text{PO}) \quad (3)$$

Notice that the last term in Eq. (3) indicates the product of an intersection; this would also be the case if instead of an "OR" operator we had an "AND" operator, so that in such a case we would have the following formula: $P(\text{NT}) = P(\text{SO} \cap \text{PO})$. To compute the probability of the intersection of events, we rely on the concept of conditional probability (e.g., assuming that politics respond directly to social concern), so that

$$P(\text{SO} \cap \text{PO}) = P(\text{SO}) P(\text{PO}/\text{SO}). \quad (4)$$

In the case that SO and PO are independent events, Eq. (4) simplifies to

$$P(\text{SO} \cap \text{PO}) = P(\text{SO}) P(\text{PO}). \quad (5)$$

This system is transferred to the evaluation of basic events in terms of those placed at a lesser level, and so on.

3.4 Computing the individual probabilities of events and the global system failure

The next step is to assign probabilities to all basic events at the bottom of the tree, and then build up (bottom-up) to assign probabilities using the Boolean rules above, until the top (full system failure) is reached. As a first step, all events were divided into four categories depending on probability of occurrence (high, medium, low, or no risk). A key point in the assessment of risk is assigning probabilities to each individual basic event, this process being quite challenging. One advantage of the fault tree approach is the possibility of assigning them at several stages of involvement, taking into

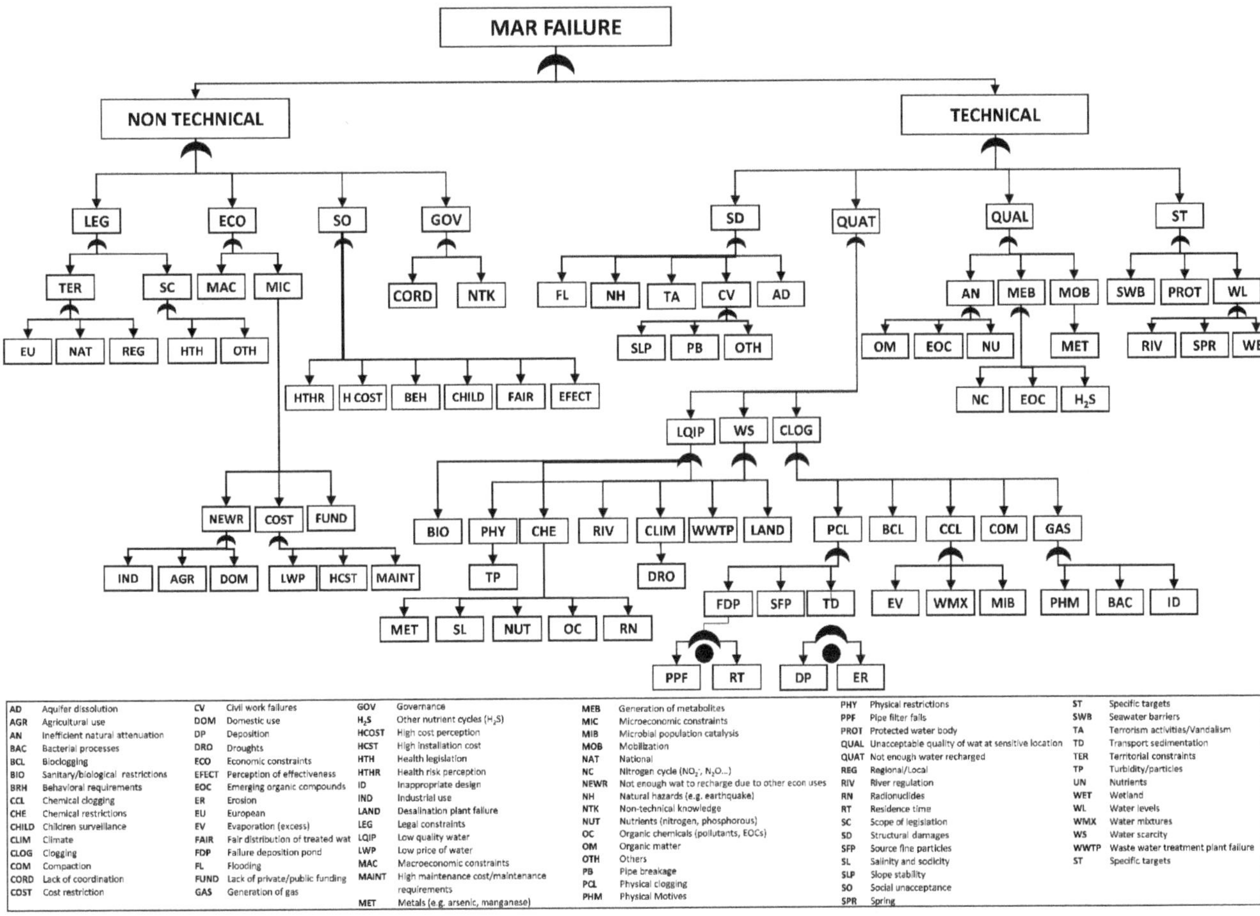

Figure 2. General fault tree for the operational phase.

account a combination of simplicity and relevance. The approach consists of first assigning a (preliminary) value to all basic events; these values may be based on the experience of the managers of the facility or experts. Such values are combined by the Boolean rules to provide a map of critical paths; i.e., events that are up in the tree and that result in a high probability of failure.

The second step would be to devote attention to these significant events and the possibility to correct or reduce their risk in order to reduce the global risk. For those significant events, whose risk contribution is largest, probabilities can be reassigned by using sophisticated approaches, based on, for example, conceptual or numerical modeling and also on changing sampling schemes or putting into operation new observation networks. From these new values, the critical events are re-assessed (including total failure). The process can be repeated as many times as needed, to arrive at an improved value of system failure.

In addition, corrective and preventive measures could be set to reduce the probability for individual events. The full procedure could go on until either economic resources are exhausted or else it is considered that further refinement cannot lead to a significant improvement in the final figure.

We provide an example for the purpose of illustration. Let us assume a surface infiltration pond located in a flood plain. We can start by assigning some probability of the MAR facility being affected by flooding (meaning that operation would have to be discontinued for a long time) using a qualitative approach provided by the facility managers (high, intermediate, low, no risk), maybe including the input of local people that would tell us about potential flooding events that took place during their lifetime. It is very relevant to state that these preliminary numbers should always be on the conservative side, meaning that the less technical the evaluation is the more caution should be included in the actual figures. Here we consider the conservative side as the one providing a larger value of probability of failure than the real one (i.e., from the administration or manager point of view). The second step would use the idea that the facility is located in a 100-year return period flood plain; if we consider the life of the facility of 30 years, and from simple statistics, we can evaluate the probability that flooding occurs during the lifetime as $1-0.99^{30}=0.26$. Now, if this number is excessively

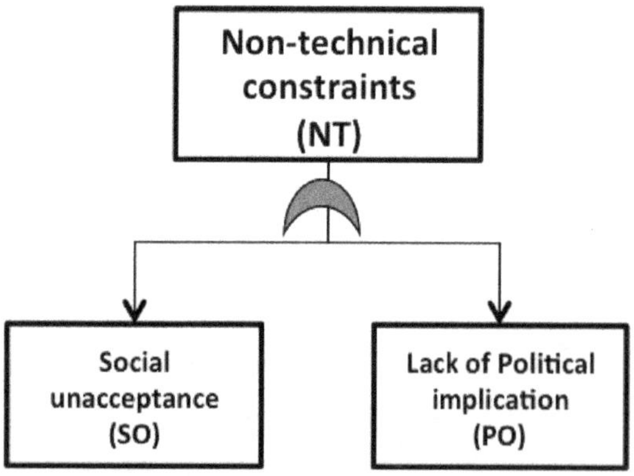

Figure 3. Simplified illustrative case for non-technical constraints involving only sociopolitical constraints. The symbol below the upper event represents an "OR" operator.

high and relevant for the evaluation of overall failure, a further step may include a full hydrological analysis through modeling to re-evaluate the probability of flooding.

Additionally, prevention measures for reducing risk or the damage caused by that risk by using protection works such as embankment construction could be included. Such a measure could then indirectly affect some events (water quality, social acceptance, etc.), leading to the need to continuously update all event risks. This step requires a deep knowledge of the system, and must be done under a local perspective and case by case. As the main goal of this paper is to develop a general methodology to evaluate MAR failure and to compare six sites, this type of analysis is out of the current scope, and we did not perform any detailed analysis of either the risk or its damage reduction of any event.

4 Description of the field sites

The PRA-FT analysis was performed for six MAR facilities, located in different parts of the Mediterranean Basin, offering a broad view of risk perception in the whole area. A summary of the characteristics and context of these sites can be found in Table 2.

The first site is located in the Algarve region (south of Portugal). It is based on an infiltration basin constructed in the Rio Seco river bed (Campina de Faro aquifer system). This MAR facility is aimed at improving the water quality of the Campina de Faro aquifer. It was constructed in 2006 (Lobo-Ferreira and Leitão, 2014). The surroundings of the MAR facility are mainly agricultural and one of its main problems is related to water quality due to agriculture diffuse water pollution, mainly by nitrate (Leitão et al., 2017). This site includes other MAR facilities in the Querença-Silves aquifer and Melides watershed, not included in this evaluation.

The second site, Los Arenales, is located in the center of Spain (Castilla y León). The MAR facility is aimed at providing enough water for the development of rural activities in the zone. Besides this, MAR is also aimed at improving the groundwater quality (reducing nitrate concentrations). In this case, the site is an ensemble of different small facilities (infiltration ponds, river bank filtration, and infiltration wells) located in the same aquifer (Los Arenales alluvial aquifer). It started its activity in 2002 and it was expanded in 2003, 2004, and in 2012.

The MAR Llobregat site is located in Catalonia (northeast Spain), in an alluvial aquifer placed some 10 km southwest of Barcelona city. It is composed of one settling pond and one for infiltration. The recharge water comes from the Llobregat River and the main goal of the facility is to increase the water storage in the aquifer, as well as to improve the quality of the recharged water. A reactive layer was placed at the bottom of the infiltration pond to improve the degradation of both traditional pollutants and emerging organic compounds (e.g., pharmaceuticals; Valhondo et al., 2015). The area surrounding the facility involves agricultural, industrial, and urban uses. It started its activity in 2009.

The induced riverbank filtration scheme along the Serchio river in Sant'Alessio (Lucca, Italy) provides continuous availability of water with good chemical quality for drinking uses to the people of the coastal Tuscany (Rossetto et al., 2015). The MAR scheme is located in a peri-urban or rural area. This facility supplies 15 $Mm^3\ year^{-1}$ and started its activity in the 1960s; it was further improved by building a river weir to increase aquifer storage at the beginning of 2000. The filtration process is so effective that only light treatment using sodium hypochlorite is needed before water is distributed by the local water utility.

The Menashe site is located in Israel. Constructed in 1967, the site includes a settling pond and three infiltration ponds and a canal in which storm water flowing in ephemeral streams are diverted for infiltration in sand dunes overlaying the northern part of the Israeli coastal aquifer. Since 2013, the facility is also used for infiltration of desalinated seawater (1–3 % of production) from the nearby Hadera desalination plant on the Mediterranean coast (Ganot et al., 2017; Ronen-Eliraz et al., 2017). Freshwater is recovered from wells surrounding the infiltration ponds mainly for residential and industrial consumers.

The Malta site is located in the South Malta Coastal Aquifer. The main objective of this site is the implementation of a MAR facility to act as a seawater intrusion barrier and to minimize the salinization risk of the aquifer using a series of infiltration boreholes. The site is located on the coastal margin of a predominantly agricultural region in a limestone aquifer. The activity started in 2016, and is considered as a pilot site to guide the future implementation of MAR in the Maltese islands.

Table 2. Events of the fault tree divided by categories for non-technical and technical issues.

Site	Algarve	Los Arenales	Llobregat	Serchio	Menashe	Malta
Location	Algarve (Portugal)	Los Arenales (Spain)	Sant Vicenç dels Horts (Spain)	Lucca (Italy)	Menashe (Israel)	south Malta (Malta)
Type of recharge	Surface infiltration basins and large wells	Surface infiltration (channels, ditches, ponds) and wells	Surface infiltration basin	Induced riverbank filtration	Surface infiltration basin	Deep wells
Source of recharge water	River water and WWTP water	River water	River water	River water	Storm water and desalinated seawater	WWTP water
Use of recharged water	Improving aquifer water quality and aquifer storage to prevent seawater intrusion	Agriculture	Improve aquifer quantity and quality	Increase groundwater availability and quality for drinking water supply	Store excess of storm and desalinated water	Coastal barrier for seawater intrusion, increase water quantity and quality
Surrounding	Farmland	Farmland	Farmland and industrial park	Farmland/ peri-urban areas	Farmland, industrial, and urban areas	Coastal zone, farmland
Aquifer geology	Alluvial	Aeolian sandy	Alluvial	Sand and gravel alluvial	Interlayered sands calcareous-sandstone and clays	Coastal, floating-lens aquifer
Supporting/managing institute	Águas do Algarve	Spanish Ministry of Agriculture, Fishing, Food and Environment	Catalan Water Agency	GEAL spa	Mekorot National Water Company	Malta Resources Authority and Water Services Corporation
Social setup	Farmers irrigation associations willing to contribute to financing MAR	Farmers, small industry presence and local public administrations	Water Users Community (Farmers and industry presence)	Drinking water needed for the town of Pisa, Lucca, and Livorno (Italy)	Pressure on land-use from industrial sector versus water sector	Coastal barrier for seawater intrusion in agricultural region

5 Probability assignment and global risk computation

5.1 Risk probability assignment

The probabilities for the basic events were defined by the personnel in charge of each MAR facility, according to their own experience (expert criteria). These values reflected the probability that the MAR facility failed due to the occurrence of one of these basic events, considering a time period of approximately 2–6 years. The expert criteria included values of each risk category (high, medium, and low risk) for each fault tree basic event. Then the users filled a qualitative questionnaire, where for each basic event one of the four risk categories had to be chosen. This questionnaire reproduced the same events of the fault tree but sorted as a list at different levels. Then, the qualitative answers were translated to absolute values of probabilities using the values of the expert criteria and the rules of Boolean algebra.

5.2 Global risk computation: MAR-RISKAPP

Once the probability values for each basic event were defined, and the questionnaires filled, global risk values for each facility were computed using a visual application tool, the MAR-RISKAPP. This tool was carried out in a friendly interface, aimed at being used by the managers of MAR facilities worldwide. The tool allows the user to assign one of the four risk categories to each basic event. A value of probability is then assigned by default to each event and category. The user can then manually modify each one of the probability values to keep updating the values of the full tree. The global probability of system failure is then computed internally.

The MAR-RISKAPP is an open-source application which can be downloaded from the website (http://marsol.eu/35-0-Results.html, last access: 6 June 2018). The main flowchart of the app is shown in Fig. 4 and the manual of the app is summarized in Sect. S3 of the Supplement.

6 Results and discussion

6.1 Comparison of risk probabilities defined by expert criteria

As a first step, we have compared the differences between the values provided by the facility managers, internally incorporating personal knowledge and technical expertise, for the six sites. Notice that this way we compare the "perception of risk" rather than actual risk. The results, presented as a box plot of all the values reported by categories, showed that the larger values of risk perceived corresponded to events classified as non-technical (Fig. 5). The risk values (in probability terms) in decreasing order were: legal constraints, social aspects, and economic constraints. On the other hand, for the technical part, the order was (also in decreasing order): water quantity, structural damage, and water quality. Therefore, the perception of risk of the managers of each MAR facility, based on their knowledge and experience, indicates that non-technical aspects are critical and can eventually lead to the facility failing to operate; it might imply that during operation and when the facility has been located in a technically appropriate site, much more uncertainty is expected bz non-technical issues than by technical ones. Legislation was the category with highest risk perception. In general terms, this is explained by the lack of or extremely new (such as in Italy with Decreto Ministeriale 100/2016) pieces of specific legislation about MAR at the European level. The existing European water directives only provide little guidance for authorizing aquifer recharge schemes (Hochstrat et al., 2010). Consequently, MAR regulation is covered by different institutions and authorities dealing with water, environmental, and health legislation. For example, in Menashe, the water recharged is to be used as drinking water, therefore the health legislation risk exceeds other associated legal risks. On the other hand, in Algarve, the infiltration zone is inside a nitrate vulnerable zone and a coastal nature reserve, leading to highest environmental regulation risks. In the Malta case, highest legal risks are associated to the potential for saline intrusion.

The following category in terms of risk perception was social aspects, related to the unacceptance of recharge technology by society. We believe that this could be related, again, to the lack of a concise legislation, which creates social uncertainty and a lack of awareness. We observed that social issues were mostly present in those sites with strong political implications, where the public administrations or the agricultural users participate in the management (and even in the construction) of the MAR facility, such as Llobregat, Los Arenales, or Malta.

Regarding the water quantity aspects, their relatively high importance could be explained because the sites are located in a Mediterranean climate (floods and droughts are typical in such environments). Besides this, it is related to the infiltration capacity of the system (especially in infiltration basins like Los Arenales and Llobregat) and this is traditionally one of the main technical issues in MAR (Fig. 1). Nevertheless, in general terms, infiltration capacity of the system was not an important category in risk perception, probably because the sites were located in highly permeable zones, suitable for recharge, and most of the sites included actions in their maintenance tasks to minimize its importance.

The risk perception on the quality issues could include the following topics: recovered-water use and water source. For example, Serchio recharged water is used as drinking water; consequently, quality plays a higher role than, for example, in Malta, where water is used as a water barrier to salt intrusion. On the other hand, quality is also important in sites where quality aspects existed independently of recharge, like Arenales (high presence of nitrate in groundwater due to farming activity in the zone; see San-Sebastián et al., 2018) and Llobregat (quality problems related to industrial and urban activities; see Valhondo et al., 2015).

This structural damages category is non-negligible, but in general it is not perceived as critical, probably because we are considering sites already in operation. This issue could be more significant in the design process of a facility, and also during construction.

6.2 Comparing risk in the different sites

The risk values for the six sites studied are summarized in Fig. 6. We can observe that the total risk of three of the six sites (Los Arenales, Algarve, and Llobregat) is very large (equal or above 0.9), indicating that facility failure is almost

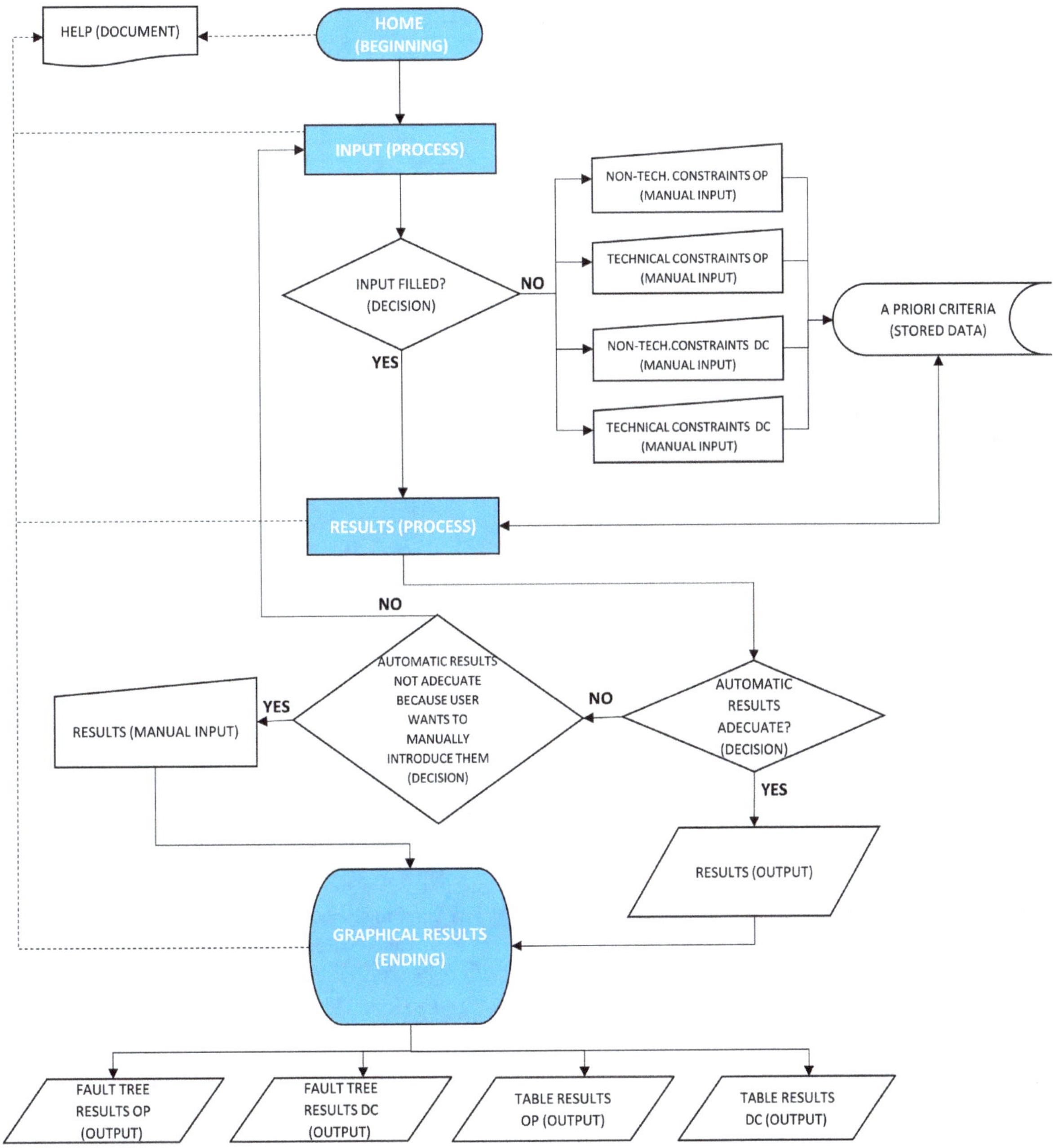

Figure 4. Flowchart for the main program of the MAR-RISKAPP.

certain during a 2- to 6-year period. This indicates that the system will most probably have to discontinue operation; however, this does not imply that the system cannot be put back into operation again. Lack of specific legislation, economic constraints, social issues, and quantitative aspects are the most probable cause for failure of these three facilities. Regarding water quantity being a potential cause of failure, the fact that all three sites involve surface infiltration with river water promote the importance of quantity and clogging aspects, as Mediterranean rivers display low flows and high suspended solid content. Besides this, these three sites are quite young (around 10 years old), which could imply that the are not completely optimized.

On the other hand, Serchio, Menashe, and Malta have lower risk values (0.18, 0.29, and 0.75, respectively). The cases of Serchio and Menashe can be explained because they are the oldest sites, with large experience in the operation of the facilities (therefore lowering the technical risks to below 0.1). Furthermore, the hydraulic characteristics of Serchio added to the riverbed recharging area produces negligible effects of clogging. The case of Menashe is justified by the use of desalinated and storm water for recharge. In the

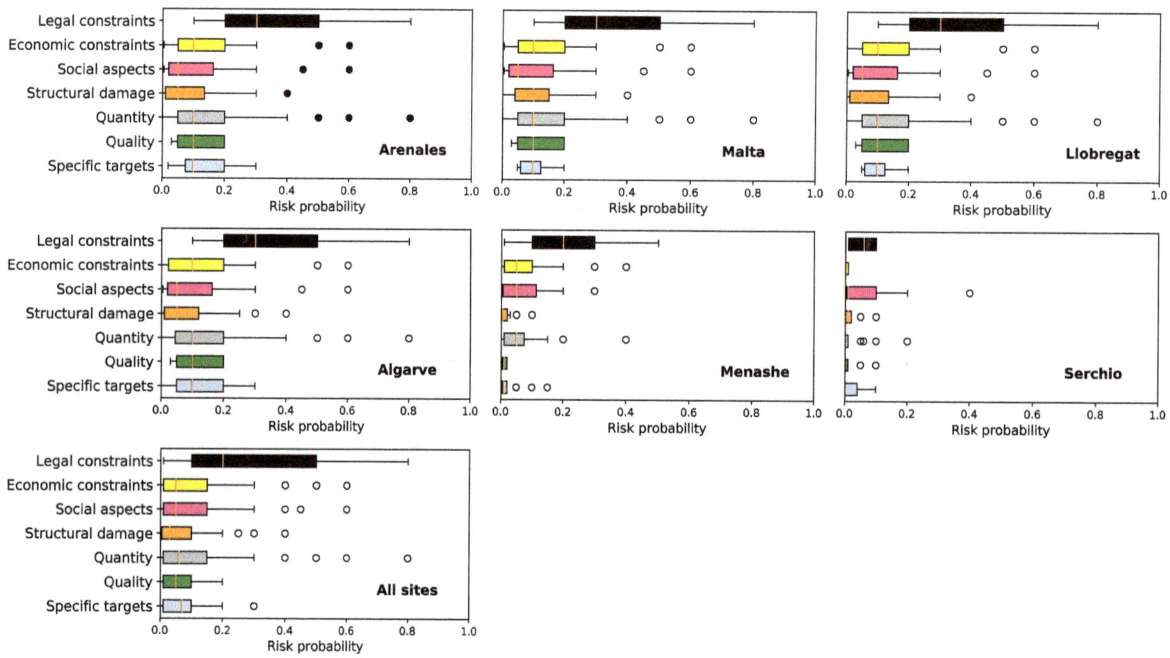

Figure 5. Distribution of the expert criteria by category and for the three levels (low, medium, high) grouped by categories.

case of Malta, the low risk value could be just perception, based on the fact that the site just recently started operation.

From all technical constrains, the one with the highest risk is related to water quantity. Half of the sites showed significant risk in terms of quantity, somewhat correlated with the sources of water for infiltration (being largest for those relying on surface water: Algarve, Llobregat, and Arenales). In the case of Malta, quantitative problems are related to the need to produce good quality water from wastewater. The Algarve site is a particular case as its aim is to improve groundwater quality with recharging water from a non-perennial stream (surface water flows only 60–70 days per year). Therefore, the lack of water is already considered in the MAR scheme. In terms of quality, again, the three sites supplied with river water are those showing the highest risks. That could respond to the variability in river water quality along the year. The geological and hydrogeological context does not seem to have any effect on the technical risk values despite it being very significant to define the site location to construct the MAR facilities.

Individually, the main risk issues for the technical issues in the Llobregat site were quantitative aspects, mainly clogging due to fine particles (probability = 0.4) and recharge water turbidity (0.4). The non-technical issues were mostly related to social aspects: lack of coordination amongst stake holders (0.4), children surveillance (0.3), and fair distribution of water (0.3). These three social aspects are aligned with the indicators of acceptance of Mankad and Walton (2015). For the Algarve site, the technical issues were mostly potential flooding (0.3), droughts (0.3), specific target (protected water body) (0.3), vandalism or terrorism (0.2), and clogging (0.2). Non-technical issues were mainly related to regional or local legislation (0.3). Non-technical issues for the Menashe site include domestic water use (0.15), perception of effectiveness (0.05), and high cost perception (0.05); for the technical ones, terrorism activities (0.02) and clogging due to compaction (0.02) were the most significant.

The Serchio site had the lack of knowledge on MAR activities (0.05) and health legislation (0.01) for the non-technical main risks. For the technical aspects, quality aspects related to organic compounds were the largest (0.01). The Los Arenales site had very large risk values associated with national (0.6), regional or local (0.8), and other legislation (0.5); agricultural water use (0.6); and fair distribution of water (0.45); the most significant technical issues were flooding (0.1), nutrients in the recharge water (0.5), droughts (0.8), generation of gas – physical clogging (0.2), nitrogen metabolites (0.2), river (0.1), and wetland water levels (0.2). The Malta site identified the European legislation (0.2) and lack of coordination (0.1) as the main non-technical risk drivers. For the technical aspects, largest risks included pipe breakage (0.05) and different specific targets: seawater barriers (0.4), protected water body (0.1), and groundwater levels (0.3).

The risk values obtained are mostly correlated to the expert criteria values. This correlation was evaluated by applying a Pearson product-moment correlation between the expert criteria basic events (considered the medium risk values) and the results basic events for each site. There were in total 65 basic events, leaving a total of 63 degrees of freedom and considering a p value of 0.05 as the confidence limit of ac-

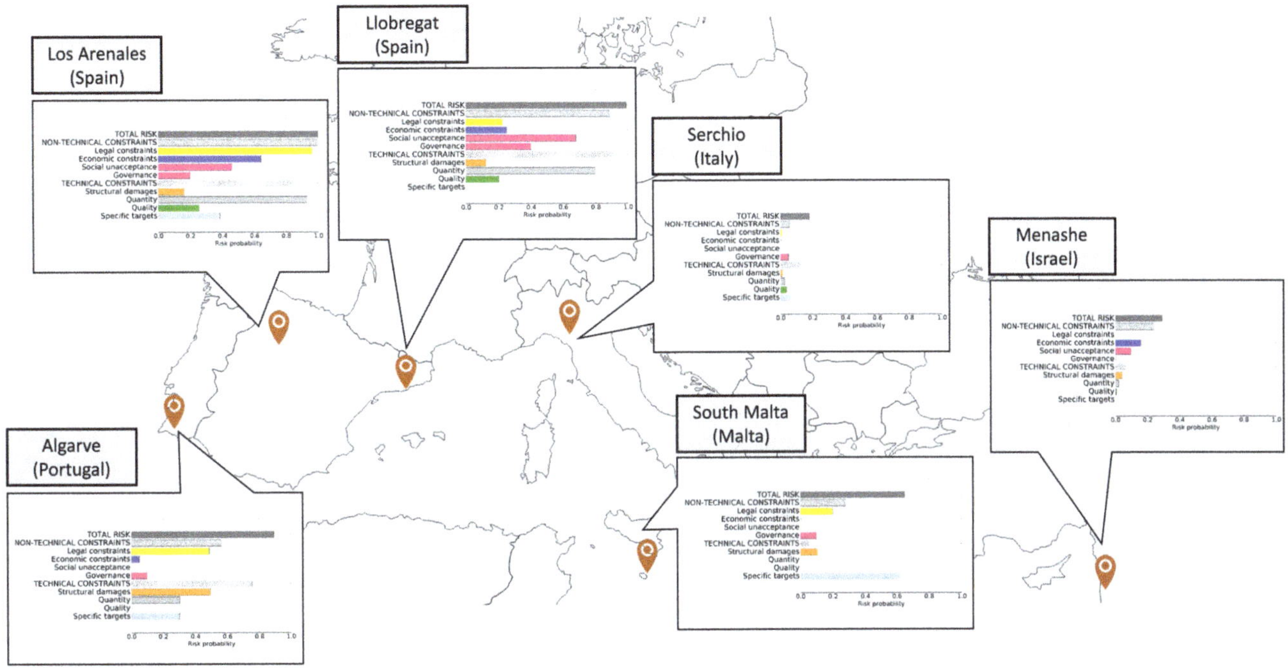

Figure 6. Risk in the different MAR sites.

ceptance. It was observed that, in general terms, a correlation between the perception of risk (expert criteria) and actual risk (results) existed. This mainly means that the facility managers know the main problems of the sites and thus define the expert criteria values accordingly. This indicates the relevance of using such a simplified method for preliminary risk assessment.

The actual results of the analyses showed correlation (in terms of p value) in the cases of Llobregat ($p = 0.026$), Los Arenales (7.38×10^{-13}), Malta (3.2×10^{-7}), Algarve (0.048), and Serchio (2.39×10^{-12}), while for Menashe site (0.6), that correlation could not be observed. Looking at the data, Menashe expert criteria values lack absolute zero values (0 from 65); however, the result values obtained show a high proportion of NO RISK (risk = 0) values (52 from the total of 65). Then the difference between both Expert criteria and Results become apparent, probably related to the knowledge of the personnel in charge of the Menashe site and their confidence in the lack of risk of their operations. A significant point to make is that there is a discrepancy between the literature review and our results. The most significant events leading to failure that we found in the literature were the technical ones, rather than the non-technical (recall Fig. 1). Our analysis on the six Mediterranean sites concluded the opposite. We attribute this discrepancy to a bias in the scientific literature towards technical issues. We could not find information on MAR facilities failure in journals devoted to social or economic sciences.

7 Conclusions

In this paper, we have developed a methodology to evaluate the risk of failure of managed aquifer recharge (MAR) facilities, and we have applied it to six different facilities located in the Mediterranean Basin. The methodology was based on the development of a probabilistic risk assessment based on fault trees. The PRA-FT methodology considered different categories affecting the operation of the facility. We further defined 65 basic events that individually or properly combined can produce global failure of the MAR facility. These events were compiled from a literature review of 51 MAR facilities, and then extended with the results of the European project MARSOL ("Demonstrating Managed Aquifer Recharge as a Solution to Water Scarcity and Drought").

The methodology consists of providing probability values to all basic events to take place in a window of time. Then, events at an upper level are computed from Boolean algebra until the top of the tree (total failure) is quantified. The initial step is to provide values based on expert criteria, assigned from the four risk categories: low, medium, high, and no risk. All values can be updated sequentially and probabilities are recalculated, until the values converge. The basic events include both technical and non-technical events.

A full preliminary (without updating) assessment of risk was developed for six sites located in the Mediterranean Basin. It was found that the non-technical aspects can be the most significant ones, contributing more than the technical issues to the overall assessment of risk. This is despite the fact that we are only considering facilities under operation, so that some issues are supposed to be already resolved. In short, the combination of legal, social, and economic factors can provide a really strong contribution to global risk. Consequently, future risk works based in managed aquifer recharge should consider how to lower risks by non-technical factors.

All events considered, we found that in the facilities analyzed, the major contributors to overall risk were the following: in decreasing order of importance, quantity issues (0.35), legal constraints (0.32), specific targets (0.23), social aspects (0.21), economic constraints (0.18), structural damages (0.16), governance (0.14), and quality (0.09). In particular, when the recharge water is supplied by a river, quantity aspects increase their relevance, due to the uncertainty in the current and future potential capacity for supplying in a dry and variable climate such as the Mediterranean Basin.

The site-specific results were obtained from a questionnaire, and so they provide "perception of risk" rather than "actual risk", and thus could and should be amended. The PRA-FT methodology allows us to now concentrate on the specific issues that individually, or combined, lead to the largest probability of failure, and concentrate the efforts in updating such values by means of detailed evaluations or specific projects of rehabilitation. The system can go on using any number of re-evaluations until an acceptable value or either until no further improvement can be obtained.

Regarding the results on perception of risk for the individual sites, it was surprising to get three of them (Los Arenales, Algarve, and Llobregat) above 0.90 in a 2- to 6-year period. The main contributors to failure were related to non-technical reasons and to quantity aspects. Actually, in recent years all three facilities had to discontinue operations at least one, indicating that the evaluations provide reasonable estimations. The Malta site is a very recent one, with little history behind, and this makes not possible to evaluate whether the perception of risk of 75 % is high or low.

On the other hand, the risks perceived for the other two sites, Serchio and Menashe, can be considered low (18 and 29 %, respectively). A potential reason is that they are the oldest facilities, so experience has been accumulated for decades. Also, the facilities have been able to adapt to evolving regulations (both local and at the European levels). In both cases, low risk values correlated also with low perception of availability risk (mainly related to clogging), due to the mechanisms and sources of recharge water in both cases (river bank filtration in Serchio and desalinated and storm water in Menashe).

Competing interests. The authors declare that they have no conflict of interest.

Acknowledgements. Financial support was provided by the project MARSOL grant agreement no. 619120, FP7-ENV-2013-WATER-INNO-DEMO, by the European Union, project ACWAPUR, PCIN-2015-239, project INDEMNE, CGL2015-69768-R (MINECO/FEDER) by the Spanish government. Arnau Canelles thanks to the Government of Catalonia for financial support (AGAUR, FI2016). Rudy Rossetto also thanks the technical support of GEAL S.p.a.

Edited by: Alberto Guadagnini

References

Aiken, G. and Kuniansky, E.: Mobilization of arsenic and other trace elements during aquifer storage and recovery, southwest Florida, US Geological Survey, 47–50, 2002.

Alazard, M., Boisson, A., Maréchal, J.-C., Perrin, J., Dewandel, B., Schwarz, T., Pettenati, M., Picot-Colbeaux, G., Kloppman, W., and Ahmed, S.: Investigation of recharge dynamics and flow paths in a fractured crystalline aquifer in semi-arid India using borehole logs: implications for managed aquifer recharge, Hydrogeol. J., 24, 35–57, https://doi.org/10.1007/s10040-015-1323-5, 2016.

Asano, T. and Cotruvo, J. A.: Groundwater recharge with reclaimed municipal wastewater: Health and regulatory considerations, Water Res., 38, 1941–1951, https://doi.org/10.1016/j.watres.2004.01.023, 2004.

Assmuth, T., Simola, A., Pitkänen, T., Lyytimäki, J., and Huttula, T.: Integrated frameworks for assessing and managing health risks in the context of managed aquifer recharge with river water, Integr. Environ. Asses., 12, 160–173, https://doi.org/10.1002/ieam.1660, 2016.

Ayuso-Gabella, N., Page, D., Masciopinto, C., Aharoni, A., Salgot, M., and Wintgens, T.: Quantifying the effect of Managed Aquifer Recharge on the microbiological human health risks of irrigating crops with recycled water, Agr. Water Manage., 99, 93–102, https://doi.org/10.1016/j.agwat.2011.07.014, 2011.

Bedford, T. and Cooke, R.: Probabilistic Risk Analysis: Foundations and Methods, Cambridge University Press, New York, 2003.

Bhusari, V., Katpatal, Y. B., and Kundal, P.: An innovative artificial recharge system to enhance groundwater storage in basaltic terrain: example from Maharashtra, India, Hydrogeol. J., 24, 1273–1286, https://doi.org/10.1007/s10040-016-1387-x, 2016.

Bolster, D., Barahona, M., Dentz, M., Fernandez-Garcia, D., Sanchez-Vila, X., Trinchero, P., Valhondo, C., and Tartakovsky, D. M.: Probabilistic risk analysis of groundwater remediation strategies, Water Resour. Res., 45, 1–10, https://doi.org/10.1029/2008WR007551, 2009.

Bouwer, H.: Artificial recharge of groundwater: Hydrogeology and engineering, Hydrogeol. J., 10, 121–142, https://doi.org/10.1007/s10040-001-0182-4, 2002.

Bouwer, H.: Environment Integrated water management for the 21st century : Problems and Solutions, J. Irrig. Drain. E., 128, 118–152, https://doi.org/10.1061/(ASCE)0733-9437(2002)128:4(193), 2003.

Chaoka, R., Alemaw, B., and Molwalelfhe, L.: Investigating the causes of water-well failure in the Gaotlhobogwe wellfield in southeast Botswana, J. Appl. Sci. Environ. Manag., 10, 59–65, 2006.

De Barros, F. P., Bolster, D., Sanchez-Vila, X., and Nowak, W.: A divide and conquer approach to cope with uncertainty, human health risk, and decision making in contaminant hydrology, Water Resour. Res., 47, W05508, https://doi.org/10.1029/2010WR009954, 2011.

De Barros, F. P., Fernàndez-Garcia, D., Bolster, D., and Sanchez-Vila, X.: A risk-based probabilistic framework to estimate the endpoint of remediation: Concentration rebound by rate-limited mass transfer, Water Resour. Res., 49, 1929–1942, https://doi.org/10.1002/wrcr.20171, 2013.

Dillon, P.: Future management of aquifer recharge, Hydrogeol. J., 13, 313–316, https://doi.org/10.1007/s10040-004-0413-6, 2005.

Dillon, P., Pavelic, P., Page, D., Beringen, H., and Ward, J.: Managed aquifer recharge : An Introduction, Tech. Rep. 13, National water Comission, Canberra, 2009.

Dillon, P., Vanderzalm, J., Page, D., Barry, K., Gonzalez, D., Muthukaruppan, M., and Hudson, M.: Analysis of ASR clogging investigations at three Australian ASR sites in a Bayesian context, Water, 8, 442, https://doi.org/10.3390/w8100442, 2016.

European Community: Directive 2000/60/EC of the European Parliament and of the Council of 23 October 2000 establishing a framework for Community action in the field of water policy, Official Journal of the European Parliament, L327, 1–82, https://doi.org/10.1039/ap9842100196, 2000.

Fader, M., Shi, S., von Bloh, W., Bondeau, A., and Cramer, W.: Mediterranean irrigation under climate change: more efficient irrigation needed to compensate for increases in irrigation water requirements, Hydrol. Earth Syst. Sci., 20, 953–973, https://doi.org/10.5194/hess-20-953-2016, 2016.

Flint, A. L. and Ellett, K. M.: The Role of the Unsaturated Zone in Artificial Recharge at San Gorgonio Pass, California, Vadose Zone J., 3, 763–774, https://doi.org/10.2136/vzj2005.0222, 2005.

Gale, I. N., Macdonald, D. M. J., Calow, R. C., Neumann, I., Moench, M., Kulkarni, H., Mudrakartha, S., and Palanisami, K.: Managed Aquifer Recharge : an assessment of its role and effectiveness in watershed management, Tech. rep., British Geological Survey, Nottingham, 2006.

Ganot, Y., Holtzman, R., Weisbrod, N., Nitzan, I., Katz, Y., and Kurtzman, D.: Monitoring and modeling infiltration-recharge dynamics of managed aquifer recharge with desalinated seawater, Hydrol. Earth Syst. Sci., 21, 4479-4493, https://doi.org/10.5194/hess-21-4479-2017, 2017.

Hochstrat, R., Wintgens, T., Kazner, C., Jeffrey, P., Jefferson, B., and Melin, T.: Managed aquifer recharge with reclaimed water: Approaches to a European guidance framework, Water Sci. Technol., 62, 1265–1273, https://doi.org/10.2166/wst.2010.386, 2010.

Izbicki, B. J. A., Metzger, L. F., Mcpherson, K. R., Everett, R. R., and Benett, G. L.: ources of High-Chloride Water to Wells, Eastern San Joaquin Groundwater Subbasin, California. Tech. rep., U.S. Geological, Survey, New York, available at: https://pubs.usgs.gov/of/2006/1309/pdf/ofr2006-1309.pdf (last access: 6 June 2018), 2006.

Ji, H. and Lee, S.-I.: Comparison of Potential Risk on Two Managed Aquifer Recharge Sites from River Basin, Water, 9, 674, https://doi.org/10.3390/w9090674, 2017.

Ji, H. W. and Lee, S. I.: Assessment of risk due to chemicals transferred in a watershed: A case of an aquifer storage transfer and recovery site, Water, 8, 242, https://doi.org/10.3390/w8060242, 2016.

Juntunen, J., Meriläinen, P., and Simola, A.: Public health and economic risk assessment of waterborne contaminants and pathogens in Finland, Sci. Total Environ., 599–600, 873–882, https://doi.org/10.1016/j.scitotenv.2017.05.007, 2017.

Jurado, A., De Gaspari, F., Vilarrasa, V., Bolster, D., Sánchez-Vila, X., Fernàndez-Garcia, D., and Tartakovsky, D. M.: Probabilistic analysis of groundwater-related risks at subsurface excavation sites, Eng. Geol., 125, 35–44, https://doi.org/10.1016/j.enggeo.2011.10.015, 2012.

Leitão, T., Lobo-Ferreira, J., Martins, T., Oliveira, M. M., and Henriques, M.: Demonstrating Managed Aquifer Recharge as a Solution to Water Scarcity and Drought – MAR to Improve the Groundwater Status in South Portugal, Tech. Rep. 619120, European Commission, 2017.

Leviston, Z., Nancarrow, B. E., Tucker, D. I., and Porter, N. B.: Predicting Community Behaviour : Indirect Potable Reuse of Wastewater through Managed Aquifer Recharge, Land and Water Science Report 2906, 2006.

Lobo-Ferreira, J. P. L. and Leitão, T. E.: Demonstrating managed aquifer recharge as a solution for climate change adaptation; results from Gabardine project and ASEMWaterNet coordination action in the Algarve region (Portugal), Acque Sotterranee Testo Stampato, 3, 15, https://doi.org/10.7343/as-080-14-0106, 2014.

Maliva, R. G. and Missimer, T. M.: Arid Lands Water Evaluation and Management, vol. 3, Springer, Berlin Heidelberg, https://doi.org/10.1007/978-3-540-88258-9, 2012.

Mankad, A. and Walton, A.: Accepting managed aquifer recharge of urban storm water reuse: The role of policy-related factors, Water Resour. Res., 51, 9696–9707, https://doi.org/10.1002/2015WR017633, 2015.

Masetti, M., Pedretti, D., Sorichetta, A., Stevenazzi, S., and Bacci, F.: Impact of a Storm-Water Infiltration Basin on the Recharge Dynamics in a Highly Permeable Aquifer, Water Resour. Manag., 30, 149–165, https://doi.org/10.1007/s11269-015-1151-3, 2016.

Murray, R. and Ravenscroft, P.: Strategy and Guideline Development for National Groundwater Planning Requirements. Potential Artificial Recharge Schemes: Planning for Implementation, Tech. rep., Department of Water Affairs, Pretoria, 2010.

Nandha, M., Berry, M., Jefferson, B., and Jeffrey, P.: Risk assessment frameworks for MAR schemes in the UK, Environ. Earth Sci., 73, 7747–7757, https://doi.org/10.1007/s12665-014-3399-y, 2015.

Navarro-Ortega, A., Acuña, V., Batalla, R. J., Blasco, J., Conde, C., Elorza, F. J., Elosegi, A., Francés, F., La-Roca, F., Muñoz, I., Petrovic, M., Picó, Y., Sabater, S., Sanchez-Vila, X., Schuhmacher, M., and Barceló, D.: Assessing and forecasting the impacts of global change on Mediterranean rivers. The SCARCE Consolider project on Iberian basins, Environ. Sci. Pollut. R., 19, 918–933, https://doi.org/10.1007/s11356-011-0566-5, 2012.

Pachauri, R., Allen, M., Barros, V., Broome, J., Cramer, W., Christ, R., Church, J., Clarke, L., Dahe, Q., Dasgupta, P., and Dubash, N.: Climate change 2014: synthesis report, Contribution of Working Groups I, II and III to the fifth assessment report of the Intergovernmental Panel on Climate Change, IPCC, https://doi.org/10.1017/CBO9781107415324, 2014.

Page, D., Dillon, P., Vanderzalm, J., Toze, S., Sidhu, J., Barry, K., Levett, K., Kremer, S., and Regel, R.: Risk Assessment of Aquifer Storage Transfer and Recovery with Urban Stormwater for Producing Water of a Potable Quality, Journal of Environment Quality, 39, 2029, https://doi.org/10.2134/jeq2010.0078, 2010.

Petersen, C. E. and Glotzbach, K.: Aquifer Storage and Recovery for the City of Roseville: a Conjunctive Use Pilot Project, Proceedings of the Water Environment Federation, WEFTEC 2005, 8634–8661, https://doi.org/10.2175/193864705783812945, 2005.

Ronen-Eliraz, G., Russak, A., Nitzan, I., Guttman, J., and Kurtzman, D.: Investigating geochemical aspects of managed aquifer recharge by column experiments with alternating desalinated water and groundwater, Sci. Total Environ., 574, 1174–1181, https://doi.org/10.1016/j.scitotenv.2016.09.075, 2017.

Rossetto, R., Barbagli, A., Borsi, I., Mazzanti, G., Vienken, T., and Bonari, E.: Site investigation and design of the monitoring system at the Sant'Alessio Induced RiverBank Filtration plant (Lucca, Italy), Rendiconti online della Società Geologica Italiana, 35, 248–251, https://doi.org/10.3301/ROL.2015.112, 2015.

San-Sebastián, J., Fernández-Escalante, E., Calero-Gil, R., Carvalho, T., and Rodríguez-Escales, P.: Characterization and benchmarking of seven managed aquifer recharge systems in south-western Europe, Sustain. Water Resour. Manag., 1–23, https://doi.org/10.1007/s40899-018-0232-x, 2018.

Schneider, B. J., Ku, H. F., and Oaksford, E. T.: Hydro-logic effects of artificial-re charge experiments with reclaimed water at east meadow, long island, new york, Tech. rep., U.S. Geological Survey, New York, available at: https://pubs.usgs.gov/wri/1985/4323/report.pdf (last access: 6 June 2018), 1987.

Sprenger, C., Hartog, N., Hernández, M., Vilanova, E., Grützmacher, G., Scheibler, F., and Hannappel, S.: Inventory of managed aquifer recharge sites in Europe: historical development, current situation and perspectives, Hydrogeol. J., 25, 1–14, https://doi.org/10.1007/s10040-017-1554-8, 2017.

Stanhill, G., Kurtzman, D., and Rosa, R.: Estimating desalination requirements in semi-arid climates: A Mediterranean case study, Desalination, 355, 118–123, https://doi.org/10.1016/j.desal.2014.10.035, 2015.

Sultana, S. and Ahmed, K. M.: Assessing risk of clogging in community scale managed aquifer recharge sites for drinking water in the coastal plain of south-west Bangladesh, Bangladesh Journal of Scientific Research, 27, 75, https://doi.org/10.3329/bjsr.v27i1.26226, 2016.

Tartakovsky, D. M.: Probabilistic risk analysis in subsurface hydrology, Geophys. Res. Lett., 34, L05404, https://doi.org/10.1029/2007GL029245, 2007.

Toze, S., Bekele, E., Page, D., Sidhu, J., and Shackleton, M.: Use of static Quantitative Microbial Risk Assessment to determine pathogen risks in an unconfined carbonate aquifer used for Managed Aquifer Recharge, Water Res., 44, 1038–1049, https://doi.org/10.1016/j.watres.2009.08.028, 2010.

Tredoux, G. and Cain, J.: The Atlantis Water Resource Management Scheme: 30 years of Artificial Groundwater Recharge, Tech. rep., Department of Water affairs, Cape Town, PRSA 000/00/11609/10 - Activity 17, http://www.artificialrecharge.co.za/casestudies/Atlantis_final_10August2010_reduceddpi.pdf (last access: 6 June 2018), 2010.

Tredoux, G., Van Der Merwe, B., and Peters, I.: Artificial recharge of the windhoek aquifer, namibia: water quality considerations, Boletin Geologico y Minero, 120, 269–278, available at: http://www.igme.es/Boletin/2009/120_2_2009/269-278.pdf (last access: 6 June 2018), 2009.

Tripathi, M.: In the Face of Changing Climate: Groundwater Development through Artificial Recharge in Hard Rock Terrain of Kumaun Lesser Himalaya, in: Geostatistical and Geospatial Approaches for the Characterization of Natural Resources in the Environment, 937–947, Springer International Publishing, Cham, https://doi.org/10.1007/978-3-319-18663-4_145, 2016.

Valhondo, C., Carrera, J., Ayora, C., Tubau, I., Martinez-Landa, L., Nödler, K., and Licha, T.: Characterizing redox conditions and monitoring attenuation of selected pharmaceuticals during artificial recharge through a reactive layer, Sci. Total Environ., 512–513, 240–250, https://doi.org/10.1016/j.scitotenv.2015.01.030, 2015.

Vesely, W. E., Goldberg, F. F., Roberts, N. H., and Haasl, D. F.: Fault Tree Handbook, Tech. Rep. NUREG-0492, available at: https://www.nrc.gov/docs/ML1007/ML100780465.pdf (last access: 6 June 2018), 1981.

5

Stochastic hydrogeology's biggest hurdles analyzed and its big blind spot

Yoram Rubin[1], **Ching-Fu Chang**[1], **Jiancong Chen**[1], **Karina Cucchi**[1], **Bradley Harken**[1], **Falk Heße**[2], **and Heather Savoy**[1]

[1]Civil and Environmental Engineering, University of California, Berkeley, 94720, USA
[2]Computational Hydrosystems, Helmholtz Centre for Environmental Research (UFZ), 04318 Leipzig, Germany

Correspondence: Yoram Rubin (rubin@ce.berkeley.edu)

Abstract. This paper considers questions related to the adoption of stochastic methods in hydrogeology. It looks at factors affecting the adoption of stochastic methods including environmental regulations, financial incentives, higher education, and the collective feedback loop involving these factors. We begin by evaluating two previous paper series appearing in the stochastic hydrogeology literature, one in 2004 and one in 2016, and identifying the current thinking on the topic, including the perceived data needs of stochastic methods, the attitude in regulations and the court system regarding stochastic methods, education of the workforce, and the availability of software tools needed for implementing stochastic methods in practice. Comparing the state of adoption in hydrogeology to petroleum reservoir engineering allowed us to identify quantitative metrics on which to base our analysis. For impediments to the adoption of stochastic hydrology, we identified external factors as well as self-inflicted wounds. What emerges is a picture much broader than current views. Financial incentives and regulations play a major role in stalling adoption. Stochastic hydrology's blind spot is in confusing between uncertainty with risk and ignoring uncertainty. We show that stochastic hydrogeology comfortably focused on risk while ignoring uncertainty, to its own detriment and to the detriment of its potential clients. The imbalance between the treatment on risk on one hand and uncertainty on the other is shown to be common to multiple disciplines in hydrology that interface with risk and uncertainty.

1 Introduction

A key element in the discussion on adoption of stochastic hydrogeology (SH), or the lack of it, is the apparent disconnect between theory and practice. The root causes for the perceived failure in adoption have been discussed in several papers (published in 2004 and 2016; for a complete list see Sect. 2). In this study, we are set to take yet another look at this issue and draw some conclusions based on a factual basis. Regardless of whether true or not, this apparent gap raises questions such as the following: first, what does "practical" mean? Does it mean "has no use in real life" or "useful but hard to apply, and would you please simplify it and make it more accessible"? How real is the perceived gap and how much of it depends on the skills of the commentator? Is there a reasonable delineation between theory and practice? For example, would it take a mobile app with all-inclusive user interfaces in order for SH to be viewed as practical? These and other questions will be addressed here.

There is an important issue looming in the background of this conversation. SH intends to model uncertainty, and uncertainty exists and must be dealt with, regardless of any perceived shortcoming. One can argue about strengths and weaknesses of SH, e.g., the practicality, but there is no argument about the need to address uncertainty. But who is to blame for ignoring uncertainty? Are there some external factors not related to the science? And what has SH missed, if anything, that could have made a difference in adoption? This paper reviews the evolution of the thoughts on these topics. It evaluates the strengths and weaknesses of arguments made and provides additional perspectives by looking at factors not considered before.

The motivation for this paper originated from the need to get to the root causes of some of the issues that we view as critical and from the growing recognition that there is a need to both go beyond the scientific and technical issues associated with SH and to consider issues related to education, public policy, economics, and the law. However, this paper is not just about external factors: we will also look at SH's biggest omissions.

The remainder of the paper is organized as follows: Sect. 2 summarizes and compares the 2004 SERRA (Stoch. Env. Res. Risk A.) and 2016 WRR (Water Resour. Res.) series; Sect. 3 reviews available SH-related software; Sect. 4 compares the role of stochastic methods in hydrogeology and petroleum engineering; and Sect. 5 discusses legal and regulatory perspectives on stochastic methods; Sect. 6 discusses the perception of uncertainty broadly and in SH; and Sect. 7 concludes with recommendations and outlook for the field of SH in practice.

2 Summary of perspectives from the 2004 and 2016 paper series

To begin the SH-specific discussion, we provide a brief summary of the comments made in two series of articles addressing SH in practice. These series were published in SERRA in 2004 and in WRR in 2016. In our subsequent discussion we shall refer to the "2004 series" (Christakos, 2004; Dagan, 2004; Freeze, 2004; Ginn, 2004; Molz, 2004; Neuman, 2004; Rubin, 2004; Sudicky, 2004; Winter, 2004; Zhang and Zhang, 2004) and to the "2016 series" (Cirpka and Valocchi, 2016; Fiori et al., 2016; Fogg and Zhang, 2016; Rajaram, 2016; Sanchez-Vila et al., 2016). The 2004 series asked nine researchers the following two specific questions: (1) "Why have there not been many real-world applications of stochastic theories and approaches, despite the significant progress in developing such rigorous theories and approaches for studying fluid flow and solute transport in heterogeneous media?" and (2) "In your opinion, what must be done in order to render stochastic theories and approaches as routine tools in hydrogeologic investigation and modeling?". The 2016 series is more open-ended, asking four teams of researchers to debate on "stochastic subsurface hydrology from theory to practice". We chose these two series as a baseline for discussion because they address similar questions, they combine multiple perspectives from a variety of researchers, and the 12-year timespan between them allows us to consider the evolution of SH. However, due to the more open nature of the 2016 series, not all of the discussion points can be compared between the two series. Of note is that we refrain from considering issues related to transport modeling specifics in as much detail as the 2016 WRR series due to the wider scope of our questions regarding stochastic methods in general.

2.1 A brief summary of the issues

The themes pursued by both paper series in trying to explore the gap between academia and practitioners regarding the adoption of SH are as follows:

1. Regulations and litigation. There is a broad agreement (but not a consensus) that institutional attitudes (in the form of regulations, public policy, and the legal system) are major factors to consider, and that in the case of SH, the institutional attitudes have played a negative role. In a departure from this broadly held view, Christakos (2004) opines that no blame should be assigned on the court system when other fields use models with uncertainty, e.g., biomedical models for DNA analysis, under legal scrutiny. We will address the following questions. What is the evidence to support the claim about institutional resistance? Is there evidence to the contrary?

2. The role of higher education. Authors criticized the lack of academic preparation in the theoretical fundamentals of stochastic theory. Suggestions were made that SH theory is too complex to be taught. We will address the following questions. What is the factual basis for this statement? Is there evidence to the contrary? Who drives the development of teaching curricula, and what can be learned about it from other disciplines?

3. Lack of appropriate measurement technology and/or data. A major theme in the 2004 series is the apparent lack of technologies needed for providing measurements attributable to the parameters and scales of stochastic models. Additionally, it is suggested that the application of SH requires large amounts of data. We will address the following questions. What are the elements of SH that require lots of measurements? Are there statistical tools to handle situations with a severe shortage of measurements?

4. Lack of user-friendly software that applies theory. This topic mainly appears as the call for user-friendly software that integrates multiple forms of information and that is computationally efficient enough for practitioners to use. We will address the following questions. What is the factual basis for this assertion?

5. SH theory is out of touch with reality. Arguments here were either directed to the theoretical research as not applicable in real-world problems (e.g., oversimplification of spatial structure, or minimizing uncertainty in parameters when other uncertainties need to be minimized) or to the lack of applications to showcase the applicability of theories in real-world problems. We will address the following questions. Is there a factual basis for these arguments? Is there any particular aspects of SH that gives rise to such arguments?

The topics most prevalently mentioned in the 2016 WRR series are Issue 4 (user-friendliness of concepts) and Issue 5 (theory too limited or not showcased in applications). Concerns regarding the availability of useful software tools remain (Cirpka and Valocchi, 2016); however there is some recognition that a few stochastic modules were introduced into forward models (Fiori et al., 2016).

2.2 Possible misconceptions and general observations

In assessing the root causes of the so-called gap, we need to distinguish between real and perceived obstacles. Perceived obstacles reside in myths and misconceptions. Real obstacles are everything else.

It is often stated that SH requires lots of data for real-life applications. As is the case with most persistent myths, there is a grain of truth here, and this is the observation that getting a well-defined variogram (or for that matter, other space random function (SRF) models such as multi-point spatial statistics) requires dozens of measurements. Whereas this observation is true, everything else that is added on top of it is not true. The variogram is a stochastic method for representing the spatial variability structural patterns with a model that is parsimonious compared to a field of unknown values but allows for the uncertainty in the local variability in un-sampled locations. A shortage of field data does make accurate modeling of variograms difficult, but that is not a gap in SH. In fact, SH is amply equipped to deal with that challenge. For example, instead of wrongly using (and wrongly assuming) the existence of a deterministic variogram model, one can consider simultaneously several potential variogram models instead of only one. Additionally, one can introduce parametric uncertainty into the estimation process (Rubin, 2003).

Fiori et al. (2016) noted the significant advancement of measurement technologies. And while this is commendable and most welcome, it does not appear to have made a significant change in closing the gap. This suggests that measurement technologies, although they advance the field of hydrogeology, should not be viewed as a critical factor in closing the gap for SH. SH's primary goal is the modeling of uncertainty, and uncertainty exists (and must be modeled) regardless of the limited availability of measurements and the limitations of measurement technologies. The data help in reducing uncertainty, but they cannot make it disappear (at least not in field applications). The need for a large amount of data in this context is somewhat of an oxymoron: having lots of data is useful, but a lack of data does not make uncertainty go away, and in fact, it only accentuates the need for stochastic methods. Should theories be accepted or rejected based on the availability of appropriate measurement technologies? If that were true, then the theory of relativity would have been kept in the drawer until the 1970s.

As noted by Cirpka and Valocchi (2016) and Li et al. (2018), Bayesian methods can use a variety of forms of information including prior knowledge and indirect data to help constrain the uncertainty in these parameters. Variograms can be used, even in the absence of a single measurement based on prior information. This would result in large uncertainty in the predictions, but (1) that is not a problem, but just a reflection of reality, and (2) that is the role of SH: to predict uncertainty, not to make it go away. The option of giving up on modeling uncertainty on the grounds that there are not enough data is not justified.

Moving forward, we will refrain from focusing on data needs based on our understanding that scarcity of data does not contribute to the gap. For the same reason, we will also avoid discussing the (always welcome) effort towards improving measurement technologies. Rather, we should focus on Issues 1, 2, and 4. We will show that there is a factual base that allows evaluating the veracity of these issues. We will also address what we call the "big miss" of SH.

3 Software issues

The 2004 and 2016 paper series both identified the lack of SH software solutions as a critical hindrance on the road to acceptance of stochastic methods by practitioners. Neuman (2004) mentioned the lack of ready-to-use and user-friendly software packages. Dagan (2004) and Winter (2004) suggested that most software solutions tend to oversimplify hydrogeologic complexity and thus are of limited appeal. Cirpka and Valocchi (2016) pointed out that stochastic software solutions for reactive transport are in their infancy. Fiori et al. (2016) mentioned that "Practical tools guided by the stochastic paradigm are very few, although the last few years have seen a promising increase of them". Here, we wish to explore the software aspect of SH further through a limited survey of the software landscape. Toward this goal, we developed a list of stochastic software solutions and related them to various components of SH studies. We investigated the extent to which these solutions are capable of handling a variety of SH problems. We do not intend to provide here an exhaustive summary of all SH software tools. Our plan is to identify whether (or not) a critical mass of software tools exists, such that most geologies and most aspects of SH studies could be addressed using an existing software solution.

Critical mass for software is a debatable concept. If one defines critical mass as having readily available software for all situations, critical mass would forever stay an elusive goal. A less demanding and possibly more realistic definition would be to consider readily available solutions as well as source codes that could be modified to address special situations. There is an ever-increasing library of open-source software solutions and modular, adjustable solutions requiring minimal programming skills for applications and it makes significant contributions to the critical mass. Hence, we will avoid making calls about whether or not critical mass is available, but we will show that we are possibly very close and way beyond the common perception.

3.1 Classification of stochastic software with examples

There is a wide range of software solutions relating to SH that we could mention. In this study, we focused on software solutions consistent with the definition of SH given in the introduction. We therefore limit our search to software solutions that are guided by complete statistical theories or that make substantial use of statistical concepts.

The three application areas covered in our review include (1) the simulation of random fields, (2) stochastic forward modeling, and (3) stochastic inverse modeling. The first area focuses on the simulation of spatially variable flow and transport domains using geostatistical models. The second area focuses on simulating dependent hydrogeological variables, such as pressure or concentration, in spatially variable domains. The third area focuses on identifying the SRF models that could then be used in application under areas 1 and 2.

Table 1 provides a list of software packages and their primary focus area. Our selection does not imply best of class and should not be viewed as a recommendation. We recognize that forward and inverse modeling tools utilize (and may even have built-in) random field generators (RFGs), but in general, random field generation is not their primary focus. Hence, we identified each software package only by its primary application area(s).

3.1.1 Random field generation and data exploration

The first application area refers to the generation of spatially variable fields and field data exploration. Available software packages cover a wide range of SRF models that can be, and have been, used to characterize a diverse range of geological environments, from unlithified sediments to sedimentary bedrock aquifers to crystalline bedrocks (see Eaton, 2006, for a more detailed review of type of geologies).

A partial review of the types of SRF models covered by software packages listed in Table 1 is provided in Table 2. In many cases, the SRF models used in the software packages were modified to meet the needs of specific applications, and, with that modification, expanded the range of applicability. For example, when investigating the karst carbonate Biscayne aquifer in southeastern Florida, Sukop and Cunningham (2014) used GSLIB to simulate the spatial distribution of porosity in the carbonated matrix of the karst aquifer after adding capability for simulating nested variogram structures. Castilla-Rho et al. (2014) used Gaussian random field generators (GRFGs) as implemented in SGeMS to simulate small-scale variability in a paleo-valley-type aquifer in northern Chile, where SGeMS is considered as an evolution of GSLIB (Bianchi and Zheng, 2009).

GRFGs represent a class of SRFs useful for modeling continuous variables, such as hydraulic conductivity, due to the parsimony of the Gaussian models. Jankovic et al. (2017) used a three-dimensional version of HYDRO_GEN in order to simulate three-dimensional SRFs of log-normally distributed hydraulic conductivity characterized by exponential and spherical variograms and their influence on breakthrough curves (Bellin and Rubin, 1996). Cvetkovic et al. (1999) used HYDRO_GEN to generate SRFs of a log-normally distributed variable characterized by an exponential correlation structure to model fracture aperture controlling fluid flow and tracer transport in a crystalline rock. While dominant in the early days of SH, GRFGs represent just one of many options currently available.

The development of multi-point statistics (MPS) allowed users to easily simulate non-Gaussian fields (Mariethoz et al., 2010). Although the concept of MPS was explored in the 1990s (Guardiano and Srivastava, 1993), it was Mariethoz et al.'s (2010) multi-point direct sampling (MPDS) algorithm that opened the door for broad use. Pirot et al. (2014) used MPDS as implemented in DeeSse to simulate topographies in a braided river system (Straubhaar, 2015). Savoy et al. (2017) used a DeeSse-generated dataset to explore transport through complex geometry that GRFs could not describe.

Another method for simulating non-Gaussian geostatistical simulations is the transition probability–Markov chain (TP–MC) approach to the random field generation of categorical variables implemented in TPROGS (Carle, 1999). The TP–MC approach focuses on modeling the spatial arrangement of lithofacies with linkage to basic geological attributes such as volumetric lithofacies proportions, mean lengths, and juxtaposition tendencies. He et al. (2014) used TPROGS to simulate structural heterogeneity for glacial deposits in a headwater catchment in Denmark. Engdahl et al. (2010) used TPROGS to simulate shallow fluvial aquifer that contain representations of sedimentary bounding surfaces. Fleckenstein et al. (2006) used TPROGS to simulate gravel, sand, and muds in the alluvial fan of the Cosumnes River in California, USA. Indicator and categorical variables have been implemented in GSLIB for simulating geologic layers (Deutsch and Journel, 1998; Marinoni, 2003). Jackson et al. (2000) used Isim3d to simulate a heterogeneous continuum porous medium to represent fractured rocks (Gómez-Hernández and Srivastava, 1990). In conclusion, a variety of software implementing different theoretical models of spatial variability are available. Our table does not aim to provide a full review of all methods and associated software for spatial variability, but rather to show that implementations are readily available for multiple types of geologies represented by the different statistical models of spatial variability. We also note that programming capability goes a long way towards expanding the original, stated capabilities of current software packages.

3.1.2 Forward modeling

Stochastic software for forward modeling refers to software tools for simulating hydrogeological processes using spatial distributions of hydrogeological parameters. A forward model, in itself, may not be an SH item, but when coupled with a random field and/or parameter generator, it becomes a bona fide SH application. There are numerous SH studies that couple process simulators with random field and param-

Table 1. Application areas of reviewed stochastic hydrogeology software.

Software	Main focus area			Reference
	Random field generation and data exploration	Forward modeling	Inverse modeling	
DeeSse	✓			Straubhaar (2015)
DREAM			✓	Vrugt (2016)
FIELDGEN	✓			Doherty and Hunt (2010)
GMS	✓	✓	✓	Aquaveo LLC (2012)
GSLIB	✓			Deutsch and Journel (1998)
GSTAT	✓			Pebesma (2004)
Isatis	✓			Bleines et al. (2004)
PEST			✓	Doherty (2005)
iTough2			✓	Finsterle (2011)
MAD			✓	Osorio-Murillo et al. (2015)
SGeMS		✓	✓	Remy et al. (2009)
TPROGS	✓			Carle (1999)
spMC			✓	Sartore (2013)
HYDRO_GEN	✓			Bellin and Rubin (1996)
Isim3D	✓			Gómez-Hernández and Srivastava (1990)

Table 2. Popular methods for simulating spatial variability of hydrogeological properties and their implementation in software packages (area 1). Additional options for random field generation are discussed here: https://petrowiki.org/Geostatistical_conditional_simulation\T1\textbackslash#Sequential_simulation (last access: 12 December 2017).

Software package	Geostatistical models of spatial variability			
	Gaussian	Non-Gaussian		
		Multi-point statistics	Transition probability	Indicator/ categorical
DeeSse		✓		✓
FIELDGEN	✓			
GMS	✓		✓	✓
GSLIB	✓			✓
GSTAT	✓			✓
Isatis	✓			
SGeMS	✓	✓		✓
TPROGS			✓	✓
Isim3d				✓

eter generators. Most commonly, investigators link between one of the software packages listed above for generating realizations of the SRF with a process simulator of their choice. Several of the software packages listed embed both RFGs and forward models into a single software package. Jones et al. (2005) presented an algorithm implemented within GMS that combines the stochastic simulation of hydraulic conductivity using TPROGS with MODFLOW 2000, a popular numerical model for flow and transport modeling (Harbaugh et al., 2000). Renard and Jeannée (2008) combined geostatistical simulations implemented using Isatis with flow simulations using METIS (Goblet, 1989). Most software packages are able to perform in two dimensions and three dimensions. The important point is that the RFGs are stand-alone and thus can be easily coupled with forward simulators, and this opens the door for a large number of situations amenable to SH treatment.

3.1.3 Inverse modeling

The third application area covers inverse modeling. Inverse modeling is a study in assimilating Type A (i.e., local data, correlated with a collocated target variable) and Type B data (non-local, usually related to a subdomain of the target variables). It is intended to produce an image of the subsurface more coherent that what could be achieved with Type

A data alone. As an example, Murakami et al. (2010) used the Method of Anchored Distributions (MAD) algorithm (Rubin et al., 2010) to condition a three-dimensional hydraulic conductivity field at the Hanford 300 Area, Washington, USA. MAD allowed the combination of large-scale injection rate measurements with small-scale electromagnetic borehole flowmeter measurements to derive posterior probability density functions (pdfs) for geostatistical parameters. In another example, Reeves et al. (2014) combined Type A data obtained from boreholes with soil deformation data. Chen and Hoversten (2012) and Hou et al. (2006) used seismic AVA and CSEM data to estimate reservoir parameters based on statistical rock physics models through Markov random fields. Kowalsky et al. (2004, 2005) combined ground-penetrating radar (GPR) data (for Type B data) with Type A hydrologic data to estimate the flow parameters in the vadose zone. Bellin and Rubin (2004) and Woodbury and Rubin (2000) used concentration for Type B data. Chen et al. (2004) used geophysical data for Type B data for geochemical characterization. Hubbard et al. (1999) used GPR for Type B data used for identifying the spatial correlation structure. Ezzedine et al. (1999) used geophysical data for Type B data in a characterization study of a superfund site, and Hubbard et al. (2001) used geophysical data for Type B data in a characterization study of a large-scale experimental site.

Published inverse modeling software packages cover a wide range of statistical concepts. It is not our goal here to rank and compare inverse modeling strategies, but rather, to provide a glimpse into the diversity for ideas employed and implemented. Detailed reviews are provided in McLaughlin and Townley (1996), Ginn and Cushman (1990), and in Rubin (2003, chap. 13). There are, of course, different ways to classify inverse modeling concepts, but the most common one is to define the inverse models by the way they treat the hydrological variables. One approach here is to consider the parameters as deterministic yet unknown, which translates into some sort of optimization- or regression-based procedure (e.g., PEST, iTough2, TPROGS, and spMC). The goal of the inverse modeling here would be to define the uncertainty associated with estimating these parameters.

Another approach is to consider the model parameters as random variables, with inverse modeling targeting their pdfs (e.g., MAD and DREAM). Here, we can identify strategies employing maximum likelihood (Kitanidis and Lane, 1985; Rubin and Dagan, 1987b), maximum a posteriori (McLaughlin and Townley, 1996), and Bayes' theorem (Osorio-Murillo et al., 2015; Rubin et al., 2010; Vrugt and Ter Braak, 2011). In the Bayesian approach, parameters are described by a pdf, and the estimation process consists in updating the prior pdf with in situ measurements. Bayesian methods offer a more complete treatment of uncertainty than optimization methods, as optimization discards nonoptimal yet realistic parameter sets, whereas in the Bayesian framework each of the parameter sets is associated with probability and they are all recognized, when assigned with the tiniest probability.

3.2 Description of properties of software packages

Several software solutions come with built-in forward model(s) and/or random field generator(s). For example, MODFLOW includes the possibility of importing SRFs simulated by TPROGS (ModelMuse Help, 2009), and the forward model GMS contains modules for the generation of conditional GRF based on FIELDGEN (Aquaveo LLC, 2012). This approach provides an easy gateway into SH applications, but at the price of limiting the range of applications. Others software solutions are designed to accommodate user-supplied computational modules (e.g., PEST and GSLIB). MAD is a hybrid between these two approaches, as it comes with built-in modules, but it is also capable of accommodating user-supplied modules. This flexibility opens the door for a wide range of applications, but it could be challenging for users lacking some minimal programming skills.

The modularity of SRF software solutions is a key aspect for their wide adoption. This modularity, implemented for example in GSLIB as a set of Fortran utility routines, allows users to identify specific code sections of relevance and use them in the data analysis workflow. Other examples of modularity include R packages, such as GSTAT or spMC, which respectively implement parameter estimation based on variograms and transitional probabilities within R (Pebesma, 2004; Sartore, 2013). This allows R users to adopt spatial statistics concepts seamlessly in their workflow. Finally, MAD has been implemented such that it can easily be linked to any RFG or forward model. Even though links to popular models such as MODFLOW and GSTAT are readily available within MAD, using other specific models is made easy thanks to the existence of built-in links that require minimal coding from the user. Allowing for flexibility in the form of packages that can be integrated into the user's workflow, either as a toolbox within a forward numerical model, or as a package that can be used within a wider programming environment, should facilitate the adoption of stochastic concepts by users.

One final important issue to consider relates to the user-friendliness of software solutions. While some users prefer to work from shell command lines providing flexibility and modularity and allowing the construction of reproducible and automated workflow, others might prefer a graphical user interface (GUI), designed to ease the use of the software by providing user-friendly options for inputting data to the statistical algorithms and visualizing results for interpretation. In our review, all software provided shell command line versions, thereby allowing practitioners and researchers to randomize parameter fields using text input files. On the other hand, GUIs were not available for all packages. For a few geostatistical programs, a GUI was added at later stages of the development as an interface to the geostatistical programs. For example Batgam© is a GUI developed for GSLIB (Buxton et al., 2005). Our review recognizes that different audiences require different capabilities in terms of user-friendliness and therefore reports available forms for using the software solution (Table 3).

3.3 Discussion

In this section, we aim to provide further thoughts about other statistical software and toolkits that were not designed directly for SH applications, but could nevertheless be easily connected to SH. In addition, we will also identify focus areas in need for additional attention and development effort.

Statistical concepts, such as Markov chain Monte Carlo simulation, transitional probability, GRF, and random walks, have been widely used in SH. In addition to a few of the existing SH packages that have already incorporated those modules, there are other statistical packages that are available, which can be easily linked with hydrogeological models. For example, Nimble provides a way to incorporate Bayesian hierarchical modeling into our subsurface parameter estimation (de Valpine et al., 2016), and the R package forecast offers a way to build a random walk to help solve reactive transport models (Hyndman and Khandakar, 2008). The geostatistical MATLAB toolbox mGstat provides a MATLAB interface to GSTAT and SGeMS, allowing its users to call GSTAT or SGeMS functions directly from MATLAB and have the output returned seamlessly (Hansen, 2004). These stand-alone statistical packages or software are not designed for SH, but could be easily connected to either existing hydrogeology models or SH toolkits.

There are some areas in which important gaps still exist. For example, modeling of uncertainty in karst formations is at its infancy. In another direction, stochastic modeling of reactive transport lags behind modeling of flow processes, mostly due to the complexity of the reaction networks and some ambiguity about the underlying physics. Some of the key transport processes such as diffusion-dominated reactive transport and transverse reactive transport (see the 2016 debate papers, as listed in Sect. 2) are still debatable. The challenge here, it might be argued, is not in the SH side. If a clearer picture of the physics and chemistry in complex systems were to emerge, the corresponding stochastic software could be used to account for uncertainties due to spatial variability. We, in fact, disagree with such an argument because uncertainty must be dealt with whether or not an underlying physical model does exist. We will expand on that in a section devoted to the shortcomings of SH.

In summary, a wide variety of stochastic software packages and toolkits are available for practitioners and researchers and have proven capability for modeling and analyzing a wide range of conditions. Despite significant efforts in linking between the software packages representing the various aspects of SH, we note that in many cases some minimal programming skills are needed in order to implement numerical solutions targeted at the specific needs of the modeling exercise. One could always aspire to obtain seamless links between all packages, but the need of programming skills is not a credible reason for explaining why practitioners in environmental engineering and groundwater hydrology are ignoring opportunities offered by the stochastic approach. We propose that concerns about limitations of software solutions may be rooted in having (or not having) programming skills. It is hard to believe that a sophisticated application in petroleum reservoirs would be scuttled because of difficulties in linking software packages, and so, it is not the lack of programming skills but the availability to pay for programming skills. And for this, you must have clients who are willing to pay. We will explore this question in the subsequent section.

4 A comparative study: hydrogeology and petroleum

Several of the contributors to the 2004 and 2016 paper series have pointed to a lack of education in SH, along with stochastic analyses being too costly and not providing enough benefit, as main reasons for the lack of adoption of these methods in practice (Cirpka and Valocchi, 2016; Dagan, 2004; Fiori et al., 2016; Molz, 2004; Neuman, 2004; Sudicky, 2004; Winter, 2004). In response, we aim to compare the factors affecting the adoption of new methods in hydrogeology to those in petroleum engineering, for which stochastic methods for characterizing, mapping, and modeling the subsurface have become commonplace (e.g., Floris et al., 2001; Jonkman et al., 2000; Liu and Oliver, 2003; Oliver and Chen, 2011; Rwechungura et al., 2011). The factors we aim to compare include incentive structure, education, and the interaction between academic and industrial organizations.

4.1 Incentive

To compare the two industries, we start by recognizing the fundamental difference in the corporate incentive structure, which in turn leads to a difference in operations and adoption of new practices. In the petroleum industry, any improvement in subsurface characterization and mapping capabilities would ultimately lead to increase in profits, whether in the form of enhanced productivity or reduction in risk. Reservoir characterization and reserve estimates have a major impact on the bottom line of a company. They affect interactions with Wall Street (Misund and Osmundsen, 2015), and they are strictly regulated by the Securities and Exchange Commission (SEC). This relationship between petroleum engineering and the financial markets relies on guidelines set by both the SEC and the Society for Petroleum Engineers on how to communicate the certainty in the inherently uncertain reserve estimates (Harrell and Gardner, 2003). By having both professional and regulatory bodies dictating the use of probabilistic methods, the petroleum engineering field has significant incentive to stay at the forefront of innovation and

Table 3. Properties of reviewed stochastic software solutions. Third-party GUI indicates that one or more graphical user interface solutions are available for the listed program, but implemented in a separate solution than the listed program.

Software	User-friendliness		Open source	Free/commercial	
	GUI	Third-party GUI		Free	Commercial
DeeSse	✓				✓
FIELDGEN			✓	✓	
GMS	✓	✓	✓	✓	
GSLIB		✓			✓
GSTAT		✓	✓	✓	
iTough2	✓	✓			✓
Isatis	✓				✓
MAD	✓		✓	✓	
PEST	✓		✓	✓	
SGeMS	✓	✓			✓
TPROGS	✓				✓
spMC			✓		✓
Isim3D		✓	✓	✓	

application regarding stochastic methods. In addition, the enormous capital costs associated with new and continued wellfield operations motivates the use of the most informative, risk-based predictions of costs versus benefits. In summary, petroleum firms are highly driven to create the most accurate depiction of underground resources in order to both accurately report reserve estimates and to best allocate costly drilling resources. Simply put, there is a direct link between improved characterization and improved profitability.

In hydrogeology, on the other hand, improvements in characterization and modeling do not always directly lead to improvements in profitability for firms involved with groundwater investigation and contaminant remediation. There are several factors that affect the perceived costs and benefits of adopting stochastic methods in hydrogeology, thus obfuscating the incentive structure. Very often, hydrogeologists involved in groundwater contamination projects must explain and justify steps taken, methods used, and conclusions reached to a wide range of people, including scientists in other fields, site owners, regulators, attorneys, and the general public. In light of these factors, it may seem more justifiable to adhere to established, commonly held concepts because adopting new methods could invite criticism. Furthermore, as will be elaborated upon in Sect. 5, it may often be the case that hydrogeologists are eager to use stochastic methods for uncertainty quantification, while their clients are reluctant to do so due to the perception that admitting any uncertainty may be construed as ignorance, thus leaving the client vulnerable to penalties from regulatory agencies.

There are no clear tangible financial benefits for improved modeling of uncertainty on the hydrogeology side. Environmental degradation, in simple economic terms, is a market failure and requires external intervention, e.g., regulations, to prevent. In terms of innovation with respect to environmental protection, the incentive must come from regulation, which heretofore has been lacking (see Sect. 5). Consider, however, a (hypothetical) situation in which owners of contaminated sites are held liable for not reporting the financial liabilities implied by their contaminated sites accurately, much the same way that energy companies are required to provide accurate estimates of their oil and gas reserves. Private companies would be required to include such liabilities in their balance sheets, whereas local and state governments would be required to set aside the resources needed for remediation. This would bring hydrogeology on par with petroleum engineering in the utilization of stochastic methods, since both would be equally motivated to find the oil, both metaphorically and literally. As things stand now, petroleum companies make money from finding oil. But that is not true for anyone else.

4.2 Education

Several contributions to the 2004 and 2016 series identified the lack of university courses in stochastic methods and, in consequence, a lack of hydrogeologists trained in stochastic methods as one of the setbacks to widespread adoption of stochastic methods in practice (Neuman, 2004; Winter, 2004). The lack of courses is driven, as explained in these two paper series, by what appears to be a high level of math and abstraction and seemingly, lack of practicality. One way to investigate this hypothesis is to look at differences between the education of hydrogeologists and reservoir engineers, as it is reasonable to expect that academic education would respond to the demands of industry.

Continuing the comparative analysis of these two industries, we compared the prevalence of courses in these subjects in the disciplines most likely to be studied by future petroleum engineers and practicing hydrogeologists. Three academic disciplines were selected for analysis: petroleum engineering, civil/environmental engineering, and earth sciences. We surveyed the top 10 schools in the US News & World Report 2016 rankings for graduate programs in petroleum engineering and earth sciences. Rankings for civil engineering are done separately from environmental engineering, and the top 10 of the two rankings produced 13 unique universities, which were the universities included for this survey.

Gathering information is made complicated because the amount of information about course offerings was variable from school to school. For some schools, course titles and course descriptions were available, while for others only course titles were available. For a couple of programs, no course information was publicly available online. While rankings exist for earth sciences as a subject, it is difficult to make direct comparisons due to the various organizational structures.

Our goal was to look for courses focused on stochastic methods in characterization and mapping. While utilization of geostatistical methods is not always necessary for incorporation of stochastic concepts, they are often used in conjunction, so our search included courses in geostatistics. For schools for which course information was found, seven out of nine petroleum engineering schools had a course in which geostatistics is the main focus (i.e., in the course title), and one school had a course where geostatistics is mentioned as part of the course. On the other hand, 4 out of 12 of the civil/environmental programs had a course for which geostatistical methods were the main focus. Two programs had a course that included geostatistics or SH in the course description, but were seemingly not a main focus. Only one earth sciences program had a course focusing on SH. Many civil/environmental engineering or earth sciences programs have at least one course on the topics of probability, statistics, or uncertainty analysis in some regard, but not specifically related to hydrology.

Note that this survey does not include information about how often these courses are taught, or what percentage of students take these courses. This survey also does not account for the possibility that these courses are being offered in other departments and taken by students of the departments in question. Despite these limitations, the results are clear: courses in geostatistics and stochastic methods are much more prevalent in petroleum engineering departments than in civil/environmental engineering or earth sciences departments. This finding suggests to us that there is no fundamental pedagogical difficulty in teaching stochastic methods. The demand for classes is not there, and it is very likely curbed by the limited marketability of the SH skill set in groundwater-related industries. Class offerings are driven by demand and are rarely curbed by intellectual challenges, especially in top academic programs. Of course, classroom teaching is just one mode of instruction and training, the other being research. This brings us to our next subsection, in which we will take a closer look at this aspect of teaching.

4.3 Research culture/collaboration

The vast majority of research in any scientific field comes from institutions belonging to one of three categories – academic, governmental, and industrial. Significant collaboration in research between these three sectors of the economy can act as a catalyst for adoption of novel methods, due to focusing of research goals, improved funding, and faster transfer of research-based knowledge from lab to field.

Since participation by industrial institutions in research is one avenue for adoption of new methods in practice, we aim to compare this activity in the two fields. To accomplish this, we performed a bibliometric survey using the Web of Science similar to the one performed by Rajaram (2016). The goal of the survey was to quantify the extent of collaboration in research between academia, government, and industry in each of the two disciplines.

The bibliometric survey was performed in March 2018 by searching matching keywords related to stochastic methods in the two fields for all articles. The author affiliations of the resulting articles were recorded, and ordered by number of articles for each institution. All institutions that appeared in at least three articles were recorded and categorized as being academic, governmental, or industrial. The number of appearances of each category was calculated for each of the two disciplines, with results presented in Fig. 1. Details of the survey and processing are provided in Appendix A.

For both hydrogeology and petroleum engineering, the majority of affiliations are academic, which may be expected for any field, but the ratio of academic to industrial affiliations is approximately 211 : 1 for hydrogeology, whereas it is only approximately 23 : 1 for petroleum engineering, a 10-fold difference in industry involvement. Hydrogeology has approximately 10 times as many publications as petroleum engineering overall, but petroleum engineering still has more industry appearances in 15 affiliations (e.g., PetroChina, DWA Energy Ltd., Chevron, and Schlumberger,) compared to hydrogeology's 12 industry appearances (e.g., Aquanty Inc., Schlumberger, and Tetra Tech Inc.). While this affiliation search is not completely exhaustive for all publications, languages, and possible search terms, the results are conclusive: there is clearly a larger presence of industrial institutions contributing to research in the petroleum field compared to hydrogeology. These results confirm the hypothesis that there is much greater participation in research by industrial institutions in the petroleum field than in hydrogeology. In turn, it can be argued that this indicates a stronger connection between theoretical development and practice: advances in theory are both motivated by practice and smoothly adopted by practice. These observations are closely aligned with our earlier observations on course offerings in the respective academic programs, for which we

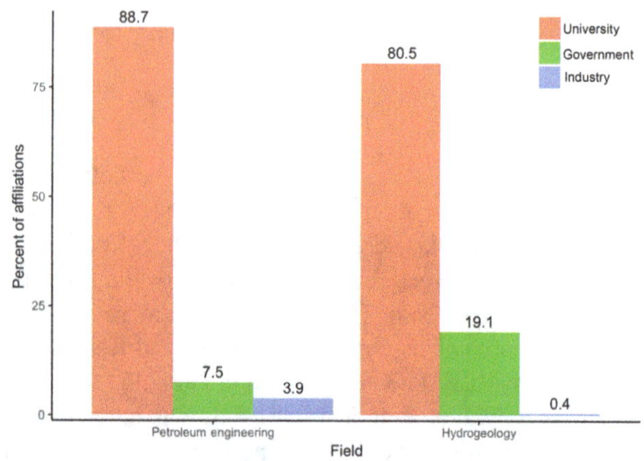

Figure 1. Bibliometric survey results for university, government, and industry author affiliations for petroleum engineering (left) and hydrogeology (right). The numbers on top of the columns indicate the percent of affiliations per category. The percent of affiliations reflect the proportion of total unique affiliations per document returned by the Web of Science database for each category (see Appendix A for full survey details). The total unique affiliation is a weighted sum of unique affiliations returned by the Web of Science with weights being the number of documents containing that affiliation at least once; i.e., two authors from the same institution on the same document are only counted once.

noted a much stronger emphasis on stochastic methods in petroleum-related schools, driven by a strong interest from industry.

4.4 Comparison summary

Despite being closely aligned in theoretical bases, hydrogeology and petroleum fields are fundamentally different in their incentive structures. The direct nature of the relationship between methodological improvements on the one hand and increased profitability plus regulatory constraints on the other provides impetus for advancement in the petroleum industry, which translates into funding opportunities and demand for skilled graduates. Schools are easy to adapt to market forces through course offerings and faculty recruitment. However, such relationships are not present in hydrogeology, leading to stagnation in industry and anemic interest in academia. When considering how often industry collaborates with academia, as quantified by journal article affiliations belonging to industry, the pattern remains: the petroleum engineering field has closer ties between industry and academia than the hydrogeology field does when it comes to SH research.

5 The law, the courts, and regulations

Previous publications opined that the regulatory and legal systems are not receptive to stochastic concepts and viewed this as one of the factors contributing to the gap. Early work on environmental regulations in the context of stochastic hydrology, e.g., Sudicky (2004), and more recent works, e.g., Fiori et al. (2016), noted that regulations do not require quantification of uncertainty, which would of course limit the demand for SH applications. Fogg and Zhang (2016) and Rajaram (2016) presented a somewhat different point of view and identified a drive towards a more explicit treatment of uncertainty.

5.1 Case study – risk-based corrective actions

Some perspective on the regulators' approach for managing uncertainty could be gained by looking at the regulations associated with risk-based corrective action (RBCA; US EPA, 1989) as a case study. RBCA focuses on cancer risk, defined as the upper bound lifetime probability of an individual developing cancer as a result of exposure to a particular level of a potential carcinogen (US EPA, 1989). For example, following Smalley et al. (2000), a risk of 10^{-6} represents an increased probability of 1 in 1 million of developing cancer (De Barros et al., 2009; De Barros and Rubin, 2008; Maxwell et al., 1999). Assessment of the exposure risk requires analysts to consider all the elements contributing to this risk, which would include the source, contaminant transport pathways, exposure pathways, and toxicology parameters. Thus, by requesting and expecting risk assessment, specifically formulated in probabilities, RBCA specifically and explicitly recognizes the stochastic nature of making predictions under uncertainty. The rationale underlying RBCA is extended to other focus areas of the US Environmental Protection Agency (EPA), such as risk-based decision-making in underground storage tank corrective action programs (US EPA, 1995). As a starting point for our conversation, we should note that risk-based assessments, and hence the underlying probabilistic approach, are endorsed by regulators.

The document shown in Fig. 2 states that "A risk-based approach is consistent with the Administrator's efforts to ensure that our environmental cleanup programs are based on the application of sound science and common sense and are flexible and cost-effective". With the link between risk and probability clearly defined, such as in RBCA, the regulator's support of stochastic concepts is clear and compelling. A rationale similar to RBCA's was pursued in the development of protective risk-based levels for the contaminants subject to the EPA's Resource Conservation and Recovery Act (RCRA). What could possibly be confused as being a hindrance is the language (used, for example, in Directive 9610.17 in Fig. 2 and US EPA, 1995) that encourages the use of RBCAs but does not mandate them. This, however, would not be an accurate interpretation, as the EPA is a federal agency, which puts it in a position to make recommendations and leave enforcement to states. EPA recommendations may be ignored by states, but this ignorance has consequences. Indeed, follow-up studies (e.g., Rifai and Suarez, 2000) confirm that most states in the USA have in fact accepted the EPA recommendations and formalized them as regulations, and that these regulations were enforced in a large number of field sites. With all this, it can be stated that regulations, if and when they exist, do not in principle pose a hindrance to the application of stochastic concepts. With such a broad embrace of stochastic concepts by the EPA, one would expect a broad adoption of SH by practitioners. This obviously is not the case.

5.2 A view on the courts, the judiciary, and administrative agencies

Administrative agencies do not operate in a vacuum, but rather in a public policy continuum that also includes the judiciary and the legislation, and the politicians who connect them. It is the interaction between these institutions that dominates the attitude towards uncertainty, and this is where agencies and courts get their cues. In this section, we seek to explore this continuum and assess the attitude towards probabilistic concepts and the modeling of uncertainty. The main question we wish to address is whether SH has experienced limited adoption outside academia due to some structural constraints rooted in the public policy continuum.

The view of the judiciary towards issues of uncertainty has been reviewed and heavily criticized by legal scholars, and the picture that emerges is that of inconsistency and confusion. There appears to be no fundamental difficulty with probabilistic concepts. There is no law, to the best of our knowledge, that forbids probabilistic concepts being presented. Furthermore, there are several cases in which courts considered and evaluated probabilities. Courts made some astute comments on probabilities. For example, when reviewing a low-probability (18 % chances of success!) plan proposed by a federal agency (NOAA Fisheries), the US Court of Appeals famously stated that "Only in Superman Comics' Bizarro world where reality is turned upside down could the [National Marine Fisheries] Service reasonably conclude that a measure that is at least four times likely to fail as to succeed offers a 'fairly high level of confidence'" (see Owen, 2009). Currell et al. (2017), commenting on the legal system in Australia, noted that courts could deal with evidence in whatever form it is presented, and they may accept it, reject it, misinterpret it, request clarifications and advice, ignore it, refer cases to other courts, or "kick the can down the road". Rejection of probabilistic concepts in the judiciary, if and when it occurs, may be driven not by resistance to probabilistic concepts but possibly as a matter of convenience. One way of avoiding hard decisions, sometimes adopted by the courts, is relying on what is euphemistically called "adaptive management", which assumes that management will be able to respond to changes. Although Currell et al. (2017) refer to the Australian judiciary, their findings seem to apply also in the USA.

At the same time, courts have been sending confusing messages. Farber (2011, p. 907) noted that "courts have sometimes reproved agencies for ignoring unquantified hazards, sometimes given their blessing as agencies buried their heads in the sand, and sometimes even forbidden agencies to act in the absence of quantification". This inconsistency leads to ambiguity, and it is reasonable to expect that professionals may want to avoid this uncharted territory. What are the root causes of this ambiguity? Legal scholars see three main factors contributing to this ambiguity: Congress, politicians, and the lack of probability standards. Although we present them as separate items, they are obviously interrelated.

The US Congress underestimated and ignored environmental uncertainties because "Uncertainty management may seem a 'politically unappealing topic'" (Owen, 2009). European legislature is even less prescriptive, opting for the very vague precautionary principle as a guideline. To understand Congress, we need to look at politicians. In the absence of clear guidance from the legislature, agencies opted for nonbinding guidance. A few environmental statutes contain signals about managing uncertainty, but these signals are often vague and indirect, leaving agencies with minimal guidance and a significant amount of discretion (Owen, 2009). To fill in the void left by Congress, agencies have rarely promulgated rules or set standards for dealing with uncertainty. More frequently, they opted for providing nonbinding guidance, which, as stated by Owen (2009), has very little visibility and often leaves major issues unresolved.

Politicians prefer to avoid stating the uncertainty associated with the stated goals of their proposed legislation and "it probably is unrealistic to expect legislators to candidly acknowledge the reality that their prescribed planning approaches will not always work out" (Owen, 2009). The attitudes by politicians are also driven by the public's view of uncertainty being a consequence of disagreement between scientists and science disciplines, and as such might as well be ignored.

The third factor to consider is the lack of probability standards. Probability standards are defined by Owen (2009) as "standards addressing plans' chances of achieving mandated outcomes". The lack of probability standards is, for example, what allowed NOAA Fisheries to submit a plan with 18 % probability of success (Owen, 2009). The lack of probability standards is possibly the most critical of the three factors because such standards, were they to exist, would promote "more systematic and transparent confrontations with the uncertainties inherent in environmental planning" and would instill "somewhat more cautious approaches to management of those uncertainties" and "reduce the persistent gaps between written environmental standards and actual performance" (Owen, 2009).

The negative consequences due to the lack of probability standards are numerous. It enhances risk-taking. It encourages making decisions on an ad hoc basis, which means that once a particular site is chosen, political pressure naturally will build to make the evaluative system fit the site rather than the underlying public safety goal" (Owen, 2009). It

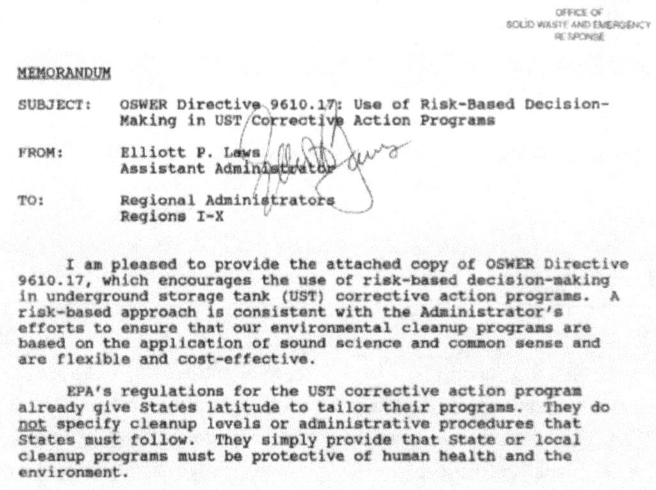

OFFICE OF
SOLID WASTE AND EMERGENCY
RESPONSE

MEMORANDUM

SUBJECT: OSWER Directive 9610.17: Use of Risk-Based Decision-Making in UST Corrective Action Programs

FROM: Elliott P. Laws
Assistant Administrator

TO: Regional Administrators
Regions I-X

I am pleased to provide the attached copy of OSWER Directive 9610.17, which encourages the use of risk-based decision-making in underground storage tank (UST) corrective action programs. A risk-based approach is consistent with the Administrator's efforts to ensure that our environmental cleanup programs are based on the application of sound science and common sense and are flexible and cost-effective.

EPA's regulations for the UST corrective action program already give States latitude to tailor their programs. They do not specify cleanup levels or administrative procedures that States must follow. They simply provide that State or local cleanup programs must be protective of human health and the environment.

Figure 2. EPA's letter encouraging (but not requiring!) the use of RBCA (US EPA, 1995). This document endorses a probability-based concept and refers to it as "sound science and common sense".

encourages individualized judgment and improvisation. The lack of probability standards "can turn stringent protection standards into a perpetually unattained aspiration, lowering short-term compliance costs but also creating persistent slippage between legal requirements and actual performance" (Owen, 2009). Furthermore, without clear probability standards, assessments undertaken with different methodologies lead to different conclusions. Such challenges are particularly difficult to address when several regulatory agencies are involved, with disparate resources and levels of expertise (Rothstein et al., 2006). It resulted in "a patchwork response to uncertainties about plan success", and finally, the lack of probability standards closes the door for SH analyses and applications.

There are plenty of reasons why probability standards are avoided, a few justified and others opportunistic, but in any case they are not related to strengths or weaknesses of statistical theories. First, it is often argued that setting standards would limit the dialog, and hence the latitude for political maneuvering, on the trade-offs between the level of protection on one hand, and compliance costs on the other. Probability standards therefore have substantial environmental and economic implications.

As already alluded to, there is the "all politics is local" factor (O'Neill and Hymel, 1995). The Interstate Technology & Regulatory Council (ITRC) (2005) points out that "Risk-based decisions are sometimes met with intense and divisive criticism by industry and environmental stakeholders, as well as the public" due to normative conflicts concerning how risks of adverse events and the associated impacts are viewed technically, socially, and politically by the affected constituencies. National or international regulatory standards may not be accepted by local communities or businesses if those actors assess and balance costs and benefits in different ways. Some stakeholders see risk-based approaches as residing in the "world of specialists" and essentially closing off debate about broader historic, social, and economic issues (Rothstein et al., 2006). Regulators, civic groups, and business interests apparently prefer to settle environmental issues without being constrained by federal "interventionist" policies. In addition, there is the issue of financial incentives. As noted in Bullard and Johnson (2000), the current regulatory environment "creates an industry around risk assessment and risk management", implying a satisfying status quo for some of the stakeholders, and hence a lack of financial incentives for change by the professional community. And this occurs despite "faulty assumptions in calculating, assessing, and managing risk".

The case in support of setting probability standards is not without its criticism (Rothstein et al., 2006), but it may also have the unintended consequence of focusing policy attention on problems that carry high institutional risks for the regulator at the expense of those that carry high risks to society. Against this background, it does not seem that setting probability standards is imminent. The courts acceptance, or not, of probabilistic concepts is perhaps not the key question to ask in this context. A more fundamental question, dated possibly to the age of Galileo, is whether courts should be the arbiters of good science. A thorough review of the interaction between courts and groundwater hydrologists is provided by Currell et al. (2017). They concluded that, whereas courts are not prevented from dealing with uncertainty, they are not well equipped to do so. Hydrogeologists should not expect courts to take an activist approach for promoting SH methods. This is not within their purview. Unfortunately, professional hydrologists generally follow the legal theories set by attorneys. Litigators may not be receptive to the modeling of uncertainty because uncertainty means ambiguity, which may translate into unfavorable judgment. Attorneys claim that their clients are innocent, not "maybe innocent". And similarly, prosecutors declare the defendants to be guilty, not "maybe guilty". In other words, "Uncertainty? What uncertainty? There is no uncertainty!", and so, who needs SH?

Realizing that courts should not be the arbiters of science, an important outcome from the extensive Australian experience in litigating groundwater hydrology (Currell et al. 2017) has been the establishment of independent expert scientific committees to advise the courts in all matters scientific. An example in which such involvement opened the door for SH can be found in the Environmental Radiation Protection Standards for Management and Disposal of Spent Nuclear Fuel and Transuranic Radioactive Wastes (40 CFR Part 190). The standards require applicants to perform a probabilistic assessment of long-term risk to the public due to accidental releases. An example of analyses based on these standards is provided in Zimmerman et al. (1998). This publication demonstrates that clear guidelines and quick responses are certainly possible. Why then is there such a difference between nuclear waste situations and what could possibly be viewed as the "lesser evil"? Legal scholars associate this special case with the involvement of a high-caliber scientific board in the form of the National Academy of Sciences (Farber, 2011; Owen, 2009).

In summary, courts are not prohibited from dealing with probability, but they have sent confusing messages with inconsistent rulings. Agencies lack probability standards for addressing their "plans' chances of achieving mandated outcomes". The guidance provided in lieu of regulations is nonbinding and fraught with problems. Congress, in the USA, underestimated and ignored environmental uncertainties. European countries rely in the precautionary principle when dealing with uncertainty, which is vague and it is even less prescriptive. Politicians refrain or hesitate acknowledging that their plans are associated with uncertainty. And as a consequence, SH is facing formidable headwinds on its way to become applicable, and this has nothing to do with the strength of the science. Unfortunately, there is no indication of any dynamics that will pull SH into a more favorable position, at least not from this angle.

6 The big miss – SH blind spot

Previous sections explored the challenges facing SH on its road to becoming broadly accepted by practitioners, regulators, and administrators. These challenges do not reflect weaknesses in the underlying SH science. There is, however, at least one challenge that is real and urgent, that has been around for a long time and that has been ignored by SH to its own detriment. This issue can be described by quoting former US Secretary of Defense Donald Rumsfeld, who, when addressing the challenges of making decisions under uncertainty, famously distinguished between "known knowns", "known unknowns", and "unknown unknowns". SH has been dealing primarily with the known unknowns and ignored the most challenging part of uncertainty management, which covers the unknown unknowns.

What are known unknowns? As explained by Rumsfeld, known unknowns are things we know that we do not know. In hydrogeology, this would include examples such as the spatial distributions of the hydraulic conductivity and almost any other soil property and, by way of consequence, state variables such as hydraulic head, concentrations, and solute travel times. It can also include forcing terms such as recharge (see Rubin and Dagan, 1987a, b).

What did Rumsfeld mean by unknown unknowns? Following Rumsfeld's definitions, these are the things we do not know that we do not know. These are the situations in which a variable is intentionally or unintentionally ignored. An example of this can be drawn from the case of nuclear waste disposal. What makes this problem unique is the life span of compounds such as plutonium. With a life span of 250 000 years, the list of unknown unknowns could include, for example, climate change and possible changes in human behavior, societal norms, and political systems. Farber (2011) noted in this context that "the calculation of the physical likelihood of leakage should (hopefully) be reasonably accurate, the assumptions about human presence and activities in the area (and therefore about exposure) are speculative, and the likelihood of human interference with the integrity of the site is completely unknown". Recalling the adaptive management concept that is sometimes employed for dealing with uncertainty, Owen (2009) noted that given the time span of many hydrological processes, "One therefore cannot reasonably assume that present social or political institutions will survive long enough to adjust waste containment systems", thus rendering any semblance of planning unreasonable.

In the case of nuclear waste, as in almost any case, it is hard to fathom the prediction of 250 000 years down the road. For some perspective, consider that the pyramids were built only 5000 years ago. The ancient Egyptians did a good job at hiding the mummies and treasures for thousands of years, but who could have predicted ground-penetrating radar? Long-term prediction is not a simple question, but possibly the most relevant question. A confirmation that the unknown unknowns are of topmost significance for regulators is given in the National Research Council (2014). This publication, dealing with risk-informed decision-making in contaminated sites, identified the unknown unknowns, i.e., the drivers of uncertainty, as the major and yet unanswered challenges in the decision-making process, including, for example, (1) "uncertainties that contamination may exist at a site", (2) estimation of the "quantities of contamination that may exist in the future" (p. 91), (3) "low probability, high consequence events" (p. 92; authors' note: these are the National Research Council's code words for human-related scenarios), and (4) uncertainty about how to predict failures of treatment and containment systems (p. 123).

Yucca Mountain, a central activity for hydrogeologists for more than 40 years, is a paramount example of SH failure. How did SH address the uncertainty challenges? As noted earlier, Farber (2011) concluded that the "calculation of the physical likelihood of leakage should (hopefully) be reasonably accurate" (our note: covered by SH; see Zimmerman et al., 1998), "the assumptions about human presence and activities in the area (and therefore about exposure) are speculative" (our note: not covered by SH), "and the likelihood of human interference with the integrity of the site is completely unknown" (our note: not covered by SH). Notably, there was no reference to issues such as the modeling of contaminant transport being showstoppers. SH opted to deal with the physics of flow (known unknowns, using Rumsfeld's terminology and otherwise known as "risk analysis' blind spot", using Farber's terminology), but ignored the societal risks residing in the interface between physics and the human environment (unknown unknowns, otherwise known as uncertainty). This omission did not escape the attention of the courts: in the absence of appropriate tools, the site planners chose to ignore "human activities and economic imperatives", opting instead to assume current conditions, a decision which was referred to by the court as "odd" (Farber, 2011, p. 951).

Modeling the unknown unknowns is a daunting task, but ignoring it has grave consequences, as Rumsfeld himself attested. It also seems that ignoring it is not an option, as the legal stance is very clear. Farber (2011, p. 910) quotes the court as saying that "the inability to quantify a risk does not justify failure to discuss it if there are other grounds for considering it significant".

Returning to our energy industry comparisons, petroleum engineering does not face such issues because most projects are of short duration and are well defined by scientific and engineering considerations, with minimal or no interfacing with non-engineering issues. But petroleum engineering also fails in situations in which human factors are involved, such as in the case of British Petroleum's Deepwater Horizon well failure and oil spill in the Gulf of Mexico, and also the now widely refuted peak oil theory.

Having therefore outlined the challenge, what sorts of solutions have been developed in the field of uncertainty characterization and uncertainty reasoning for dealing with unknown unknowns, and can these solutions be transferred to the field of hydrogeology? An important milestone here is the groundbreaking work of Savage (1954), which provided probability theory with the necessary link to decision theory. So strong was the impact of Savage's work that it not only became the mainstay of decision theory everywhere (Baron, 2004; Berger, 1985; Bernardo and Smith, 1994; Gilboa, 2009; Jeffrey, 1992; Kahneman et al., 1982; Peterson, 2009), that even when Nobel laureates (Kahneman and Tversky, 1973) showed how decision-making in real life often diverges from the axioms described by Savage (1954), their ideas were largely ignored (initially). On the contrary, probability theory and probabilistic decision theory was simply declared to be the rational norm and that anything else is irrational.

This harsh repudiation of people's intuition began to slowly change with the growing recognition that real-world situations are ripe with unknown unknowns due to ignorance, ambivalence, or ambiguity. In such situations, it is not easy to see how a unique prior distribution can be defined by an agent as demanded by classic decision theory. To account for unknown unknowns, two lines of thought can be identified. The first approach stays within the well-defined confines of probability theory and accounts for possible unknown unknowns by using a whole set of prior distributions instead of just a single unique one. Updating of beliefs is then performed by Bayes' rule and the common rules of decision theory, i.e., the maximization of expected utility (see, e.g., Berger, 1990; Wasserman and Kadane, 1990). The second approach goes beyond probability and is based on the works of Dempster (1968) and Shafer (1976). Here, beliefs are no longer represented by a probability function but by a so-called belief function, and for updating, Bayes' rule is replaced by the Dempster–Shafer rule. This Dempster–Shafer theory has historically been criticized for its failure to connect with decision theory (Pearl, 1990), but more recent developments by Gilboa and Schmeidler (1993) have provided solid foundations for deriving decision rules from Dempster–Shafer belief functions, too.

Whereas we think that in science alone, SH included, the importance of unknown unknowns is constantly decreasing with the accumulation of knowledge through research, the prediction of societal, political, economic, technological, and demographic changes will remain a source of unknown unknowns for the foreseeable future. Since scientific results constantly interface with these fields through the perspective of decision makers, we should also expand the applicability of SH into this domain.

7 Summary

Several publications have attempted to explain the apparent failure of SH to gain broad acceptance in application. In this paper, we reviewed the various hypotheses raised and evaluated them by establishing and analyzing a factual basis. We found a large disparity between several of the hypotheses and the facts on the ground as we see them. For example, we show that it is not justified to view SH education or software as roadblocks on the way to broad acceptance.

The roadblocks on the way to making SH broadly accepted are primarily twofold. One roadblock is external, meaning that it is not related to SH fundamentals. The other factor is in fact internal as it is related to SH self-definition: the topics it chose to pursue, and more important, those that it chose, knowingly or not, to ignore; it is what we call SH's blind spot, and it is related to SH fundamentals.

The external factor resides in the continuum defined by the legislature, the judiciary, and government. Legal scholars summarize it succinctly as a lack of probability standards. Probability standards are defined by Owen (2009) as "standards addressing plans' chances of achieving mandated outcomes" (see discussion in Sect. 5). The lack of probability standards institutionalizes the fragmentation between the three branches of government often asked to deal with uncertainty and eliminates the financial incentive needed for industry to move in and address the challenges.

The internal roadblock is the focus on risk (Rumsfeld's known unknowns) and ignoring uncertainty (the unknown unknowns). Uncertainty resides in the interface between hydrogeology and related disciplines such as climate science and sociology and political science. It is noted that Yucca Mountain project failed because of uncertainty, not because of any doubt about the values of the conductivity anywhere, but because of the inability to make long-term predictions regarding climate change and to model the possible changes in human behavior, societal norms, and political systems. These are by no means easy challenges, but work on these topics is already under way, alas, not by hydrologists. Expanding SH to account for risk as well as uncertainty is a key for making an impact. This lacuna is not the privilege of SH: as noted by several authors, there are other environmental problems needing to address uncertainty, such as fisheries, air quality, endangered species, water quality, and climate change (see Owen, 2009). And of course outside the environmental sciences we have financial risks and other catastrophes. Publications on the interface between environmental science and the social sciences are available (Liu et al., 2007; Torn and Harte, 2006) and there are quite a few from legal scholars (reviewed earlier), but SH is still lagging.

Appendix A

The bibliometric survey comparing the collaboration between academia and industry for the two fields of hydrogeology and petroleum engineering consisted of first compiling information from the Web of Science database at http://apps.webofknowledge.com/ (last access: 8 March 2018) (v.5.27.2 during March 2018) and then processing that information into useable metrics. This database allows users to first query certain search terms to generate a list of documents and then access summaries about that list such as the number of articles published per year or how often organizations are included in the author affiliations. To perform the analysis of comparing the collaboration of academia and industry in the two fields of interest, further data summary metrics were calculated with the results downloaded from the Web of Science. The two lists below detail the steps taken for these two procedures.

Steps for compiling data from Web of Science:

1. Search terms were defined to capture relevant documents for the two fields of hydrogeology and petroleum engineering. See Table 1 for the search terms used. There were five sets of iterative search terms used per field to confirm that the resulting metrics comparing the two fields were consistent.

2. Each of the search term sets was queried with the Web of Science database using the advanced search option. A time span of the last 5 years (2013–2018) was selected, and all documents types were used; i.e., the search was not limited to just journal articles in order to capture collaborations in conference posters and presentations or other documents.

3. For each search set, the Analyze Results feature was used to generate information regarding the affiliations listed for the documents returned by the query. The Organizations field was selected with a minimum record threshold set to 3; i.e., an organization had to be listed on at least three documents, in order to make the volume of unique affiliations manageable for manual classification. The data returned include organization names and both the raw number and percent of articles in the search containing those organizations as an affiliation. These numbers reflect how many documents in which the organization is listed as an affiliation for at least one author. The results table was downloaded as a text file for each search set.

Steps for calculating metrics:

1. Each organization returned by Web of Science was manually categorized as a degree-granting academic institution ("university"), a government agency or research organization primarily funded by government agencies ("government"), or, if neither of those, then "industry". This was accomplished by searching for certain text patterns in the organizations' names. For example, if the pattern "univ" or "college" was in the organization name, then it was placed in the university category. Or if "geological survey" or "national research council" was in the organization name, then it was categorized as government. Additional patterns were created until all organizations were categorized. When the name of the organization was not sufficient to assign a category, the organization was investigated to ascertain the most appropriate affiliation.

2. For each field, category, and search set, the number of records across organizations were summed and a percentage was calculated. This percentage represents the proportion of affiliations on documents belonging to the three categories.

3. The relative relationship between the three categories per field was confirmed to remain consistent across the iterative search terms. The fifth set was used for the rest of the analysis.

4. The categorical percentages were compared between the two fields to ascertain if the two fields of hydrogeology and petroleum engineering had more participation of industry in research collaborations leading to publications or conference abstracts.

Table A1. Web of Science search terms used to compare hydrogeology and petroleum engineering documents. The stochastic terms (leftmost column) were used for all search sets. For hydrogeology, the searches were limited to the Web of Science categories listed in the second leftmost column. The five search sets for hydrogeology are listed in the middle column and are composed of the search terms listed. The search sets vary by incrementally more search terms related to hydrogeology. Similarly, the Web of Science categories and search term sets for petroleum engineering are listed in the two rightmost columns. Each search set consists of combining the stochastic terms, the Web of Science categories, and field terms with a logical AND operator via the Web of Science.

Stochastic terms	Hydrogeology		Petroleum engineering	
	Web of Science categories	Field terms	Web of Science categories	Field terms
stochastic OR geostatist*	Water Resources OR Environmental Sciences OR Limnology OR Engineering Environmental OR Science	1: porous OR soil OR aquifer OR subsurface OR hydrogeology OR geology	Petroleum OR Energy	1: porous OR soil OR subsurface OR geology OR reservoir OR oil
		2: 1+ OR groundwater		2: 1+ OR petrol*
		3: 2+ OR remediation		3: 2+ OR exploration
		4: 3+ OR hyporheic		4: 3+ OR natural gas
		5: 4+ OR phreatic		5: 4+ OR hydraulic fracturing

Note the asterisk * denotes the wildcard search feature; e.g., geostatist* matches geostatistics and geostatistical.

Author contributions. YR defined the scope of the manuscript. CC researched legal case examples. JC and KC researched and described software availability. BH researched education and the petroleum engineering industry. FH contributed to the discussion on the perception of uncertainty. HS performed the bibliometric analysis. The manuscript was written through contributions of all authors. All authors have given approval to the final version of the manuscript.

Competing interests. The authors declare that they have no conflict of interest.

Edited by: Bill X. Hu

References

Aquaveo LLC: GMS User Manual: The Groundwater Modeling System, 558 pp., 2012.

Baron, J.: Blackwell Handbook of Judgment and Decision Making, edited by: Koehler, D. J. and Harvey, N., Blackwell Publishing Ltd., Oxford, UK, 19–37,2004.

Bellin, A. and Rubin, Y.: HYDRO_GEN: A spatially distributed random field generator for correlated properties, Stoch. Hydrol. Hydraul., 10, 253–278, https://doi.org/10.1007/BF01581869, 1996.

Bellin, A. and Rubin, Y.: On the use of peak concentration arrival times for the inference of hydrogeological parameters, Water Resour. Res., 40, W07401, https://doi.org/10.1029/2003WR002179, 2004.

Berger, J.: Robust Bayesian analysis: sensitivity to the prior, J. Stat. Plan. Infer., 25, 303–328, 1990.

Berger, J. O.: Statistical Decision Theory and Bayesian Analysis, Springer, New York City, U.S.A., 1985.

Bernardo, J. M. and Smith, A. F. M.: Bayesian Theory, Wiley, Hoboken, NJ, U.S.A., 1994.

Bianchi, M. and Zheng, C.: SGeMS: A free and versatile tool for three-dimensional geostatistical applications, Ground Water, 47, 8–12, https://doi.org/10.1111/j.1745-6584.2008.00522.x, 2009.

Bleines, C., Perseval, S., Rambert, F., Renard, D., and Touffait, Y.: ISATIS. Isatis software manual, 5th edn., Geovariances & Ecole Des Mines De Paris, Paris, 2004.

Bullard, R. D. and Johnson, G. S.: Environmental justice: Grassroots activism and its impact on public policy decision making, J. Soc. Issues, 56, 555–578, https://doi.org/10.1111/0022-4537.00184, 2000.

Buxton, B., Pate, A., and Morara, M.: Batgam©geostatistical software based on GSLIB, in: Geostatistics Banff 2004, edited by: Leuangthong, O. and Deutsch, C. V., Quant. Geo. G., 14, 1131–1136, 2005.

Carle, S.: T-PROGS: Transition probability geostatistical software, University of California, Davis, CA, 1999.

Castilla-Rho, J. C., Mariethoz, G., Kelly, B. F. J. and Andersen, M. S.: Stochastic reconstruction of paleovalley bedrock morphology from sparse datasets, Environ. Modell. Softw., 53, 35–52, https://doi.org/10.1016/j.envsoft.2013.10.025, 2014.

Chen, J. and Hoversten, G. M.: Joint inversion of marine seismic AVA and CSEM data using statistical rock-physics models and Markov random fields, Geophysics, 77, R65–R80, https://doi.org/10.1190/geo2011-0219.1, 2012.

Chen, J., Hubbard, S., Rubin, Y., Murray, C., Roden, E., and Majer, E.: Geochemical characterization using geophysical data and Markov Chain Monte Carlo methods: A case study at the South Oyster bacterial transport site in Virginia, Water Resour. Res., 40, W12412, https://doi.org/10.1029/2003WR002883, 2004.

Christakos, G.: A sociological approach to the state of stochastic hydrogeology, Stoch. Env. Res. Risk A., 18, 274–277, https://doi.org/10.1007/s00477-004-0197-1, 2004.

Cirpka, O. A. and Valocchi, A. J.: Debates – Stochastic subsurface hydrology from theory to practice – Does stochastic subsurface hydrology help solving practical problems of contaminant hydrogeology?, Water Resour. Res., 52, 9218–9227, https://doi.org/10.1002/2016WR019087, 2016.

Currell, M. J., Werner, A. D., McGrath, C., Webb, J. A., and Berkman, M.: Problems with the application of hydrogeological science to regulation of Australian mining projects: Carmichael Mine and Doongmabulla Springs, J. Hydrol., 548, 674–682, https://doi.org/10.1016/j.jhydrol.2017.03.031, 2017.

Cvetkovic, V., Selroos, J.-O., and Cheng, H.: Transport of reactive tracers in rock fractures, J. Fluid Mech., 378, 335–356, https://doi.org/10.1017/S0022112098003450, 1999.

Dagan, G.: On application of stochastic modeling of groundwater flow and transport, Stoch. Env. Res. Risk A., 18, 266–267, https://doi.org/10.1007/s00477-004-0191-7, 2004.

De Barros, F. P. J. and Rubin, Y.: A risk-driven approach for subsurface site characterization, Water Resour. Res., 44, W01414, https://doi.org/10.1029/2007WR006081, 2008.

De Barros, F. P. J., Rubin, Y., and Maxwell, R. M.: The concept of comparative information yield curves and its application to risk-based site characterization, Water Resour. Res., 45, W06401, https://doi.org/10.1029/2008WR007324, 2009.

Dempster, A. P.: A Generalization of Bayesian Inference, J. Roy. Stat. Soc. B Met., 30, 205–247, 1968.

Deutsch, C. V. and Journel, A. G.: Geostatistical software library and user's guide, Oxford University Press, New York, 1–369, 1998.

de Valpine, P., Turek, D., Paciorek, C. J., Anderson-Bergman, C., Lang, D. T., Bodik, R., Valpine, P. De, Turek, D., Paciorek, C. J., Lang, D. T., and Bodik, R.: Programming with models: writing statistical algorithms for general model structures with NIMBLE, J. Comput. Graph. Stat., 26, 403–413, https://doi.org/10.1080/10618600.2016.1172487, 2016.

Doherty, J.: PEST User Manual, 5th edn., Watermark Numerical Computing, Brisbane, Australia, 2005.

Doherty, J. and Hunt, R.: Approaches to highly parameterized inversion – A guide to using PEST for groundwater-model calibration, U.S. Geological Survey, Scientific Investigations Report 2010-5169, 59 pp., 2010.

Eaton, T. T.: On the importance of geological heterogeneity for flow simulation, Sediment. Geol., 184, 187–201, https://doi.org/10.1016/j.sedgeo.2005.11.002, 2006.

Engdahl, N. B., Weissmann, G. S., and Bonal, N. D.: An integrated approach to shallow aquifer characterization: Combining geophysics and geostatistics, Comput. Geosci., 14, 217–229, https://doi.org/10.1007/s10596-009-9145-y, 2010.

Ezzedine, S., Rubin, Y., and Chen, J.: Bayesian Method for hydrogeological site characterization using borehole and geophysical survey data: Theory and application to the Lawrence Livermore National Laboratory Superfund Site, Water Resour. Res., 35, 2671, https://doi.org/10.1029/1999WR900131, 1999.

Farber, D. A.: Uncertainty, Georgetown Law J., 99, 901–959, 2011.

Finsterle, S.: iTOUGH2 User's Guide, Lawrence Berkeley National Lab (LBNL), Berkeley, 123, 407–408, 2011.

Fiori, A., Cvetkovic, V., Dagan, G., Attinger, S., Bellin, A., Dietrich, P., Zech, A., and Teutsch, G.: Debates – Stochastic subsurface hydrology from theory to practice: The relevance of stochastic subsurface hydrology to practical problems of contaminant transport and remediation. What is characterization and stochastic theory good for?, Water Resour. Res., 52, 9228–9234, https://doi.org/10.1002/2015WR017525, 2016.

Fleckenstein, J. H., Niswonger, R. G., and Fogg, G. E.: River-aquifer interactions, geologic heterogeneity, and low-flow management, Ground Water, 44, 837–852, https://doi.org/10.1111/j.1745-6584.2006.00190.x, 2006.

Floris, F. J. T., Bush, M. D., Cuypers, M., Roggero, F., and Syversveen, A.-R.: Methods for quantifying the uncertainty of production forecasts: a comparative study, Petrol. Geosci., 7, S87–S96, https://doi.org/10.1144/petgeo.7.S.S87, 2001.

Fogg, G. E. and Zhang, Y.: Debates–Stochastic subsurface hydrology from theory to practice: A geologic perspective, Water Resour. Res., 52, 5375–5377, https://doi.org/10.1002/2016WR019699, 2016.

Freeze, R. A.: The role of stochastic hydrogeological modeling in real-world engineering applications, Stoch. Env. Res. Risk A., 18, 286–289, https://doi.org/10.1007/s00477-004-0194-4, 2004.

Gilboa, I.: Theory of Decision Under Uncertainty, Cambridge University Press, Cambridge, UK, 2009.

Gilboa, I. and Schmeidler, D.: Updating ambiguous beliefs, J. Econ. Theory, 59, 33–49, 1993.

Ginn, T. R.: On the application of stochastic approaches in hydrogeology, Stoch. Env. Res. Risk A., 18, 282–284, https://doi.org/10.1007/s00477-004-0199-z, 2004.

Ginn, T. R. and Cushman, J. H.: Inverse methods for subsurface flow: A critical review of stochastic techniques, Stoch. Hydrol. Hydraul., 4, 1–26, https://doi.org/10.1007/BF01547729, 1990.

Goblet, P.: Programme METIS, Simulation d'écoulement et de transport miscible en milieu poreux et fracturé, notice de conception, Ecole des Mines de Paris, Paris, 1989.

Gómez-Hernández, J. J. and Srivastava, R.: Isim3D – an Ansi-C 3-Dimensional Multiple Indicator Conditional Simulation Program, Comput. Geosci., 16, 395–440, 1990.

Guardiano, F. and Srivastava, R.: Multivariate geostatistics: beyond bivariate moments, Geostatistics Troia'92, 133–144, https://doi.org/10.1007/978-94-011-1739-5_12, 1993.

Hansen, T. M.: mgstat: A geostatistical matlab toolbox, Online web resource, available at: http://mgstat.sourceforge.net. (last access: 22 December 2017), 2004.

Harbaugh, A. W., Banta, E. R., Hill, M. C., and McDonald, M. G.: MODFLOW-2000, The U.S. Geological Survey modular ground-water model — User guide to modularization concepts and the ground-water flow process, U.S. Geological Survey, 130 pp., 2000.

Harrell, D. R. and Gardner, T. L.: Significant Differences in Proved Reserves Volumes Estimated Using SPE/WPC Reserves Compared to United States Securities and Exchange Commission (SEC) Definitions, SPE Annual Technical Conference and Exhibition, 5–8 October 2003, Denver, Colorado, SPE-84145-MS, https://doi.org/10.2118/84145-MS, 2003.

He, X., Koch, J., Sonnenborg, T. O., Flemming, J., Schamper, C., and Refsgaard, J. C.: Transition probability-based stochastic geological modeling Using Airborne Geophysical Data and Borehole Data, Water Resour. Res., 50, 3147–3169, https://doi.org/10.1002/2013WR014593, 2014.

Hou, Z., Rubin, Y., Hoversten, G. M., Vasco, D., and Chan, J.: Reservoir-parameter identification using minimum relative entropy-based Bayesian inversion of seismic AVA and marine CSEM data, Geophysics, 71, O77–O88, 2006.

Hubbard, S. S., Rubin, Y., and Majer, E.: Spatial correlation structure estimation using geophysical and hydrogeological data, Water Resour. Res., 35, 1809–1825, https://doi.org/10.1029/1999WR900040, 1999.

Hubbard, S. S., Chen, J., Peterson, J. E., Mayer, E. L., Williams, K. H., Swift, D. J., Mailloux, B., and Rubin, Y.: Hydrogeological characterization of South Oyster Bacterical Transport Site using geophysical data, Water Resour. Res., 37, 2431–2456, https://doi.org/10.1029/2001WR000279, 2001.

Hyndman, R. J. and Khandakar, Y.: Automatic time series forecasting: The forecast package for R, J. Stat. Softw., 27, 22 pp., https://doi.org/10.18637/jss.v027.i03, 2008.

ITRC: Examination of Risk-Based Screening Values and Approaches of Selected States, Interstate Technology & Regulatory Council (ITRC), Washington, D.C., 115 pp., 2005.

Jackson, C. P., Hoch, A. R., and Todman, S.: Self-consistency of a heterogeneous continuum porous medium representation of a fractured medium, Water Resour. Res., 36, 189–202, https://doi.org/10.1029/1999WR900249, 2000.

Jankovic, I., Maghrebi, M., Fiori, A., and Dagan, G.: When good statistical models of aquifer heterogeneity go right: The impact of aquifer permeability structures on 3D flow and transport, Adv. Water Resour., 100, 199–211, 2017.

Jeffrey, R.: Probability and the Art of Judgment, Cambridge University Press, Cambridge, UK, 1992.

Jones, N. L., Walker, J. R., and Carle, S. F.: Hydrogeologic unit flow characterization using transition probability geostatistics, Ground Water, 43, 285–289, https://doi.org/10.1111/j.1745-6584.2005.0007.x, 2005.

Jonkman, R. M., Bos, C. F. M., Breunese, J. N., Morgan, D. T. K., Spencer, J. A., and Sondena, E.: Best practices and methods in hydrocarbon resource estimation, production and emissions forecasting, uncertainty evaluation and decision making, Proceedings of the European Petroleum Conference, SPE European Petroleum Conference, Paris, France, 24–25 October 2000.

Kahneman, D. and Tversky, A.: On the psychology of prediction, Psychol. Rev., 80, 237–251, https://doi.org/10.1037/h0034747, 1973.

Kahneman, D., Slovic, P., and Tversky, A. (Eds.): Judgment under Uncertainty: Heuristics and Biases, Cambridge University Press, Cambridge, UK, 1982.

Kitanidis, P. K. and Lane, R. W.: Maximum likelihood parameter estimation of hydrologic spatial processes by the Gauss-Newton method, J. Hydrol., 79, 53–71, https://doi.org/10.1016/0022-1694(85)90181-7, 1985.

Kowalsky, M. B., Finsterle, S., and Rubin, Y.: Estimating flow parameter distributions using ground-penetrating radar and hydrological measurements during transient flow in the vadose zone, Adv. Water Resour., 27, 583–599, https://doi.org/10.1016/j.advwatres.2004.03.003, 2004.

Kowalsky, M. B., Finsterle, S., Peterson, J., Hubbard, S., Rubin, Y., Majer, E., Ward, A., and Gee, G.: Estimation of field-scale soil hydraulic and dielectric parameters through joint inversion of GPR and hydrological data, Water Resour. Res., 41, W11425, https://doi.org/10.1029/2005WR004237, 2005.

Li, X., Li, Y., Chang, C. F., Tan, B., Chen, Z., Sege, J., Wang, C., and Rubin, Y.: Stochastic, goal-oriented rapid impact modeling of uncertainty and environmental impacts in poorly-sampled sites using ex-situ priors, Adv. Water Resour., 111, 174–191, https://doi.org/10.1016/j.advwatres.2017.11.008, 2018.

Liu, J., Dietz, T., Carpenter, S. R., Alberti, M., Folke, C., Moran, E., Pell, A. N., Deadman, P., Kratz, T., Lubchenco, J., Ostrom, E., Ouyang, Z., Provencher, W., Redman, C. L., Schneider, S. H., and Taylor, W. W.: Complexity of Coupled Human and Natural Systems, Science, 317, 1513–1516, https://doi.org/10.1126/science.1144004, 2007.

Liu, N. and Oliver, D. S.: Evaluation of Monte Carlo Methods for Assessing Uncertainty, SPE J., 8, 188–195, https://doi.org/10.2118/84936-PA, 2003.

Mariethoz, G., Renard, P., and Straubhaar, J.: The direct sampling method to perform multiple-point geostatistical simulations, Water Resour. Res., 46, W11536, https://doi.org/10.1029/2008WR007621, 2010.

Marinoni, O.: Improving geological models using a combined ordinary-indicator kriging approach, Eng. Geol., 69, 37–45, https://doi.org/10.1016/S0013-7952(02)00246-6, 2003.

Maxwell, R. M., Kastenberg, W. E., and Rubin, Y.: A methodology to integrate site characterization information into groundwater-driven health risk assessment, Water Resour. Res., 35, 2841–2855, https://doi.org/10.1029/1999WR900103, 1999.

McLaughlin, D. and Townley, L. R.: A reassessment of the groundwater inverse problem, Water Resour. Res., 32, 1131–1161, https://doi.org/10.1029/96WR00160, 1996.

Misund, B. and Osmundsen, P.: Probable Oil and Gas Reserves and Shareholder Returns: The Impact of Shale Gas, CESifo Working Paper Series 5687, CESifo Group Munich, 2015.

ModelMuse Help: Import T-PROGS File Dialog Box, USGS Publications, https://water.usgs.gov/nrp/gwsoftware/ModelMuse/Help/index.html?import_t_progs_file_dialog_box.htm (last access: 21 October 2018), 2009.

Molz, F.: A rational role for stochastic concepts in subsurface hydrology: a personal perspective, Stoch. Env. Res. Risk A., 18, 278–279, https://doi.org/10.1007/s00477-004-0195-3, 2004.

Murakami, H., Chen, X., Hahn, M. S., Liu, Y., Rockhold, M. L., Vermeul, V. R., Zachara, J. M., and Rubin, Y.: Bayesian approach for three-dimensional aquifer characterization at the Hanford 300 Area, Hydrol. Earth Syst. Sci., 14, 1989–2001, https://doi.org/10.5194/hess-14-1989-2010, 2010.

National Research Council: Best Practices for Risk-Informed Decision Making Regarding Contaminated Sites: Summary of a Workshop Series. Washington, DC: The National Academies Press, https://doi.org/https://doi.org/10.17226/18747, 2014.

Neuman, S. P.: Stochastic groundwater models in practice, Stoch. Env. Res. Risk A., 18, 268–270, https://doi.org/10.1007/s00477-004-0192-6, 2004.

O'Neill, T. and Hymel, G.: All Politics is Local: And Other Rules of the Game, edited by: Adams, B., Inc, Holbrook, M, 1995.

Oliver, D. S. and Chen, Y.: Recent progress on reservoir history matching: A review, Comput. Geosci., 15, 185–221, https://doi.org/10.1007/s10596-010-9194-2, 2011.

Osorio-Murillo, C. A., Over, M. W., Savoy, H., Ames, D. P., and Rubin, Y.: Software framework for inverse modeling and uncertainty characterization, Environ. Modell. Softw., 66, 98–109, https://doi.org/10.1016/j.envsoft.2015.01.002, 2015.

Owen, D.: Probabilities, Planning Failures, and Environmental Law, Tulane Law Review, 84, 265–335, 2009.

Pearl, J.: Reasoning with belief functions: An analysis of compatibility, Int. J. Approx. Reason., 4, 363–389, https://doi.org/10.1016/0888-613X(90)90013-R, 1990.

Pebesma, E. J.: Multivariable geostatistics in S: the gstat package, Comput. Geosci., 30, 683–691, https://doi.org/10.1016/j.cageo.2004.03.012, 2004.

Peterson, M.: An Introduction to Decision Theory, Cambridge University Press, Cambridge, UK, 2009.

Pirot, G., Straubhaar, J., and Renard, P.: Simulation of braided river elevation model time series with multiple-point statistics, Geomorphology, 214, 148–156, https://doi.org/10.1016/j.geomorph.2014.01.022, 2014.

Rajaram, H.: Debates –Stochastic subsurface hydrology from theory to practice: Introduction, Water Resour. Res., 52, 9215–9217, https://doi.org/10.1002/2016WR020066, 2016.

Reeves, J. A., Knight, R., and Zebker, H. A.: An analysis of the uncertainty in InSAR deformation measurements for groundwater applications in agricultural areas, IEEE J. Sel. Top. Appl., 7, 2992–3001, https://doi.org/10.1109/JSTARS.2014.2322775, 2014.

Remy, N., Boucher, A., and Wu, J.: Applied Geostatistics with SGeMS, Cambridge University Press, Cambridge, UK, 2009.

Renard, F. and Jeannée, N.: Estimating transmissivity fields and their influence on flow and transport: The case of Champagne mounts, Water Resour. Res., 44, W11414, https://doi.org/10.1029/2008WR007033, 2008.

Rifai, H. S. and Suarez, M.: The RBCA Success Story, Environmental Protection, 1–8, 2000.

Rothstein, H., Irving, P., Walden, T., and Yearsley, R.: The risks of risk-based regulation: Insights from the environmental policy domain, Environ. Int., 32, 1056–1065, https://doi.org/10.1016/j.envint.2006.06.008, 2006.

Rubin, Y.: Applied Stochastic Hydrogeology, Oxford Press, Oxford, 2003.

Rubin, Y.: Stochastic hydrogeology – challenges and misconceptions, Stoch. Env. Res. Risk A., 18, 280–281, https://doi.org/10.1007/s00477-004-0193-5, 2004.

Rubin, Y. and Dagan, G.: Stochastic identification of transmissivity and effective recharge in steady groundwater flow: 1. Theory, Water Resour. Res., 23, 1185–1192, https://doi.org/10.1029/WR023i007p01185, 1987a.

Rubin, Y. and Dagan, G.: Stochastic identification of transmissivity and effective recharge in steady groundwater flow: 2. Case study, Water Resour. Res., 23, 1193–1200, https://doi.org/10.1029/WR023i007p01193, 1987b.

Rubin, Y., Chen, X., Murakami, H., and Hahn, M.: A Bayesian approach for inverse modeling, data assimilation, and conditional simulation of spatial random fields, Water Resour. Res., 46, W10523, https://doi.org/10.1029/2009WR008799, 2010.

Rwechungura, R. W., Dadashpour, M., and Kleppe, J.: Advanced History Matching Techniques Reviewed, SPE Middle East Oil and Gas Show and Conference, 25–28 September 2011, Manama, Bahrain, SPE-142497-MS, https://doi.org/10.2118/142497-MS, 2011.

Sanchez-Vila, X. and Fernàndez-Garcia, D.: Debates – Stochastic subsurface hydrology from theory to practice- Why stochastic modeling has not yet permeated into practitioners?, Water Resour. Res., 52, 9246–9258, https://doi.org/10.1002/2016WR019302, 2016.

Sartore, L.: spMC: Modelling Spatial Random Fields with Continuous Lag Markov Chains, The R Journal, 5, 16–28, 2013.

Savage, L. J.: The Foundations of Statistics, Dover Publications, John Wiley & Sons, New York, 1954.

Savoy, H., Kalbacher, T., Dietrich, P., and Rubin, Y.: Geological heterogeneity: Goal-oriented simplification of structure and characterization needs, Adv. Water Resour., 109, 1–13, https://doi.org/10.1016/j.advwatres.2017.08.017, 2017.

Shafer, G.: A Mathematical Theory of Evidence, Princeton University Press, Princeton, 1976.

Smalley, J. B., Minsker, B. S., and Goldberg, D. E.: Risk-based in site bioremediation design using a noisy genetic algorithm, Water Resour. Res., 36, 3043–3052, 2000.

Straubhaar, J.: DeeSse User's Guide, The Centre for Hydrogeology and Geothermics (CHYN), University of Neuchâtel, Switzerland, 2015.

Sudicky, E.: On certain stochastic hydrology issues, Stoch. Env. Res. Risk A., 18, 285–285, https://doi.org/10.1007/s00477-004-0196-2, 2004.

Sukop, M. C. and Cunningham, K. J.: Lattice Boltzmann methods applied to large-scale three-dimensional virtual cores constructed from digital optical borehole images of the karst carbonate Biscayne aquifer in southeastern Florida, Water Resour. Res., 50, 8807–8825, https://doi.org/10.1002/2014WR015465, 2014.

Torn, M. S. and Harte, J.: Missing feedbacks, asymmetric uncertainties, and the underestimation of future warming, Geophys. Res. Lett., 33, L10703, https://doi.org/10.1029/2005GL025540, 2006.

US EPA: Risk Assessment, Guidance for Superfund, Volume I: Human Health, Evaluation Manual (Part A), US EPA, Washington, D.C., EPA/540/1-89/002, 1989.

US EPA: OSWER Directive 9610.17: Use of Risk-based decision-making in UST corrective action programs, US EPA, Washington, D.C., 1995.

US News & World Report: US News & World Report 2016, U.S. News & World Report L.P., New York, USA, 2016.

Vrugt, J. A.: Markov chain Monte Carlo simulation using the DREAM software package: Theory, concepts, and MATLAB implementation, Environ. Modell. Softw., 75, 273–316, https://doi.org/10.1016/j.envsoft.2015.08.013, 2016.

Vrugt, J. A. and Ter Braak, C. J. F.: DREAM$_{(D)}$: an adaptive Markov Chain Monte Carlo simulation algorithm to solve discrete, noncontinuous, and combinatorial posterior parameter estimation problems, Hydrol. Earth Syst. Sci., 15, 3701–3713, https://doi.org/10.5194/hess-15-3701-2011, 2011.

Wasserman, L. A. and Kadane, J. B.: Bayes' Theorem for Choquet Capacities, Ann. Stat., 18, 1328–1339, 1990.

Winter, C. L.: Stochastic hydrology: practical alternatives exist, Stoch. Env. Res. Risk A., 18, 271–273, https://doi.org/10.1007/s00477-004-0198-0, 2004.

Woodbury, A. D. and Rubin, Y.: A full-Bayesian approach to parameter inference from tracer travel time moments and investigation of scale effects at the Cape Cod experimental site, Water Resour. Res., 36, 159–171, https://doi.org/10.1029/1999WR900273, 2000.

Zhang, Y. K. and Zhang, D.: Forum: The state of stochastic hydrology, Stoch. Env. Res. Risk A., 18, p. 265, https://doi.org/10.1007/s00477-004-0190-8, 2004.

Zimmerman, D. A., de Marsily, G., Gotway, C. A., Marietta, M. G., Axness, C. L., Beauheim, R. L., Bras, R. L., Carrera, J., Dagan, G., Davies, P. B., Gallegos, D. P., Galli, A., Gómez-Hernández, J., Grindrod, P., Gutjahr, a. L., Kitanidis, P. K., Lavenue, A. M., McLaughlin, D., Neuman, S. P., RamaRao, B. S., Ravenne, C., and Rubin, Y.: A comparison of seven geostatistically based inverse approaches to estimate transmissivities for modeling advective transport by groundwater flow, Water Resour. Res., 34, 1373, https://doi.org/10.1029/98WR00003, 1998.

6

Using hydraulic head, chloride and electrical conductivity data to distinguish between mountain-front and mountain-block recharge to basin aquifers

Etienne Bresciani[1,2], **Roger H. Cranswick**[1,3], **Eddie W. Banks**[1], **Jordi Batlle-Aguilar**[1,4], **Peter G. Cook**[1], **and Okke Batelaan**[1]

[1]National Centre for Groundwater Research and Training, School of the Environment, Flinders University, Adelaide, SA 5001, Australia
[2]Korea Institute of Science and Technology, Seoul, 02792, Republic of Korea
[3]Department of Environment, Water and Natural Resources, Government of South Australia, Adelaide, SA 5000, Australia
[4]Kansas Geological Survey, University of Kansas, Lawrence, KS 66047, USA

Correspondence: Etienne Bresciani (etienne.bresciani@flinders.edu.au)

Abstract. Numerous basin aquifers in arid and semi-arid regions of the world derive a significant portion of their recharge from adjacent mountains. Such recharge can effectively occur through either stream infiltration in the mountain-front zone (mountain-front recharge, MFR) or subsurface flow from the mountain (mountain-block recharge, MBR). While a thorough understanding of recharge mechanisms is critical for conceptualizing and managing groundwater systems, distinguishing between MFR and MBR is difficult. We present an approach that uses hydraulic head, chloride and electrical conductivity (EC) data to distinguish between MFR and MBR. These variables are inexpensive to measure, and may be readily available from hydrogeological databases in many cases. Hydraulic heads can provide information on groundwater flow directions and stream–aquifer interactions, while chloride concentrations and EC values can be used to distinguish between different water sources if these have a distinct signature. Such information can provide evidence for the occurrence or absence of MFR and MBR. This approach is tested through application to the Adelaide Plains basin, South Australia. The recharge mechanisms of this basin have long been debated, in part due to difficulties in understanding the hydraulic role of faults. Both hydraulic head and chloride (equivalently, EC) data consistently suggest that streams are gaining in the adjacent Mount Lofty Ranges and losing when entering the basin. Moreover, the data indicate that not only the Quaternary aquifers but also the deeper Tertiary aquifers are recharged through MFR and not MBR. It is expected that this finding will have a significant impact on the management of water resources in the region. This study demonstrates the relevance of using hydraulic head, chloride and EC data to distinguish between MFR and MBR.

1 Introduction

Numerous basin aquifers in arid and semi-arid regions receive a significant portion of their recharge from adjacent mountains, largely because the latter typically benefit from higher precipitation and lower evapotranspiration (Winograd et al., 1998; Wilson and Guan, 2004; Earman et al., 2006). Two recharge mechanisms can be recognized (Wahi et al., 2008): mountain-front recharge (MFR), which predominantly consists of stream infiltration in the mountain-front zone, and mountain-block recharge (MBR), which consists of subsurface flow from the mountain towards the basin. Here the mountain-front zone is defined after Wilson and Guan (2004) as the upper zone of the basin, between the basin floor and the mountain block (Fig. 1a). The term MFR has traditionally been used to encompass the two recharge mechanisms described above, but it may be more appropriate to

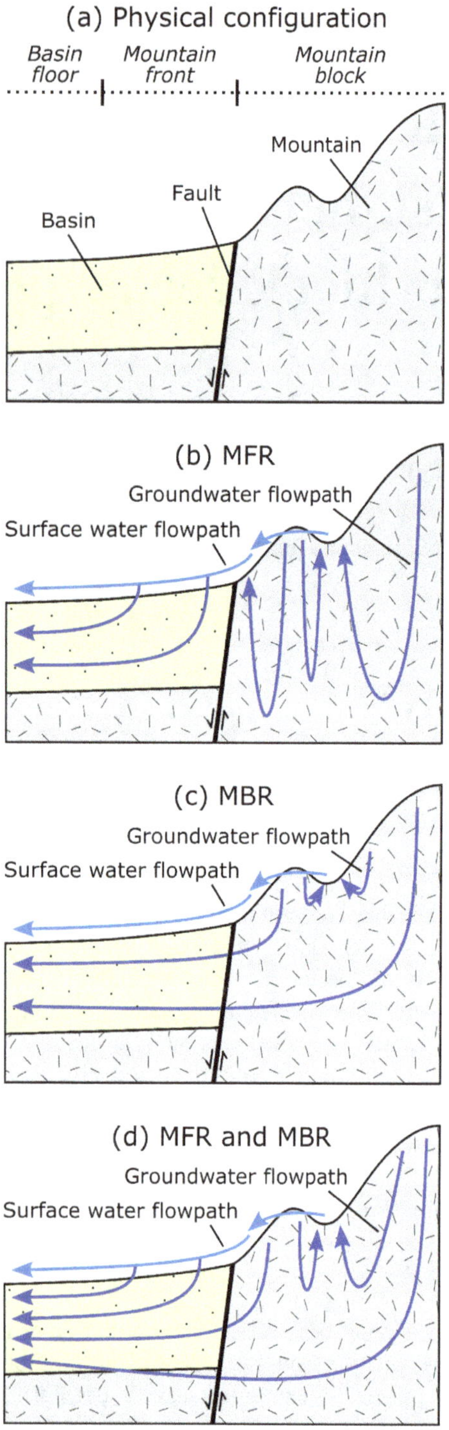

Figure 1. Conceptual models of the transition between mountain and basin: **(a)** physical configuration and **(b–c)** possible conceptualizations of MSR where **(b)** MFR dominates, **(c)** MBR dominates, and **(d)** both MFR and MBR are significant.

use it for the first one only. Following Wahi et al. (2008), the collective process of MFR and MBR is referred to as mountain system recharge (MSR).

The distinction between MFR and MBR is important. The conceptualization of a basin groundwater system critically depends on whether recharge occurs through MFR or MBR, as each of these mechanisms implies different groundwater flow paths, groundwater age and geochemical characteristics. MFR and MBR can also imply different responses to land and water resource management practices (both in the basin and the mountain) as well as to climate change. A good understanding of these mechanisms is thus essential for an effective coordinated management approach of water resources in basins and adjacent mountains (Manning and Solomon, 2003; Wilson and Guan, 2004).

While various methods exist to estimate MSR as a bulk, characterizing the individual contributions of MFR and MBR is difficult. For instance, Darcy's law calculations and inverse groundwater flow modelling typically provide bulk MSR estimates (e.g. Hely et al., 1971; Anderson, 1972; Maurer and Berger, 1997; Siade et al., 2015). It is possible to consider MFR and MBR independently in a groundwater flow model, but the solution to the inverse problem is more likely to be non-unique (e.g. Bresciani et al., 2015b). The water balance and chloride mass balance methods also provide bulk MSR estimates when the analysis is performed at the base of the mountain-front zone or further downstream in the basin (e.g. Maxey and Eakin, 1949; Dettinger, 1989). Environmental tracers such as noble gases (e.g. Manning and Solomon, 2003), stable isotopes (e.g. Liu and Yamanaka, 2012) and radioactive isotopes (e.g. Plummer et al., 2004) can help to determine which of MFR or MBR is the dominant mechanism, but their analysis remains expensive and their interpretation can be difficult. The most robust approach for characterizing MFR and MBR might be the integrated analysis of all available hydraulic, temperature and concentration data through the coupled modelling of groundwater flow, heat and solute transport in the combined basin–mountain system (e.g. Manning and Solomon, 2005) – but it is also arguably the most complex approach.

In this study, we explore alternatives to expensive and complex methods to investigate whether MSR to basin aquifers is dominated by MFR (Fig. 1b) or MBR (Fig. 1c), or if both recharge mechanisms are significant (Fig. 1d). We focus on the use of hydraulic head, chloride (Cl^-) and electrical conductivity (EC), which are inexpensive to measure and may be readily available from hydrogeological databases in many cases. The general utility of hydraulic head and Cl^- data to infer groundwater dynamics is well established (e.g. Domenico and Schwartz, 1997; Herczeg and Edmunds, 2000). Furthermore, EC values can be converted to Cl^- concentrations (as demonstrated later), and hence can be used in a similar manner to Cl^-. However, studies demonstrating the specific use of these data for the characterization of MSR mechanisms appear to be rare (Feth et al., 1966). This may reflect a traditionally low data density along mountain fronts, which are not generally the prime locations for drilling groundwater wells due to an expected lower yield than elsewhere in the basin (aquifer thickness may typically be smaller, and prevailing recharge conditions are not favourable to well yield).

After presenting a general rationale for the use of hydraulic head, Cl^- and EC data to distinguish between MFR and MBR, the Adelaide Plains (AP) basin in South Australia (Fig. 2) is used as a case study to test the relevance of the approach. This semi-arid region features a sedimentary basin bounded by a mountain range – the Mount Lofty Ranges, from which most of the recharge is believed to be derived (Miles, 1952; Shepherd, 1975; Gerges, 1999; Bresciani et al., 2015a). Groundwater in this basin has been used for over a century for industry, water supply and agriculture. Nonetheless, and despite several recent studies, the relative contributions of MFR and MBR is still subject to debate. In particular, the hydraulic role of faults (i.e. acting as barrier or conduit to flow) that run along the mountain front remains unclear (Green et al., 2010; Bresciani et al., 2015a; Batlle-Aguilar et al., 2017). As a result of the common use of groundwater, a relatively high density of wells exists in the region, including in the mountain-front zone. Therefore, this case study provides a good example of the potential of the proposed approach.

2 Rationale

In this section, a generic rationale is presented for the use of hydraulic head and Cl^- (or EC-derived Cl^-) data to distinguish between MFR and MBR to basin aquifers. Hydraulic head and Cl^- data can be used independently, but as they are of different nature, it is expected that their simultaneous use will result in a more complete and reliable characterization of the recharge mechanisms.

2.1 Using hydraulic head

Hydraulic heads directly relate to groundwater dynamics. Consequently, hydraulic head patterns could theoretically enable the identification of groundwater flow paths, both in mountains and basins. Specifically, four types of analysis are suggested that could inform the likely occurrence or absence of MFR and MBR:

- *Assessment of the correlation between hydraulic head and topography.* In the mountain block, a good correlation would suggest that groundwater flow is dominated by local flow systems as opposed to regional flow systems (Tóth, 1963). This would imply that only a small portion of the recharge occurring over the mountain would make its way towards the basin. In fact, in this case MBR would be mostly limited to the recharge occurring over triangular facets in between stream catchments at the base of the mountain block (Fig. 3). Here, the recharge is less likely to be routed towards mountain streams, and instead it may be routed towards the basin (Welch and Allen, 2012). In the mountain-front zone, a good correlation between hydraulic head and topography would suggest that groundwater discharges to streams, so that MFR from stream leakage would be limited or non-existent.
- *Analysis of the shape of hydraulic head contours adjacent to surface water features to identify losing and gaining stream conditions.* It is well known that head contours show a curvature pointing in the downstream direction where the contour lines cross a losing stream (due to the mounding induced by groundwater recharge) (Fig. 4a), whereas they show a curvature pointing in the upstream direction where the contour lines cross a gaining stream (due to the depression induced by groundwater discharge) (Fig. 4b) (Winter et al., 1998). Performing such analysis in the mountain block should indicate whether mountain groundwater appears mostly routed towards local streams, which would make it less likely for MBR to be significant. Additionally, performing such analysis in the mountain-front zone should allow for testing the occurrence or absence of MFR in the form of stream infiltration, which is the predominant form of MFR (Wilson and Guan, 2004).
- *Comparison of stream levels with nearby groundwater levels.* A stream level higher than nearby groundwater levels would indicate a potential for stream infiltration, while the opposite would indicate a potential for groundwater discharge to stream. If the data density is low, this analysis may be preferable over the previous one (list item no. 2) as it does not require head contours to be accurately determined. However, it can only provide information on a *potential* interaction: groundwater discharge or recharge would be significant only if the hydraulic conductivity of the streambed is high enough. In contrast, the previous analysis (no. 2) could give a more definite answer, because the curvature of head contours at some distance from the stream should only be visible if the groundwater–surface water interaction is significant relative to other flow components (i.e. horizontal flow).
- *Evaluation of the vertical head gradient in the mountain-front zone.* Recharge areas are associated with a decrease in hydraulic head with depth, while discharge areas are associated with an increase in hydraulic head with depth (e.g. Wang et al., 2015). Hence, in the mountain-front zone, a head decrease with depth would suggest that MFR occurs (at a rate that depends on the vertical hydraulic conductivity of the aquifer). In contrast, an absence of head decrease (or a head increase) with depth would suggest that MFR does not occur.

In cases where faults run between the basin and the mountain, it may be tempting to study the difference in hydraulic

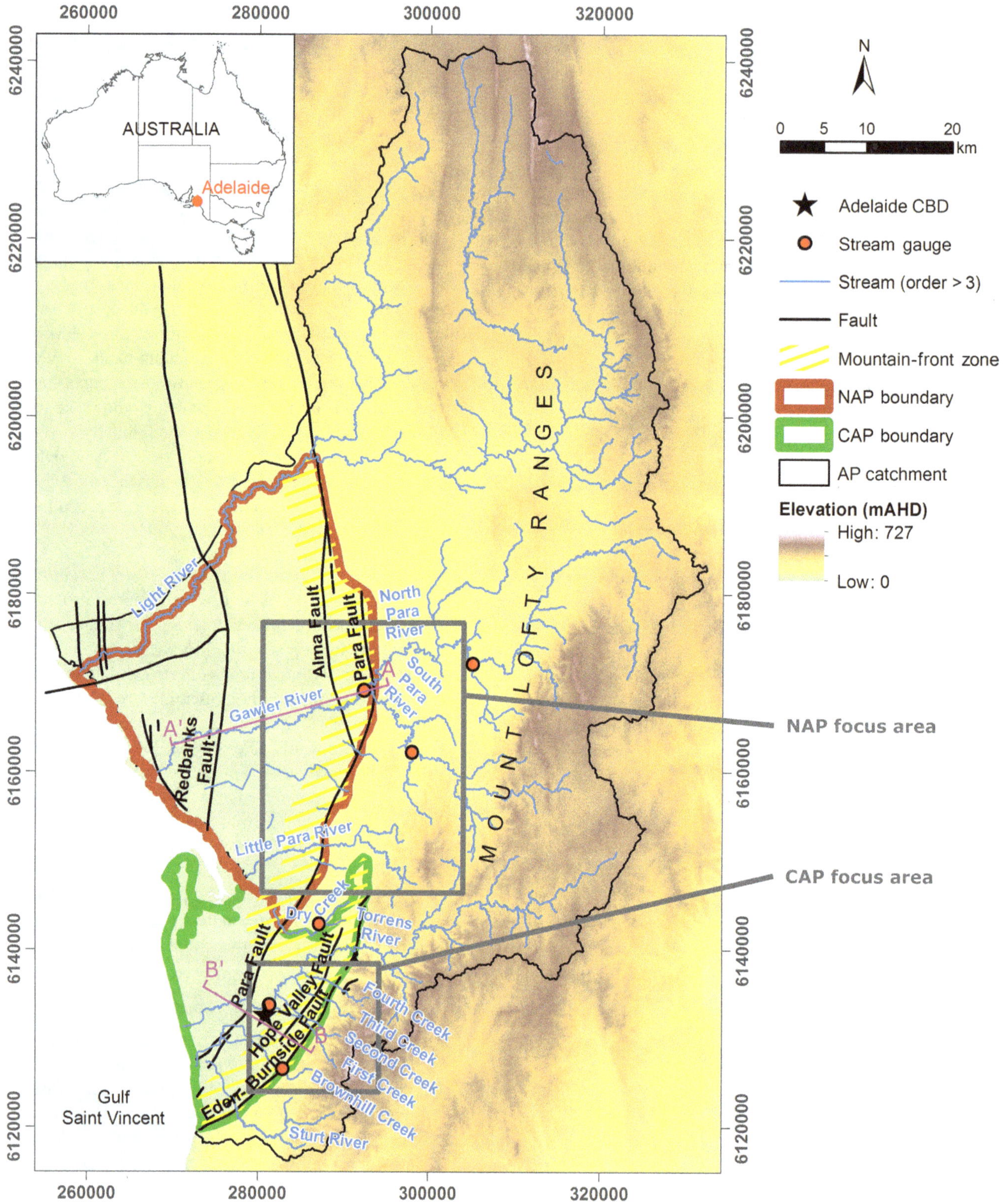

Figure 2. Situation map showing elevation (in metres according to the AHD, i.e. Australian Height Datum) and relevant features of this study.

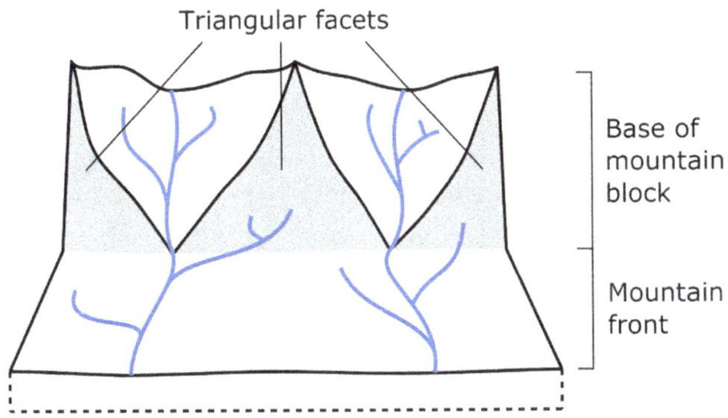

Figure 3. Schematic diagram showing triangular facets at the base of the mountain block (after Welch and Allen (2012)).

head between the two sides of the fault zones. Intuitively, a large head difference would indicate that a fault zone constitutes a barrier to flow in the direction perpendicular to it (e.g. Bense et al., 2013), and consequently that MBR would be low. However, a large head difference across a fault zone may not always imply that the fault zone constitutes a hydraulic barrier. Let us consider the hypothetical case of a sedimentary layer overlying a basement of relatively low hydraulic conductivity and that features a sharp transition in elevation as a consequence of faulting (Fig. 5a). The hydraulic conductivity of the fault zone itself is assumed to be no different from that of the embedding materials (i.e. the fault only implies a difference in basement elevation). In this simple configuration, if the groundwater level right below the fault is lower than the basement elevation above the fault (as a result of downstream hydraulic controls), the groundwater level above the fault is essentially "disconnected" from the lower part (Fig. 5b). This is because in all cases, the groundwater level above the fault has to satisfy a minimum height (i.e. transmissivity has to be large enough) for groundwater to flow there. Hence, this example demonstrates that a large difference in head can exist across the fault zone despite the fault zone itself having no particularly low hydraulic conductivity. It should also be noted that regardless of the cause, the implications of a large difference in head, in terms of the amount of flow eventually crossing the fault zone, is far from obvious, as it depends on the hydraulic conductivity of the fault or the basement (which is in either case difficult to determine). A more relevant analysis may be to investigate whether or not the hydraulic head above of the fault is so high (relative to topography) as to imply that groundwater discharges locally to mountain streams instead of flowing across the fault towards the basin. In other words, what matters is the partitioning of the mountain groundwater between these two pathways. This is precisely what the first three types of analysis presented above should contribute to determining.

2.2 Using chloride

Chloride (Cl^-) is a naturally occurring ion in groundwater that is generally considered as conservative in geochemical studies (Clark and Fritz, 1997). If there is no removal or addition of Cl^- in the aquifer, and if the effects of dispersion (i.e. mixing of water from different flow paths) can be neglected, the Cl^- concentration will be constant along each groundwater flow path (Bresciani et al., 2014). Under such conditions, if the potential MFR and MBR sources have a different Cl^- concentration, Cl^- could be an excellent tracer to distinguish between these two recharge mechanisms.

In many cases, Cl^- in groundwater primarily originates from atmospheric deposition (Allison et al., 1994). The rate of atmospheric deposition depends on a number of factors including distance to the source (oceanic or terrestrial), elevation, terrain aspect, slope, vegetation cover and climatic conditions (Hutton and Leslie, 1958; Guan et al., 2010b; Bresciani et al., 2014). Other potential sources of Cl^- in groundwater include anthropogenic inputs (e.g. salting of roads, irrigation, application of fertilizer, leakage from septic/sewage systems) and dissolution of Cl-bearing minerals. Cl^- removal from solution is unlikely as Cl does not easily adsorb onto clays or precipitate as mineral (Clark and Fritz, 1997). The Cl^- concentration in recharge water also depends on evapotranspiration, which leaves Cl^- in solution, implying its enrichment (Eriksson and Khunakasem, 1969; Allison et al., 1994) (a growing vegetation could in theory counter this effect since Cl is a nutrient for plants (White and Broadley, 2001), but in practice the uptake of Cl from soil by most plant species is insignificant (Allison et al., 1994)). Thus, depending on how variable the above controlling factors are, the Cl^- concentration in mountain groundwater – i.e. the potential MBR source – may show significant spatial and temporal variability. The Cl^- concentration in stream water entering the basin – i.e. the potential MFR source – strongly depends on the streamflow generation mechanisms. If the mountain streams are supported by large proportions of overland flow or interflow, the Cl^- concentration in stream water entering the basin should tend to be lower than that in mountain groundwater, because these streamflow generation mechanisms imply relatively little evapotranspiration. In contrast, if the mountain streams are mostly supported by groundwater discharge, the Cl^- concentration in stream water entering the basin should tend to have an integrated value of the mountain groundwater concentration. In conclusion, while no general statement can be made, chances are that the potential MFR and MBR sources have a distinct Cl^- signature.

In this study, the proposed strategy consists of analysing three types of water for Cl^-: groundwater in the basin, groundwater in the mountain, and stream water in the mountain-front zone. Assuming steady-state concentrations and conservative Cl^-, groundwater in the basin should have the same concentration as stream water in the mountain-front

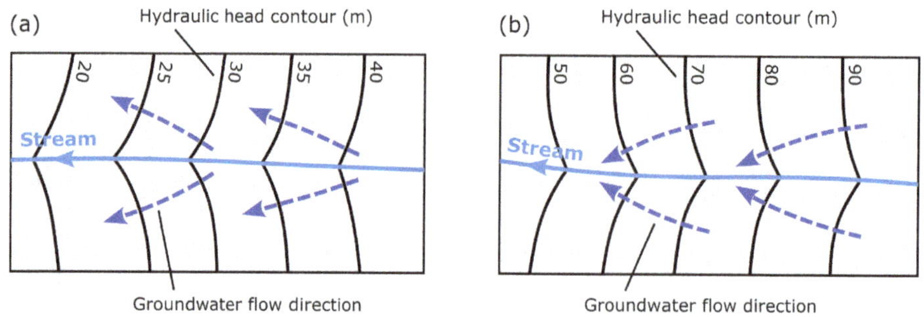

Figure 4. Schematic diagram showing hydraulic head contours and groundwater flow directions in the horizontal plane for **(a)** losing and **(b)** gaining stream conditions (after Winter et al. (1998)).

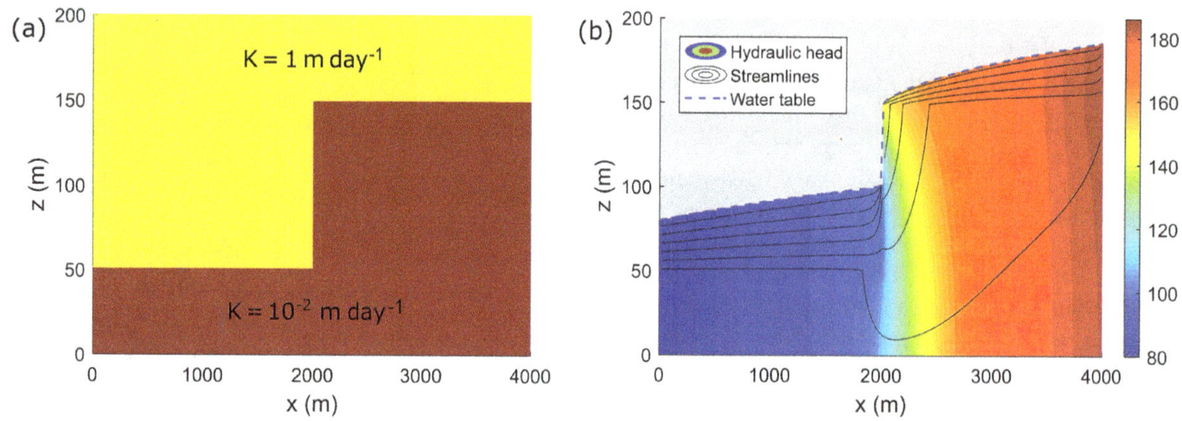

Figure 5. Impact of a difference in basement elevation induced as a result of faulting on hydraulic head in a hypothetical setting. **(a)** Hydraulic conductivity in a vertical cross section representing a sedimentary layer overlying a basement with significantly lower hydraulic conductivity, with the fault zone itself having no different hydraulic conductivity (i.e. the fault only implies the difference in basement elevation). **(b)** Results from an unconfined groundwater flow simulation in which a constant head (80 m) was specified on the left boundary and inflow was specified on the right boundary (at a rate proportional to the hydraulic conductivity). A sharp difference in hydraulic head is observed across the fault zone. The simulation was performed using MODFLOW-NWT (Niswonger et al., 2011), with 200 cells in horizontal direction and 150 cells in the vertical direction.

zone in the case of MFR (further assuming that transpiration from plants after stream infiltration and potential mixing with diffuse recharge are negligible), while it should have the same concentration as groundwater in the mountain in the case of MBR. As much as possible, the analysis should focus on comparing points that are not too far apart from one another and are along presumed flow paths, so as to reduce risks of misinterpretation caused by spatiotemporal variability in Cl^- concentration and dispersion. In particular, MBR should be most reliably assessed by comparing the groundwater Cl^- concentration in the uppermost part of the basin to that in the lowermost part of the mountain, along lines running perpendicular to the mountain front.

Electrical conductivity values are known to be strongly correlated to Cl^- concentrations (Guan et al., 2010a). Therefore, EC data can be converted to Cl^- data if a relationship between the two can be assumed. Typically, EC is more routinely measured than Cl^-, and thus this should significantly increase the dataset. Ideally, an empirical relationship between EC and Cl^- should be developed based on available pair measurements in the study area.

3 Case study

3.1 Study area and background

The Adelaide Plains basin is a coastal sedimentary embayment of 1700 km^2 in South Australia (Fig. 2). The area is bounded by the Mount Lofty Ranges to the east and south, by the Light River to the north, and by the Gulf Saint Vincent to the west. It can be split into two sub-basins: the central Adelaide Plains (CAP) sub-basin south of Dry Creek, and the northern Adelaide Plains (NAP) sub-basin north of Dry Creek. The topographic gradient is more pronounced in the CAP and adjacent mountains (regional slopes of about 0.8 and 7 %, respectively) than in the NAP and adjacent mountains (regional slopes of about 0.3 and 2.5 %, respectively). Torrens River and Gawler River are the largest rivers in the

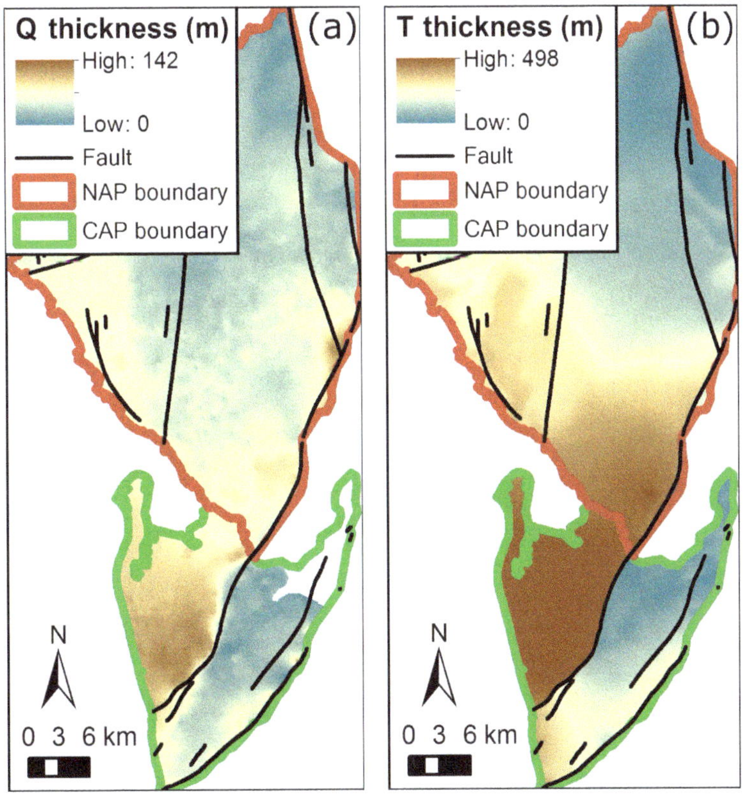

Figure 6. Total thickness of the Quaternary (Q) sediments **(a)** and Tertiary (T) sediments **(b)** in the AP basin. The colour schemes are different in **(a)** and **(b)**.

CAP and in the NAP, respectively. A number of streams run down from the Mount Lofty Ranges, either feeding these rivers or flowing directly into the ocean.

Precipitation is relatively low and potential evapotranspiration is high in this semi-arid area. The average rainfall is 445 mm yr^{-1} (no snowfall) and the annual average maximum daily temperature of 21.6 °C at Adelaide Airport station, located near the coast (station number 23034, 1970–2013; Australian Government, Bureau of Meteorology). Direct recharge from rainfall in the basin is thus expected to be relatively low. Instead, most of the recharge is believed to be derived from the adjacent Mount Lofty Ranges. The latter receive an average rainfall of 983 mm yr^{-1} and negligible snowfall (i.e. only exceptionally and in insignificant quantities) at Mount Lofty Cleland Conservation Park (station number 23810, 1970–2013; Australian Government, Bureau of Meteorology), i.e. more than twice that of the basin. It also experiences cooler temperatures, with an annual average maximum daily temperature of 15.2 °C at Mount Lofty (station number 23842, 1993–2007; Australian Government, Bureau of Meteorology). The majority (77 %) of the rainfall in the Mount Lofty Ranges occurs during autumn and winter (May–September) (station number 23810, 1970–2013), suggesting a strong seasonality of the recharge.

The basin comprises complex, spatially dependent sequences of Quaternary and Tertiary sedimentary deposits (Gerges, 1999). The Quaternary sediments are dominated by fluvio-lacustrine clay interbedded with sand and gravel. The Tertiary sediments are dominated by sand, sandstone, limestone, chert, marl and shell remains interbedded with clay. A number of faults dissect the basin. Among these, the Eden–Burnside Fault and the Para Fault are of primary interest in this study since these faults run along the foothill, almost at the margin of the CAP and the NAP sub-basins, respectively (Fig. 2). The total thickness of the sedimentary units increases sharply on the downthrown side of the major faults (up to 400 m in places). The thickness of the Quaternary sediments ranges from 0 to about 140 m across the basin (Fig. 6a), while that of the Tertiary sediments ranges from 0 to about 500 m (Fig. 6b). The Tertiary sediments are directly outcropping in the northeast part of the CAP. The basement of the basin and the Mount Lofty Ranges are mostly comprised of Proterozoic fractured rocks of various lithologies including slate, phyllite, quartzite, limestone and dolomite. Superficial sedimentary deposits (typically less than 20 m in thickness) also exist locally in the Mount Lofty Ranges.

Up to six semi-confined aquifers (named Q1 to Q6) are recognized within the Quaternary sediments from the central to western side of the basin (Gerges, 1999) (i.e. downstream

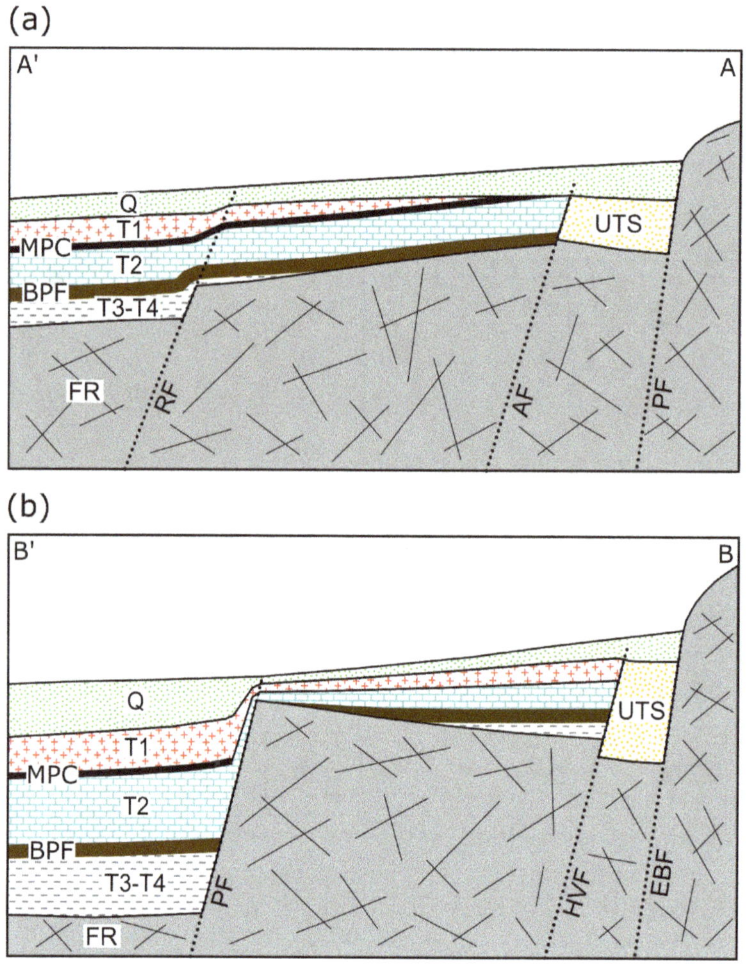

Figure 7. Schematic hydrogeological cross sections in **(a)** the NAP and **(b)** the CAP sub-basins (see Fig. 2) (after Shepherd, 1975; Gerges, 1999; Zulfic et al., 2008; Baird, 2010). Q: Quaternary aquifers (lumped together; in reality there are up to six different aquifers depending on location, separated by clay layers); T1, T2, T3 and T4: Tertiary aquifers; UTS: undifferentiated Tertiary sand aquifers; MPC: Munno Para clay aquitard; BPF: Blanche Point Formation aquitard; FR: fractured-rock aquifers; RF: Redbank Fault; AF: Alma Fault; PF: Para Fault; HVF: Hope Valley Fault; EBF: Eden–Burnside Fault.

of the mountain-front zone). These aquifers contain water of variable salinity with a median value of around 1300 mg L^{-1}. The underlying Tertiary sediments are generally subdivided into four aquifers (named T1 to T4) over a large part of the basin. However, there is no clear hydrogeological distinction between the various Tertiary sediments along most of the mountain front in both sub-basins, and thus in this area they are considered to form a single undifferentiated Tertiary aquifer (Gerges, 1999; Zulfic et al., 2008; Baird, 2010). Simplified cross sections of the aquifers in the NAP and CAP sub-basins are shown in Fig. 7a and b, respectively. Salinity is relatively low in the upper aquifer (T1) with a median value of around 600 mg L^{-1}, slightly higher in the T2 aquifer with median values of around 1000 mg L^{-1}, and significantly higher in deeper aquifers with median values of 8400 and 40 000 mg L^{-1} in T3 and T4, respectively (but note that very few data are available from the latter two aquifers).

Because they present large areas of good salinity and yield, the T1 and T2 aquifers have been used since 1914 for occasional water supply, irrigation and industrial activities and are currently the main targets of groundwater extraction in the AP (Gerges, 1999; Zulfic et al., 2008). Long-term, large cones of depression in both of these aquifers and forecasted increases in groundwater demand raise concerns about the sustainability of extraction in the coming years (Bresciani et al., 2015a). Risks are related to both potential depletion of the resource and rise in salinity from the migration of higher-salinity groundwater, which could make groundwater unusable. To better assess these risks, a thorough understanding of the recharge mechanisms to these aquifers is necessary.

Early investigations suggested that the natural (i.e. predevelopment) recharge to the Tertiary aquifers of the basin was dominated by stream infiltration along the mountain front (i.e. MFR) (Miles, 1952; Shepherd, 1975). In contrast,

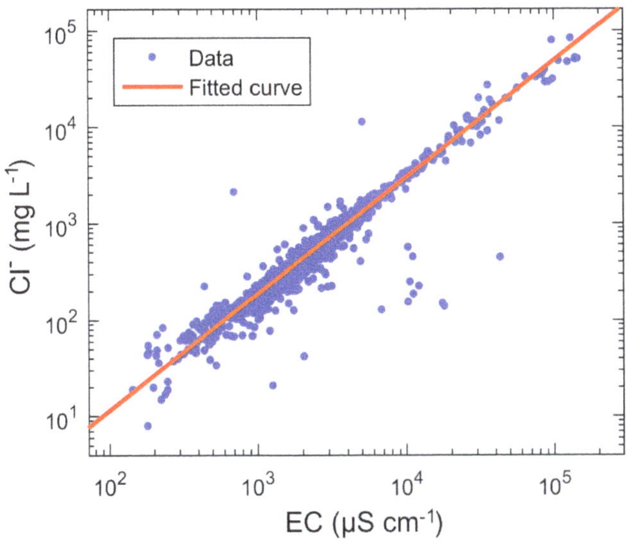

Figure 8. Cl^- versus EC data in the AP catchment. The fitted function used to describe the relationship is $[Cl^-] = 004411 \times [EC]^{1.208}$, where $[Cl^-]$ and [EC] are in units of milligrams per litre ($mg\,L^{-1}$) and microsiemens per centimetre ($\mu S\,cm^{-1}$), respectively (coefficient of determination: $R^2 = 0.9996$).

subsequent investigations suggested that the natural recharge of the Tertiary aquifers was dominated by subsurface flow from the Mount Lofty Ranges (i.e. MBR) (Gerges, 1999, 2006). The latter conceptual model has formed the basis of most investigations of the Tertiary aquifers since its presentation and has underpinned the development of a number of groundwater flow and transport models of the basin aquifers (Jeuken, 2006a, b; Zulfic et al., 2008; Baird, 2010; Georgiou et al., 2011; Bresciani et al., 2015b). However, studies from Green et al. (2010) and Bresciani et al. (2015a) produced results supporting the hypothesis that MFR could also be significant. To further investigate this question, the present study provides a re-appraisal of available hydraulic head, Cl^- and EC data through application of the above rationale.

3.2 Datasets

3.2.1 Hydraulic head dataset

Hydraulic head data in the AP catchment (i.e. the area including both the basin and contributing mountain areas based on surface topography) were retrieved from the WaterConnect database (http://www.waterconnect.sa.gov.au, Government of South Australia) on 4 November 2016. The collection dates span more than a century, the earliest measurements being from 1906 and the latest from 2016. The data were filtered out for unsuitable measurements such as measurements taken during pumping, aquifer testing or drilling. After filtering, 111 538 hydraulic head measurements from 9561 wells were obtained.

The data were subsequently split according to three aquifer groups: the AP Quaternary aquifers, the AP Tertiary aquifers ("AP" in these expressions will be omitted in the remaining text) and the Mount Lofty Ranges aquifers. This grouping is relevant in view of the hydrogeological characteristics of the system and the objective of the study. In particular, we did not distinguish between the T1 and T2 aquifers (i.e. the two main aquifers of the AP basin) because, as mentioned earlier, they are undifferentiated along most of the mountain front. In the Mount Lofty Ranges, the presence of complex fracture networks and high relief can induce the blurring of otherwise depth-dependent hydraulic signals, and so splitting the data according to depth in this area may not be very meaningful and it would reduce data density. The name of the aquifer into which the wells were screened was informed in the database for about two-thirds of the wells (6209). This allowed for assignment of these wells to one of the above aquifer groups. For the remaining one-third of wells, the aquifer group for the wells located in the basin was determined by comparing the well mid-screen elevation to the bottom elevation of the Quaternary sediments and to the top elevation of the basement. The largest number of wells was from the Quaternary aquifers (3964), followed by the Mount Lofty Ranges aquifers (3589) and the Tertiary aquifers (1768). Wells screened into the basement of the basin were disregarded (240 wells).

Groundwater-level fluctuations can be an issue for data interpretation. In particular, as this study focuses on natural recharge mechanisms, the impact of pumping constitutes a potentially important bias. It should be noted that the density of hydraulic head data is higher in areas of lower groundwater salinity, which coincides with areas that have experienced greater changes due to pumping. The measurements made before the main development period (i.e. before 1950) may have been less affected by pumping than more recent measurements, but limiting the analysis only to these measurements would dramatically reduce the data density. In addition, even the earliest measurements may not be free of pumping influence, since it is likely that these were precisely taken to monitor the impact of pumping. Hence, instead of subjectively fixing an arbitrary date beyond which the data would be excluded, all data were retained regardless of the measurement date. For each of the wells that had multiple measurements, the temporal mean hydraulic head was calculated in an effort to smooth out the measurement errors and temporal fluctuations. The analysis focuses on these mean values.

3.2.2 Chloride dataset

Groundwater Cl^- data in the AP catchment were also retrieved from the WaterConnect database on 4 November 2016. This dataset was extended using the more commonly available EC data from the database. A strong relationship between EC and Cl^- was found from 1,559 pair measure-

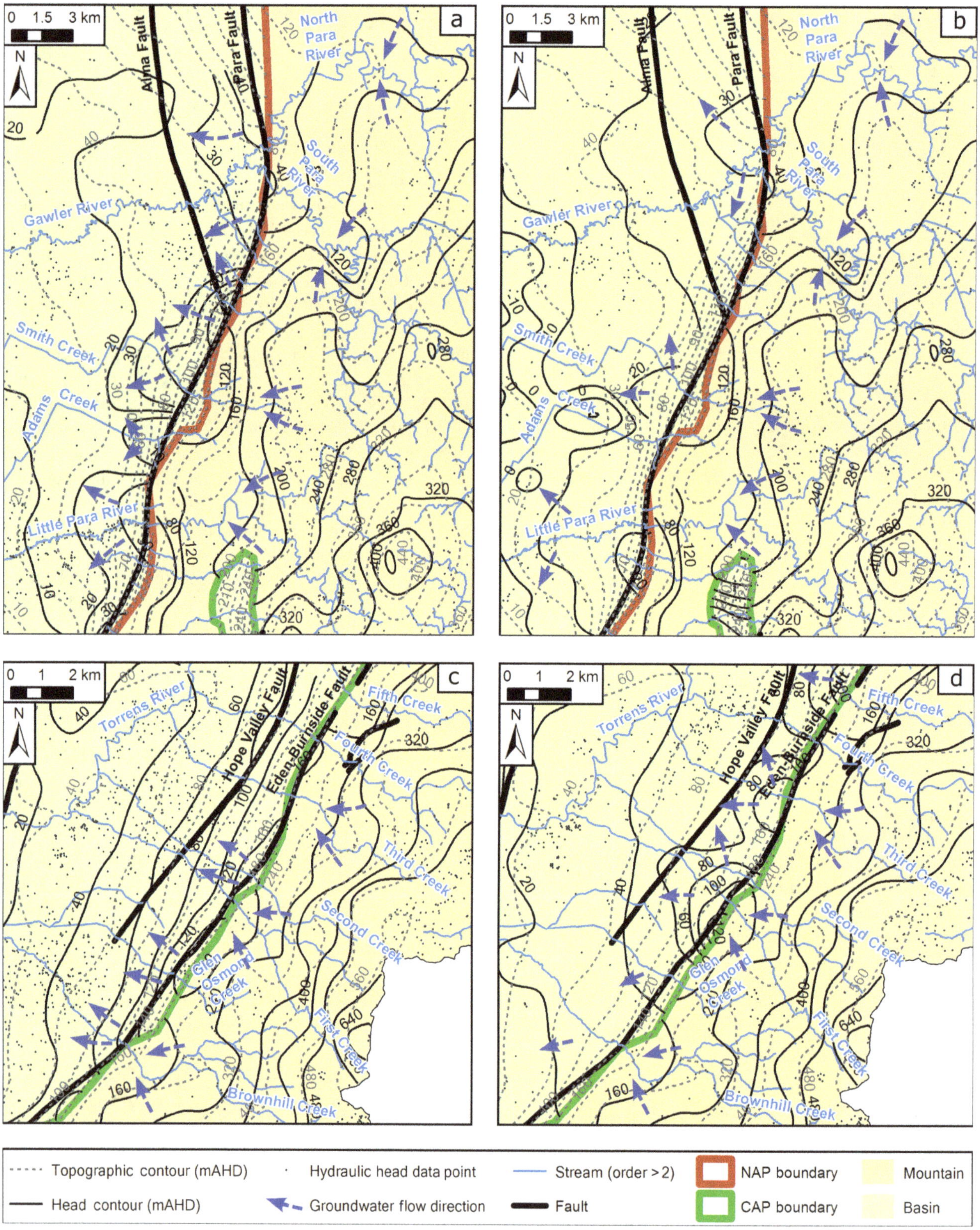

Figure 9. Hydraulic head and topographic contours in the NAP **(a–b)** and CAP **(c–d)** focus areas. On the basin side, (a) and (c) show the head in the Quaternary aquifers, while **(b)** and **(d)** show the head in the Tertiary aquifers. On the mountain side, all four sub-figures show the head in the Mount Lofty Ranges aquifers. In **(a–b)**, the contour interval is 10 m in the basin and 40 m in the mountain (both for head and topography), while in **(c–d)**, it is 20 m in the basin and 80 m in the mountain. Selected groundwater flow directions highlight apparent gaining stream conditions in the lower part of the mountain and loosing stream conditions in the mountain-front zone.

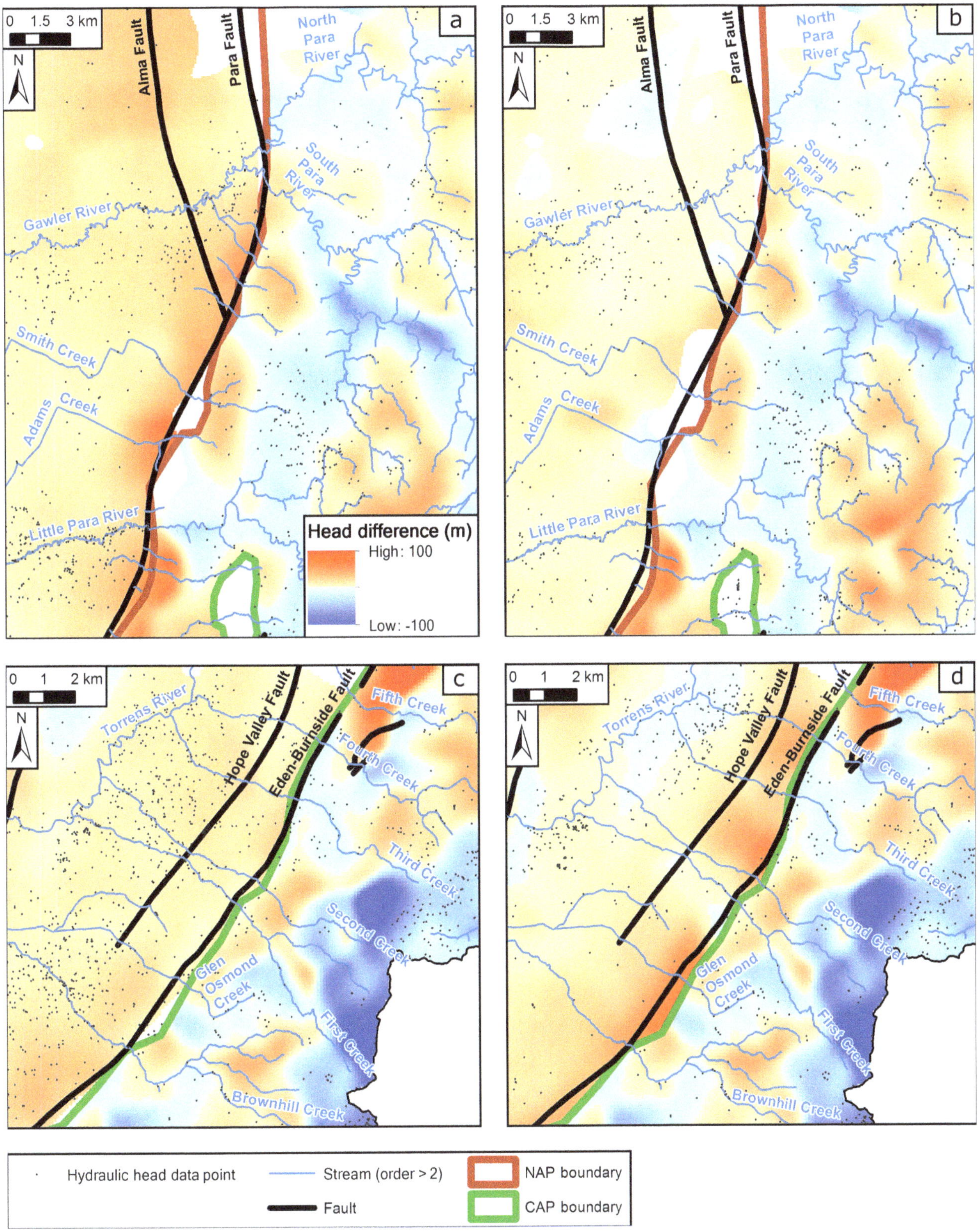

Figure 10. Head difference in the NAP **(a–b)** and CAP **(c–d)** focus areas. On the basin side, **(a)** and **(c)** show the nearest river head (approximated by topography) minus the head in the Quaternary aquifers, while **(b)** and **(d)** show the head in the Quaternary aquifers minus the head in the Tertiary aquifers. On the mountain side, all four sub-figures show the nearest river head minus the head in the Mount Lofty Ranges aquifers. The same colour scheme is applied everywhere and is indicated in **(a)**. Reddish colours are for positive values, indicating a potential for downward flow (i.e. groundwater recharge). Blueish colours are for negative values, indicating a potential for upward flow (i.e. groundwater discharge).

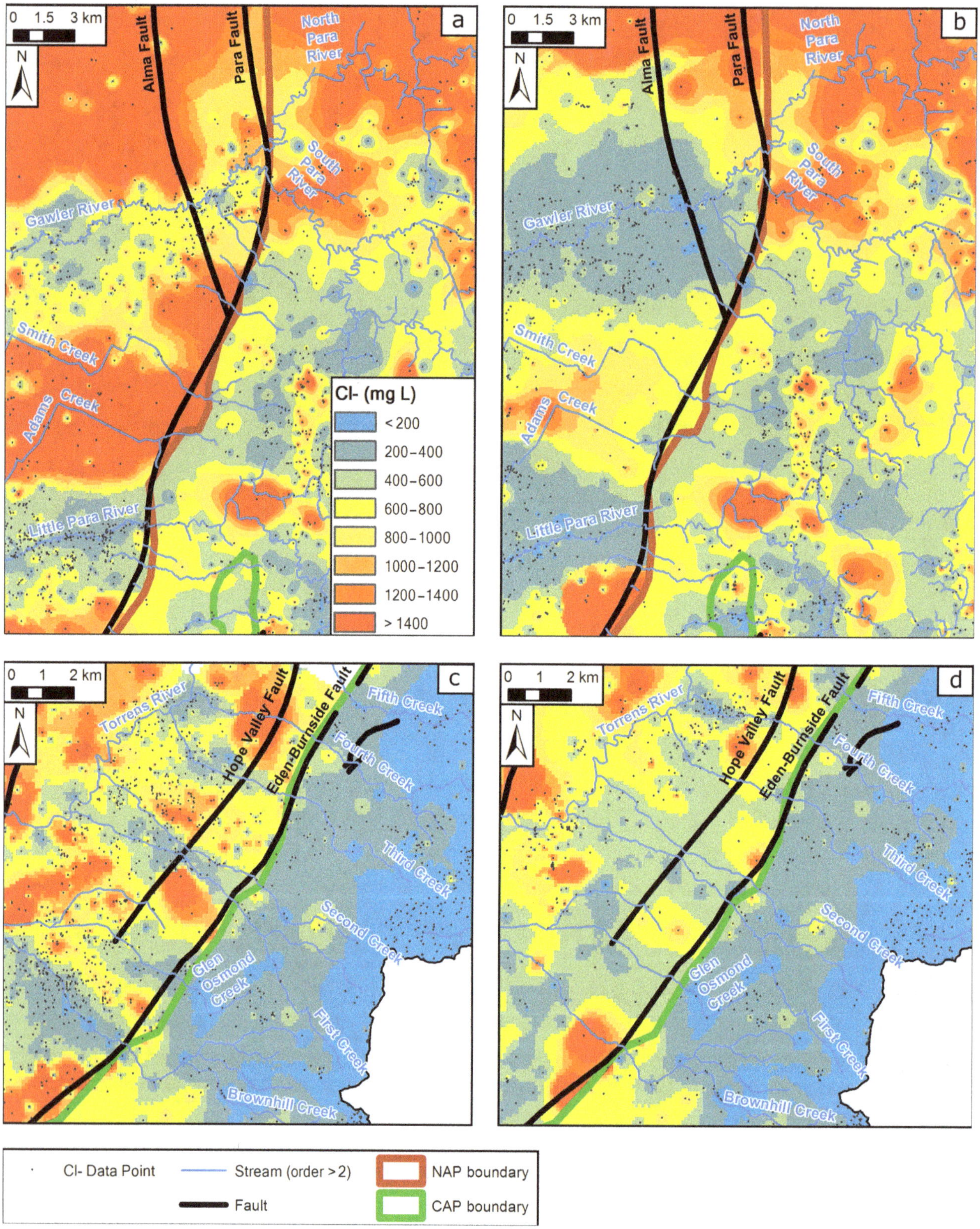

Figure 11. Cl^- concentration in the NAP **(a–b)** and CAP **(c–d)** focus areas. On the basin side, **(a)** and **(c)** show the Cl^- concentration in the Quaternary aquifers, while **(b)** and **(d)** show the Cl^- concentration in the Tertiary aquifers. On the mountain side, all four sub-figures show the Cl^- concentration in the Mount Lofty Ranges aquifers. The same colour scheme is applied everywhere and is indicated in **(a)**. The latter is chosen to be favourable to the study of relatively low-salinity zones; i.e. all Cl^- concentrations larger than 1400 mg L^{-1} are included in the same class (red colour), but in reality much higher values exist.

ments ($R^2 = 0.9996$, Fig. 8). In wells where only EC data were available, EC values were converted into Cl^- concentrations using this relationship. In total, 34 145 Cl^- or EC-converted Cl^- data (simply referred to as Cl^- data) from 12 660 wells were obtained (i.e. slightly more than for hydraulic heads, partly due to a less-restrictive filtering, i.e. keeping measurements taken during pumping or aquifer testing). The collection dates span the same period as for the hydraulic head data.

The Cl^- data were subsequently split according to the same three aquifer groups as indicated above for the hydraulic head data. The same procedure was also applied to determine the aquifer group to which the wells belong. The largest number of wells was from the Quaternary aquifers (4963), followed by the Mount Lofty Ranges aquifers (4395) and the Tertiary aquifers (2963).

Pumping may also have impacted Cl^- concentrations by inducing migration of the original groundwater, whose concentration is spatially variable (although this effect may be seen later than on hydraulic heads since solute travel times are typically longer than pressure travel times). Also, as for hydraulic head data, the density of Cl^- data is higher in areas that have experienced pumping. Additionally, in the uppermost aquifers, irrigation may have locally influenced the Cl^- concentration. Nonetheless, following the same reasoning as for hydraulic heads, all available Cl^- data were retained regardless of the measurement date. For each of the wells that had multiple measurements, the temporal mean Cl^- concentration was calculated. The analysis focuses on these values.

Flow rate and EC data from streams running down from the Mount Lofty Ranges into the AP basin were also retrieved from the WaterConnect database. Six gauging stations were located close enough to the mountain front to be relevant to the current study. Details on this dataset are given in Table 1. The reported EC values of surface water were converted into Cl^- concentrations using the same relationship as that developed for groundwater. This approach is deemed appropriate given the common origin of these waters, even though potentially different chemical reactions might slightly affect the relationship.

3.3 Data analysis

Given the relatively large area investigated, the analysis presented below concentrates on two "focus areas" that cover the transition between the Mount Lofty Ranges and the AP basin: one at the margin of the NAP sub-basin and one at the margin of the CAP sub-basin (locations indicated in Fig. 2). Figures for the entire study area are also available in the Supplement (Figs. S1–S12). These do not call for a different interpretation.

3.3.1 Hydraulic heads

Hydraulic head maps were constructed for the three aquifer groups (Quaternary aquifers, Tertiary aquifers and Mount Lofty Ranges aquifers) (Figs. 9, 10). In constructing these maps, the choice of the interpolation method and associated parameters revealed to be critical. The classical inverse distance weighting method would produce the well-known "bull's eye" effect around individual data points. This could severely compromise the interpretation of head contours. Instead, the diffusion kernel interpolation method from the Geostatistical Analyst extension of ArcGIS 10.4.1 was used. This method allows for a more realistic interpolation when the underlying phenomenon governing the data is diffusive, as is the case for hydraulic heads. The most important parameter in this method is the bandwidth, which is used to specify the maximum distance within which data points are used for prediction. Making this parameter too small would undermine the prediction capability as many areas would remain uncovered by the interpolation, while making it too large would produce overly smoothed results. A good compromise was found by setting this parameter to 1200 m for the NAP focus area and to 800 m for the CAP focus area – reflecting a higher density of streams and data in the latter case. Topographic contours were also constructed. To facilitate the comparison with head contours, these were created after application of a circular moving-average window to the topography using a radius that matches the bandwidth used in the interpolation method for hydraulic head (i.e. 1200 m in the NAP focus area and 800 m in the CAP focus area).

Figure 9a displays hydraulic head and topographic contours in the NAP focus area, showing the Quaternary aquifers on the basin side and the Mount Lofty Ranges aquifers on the mountain side. The results are quite contrasted between the mountain and the basin. In the mountain, head contours follow the topographic contours relatively closely, and their shape is most often indicative of gaining stream conditions. Note that one should not expect to see sharp "V" shapes where head contours cross streams (i.e. as in Fig. 4) due to the limited data density. Instead, head contours are smoothly curved. In the basin, head contours do not closely follow the topographic contours, and their shape is generally indicative of losing stream conditions (especially close to the basin margin). Figure 9b also displays hydraulic head and topographic contours in the NAP focus area, but showing the Tertiary aquifers on the basin side instead of the Quaternary aquifers (on the mountain side, this figure is identical to Fig. 9a). Head contours in the Tertiary aquifers are generally indicative of focused recharge along streams, but on a somewhat larger scale, i.e. showing wider curvatures than in the Quaternary aquifers (mostly around Gawler River and Little Para River). Figure 9c and d display analogous results for the CAP focus area. Similar to above, in the mountain, head contours are relatively well correlated with topographic contours and their shape is generally indicative of gaining conditions.

Table 1. Surface water flow rate, EC values and their conversion into Cl^- concentrations for the six gauging stations located near the mountain-front zone (gauging station number in parenthesis; see Fig. 2 for site locations). Note that the North Para River and South Para River join about 1 km downstream of the Para Fault to form the Gawler River.

	North Para River (A5050502)	South Para River (A5051009)	Gawler River (A5050505)	Dry Creek (A5041051)	First Creek (A5040578)	Brownhill Creek (A5040901)
Mean EC ($\mu S\,cm^{-1}$)	3424	1414	3121	1579	1276	808
Flow-weighted mean EC ($\mu S\,cm^{-1}$)	1095	656	758	411	611	545
10th percentile Cl^- ($mg\,L^{-1}$)	214	99	183	37	76	87
90th percentile Cl^- ($mg\,L^{-1}$)	852	304	1615	184	259	191
Median Cl^- ($mg\,L^{-1}$)	1441	457	600	990	432	144
Mean Cl^- ($mg\,L^{-1}$)	846	288	769	353	257	145
Mean flow rate ($ML\,d^{-1}$)	28.05	13.63	82.32	17.20	6.24	5.26
Flow-weighted mean Cl^- ($mg\,L^{-1}$)	221	115	146	67	107	91
No. of EC record days	6064	1221	131	944	937	29
Records period	1994–2016	2003–2010	1970–1995	2013–2016	2013–2016	2012–2016

In the basin, head contours in the Quaternary aquifers are not very distinct from topographic contours, but nevertheless tend to indicate losing rather than gaining stream conditions close to the basin margin (Fig. 9c). In the Tertiary aquifers, head contours are quite distinct from topographic contours and are quite clearly indicative of focused recharge along a majority of streams (Fig. 9d). Near Glen Osmond Creek and Brownhill Creek, groundwater flow predominantly appears oriented towards the southwest, which may result from the bedrock sloping in this direction (the Tertiary sediment thickness can be seen to increase in Fig. 6b).

Figure 10a–d display the difference, in every point, between river head (approximated by the topographic elevation of the nearest river) and groundwater head. Figure 10a shows the NAP focus area, with the Quaternary aquifers on the basin side and the Mount Lofty Ranges aquifers on the mountain side. In the mountain, the results generally reveal a potential for gaining stream conditions along large portions of the main rivers (i.e. North Para River, South Para River and Little Para River). Potential losing stream conditions are indicated around the upper reaches of streams, suggesting that these are not supported by groundwater discharge, but are rather initiated by overland flow or interflow. This observation is consistent with the fact that most of the stream headwaters in the Mount Lofty Ranges are ephemeral. Potential losing stream conditions are also observed locally around a few streams in the lowest part of the Mount Lofty Ranges (e.g. South Para River and Smith Creek). This observation is not in line with the interpretation of head contours made from Fig. 9a, and this inconsistency might be an artefact of the temporal averaging of hydraulic heads, i.e. the hydraulic heads might be on average lower than the river head but the stream might still be gaining due to important groundwater discharge in some periods (but this explanation remains a hypothesis). In the basin, the Quaternary aquifers are revealed as potentially receiving water from streams everywhere, and especially close to the basin margin where the head difference is the largest. Figure 10b shows the head difference between the Quaternary aquifers and the Tertiary aquifers on the basin side, such as to investigate the vertical connection between these aquifers (on the mountain side, this figure is identical to Fig. 10a). The hydraulic head appears larger in the Quaternary aquifers than in the Tertiary aquifers over most of the area. This indicates a potential for downward groundwater leakage from the Quaternary aquifers to the Tertiary aquifers. The rate at which this leakage occurs is nonetheless difficult to estimate, since it is also a function of the effective vertical hydraulic conductivity and vertical distance between these units, which are largely unknown. Similar observations and interpretations can be made of the CAP focus area (Fig. 10c, d).

Most observations from Figs. 9 and 10 suggest that groundwater flow is dominated by local flow systems in the Mount Lofty Ranges. This indicates that only a small proportion of the recharge occurring over the mountain may make its way towards the basin. Hence, if MBR occurs, it would be probably limited to the routing of the recharge occurring over triangular facets in between stream catchments at the base of the mountain (see Sect. 2.1). Moreover, the results generally suggest that MFR through stream leakage is an important recharge mechanism for both the Quaternary aquifers and the Tertiary aquifers of the AP basin.

3.3.2 Chloride concentrations

Cl^- concentration maps were also constructed for the three aquifer groups (Quaternary aquifers, Tertiary aquifers and Mount Lofty Ranges aquifers) (Fig. 11). Here the inverse distance weighting interpolation method from the Geostatistical Analyst extension of ArcGIS 10.4.1 was used. This method is appropriate for Cl^- because its analysis does not especially make use of the shape of concentration contours, and hence the bull's eye effect is not really an issue. Furthermore, Cl^- cannot be assumed to result from a diffusive process since advection is expected to dominate on the scale of this study, and so the diffusive kernel method would be inappropriate. The inverse distance weighting interpolation method also has the advantage of being exact at the data points. The power parameter was set to 2, and a standard neighbourhood was used with 15 maximum neighbours and 10 minimum neighbours. The same parameters were used for both NAP and CAP focus areas.

Figure 11a shows the Cl^- concentrations in the Quaternary aquifers and the Mount Lofty Ranges aquifers in the NAP focus area. This figure reveals a strong correlation between low Cl^- concentration zones and the location of the main rivers in the basin (Gawler River and Little Para River). It seems unlikely that such a correlation would be observed if MBR was the main recharge mechanism. By contrast, no obvious correlation can be found in the mountain. Here, the Cl^- concentration mostly appears correlated with elevation, with lower values occurring at higher elevations. This trend is expected, since the rate of evapotranspiration – which largely controls Cl^- concentration – is expected to decrease with elevation as a result of higher rainfall and lower temperature. In line with these observations, there is a clear discontinuity in Cl^- concentration at the transition between the mountain and the basin almost everywhere along the mountain front. This suggests that little or no hydraulic connection occurs between the mountain and the basin through the subsurface (i.e. MBR is probably insignificant). Similar observations hold in Fig. 11b, where the Tertiary aquifers are shown in the basin instead the Quaternary aquifers. The zones of low Cl^- concentration around Gawler River and Little Para River are wider in these aquifers than in the Quaternary aquifers, which is consistent with the above observation that the head contours display a wider curvature around these rivers. Figure 11c and d show analogous results for the CAP focus area. The Cl^- concentration is generally lower than in the NAP focus area, especially in the mountain, most likely a result of lower evapotranspiration associated with the higher elevation of this area. A strong correlation between zones of low Cl^- concentration and stream locations can also be seen in the basin. These zones appear wider and somewhat less distinct in the Tertiary aquifers than in the Quaternary aquifers, but it should be noted that the data density is lower in these aquifers. In both cases, as in the NAP, a sharp change in Cl^- concentration can be seen at the transition between the mountain and the basin, therefore suggesting that MBR is insignificant.

The Cl^- concentration in streams running down from the Mount Lofty Ranges into the AP basin was analysed to investigate whether stream leakage can explain the groundwater Cl^- concentrations measured in the basin. A summary of available flow rate, electrical conductivity and derived Cl^- concentration data for six monitoring stations located near the transition between the mountain and the basin is presented in Table 1. The location of the stream gauges is indicated in Fig. 2. The relationship between streamflow rate and Cl^- concentration is shown from a scatter plot in

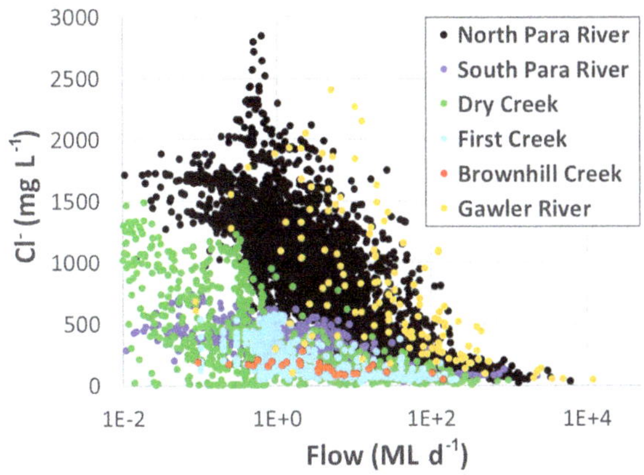

Figure 12. Scatter plot of chloride concentration versus streamflow rate for six streams running down from the Mount Lofty Ranges into the AP basin.

Fig. 12. The stream Cl^- concentration displays significant variations, with a decreasing trend as flow increases. The relationship between flow rate and Cl^- concentration nevertheless varies between different streams. This probably reflects different catchment characteristics that are likely to influence the streamflow generation mechanisms (i.e. topography, climate, geology, land use). The variations of streamflow rate and Cl^- concentration show a strong seasonality, as illustrated through selected time series for Gawler River and Brownhill Creek in Fig. 13a and b, respectively. These time series show that the periods of low Cl^- concentration and high flow coincide with the wet season (May–September). During this season, lower Cl^- concentrations in stream water can be explained by a relatively large contribution of overland flow or interflow to streamflow generation, as these processes should experience little evapotranspiration relative to subsurface flow contribution. The significance of overland flow or interflow is also supported by observations mentioned above. During high flow periods, the infiltration potential in the mountain-front zone should be enhanced due to higher stream water levels and wider wetted areas. We therefore propose to estimate a representative value of the MFR Cl^- concentration by calculating the flow-weighted average Cl^- concentration in stream water. A more rigorous approach would require knowledge of the timing and rate of stream leakage, which are not available. The flow-weighted average Cl^- concentration is 221, 115, 146, 67, 107 and 91 $mg\,L^{-1}$ in the North Para River, South Para River, Gawler River, Dry Creek, First Creek and Brownhill Creek, respectively (Table 1). These data show that streams are a plausible source for the low Cl^- concentrations observed in the basin aquifers (Fig. 11).

4 Discussion

4.1 Strengths and limitations of using hydraulic head and chloride data

One of the main strengths of hydraulic head data is that they can indicate the contemporary flow direction (if hydraulic conductivity can be considered to be isotropic). In this study, the analysis of head contours suggested that groundwater in the Mount Lofty Ranges discharges to local streams for a large part. It also allowed for the identification of losing stream conditions in the mountain-front zone, where leakage from streams appears to recharge not only the Quaternary aquifers but also the deeper Tertiary aquifers. Studying the head variation with depth in the basin gave further evidence that groundwater flows from the Quaternary aquifers to the underlying Tertiary aquifers, even though the rate of this flow is unknown.

The main limitation of hydraulic head data is probably that these data are quite sensitive to pumping. This is problematic when the objective is to study the natural (i.e. pre-development) recharge mechanisms. Pumping in the AP basin mostly affects groundwater levels in the western part of basin, where large cones of depression exist in the Tertiary aquifers due to extensive historical and ongoing pumping (Bresciani et al., 2015a). Therefore, for the purpose of this study, which focuses on the eastern part of the basin (where the mountain-front zone is located), this issue may not be as critical. However, smaller-scale pumping wells surely also exist in the mountain-front zone and in the mountain, and may affect the results to an unknown degree. This represents a non-negligible source of uncertainty.

Cl^- is potentially an effective tool to distinguish between MFR and MBR. But for this, the stream water Cl^- concentration in the mountain-front zone (i.e. a potential MFR source) needs to be significantly different from the groundwater Cl^- concentration at the base of the mountain (i.e. potential MBR source). Different processes involved in the generation of MFR and MBR imply that these two potential sources of water may indeed have different Cl^- concentrations (see Sect. 2.2). This is certainly the case in the present case study, where the Cl^- concentration at the base of the mountain is seen to be significantly different from that of the basin, thereby suggesting that MBR is insignificant. In contrast, the Cl^- concentration in streams appears to be similar to that of the low Cl^- concentration zones of the basin, which are aligned with surface water features. These observations leave little doubt regarding the recharge mechanisms in the AP basin, i.e. MFR appears to be the dominant recharge mechanism.

As for hydraulic heads, pumping can potentially distort the Cl^- concentrations from those of the natural system. In addition, irrigation may influence the Cl^- concentration in shallow wells located in agricultural areas. However, changes in solute concentrations are expected to be observed later than

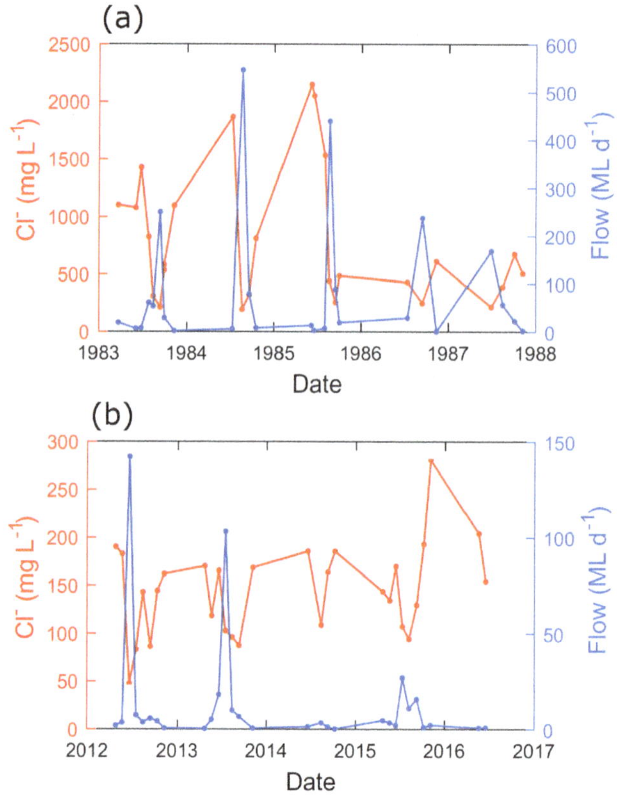

Figure 13. Stream water chloride concentration and flow rate over selected periods for **(a)** Gawler River and **(b)** Brownhill Creek. The axes in **(a)** and **(b)** have a different scale.

hydraulic head changes, and a dramatic shift in Cl^- concentrations is unlikely to be seen in Cl^- concentrations away from the main areas of pumping. Furthermore, if recharge from streams in the basin was only a result of recent pumping, groundwater should have a very modern (i.e. post-development) recharge signature. Groundwater dating shows that this is not the case (Batlle-Aguilar et al., 2017). It can also be noted that the correlation between the zones of low groundwater salinity and stream locations was already observed in the 1950s, i.e. using measurements anterior to the main development period (Miles, 1952).

The interpretation of Cl^- data also relies on the assumptions of constant Cl^- inputs. This assumption is widely accepted in the literature for hydrogeological studies and in applications of the chloride mass balance method to estimate recharge (Wood, 1999; Scanlon et al., 2006; Crosbie et al., 2010; Healy and Scanlon, 2010). This assumption may not be strictly satisfied over the AP basin because groundwater in the Tertiary aquifers can be quite old, as revealed by numerous samples showing paleo-meteoric origin (> 12 000 years) according to carbon-14 dating and noble gas measurements (Batlle-Aguilar et al., 2017). This indicates that different climatic conditions may have prevailed at the time of recharge, implying possible variations in Cl^- inputs. However, such

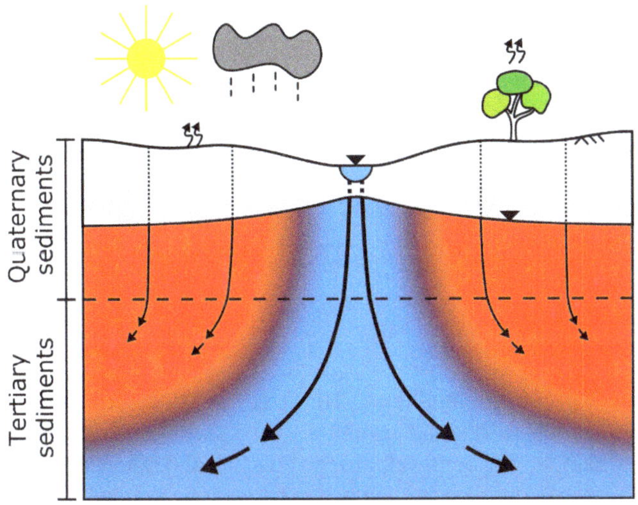

Figure 14. Conceptual model of the recharge mechanisms for the aquifers of the AP basin, as seen in a cross section perpendicular to a stream in the mountain-front zone, and how these mechanisms can explain the observed groundwater salinity. Blue to red colours indicate relatively low to high salinity.

old groundwater is mostly observed in the western part of the basin. Groundwater in the mountain-front zone is younger (in most cases < 10000 years according to carbon-14 dating), making it less likely for these to reflect drastically different paleo-climatic conditions. Furthermore, even if the Cl^- inputs did vary over time, such temporal variation would not in itself explain the correlation of groundwater Cl^- concentration with stream locations in the basin.

Finally, the assumption of conservative Cl^- is deemed reasonable in view of the geology of the study area, which is not known to bear evaporite deposits such as halite. This assumption was also tested by analysing the chloride / bromide (Cl^- / Br^-) ratio from 161 well samples distributed over the study area. The average Cl^- / Br^- ratio from these samples is 739 +/− 173 (molar) (Fig. S13). This shows that groundwater has a similar Cl^- / Br^- ratio to seawater (~ 650 molar), hence indicating that there is little water-rock interactions or dissolution of evaporite deposits (Drever, 1997; Davis et al., 1998). The low variance of the Cl^- / Br^- ratio also suggests that the dominant process for the increasing salinity in the groundwater is evapotranspiration (Cartwright et al., 2006). Furthermore, even if Cl^- was not strictly conservative, focused recharge from streams in the basin would again appear necessary to explain the fact that the zones of low groundwater Cl^- concentration are found along streams.

Compared to other methods that use noble gas, radioactive or isotopic tracers, the above approach appears simpler, more cost effective and more reliable due to the much higher data density generally achievable (i.e. given current technologies and budget constraints). Nevertheless, in contrast with some other methods (e.g. noble gases), it should be noted that this approach is only qualitative. I.e. it does not allow for the quantification of the relative proportion of MFR and MBR and their absolute rate. One way to extend the ideas presented in this study to gain more quantitative insight would be to use the data as calibration targets in a groundwater flow and Cl^- transport model. This is also the subject of ongoing efforts (Bresciani et al., 2015b).

4.2 MFR versus MBR in the AP basin

In an early study, Miles (1952) noted that the pre-development groundwater levels along the mountain front of the AP basin were reflective of unconfined conditions, and that the subsurface materials in this zone were favourable to stream infiltration. In addition, Miles (1952) already analysed the groundwater salinity distribution. He observed that salinity contours were forming fan-shaped zones of low-salinity "mushrooming" outwards from streams, with such patterns being visible up to more than 100 m below the ground surface. He concluded that stream infiltration in the mountain-front zone was a major recharge mechanism for the basin aquifers. Later, in a study of the NAP aquifers, Shepherd (1975) arrived at the same conclusion, partly using similar arguments and further noting that (i) groundwater hydrographs in the Quaternary aquifers were each year showing a rapid rise in water level shortly after Gawler River and Little Para River started to flow, and (ii) the vertical head gradient and vertical hydraulic conductivity were indicative of significant downward flow from the Quaternary to the Tertiary aquifers. Additionally, a number of studies directly measured groundwater gains and losses using differential flow-gauging along streams entering the AP basin (Hutton, 1977; Green et al., 2010; Cranswick and Cook, 2015). All found that several streams were losing a significant amount of water in the mountain-front zone. Finally, Zulfic et al. (2010) found that bore yield based on air-lift testing conducted at the time of drilling (e.g. Williams et al., 2004) in the Mount Lofty Ranges did not increase beyond 100 m depth for most geology types. This finding can be interpreted as showing that hydraulic conductivity is relatively low beyond that depth. This would promote local groundwater flow systems in the mountain instead of deep flow towards the basin, in line with the current analysis.

In contrast, Gerges (1999) and Batlle-Aguilar et al. (2017) proposed that MBR is the dominant recharge mechanism for the Tertiary aquifers of the AP basin. A major argument in these studies was based on the observation that salinity is generally higher in the Quaternary aquifers than in the T1 and T2 aquifers. From this observation, the authors suggested that the relatively fresh water found in the T1 and T2 aquifers could not be the result of downward leakage. Instead, they proposed that this water should come from the Mount Lofty Ranges (where the salinity is lower) through subsurface flow. However, along streams, the Cl^- concentration in the Quaternary aquifers is in fact very similar to that of the underlying Tertiary aquifers (Fig. 11). Furthermore, the possibility that this relatively fresh water originates from the higher elevation areas of the Mount Lofty Ranges through deep groundwater flow paths is unlikely since (i) the hydraulic head data suggest a predominance of local flow systems in the Mount Lofty Ranges (Sect. 3.2.1), (ii) there is an important mismatch between groundwater Cl^- concentrations at the base of the mountain block and those in the upper part of the basin (Sect. 3.2.2), and (iii) groundwater in the deep layers of the basin (i.e. T3 and T4 aquifers) generally shows higher salinity (Gerges, 1999), implying that deep groundwater flow paths cannot explain the observed fresh water. Another argument in Batlle-Aguilar et al. (2017) was based on the observation that relatively old groundwater was measured near the top of Tertiary aquifers (from carbon-14 dating). From this observation, the authors suggested that groundwater could not be recharged in the basin, but rather further away, in the Mount Lofty Ranges. However, most of the old groundwater samples analysed were from wells found quite some distance away from the mountain-front zone, where the Tertiary aquifers become confined – logically implying an increase in age with distance from the recharge zone. Furthermore, relatively young groundwater was found at significant depth near major faults of the basin, precisely suggesting the occurrence of focused recharge (Batlle-Aguilar et al., 2017). Finally, neither Gerges (1999) nor Batlle-Aguilar et al. (2017) proposed a mechanism to explain how the groundwater could be more saline outside of the low-salinity corridors under MBR-prevailing conditions, as was consistently observed in both the Quaternary aquifers and the Tertiary aquifers (Fig. 11). These zones of higher salinity directly contradict the MBR hypothesis, including MBR that would be derived from the recharge over triangular facets at the base of the mountain. It seems more likely that the groundwater of higher salinity in the basin originates from diffuse recharge, which would naturally imply a much higher salinity as a result of evapotranspiration than focused recharge from streams.

Hence, on the basis of robust consistent evidence given in this work (including consistent findings from hydraulic head and Cl^- analyses) and through a critical review of earlier investigations, we propose that MFR is the most plausible and predominant recharge mechanism for the relatively fresh water found in the AP basin (i.e. as in Fig. 1b). A conceptual model depicting the suggested recharge mechanisms, and how these can explain the observed Cl^- (or salinity) patterns – at least in the eastern part of the basin – is shown in Fig. 14.

5 Conclusions

We presented and demonstrated through a regional-scale example the use of hydraulic head, Cl^- and EC data to distinguish between MFR and MBR to basin aquifers. Hydraulic heads can provide information on groundwater flow directions and stream–aquifer interactions, while chloride concentrations and EC values can be used to distinguish between different water sources if these have a distinct signature. This information can provide evidence for the occurrence or absence of MFR and MBR.

In the above case study, both hydraulic head and Cl^- (equivalently, EC) analyses gave informative and consistent results (i.e. both suggested a predominance of MFR), which gives confidence in the interpretation. The Cl^- analysis was particularly straightforward and authoritative, and it further allowed for the identification of diffuse recharge in the basin.

Difficulties in the interpretation of hydraulic head and Cl^- data may arise for particular conditions such as when pumping effects are significant. However, the study of the AP basin demonstrates that even for a basin that has been subject to long-term groundwater extraction (for about a century in this case), the data can allow for the identification of natural recharge mechanisms. Cl^- signature in particular is expected to be less affected (at least, less rapidly affected) by pumping than hydraulic head.

The relevance of the presented approach lies in that the variables used (hydraulic heads, Cl^- concentrations and EC values) have been routinely measured for decades in many parts of the world; if not, their measurement is inexpensive (provided wells already exist). While these data cannot tell everything (e.g. they cannot directly provide information on the recharge rates), we expect that in many cases a significant dataset would be readily available, bearing an as-yet-unexploited potential to inform the recharge mechanisms. Such information is critical for conceptualizing and managing groundwater systems.

The AP basin serves as an example of a region where the recharge mechanisms have long been debated in the context of groundwater resources management. Application of the proposed rationale was revealed to be effective in resolving this debate. It is expected that the findings of this study will have important consequences for future hydrogeological investigations and the management of water resources in the Adelaide region.

Competing interests. The authors declare that they have no conflict of interest.

Acknowledgements. This study was supported by the Goyder Institute for Water Research through the project I.1.6 "Assessment of Adelaide Plains Groundwater Resources" (2013–2015), and by the Korea Research Fellowship Program through the National Research Foundation of Korea (NRF) funded by the Ministry of Science, ICT and Future Planning (project number 2016H1D3A1908042). The authors thank two anonymous reviewers for their constructive comments.

Edited by: Graham Fogg

References

Allison, G. B., Gee, G. W., and Tyler, S. W.: Vadose-zone techniques for estimating groundwater recharge in arid and semiarid regions, Soil Sci. Soc. Am. J., 58, 6–14, 1994.

Anderson, T. W.: Electrical-analog analysis of the hydrologic system, Tucson basin, southeastern Arizona, Geological Survey Water-Supply Paper 1939C, 1972.

Baird, D. J.: Groundwater recharge and flow mechanisms in a perturbed, buried aquifer system: Northern Adelaide Plains, South Australia, PhD thesis, Flinders University, Adelaide, South Australia, 345 pp., 2010.

Batlle-Aguilar, J., Banks, E. W., Batelaan, O., Kipfer, R., Brennwald, M. S., and Cook, P. G.: Groundwater residence time and aquifer recharge in multilayered, semi-confined and faulted aquifer systems using environmental tracers, J. Hydrol., 546, 150–165, 2017.

Bense, V. F., Gleeson, T., Loveless, S. E., Bour, O., and Scibek, J.: Fault zone hydrogeology, Earth-Sci. Rev., 127, 171–192, 2013.

Bresciani, E., Ordens, C. M., Werner, A. D., Batelaan, O., Guan, H., and Post, V. E. A.: Spatial variability of chloride deposition in a vegetated coastal area: Implications for groundwater recharge estimation, J. Hydrol., 519, 1177–1191, 2014.

Bresciani, E., Batelaan, O., Banks, E. W., Barnett, S. R., Batlle-Aguilar, J., Cook, P. G., Costar, A., Cranswick, R. H., Doherty, J., Green, G., Kozuskanich, J., Partington, D., Pool, M., Post, V. E. A., Simmons, C. T., Smerdon, B. D., Smith, S. D., Turnadge, C., Villeneuve, S., Werner, A. D., White, N., and Xie, Y.: Assessment of Adelaide Plains Groundwater Resources: Summary Report, Goyder Institute for Water Research, Adelaide, South Australia, Technical Report Series 15/31, 2015a.

Bresciani, E., Partington, D., Xie, Y., Post, V. E. A., and Batelaan, O.: Appendix K. Regional groundwater modelling. In: Assessment of Adelaide Plains Groundwater Resources: Appendices Part II – Regional Groundwater Modelling, Goyder Institute for Water Research, Adelaide, South Australia, Technical Report Series 15/33, 2015b.

Cartwright, I., Weaver, T. R., and Fifield, L. K.: Cl/Br ratios and environmental isotopes as indicators of recharge variability and groundwater flow: An example from the southeast Murray Basin, Australia, Chemical Geology, 231, 38-56, 2006.

Clark, I. D. and Fritz, P.: Environmental Isotopes in Hydrogeology, CRC Press, 352 pp., 1997.

Cranswick, R. H. and Cook, P. G.: Appendix D. Groundwater – surface water exchange. In: Assessment of Adelaide Plains Groundwater Resources: Appendices Part I – Field and Desktop Investigations, Goyder Institute for Water Research, Adelaide, South Australia, Technical Report Series 15/32, 2015.

Crosbie, R. S., Jolly, I. D., Leaney, F. W., and Petheram, C.: Can the dataset of field based recharge estimates in Australia be used to predict recharge in data-poor areas?, Hydrol. Earth Syst. Sci., 14, 2023–2038, https://doi.org/10.5194/hess-14-2023-2010, 2010.

Davis, S. N., Whittemore, D. O., and Fabryka-Martin, J.: Uses of chloride/bromide ratios in studies of potable water, Ground Water, 36, 338–350, 1998.

Dettinger, M. D.: Reconnaissance estimates of natural recharge to desert basins in Nevada, USA, by using chloride-balance calculations, J. Hydrol., 106, 55–78, 1989.

Domenico, P. A. and Schwartz, F. W.: Physical and chemical hydrogeology, 2nd ed., Wiley, Chichester, 528 pp., 1997.

Drever, J. I.: The geochemistry of natural waters: Surface and groundwater environments, Prentice Hall, New Jersey, USA, 436 pp., 1997.

Earman, S., Campbell, A. R., Phillips, F. M., and Newman, B. D.: Isotopic exchange between snow and atmospheric water vapor: Estimation of the snowmelt component of groundwater recharge in the southwestern United States, J. Geophys. Res.-Atmos., 111, https://doi.org/10.1029/2005JD006470, 2006.

Eriksson, E. and Khunakasem, V.: Chloride concentration in groundwater, recharge rate and rate of deposition of chloride in the Israel Coastal Plain, J. Hydrol., 7, 178–197, 1969.

Feth, J. H. F., Barker, D. A., Moore, L. G., Brown, R. J., and Veirs, C. E.: Lake Bonneville: Geology and hydrology of the Weber Delta district, including Ogden, Utah, Professional Paper 518, 1966.

Georgiou, J., Stadter, M., and Purczel, C.: Adelaide Plains groundwater flow and solute transport model (AP2011), RPS Aquaterra, Report A141B/R001f, 2011.

Gerges, N. Z.: The Geology and Hydrogeology of the Adelaide Metropolitan Area Volume 1 & Volume 2, Flinders University, Adelaide, Australia, 1999.

Gerges, N. Z.: Overview of the hydrogeology of the Adelaide metropolitan area, Department of Water, Land and Biodiversity Conservation, South Australia, DWLBC Report 2006/10, 2006.

Green, G., Watt, E., Alcoe, D., Costar, A., and Mortimer, L.: Groundwater flow across regional scale faults, Government of South Australia, through Department for Water, Adelaide, DFW Technical Report 2010/15, 2010.

Guan, H., Love, A. J., Simmons, C. T., Hutson, J., and Ding, Z.: Catchment conceptualisation for examining applicability of chloride mass balance method in an area with historical forest clearance, Hydrol. Earth Syst. Sci., 14, 1233–1245, https://doi.org/10.5194/hess-14-1233-2010, 2010a.

Guan, H., Love, A. J., Simmons, C. T., Makhnin, O., and Kayaalp, A. S.: Factors influencing chloride deposition in a coastal hilly area and application to chloride deposition mapping, Hydrol. Earth Syst. Sci., 14, 801–813, https://doi.org/10.5194/hess-14-801-2010, 2010b.

Healy, R. W. and Scanlon, B. R.: Estimating Groundwater Recharge, Cambridge University Press, 2010.

Hely, A. G., Mower, R. W., and Harr, C. A.: Water resources of Salt Lake County, Utah, Utah Department of Natural Resources, Technical Publication 31, 1971.

Herczeg, A. L. and Edmunds, W. M.: Inorganic ions as tracers, in: Environmental tracers in subsurface hydrology, edited by: Cook,

P. G., and Herczeg, A. L., Springer US, Boston, MA, 31–77, 2000.

Hutton, J. and Leslie, T.: Accession of non-nitrogenous ions dissolved in rainwater to soils in Victoria, Austr. J. Agric. Res., 9, 492–507, 1958.

Hutton, R. G.: Northern Adelaide Plains surface water assessment 1971–1975, Department of Mines, Adelaide, South Australia, Report 77/11, 1977.

Jeuken, B.: Tertiary aquifers of the Adelaide Coastal Plains groundwater model: Steady-state model set-up and calibration report, Resources and Environmental Management Pty Ltd, Adelaide, South Australia, Report AE-08-04-2, 2006a.

Jeuken, B.: Tertiary aquifers of the Adelaide Coastal Plains groundwater model: Transient model set-up and calibration report, Resources and Environmental Management Pty Ltd, Adelaide, South Australia, Report AE-08-04-3, 2006b.

Liu, Y., and Yamanaka, T.: Tracing groundwater recharge sources in a mountain–plain transitional area using stable isotopes and hydrochemistry, J. Hydrol., 464–465, https://doi.org/10.1016/j.jhydrol.2012.06.053, 2012.

Manning, A. H. and Solomon, D. K.: Using noble gases to investigate mountain-front recharge, J. Hydrol., 275, 194–207, 2003.

Manning, A. H., and Solomon, D. K.: An integrated environmental tracer approach to characterizing groundwater circulation in a mountain block, Water Resour. Res., 41, W12412, https://doi.org/10.1029/2005WR004178, 2005.

Maurer, D. K. and Berger, D. L.: Subsurface flow and water yield from watersheds tributary to Eagle Valley hydrographic area, west-central Nevada, Water-Resources Investigations Report 97-4191, 1997.

Maxey, G. B. and Eakin, T. E.: Ground water in White River Valley, White Pine, Nye, and Lincoln counties, Nevada, Nevada State Engineer, Carson City (Nev.), Water Resources Bulletin, 8, 61 p., 1949.

Miles, K. R.: Geology and underground water resources of the Adelaide Plains area, Department of Mines, Geological Survey of South Australia, Adelaide, Bulletin, 27, 257 p.,1952.

Niswonger, R. G., Panday, S., and Ibaraki, M.: MODFLOW-NWT, a Newton formulation for MODFLOW-2005, U.S. Geological Survey Techniques and Methods 6-A37, 2011.

Plummer, L. N., Bexfield, L. M., Anderholm, S. K., Sanford, W. E., and Busenberg, E.: Hydrochemical tracers in the middle Rio Grande Basin, USA: 1. Conceptualization of groundwater flow, Hydrogeol. J., 12, 359–388, 2004.

Scanlon, B. R., Keese, K. E., Flint, A. L., Flint, L. E., Gaye, C. B., Edmunds, W. M., and Simmers, I.: Global synthesis of groundwater recharge in semiarid and arid regions, Hydrol. Proc., 20, 3335–3370, 2006.

Shepherd, R. G.: Northern Adelaide Plains groundwater study, Stage II, 1968–74, Department of Mines, South Australia, Report Book 75/38, 1975.

Siade, A., Nishikawa, T., and Martin, P.: Natural recharge estimation and uncertainty analysis of an adjudicated groundwater basin using a regional-scale flow and subsidence model (Antelope Valley, California, USA), Hydrogeol. J., 23, 1267–1291, 2015.

Tóth, J.: A theoretical analysis of groundwater flow in small drainage basins, J. Geophys. Res., 68, 4795–4812, 1963.

Wahi, A. K., Hogan, J. F., Ekwurzel, B., Baillie, M. N., and Eastoe, C. J.: Geochemical Quantification of Semiarid Mountain Recharge, Ground Water, 46, 414–425, 2008.

Wang, J.-Z., Jiang, X.-W., Wan, L., Wörman, A., Wang, H., Wang, X.-S., and Li, H.: An analytical study on artesian flow conditions in unconfined-aquifer drainage basins, Water Resour. Res., 51, 8658–8667, 2015.

Welch, L. A. and Allen, D. M.: Consistency of groundwater flow patterns in mountainous topography: Implications for valley bottom water replenishment and for defining groundwater flow boundaries, Water Resour. Res., 48, W05526, https://doi.org/10.1029/2011WR010901, 2012.

White, P. J. and Broadley, M. R.: Chloride in soils and its uptake and movement within the plant: A review, Ann. Bot., 88, 967–988, 2001.

Williams, L. J., Albertson, P. N., Tucker, D. D., and Painter, J. A.: Methods and hydrogeologic data from test drilling and geophysical logging surveys in the Lawrenceville, Georgia, area, Open-File Report, 2004–1366, 2004.

Wilson, J. L. and Guan, H.: Mountain-block hydrology and mountain-front recharge, in: Groundwater Recharge in a Desert Environment: The Southwestern United States, edited by: Hogan, J. F., Phillips, F. M., and Scanlon, B. R., American Geophysical Union, Washington, D.C., 113–137, 2004.

Winograd, I. J., Riggs, A. C., and Coplen, T. B.: The relative contributions of summer and cool-season precipitation to groundwater recharge, Spring Mountains, Nevada, USA, Hydrogeol. J., 6, 77–93, 1998.

Winter, T. C., Harvey, J. W., Franke, O. L., and Alley, W. M.: Ground water and surface water - A single resource, U.S. Geological Survey, Denver, Colorado, U.S. Geological Survey Circular 1139, 1998.

Wood, W. W.: Use and misuse of the chloride-mass balance method in estimating ground water recharge, Ground Water, 37, 2–3, 1999.

Zulfic, D., Osei-Bonsu, K., and Barnett, S. R.: Adelaide Metropolitan Area Groundwater Modelling Project, Volume 1 – Review of Hydrogeology, and Volume 2 – Numerical model development and prediction run, Department of Water, Land and Biodiversity Conservation, South Australia, DWLBC Report 2008/05, 2008.

Zulfic, D., Wilson, T., Costar, A., and Mortimer, L.: Predicting catchment scale processes, Mount Lofty Ranges, South Australia, Government of South Australia, through Department for Water, Adelaide, DFW Technical Report 2010/17, 2010.

7

An assessment of trends and potential future changes in groundwater-baseflow drought based on catchment response times

Jost Hellwig and Kerstin Stahl

Faculty of Environment and Natural Resources, University of Freiburg, 79098 Freiburg, Germany

Correspondence: Jost Hellwig (jost.hellwig@hydrology.uni-freiburg.de)

Abstract. Drought is an important natural hazard with large impacts on society. Changes in drought characteristics have been studied for different parts of the hydrological cycle, but insights into changes of groundwater resources are obscured due to the lack of long-term observations and large heterogeneity of hydrogeological conditions. Moreover, predicted future changes in precipitation are uncertain and have a lagged effect on streamflow and groundwater. We investigated past changes and potential future changes in catchment baseflow as a reflection of groundwater drought for 338 headwater catchments across Germany based on catchments' characteristic response times. First, baseflow dynamics as a proxy of groundwater storage and outflow on a catchment scale were derived from streamflow records and related to precipitation input. Second, past trends in baseflow minima were calculated and attributed to climate and catchment controls. Last, response times and the timing of yearly baseflow minima were combined into estimates of the sensitivity to future precipitation changes. Baseflow response times of the studied headwaters are heterogenous across Germany, ranging from a few months to several years, and depend significantly on hydrogeological conditions. Few significant trends were found in past baseflow minima, and trends are highly dependent on the period of analysis. Based on the assumption of a typical regional scenario of increasing winter precipitation and decreasing summer precipitation, increases in hydrological drought hazard or no changes are projected for most parts of Germany. Catchments with longer response times can buffer interannual precipitation shifts, whereas catchments with fractured rocks are sensitive to summer precipitation decreases. These results urge for a surface water and groundwater management based on local groundwater response to precipitation and help to assess impacts of climate change on overall water supply.

1 Introduction

Drought is a natural phenomenon occurring in all compartments of the hydrological cycle. Accordingly, it is classified into meteorological drought, hydrological drought, agricultural drought, socio-economic drought and groundwater drought (Mishra and Singh, 2010). Due to the large number of people affected by drought and the high economic loss related to drought events (EC, 2007), it is important to enhance the understanding of drought processes considering projected changes in drought hazard, most importantly due to climate change. Empirical analysis of monitored hydrological time series remains an important tool with which to validate theory-based or model-derived hypotheses on these changes, since projected future changes are often uncertain. Particularly for the groundwater compartment there is a high diversity in response to climate input (Eltahir and Yeh, 1999; Green et al., 2011), making predictions even more difficult. Most of the empirical studies on long-term trends in drought hazard have focused on meteorological drought (e.g. Sheffield et al., 2012; Spinoni et al., 2017) and on hydrological drought (e.g. Stahl et al., 2010; Laaha et al., 2017). For groundwater drought, empirical trend analysis is difficult for two reasons: (i) groundwater time series are usually short or influenced by abstractions, and (ii) where long and natural time series are available, they only give point information.

Some countries and states now display groundwater level anomalies at observation wells as part of their drought monitoring (e.g. Switzerland:

https://www.hydrodaten.admin.ch/de, last access: 28 November 2018; Bavaria: https://www.nid.bayern.de/grundwasser, last access: 28 November 2018). Studies on groundwater drought have mainly used borehole time series for index generation and statistical analysis (e.g. Bloomfield and Marchant, 2013; Bloomfield et al., 2015; Kumar et al., 2016; Van Loon et al., 2017; Heudorfer and Stahl, 2017; Leelaruban et al., 2017). The groundwater response to drought and its changes over time are typically very site specific, and observations are therefore hardly scalable from point to catchment scale (Kumar et al., 2016). However, larger regional-scale estimates of groundwater resources are relevant for effective groundwater management and adaptation to climate change. Model-based approaches for groundwater drought analysis and estimation (e.g. Peters et al., 2003, 2006; Tallaksen et al., 2009; Li and Rodell, 2015; Apurv et al., 2017) are one common way to get information on the catchment scale. There are also a few studies that analyse spring data (Fiorillo and Guadagno, 2012) or baseflow time series (Brutsaert, 2008, 2010; Fendeková and Fendek, 2012), two flux variables that provide a more integrative measure of the groundwater situation during drought.

To overcome the difficulties related to borehole data, in this study we use a baseflow approach to characterize groundwater drought on a catchment scale. We analyse a large dataset of long baseflow time series in central Europe. In this region groundwater is often used as drinking water, and aquifers act as an important buffer to climatic variability. Most droughts start with a deficit in precipitation, especially when precipitation falls as rain (Van Loon and Van Lanen, 2012). For the propagation from a meteorological to a groundwater drought different processes are relevant, i.e. attenuation, delay and pooling (e.g. Peters et al., 2003; Tallaksen et al., 2009; Heudorfer and Stahl, 2017). Therefore, the drought signal in groundwater storage not only depends on current meteorological conditions, but also the previous months are important. A catchment-specific timescale for this dependence may be called the catchment's response time. Response times have been analysed by correlations between groundwater depth and time series of precipitation accumulated for different periods. Studies have found that the response times for borehole water tables (Bloomfield et al., 2015; Van Loon et al., 2017; Leelaruban et al., 2017) or spring discharge (Fiorillo and Guadagno, 2012) vary strongly. Moreover, some studies suggest time lags for the highest correlations between precipitation and groundwater time series because of delayed groundwater response (Bloomfield et al., 2015; Fiorillo and Guadagno, 2012). However, when looking at monthly scales, this lag was always found to be quite small and often non-existent (e.g. Haas and Birk, 2017).

There are two approaches to identifying drought periods in a time series. The climatological approach is based on anomalies and often used also in hydrology to track the propagation of relative seasonal water deficits through the hydrological cycle (e.g. Barker et al., 2016; Kumar et al., 2016). The traditional hydrological approach is the "threshold level approach", which defines streamflow droughts as events below a certain fixed limit and is therefore focused on actual low water availability (e.g. Yevjevich, 1967; Peters et al., 2006; Tallaksen et al., 2009). In this work we use the term drought according to the threshold level approach; thus we consider drought events as periods of low baseflow in absolute terms. If there is a distinct seasonal regime, droughts mostly occur in the dry season.

Recent work on central European low flows, i.e. periods when streamflow mostly consists of baseflow, found that climate change is expected to alter low flows (Marx et al., 2018; Forzieri et al., 2014; Van Vliet et al., 2015; Gosling et al., 2017). However, the sign and magnitude of change in central Europe are subject to model choice and emission scenario (Marx et al., 2018; Forzieri et al., 2014). Those modelling studies were focused on large river basins, and the change they predicted reflects strongly that of the precipitation change. Marx et al. (2018) found a high correlation between changes in annual precipitation sums and low flows. Stahl et al. (2012) found that hindcasting summer low flow trends with large-scale models suggests a too-homogenous spatial pattern of change compared to trends found in observations. Together with difficulties of models to capture the persistence of drought events in runoff generation found by Tallaksen and Stahl (2014), it can be assumed that some large-scale models do not necessarily resolve the heterogeneity of catchment storage and release for the hydrological response on a headwater catchment scale. However, recent drought events have demonstrated that especially headwaters are prone to groundwater-related drought impacts like shortages in water supply (Van Lanen et al., 2016). This coincides with findings that, independent from elevation, groundwater is an important catchment storage (Staudinger et al., 2017). Thus, predicting future changes in groundwater drought on a catchment scale will be a prerequisite for effective drought management.

Depending on the projected climate change, different scenarios of the future development of natural baseflow can be expected (Fig. 1). If there is an increase (decrease) in precipitation projected for the entire year, flow during the dry season is also expected to increase (decrease). However, if a seasonal shift of precipitation is expected, the future development of flow during the dry season is not that straightforward. It will depend on the timing of seasonal shift and dry period and on the catchment's characteristic response time to precipitation. Stölzle et al. (2014) found that for baseflow drought changes in precipitation are especially relevant during the recharge period, which is dependent on the hydrogeology of the catchment.

For many parts of central Europe climate projections indicate a seasonal shift of precipitation to wetter winters and drier summers rather than a consistent increase/decrease (Ja-

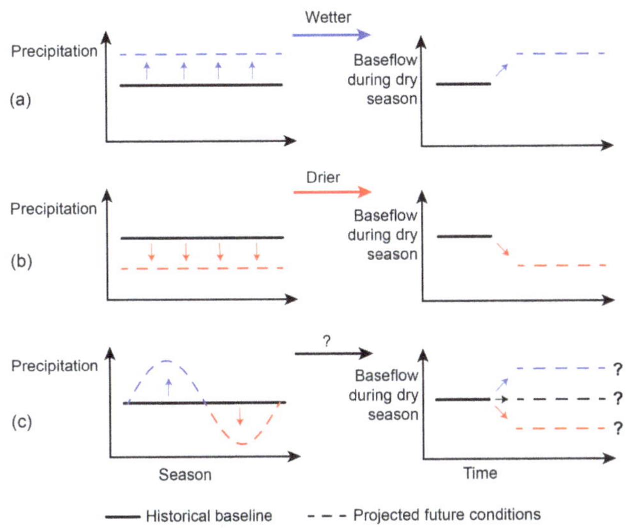

Figure 1. Schematic direction of change of natural baseflow in the dry season under different climate change scenarios: **(a)** all seasons becoming wetter, **(b)** all seasons becoming drier and **(c)** seasonal shift of precipitation.

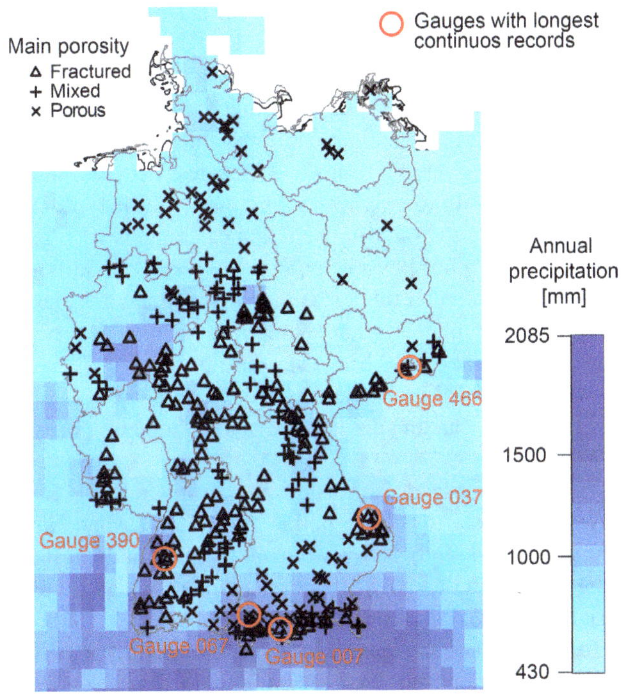

Figure 2. Location of gauges in Germany, catchments' dominant type of porosity (derived from the German hydrogeological map) and mean annual precipitation sums (calculated from European Climate Assessment and Dataset E-OBS).

cob et al., 2014), urging for statistical tools to assess the prospective changes in baseflow. Knowledge of the seasonal to multi-annual scale of the baseflow response to climatic variation and extremes is therefore particularly important in central Europe under this seasonally diverging expected climate change.

This study aims to explain differences in drought trends by catchment characteristics to allow for more accurate predictions under climate projection uncertainty on a headwater scale. Firstly, past trends in baseflow drought and catchment-relevant response times are analysed. Secondly, past trends are attributed to climatic and catchment controls. Finally, based on these statistics an estimate for future changes in baseflow drought valid for all common emission scenarios and climate models is realized.

2 Study area and data

The dataset used in this study is the same set of headwater catchments (that is, all catchment areas are below 200 km^2) that were used in Hellwig et al. (2018) to evaluate the representativeness of meteorological grid data. It comprises streamflow data of daily resolution available from the responsible environment agencies of the German federal states. The lengths of available time series differ, so there is a trade-off between selected period and the number of records with sufficient data. For this study, we only used catchments with streamflow data of the period 1970 to 2009. Records with data gaps were not considered. Even though the dataset consists of near-natural headwaters with minimal regulations, the hydrographs and their double mass curves were visually screened for signs of anthropogenic influence. In a natural catchment under relatively constant conditions, the precipitation–streamflow relationship should be similar for the entire observation period. Four of the gauges showed sudden changes in the relationship and were subsequently removed. The final dataset consisted of 338 unnested and independent gauges across Germany (Fig. 2).

The selected catchments cover the flat lowland regions in the north of Germany, the low mountain ranges in south-central Germany, and the Alpine foothills and non-glacierized front range in the south. Precipitation varies with highest annual precipitation sums in the Alpine south (> 2000 mm) and lowest sums in the northeast (< 500 mm). Climate in Germany is humid with slightly higher precipitation sums in summer than in winter for most regions. Precipitation was analysed in the form of monthly precipitation sums taken from the European Climate Assessment and Dataset (Haylock et al., 2008), Version 13.1. According to the procedure described in Hellwig et al. (2018), catchment-specific precipitation was calculated as an area-weighted mean of the intersecting grid cells. Hellwig et al. (2018) found that due to the low spatial resolution of the meteorological dataset compared to catchment size there are some biases towards products of higher resolution; however, correlations between products were found to be very high.

Information on the hydrogeology of the catchments was taken from the digital German hydrogeological map (Hydrogeologische Übersichtskarte von Deutschland 1 : 200.000, Oberer Grundwasserleiter; BGR and SGD, 2016). The catchments were classified according to the main type of porosity found in the underlying geology, either "porous" for porous aquifers such as unconsolidated alluvial and glacial fillings or "fractured" for fractured bedrock. Where less than two-thirds of the catchment's area is covered by one of these types, the class "mixed" was assigned, including catchments which are dominated by other types of porosity like karst. Additional analysis revealed that other reasonable thresholds for classifying the catchments do not change the results. According to this classification, 80 headwater catchments (23.7 %) are mainly "porous", 170 catchments (50.3 %) are mainly "fractured" and 88 catchments (26.0 %) have "mixed" porosity (Fig. 2).

The dominant land use was derived from CORINE Land Cover 2006 data (available online from the German Environment Agency (https://www.umweltbundesamt.de, last access: 28 November 2018). Each catchment is classified as agricultural (53.6 %), urban (1.2 %) or forested (45.2 %).

To differentiate the catchments regarding topography, the average height above the catchment's outlet was calculated. This metric describes the potential groundwater waterbody that can contribute to the flow at the gauge and thus characterizes the potential storage. Elevation data (1 arcsec digital elevation model over Europe) were obtained from the European Environment Agency (2013).

3 Methods

3.1 Characterizing baseflow drought

3.1.1 Baseflow as a measure of groundwater storage at the catchment scale

Baseflow Q_b is defined as the delayed component of streamflow Q, i.e. flow originating from stored sources as opposed to flow originating from event water (WMO, 2008). In groundwater-dominated catchments, where groundwater is the main water storage, Q_b is directly related to groundwater outflow (see Appendix A for a detailed rationale). Therefore, Q_b can be taken as an integrated measure of groundwater storage S as it was done in the study of Fendeková and Fendek (2012) and others. It is important to be aware that the unknown functional relationship between Q_b and S strongly differs between catchments and might be non-linear, e.g. due to temporal changes in connectivity of groundwater and surface water (Elthair and Yeh, 1999; Brunner et al., 2011). Hence, without further knowledge about this functional relationship, the observation of Q_b allows for conclusions about S solely on an ordinal scale.

Despite the clear concept of baseflow, there is no universally valid way to separate Q_b from time series of Q. Instead, a number of methods exist (e.g. WMO, 2008; Lyne and Hollick, 1979; Nathan and McMahon, 1990; Eckhardt, 2005; Mei and Anagnostou, 2015) that differ especially in baseflow estimation during high flow, when Q_b is relatively uncertain. For the type of extreme low-flow conditions considered in this study, most methods consistently assume that Q is almost completely comprised of Q_b. The World Meteorological Organization recommends a baseflow separation method (WMO, 2008) that was applied in this work using the package "lfstat" (Koffler et al., 2016) in the programming language R (R Core Team, 2016). The separation procedure consists of five steps: (i) divide the time series into non-overlapping blocks of a certain length N, where N is recommended to be 5 days for rainfall regimes which have a fast streamflow response; (ii) select the minimum value for every block; (iii) compare each minimum value with the adjacent ones (if the central value is a times smaller than the adjacent values, then it becomes a turning point; a is recommended to be 0.9 for rainfall regimes which have fast streamflow response); (iv) join the turning points by straight lines; and (v) compute the baseflow for every day by linearly interpolating between the turning points (if the computed value is higher than observed flow, use the observation instead).

3.1.2 Catchment-specific response times to precipitation

To quantify the catchments' baseflow response times to precipitation (T_R), time series of very different characteristics need to be related to each other. To enhance comparability over space, time and different parts of the hydrological cycle, it is common to standardize time series. Recent studies on drought have used either parametric (e.g. Fiorillo and Guadagno, 2012; Barker et al., 2016) or non-parametric approaches (e.g. Bloomfield and Marchant, 2013; Kumar et al., 2016) for standardization. Parametric approaches like the Standardized Precipitation Index (SPI), introduced by McKee et al. (1993), rely on the fitting of a theoretical distribution. Since time series of precipitation (P) and Q_b have distinct characteristics, entirely different distributions would have to be selected. To ensure consistency in the way of standardization, in this study a non-parametric approach was applied (Bloomfield and Marchant, 2013; Kumar et al., 2016). All time series (P in accumulation periods from 1 to 36 months, Q_b) were standardized according to the procedure of the Standardized Groundwater Index (SGI; Bloomfield and Marchant, 2013).

P (or Q_b) was transformed by using the inverse normal cumulative distribution function to calculate n equally spaced p_i values between $1/(2n)$ and $1 - 1/(2n)$, where n is the number of observations. These values were afterwards assigned to the time series by reordering them according to the ranks of the original time series. Hence, the ranks of the original time series and the new ones coincide, but the newly generated time series are normally distributed. Standardization was done separately for all 12 months of the year to account for the regime.

For all 338 catchments, correlation coefficients R between P and Q_b were calculated:

$$R = \frac{\mathrm{cov}(x, y)}{\sigma_x \sigma_y} = \mathrm{cov}(xy), \tag{1}$$

where x and y are P and Q_b, respectively, with cov() the covariance and σ the standard deviation of each time series, which equals one for standardized time series. To derive the catchments' response times to precipitation, precipitation accumulation periods from 1 to 36 months were tested as well as time lags of these indices to baseflow from 0 to 10 months. For each catchment, the combination of precipitation accumulation period and time lag that has the highest correlation coefficient was selected as the optimal accumulation period and the optimal time lag to describe the Q_b response. Longer optimal accumulation periods indicate higher groundwater attenuation of precipitation input, whereas higher time lags represent a more delayed groundwater response to precipitation. To account for seasonal variations in the response times, correlations were calculated separately for all seasons (winter: DJF; spring: MAM; summer: JJA; autumn: SON).

3.1.3 Catchment-relevant recharge period

Catchments analysed in this study follow a distinct regime in Q_b. Thus, droughts in Q_b are most likely to occur in a certain period of the year. To derive the catchment-characteristic drought season, annual minimal 7-day Q_b (Q_{b7}) were calculated. The most frequent month of Q_{b7} was assigned as the month of highest baseflow drought hazard ($M_{Q_{b7}}$). Subsequently, the derived seasonal T_R was used to determine the months influencing Q_b in $M_{Q_{b7}}$. For example, if $M_{Q_{b7}}$ in a catchment was October and T_R for that catchment in autumn was 6 months, then the catchment-relevant recharge period would be May through October.

3.2 Detection of past baseflow drought trends

Time series of Q_{b7} for the time period of 1970–2009 were used to analyse past trends in minimal baseflow. The analysis was carried out with the non-parametric Mann–Kendall (MK) test, which is a common tool used to detect monotonic increases or decreases in hydrological time series (e.g. used by Douglas et al., 2000; Lins and Slack, 1999; Lorenzo-Lacruz et al., 2012; Rennermalm et al., 2010; Asarian et al., 2016) and is robust to outliers. The test was performed after pre-whitening the time series to account for the influence of serial correlation (see Appendix B for calculation steps). A significance level of $\alpha = 0.05$ was used.

Trends only give information on the period they are calculated for. Many studies found that trends of a certain period are not part of a trend on another timescale (e.g. Stahl et al., 2010; Hannaford and Buys, 2012; Giuntoli et al., 2013; Hannaford et al., 2013). The analysis of trends for multiple periods may help to assess whether observed trends are steady or rather fluctuating. To evaluate the trends found for the period 1970–2009, we additionally calculated trends over multiple periods for the five gauges with longest continuous records. Three of these are in the porosity class "fractured", and one each in "porous" and "mixed".

3.3 Attribution of baseflow drought trends

Trends in Q_{b7} were assumed to be driven by both climatic (climate control) and catchment-specific (catchment control) factors. Thus, the results were related to a set of potential predictors: (i) trends in P for the catchment-specific relevant recharge period, (ii) T_R for the catchment-specific $M_{Q_{b7}}$, (iii) $M_{Q_{b7}}$, (iv) dominant porosity of the catchment, (v) catchment size (A), (vi) dominant land use and (vii) potential storage. Statistical significance of the factors was determined using the correlation coefficient R (Eq. 1) for the continuous variables (trends in P, A, potential storage), Spearman's rank-based correlation coefficient for the discrete variable (T_R) and an analysis of variance (ANOVA) combined with the post hoc Tukey test (for details see Appendix C) for the nominally scaled ones ($M_{Q_{b7}}$, dominant porosity, dominant land use).

3.4 Estimate of drought sensitivity to expected future precipitation change

Since past trends derived from empirical trend analysis (e.g. MK) are solely valid for the observation period, they cannot be extrapolated beyond the period of data availability. Future drought predictions therefore mostly rely on climate projections and process modelling. For Germany climate projections indicate little to no changes of annual precipitation sums but seasonal shifts to lower summer precipitation and higher winter precipitation. However, the magnitude of the shifts differs considerably for different projections (Zebisch et al., 2005; Jacob et al., 2012, 2014; Hübener et al., 2017). A common approach to dealing with such uncertainties is to use a range of possible trajectories to model hydrological change. Instead of using uncertain quantitative inputs in forward modelling, here we propose a more qualitative inverse approach. We assume that the general direction of future development is the most important piece of information for future groundwater management planning and formulate a qualitative test scenario of the consistent direction of different projections of future precipitation change. Thus, the approach is scenario-neutral regarding emission scenarios and climate models.

Trends in future baseflow drought hazard were assumed to depend not only on precipitation changes but also on T_R and $M_{Q_{b7}}$ calculated for every catchment. A particular test scenario can still have three possible outcomes regarding future changes in baseflow drought hazard (Fig. 3): (i) increased

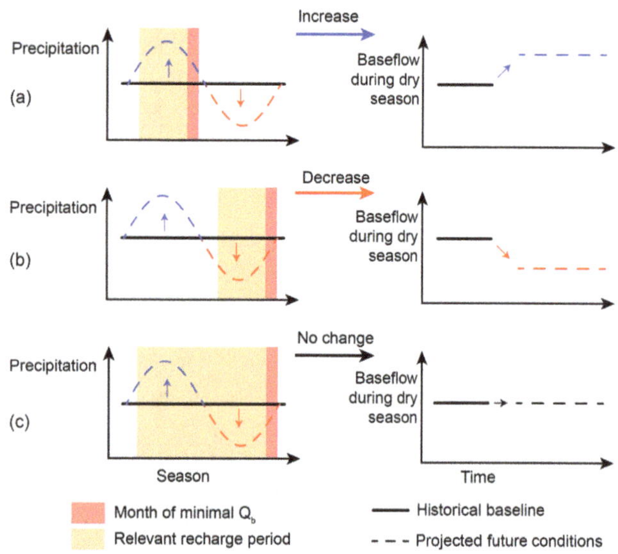

Figure 3. Potential directional changes in Q_{b7} under a constant seasonally diverging precipitation change scenario, depending on T_R and $M_{Q_{b7}}$: **(a)** increase, **(b)** decrease and **(c)** no change.

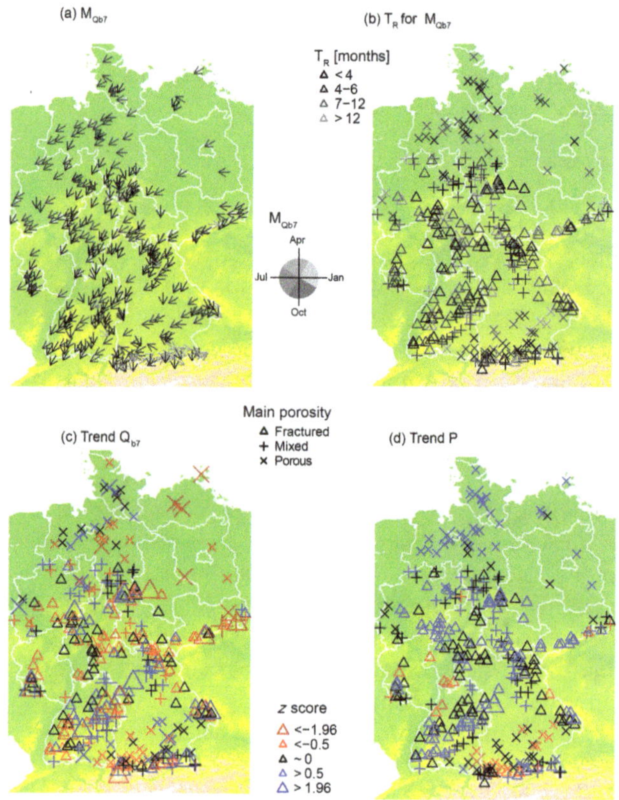

Figure 4. Patterns of Q_{b7} across Germany. **(a)** Low-flow month $M_{Q_{b7}}$ and **(b)** seasonal response time T_R. **(c)** Trend of Q_{b7} and **(d)** P in the relevant recharge period for 1970–2009.

baseflow – there are relevant months with increasing precipitation but not with decreasing precipitation; (ii) decreased baseflow – there are relevant months with decreasing precipitation but not with increasing precipitation; or (iii) no change – either there are relevant months with both increasing and decreasing precipitation or there are no relevant months with any change.

The test scenario applied in this study is a decrease of precipitation in summer (JJA) and an increase of precipitation in winter (DJF) with no change in the annual precipitation sum. To derive the future change in Q_{b7} for a catchment, we analyse whether months with increasing or decreasing precipitation belong to the catchment-relevant recharge period. For example, for a catchment with a relevant recharge period from May to October we would estimate an increased baseflow drought hazard, since there are months with decreasing precipitation in the relevant recharge period (June, July and August) but no months with increasing precipitation. This analysis was carried out for all 338 catchments.

4 Results

4.1 Past minimal baseflow trends

$M_{Q_{b7}}$ occurs in most parts of Germany in late summer or autumn with some exceptional winter low-flow catchments in the mountainous regions (Fig. 4). Seasonal response times vary across Germany, ranging from short subseasonal response times (1–3 months) to response times of over a year (Fig. 4b). Altogether 22 out of 338 catchments (~6.5 %) have a time lag in the response, though these are mainly short. All five catchments with a time lag above 1 month have response times longer than 12 months. According to the MK test, 40 out of the 338 catchments show a significant trend of Q_{b7} (Fig. 4c), corresponding to 11.8 %. Assuming independence of the baseflow observations, this is slightly more than the expected 5 % in case of no real trends. Hence, for the period of 1970–2009 there are some changes detectable for all catchments' porosity classes. For P there are even fewer significant trends than would be expected by chance (Fig. 4d). However, the results are skewed towards predominantly positive trends in P during the relevant recharge periods across Germany.

The five selected gauges for a trend analysis on multiple periods all have a negative z score in the original period: 1970–2009 (Fig. 5). However, in all cases a significant positive trend could also have been observed if another period had been chosen for analysis (Fig. 5). This reveals a high influence of the observation period on the results of the trend analysis. The high variability of trends is also independent of the porosity, T_R and $M_{Q_{b7}}$.

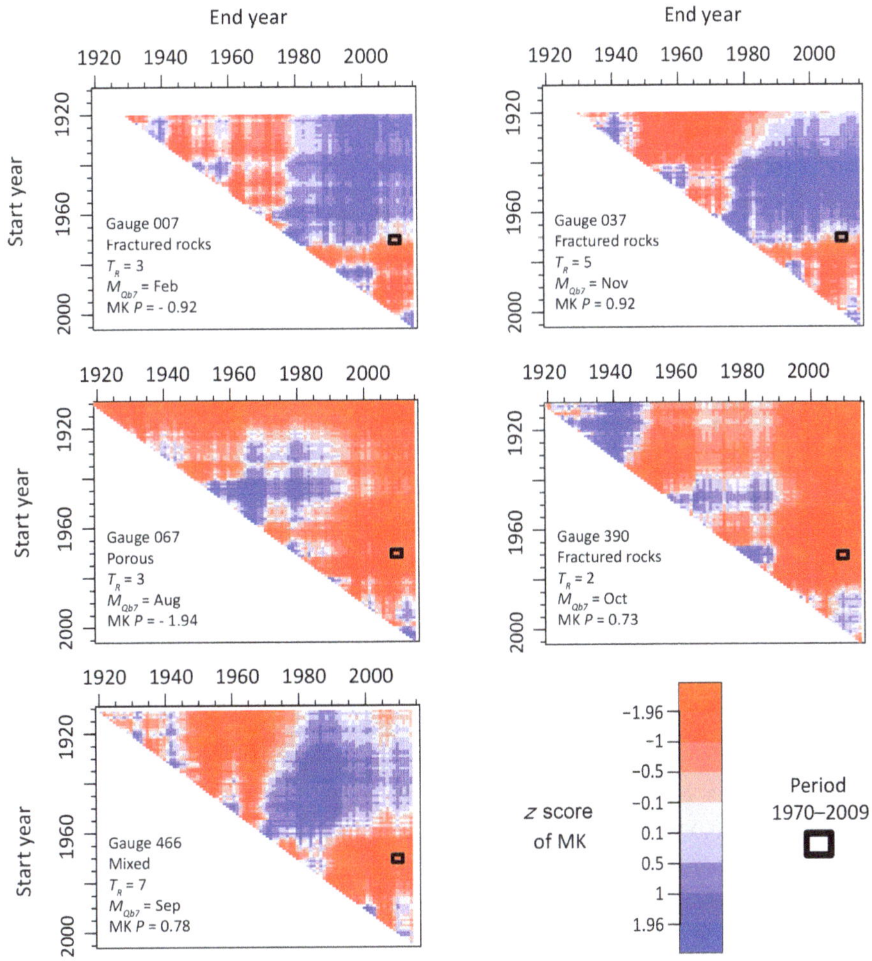

Figure 5. z-score values for all possible periods for the five catchments with longest continuous records.

4.2 Trend attribution

None of the factors tested was found to explain past trends of Q_{b7} very well. This lack of statistical relation coincides with the results that few trends are significant and trend magnitude is highly dependent on the period of analysis. However, there is a small but significant correlation between the trends in P for the catchment-relevant recharge period and trends in Q_{b7} (Fig. 6). Moreover, the ANOVA indicates a significant influence of $M_{Q_{b7}}$ on the trends in Q_{b7} even though Tukey's test does not distinguish two groups significantly. The largest differences are between July and November, so the season of low flow might be relevant for trend magnitude in Q_{b7} as well. The remaining factors (T_R, dominant porosity, A, dominant land use and potential storage) do not show any relation to the trends in Q_{b7} (Fig. 6).

4.3 Potential future drought hazard

According to the test scenario, Q_{b7} will decrease or not change for most of Germany (Fig. 7). The only catchments with estimated increases of Q_{b7} are located in the mountainous regions of southeast Germany, especially in the Alpine foothills in the catchments with annual low flows in winter. However, in the catchments with annual low flows in winter or spring, impacts might also be more severe for low flows in summer or autumn. Additional tests revealed that, for 22 (27) out of 35 catchments with $M_{Q_{b7}}$ in winter or spring, baseflows of the summer (autumn) season are expected to decrease according to the test scenario.

The changes in drought hazard are significantly related to the catchments' response times T_R (Fig. 8). Moreover, there is a strong statistical linkage to the dominant porosity (Fig. 8b). Catchments dominated by fractured rocks more frequently show estimated decreases in Q_{b7} than the other classes. This coincides with a significant relationship between T_R and the dominant porosity of the catchment (Fig. 8c). According to Turkey's test, catchments with fractured rocks have shorter response times than the other classes. Contrarily, catchments with porous aquifers or mixed hydrogeology, which have long T_R, can more often

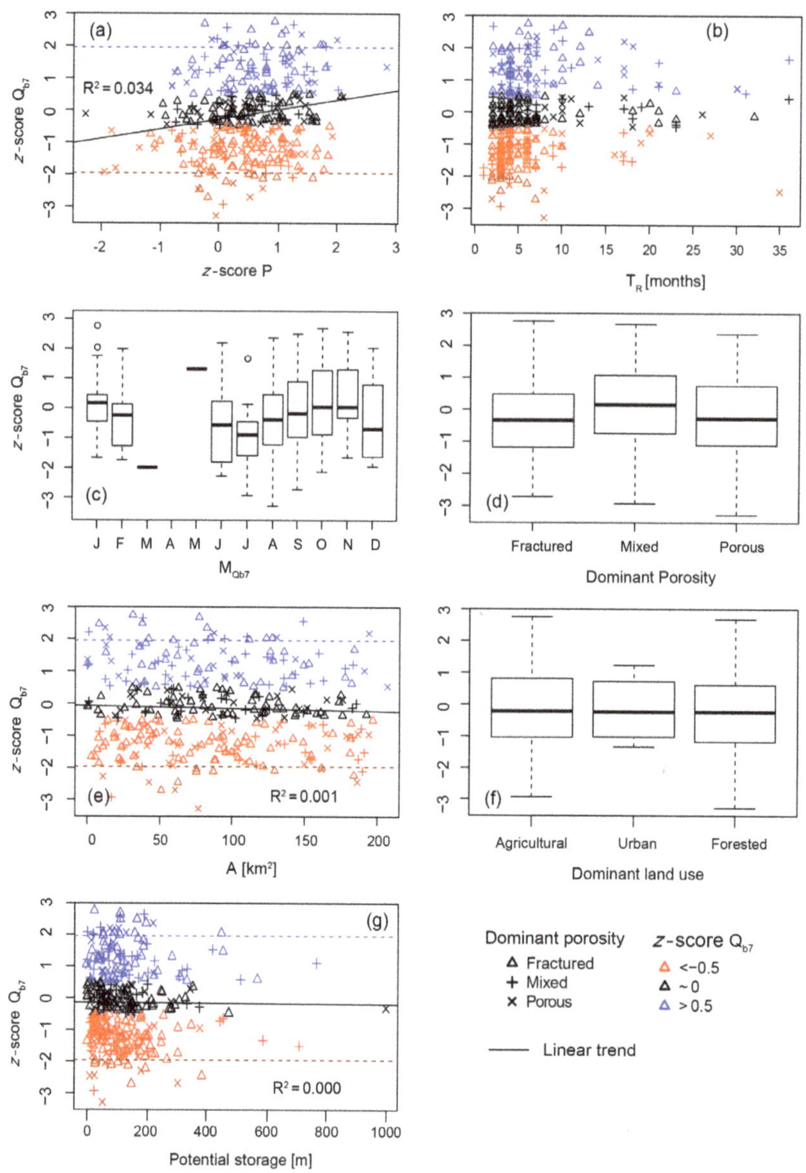

Figure 6. Attribution of trends in Q_{b7} to potential influencing factors. **(a)** Trends in precipitation P during the relevant recharge period, **(b)** length of response time T_R, **(c)** $M_{Q_{b7}}$, **(d)** dominant porosity in the catchment **(e)** catchment size A, **(f)** dominant land use in the catchment and **(g)** potential storage. Boxes represent the quartiles 0.25 and 0.75, and whiskers last to the most extreme value within a maximum distance of a 1.5-fold interquartile range.

compensate for summer precipitation decreases with winter increases and are therefore more frequently in the category "no change".

5 Discussion

5.1 The catchment baseflow response to precipitation

Catchments' response times to precipitation were found to be highly diverse across Germany, ranging from 1 month to 3 years. In general, baseflow response times determined as a proxy of groundwater response are rather short compared to other studies. Fiorillo and Guadagno (2012) found response times of 12 to 24 months for a karst region in southern Italy and highest correlations for shorter precipitation accumulation periods when adding a short time lag. Bloomfield and Marchant (2013) also found in 3 out of 14 cases a time lag for highest correlations. We found time lags to be an exception, supporting the results of Kumar et al. (2016), Barker et al. (2016) and Haas and Birk (2017). This indicates that, in the headwater catchments studied, the delay of the groundwater baseflow response to meteorological conditions

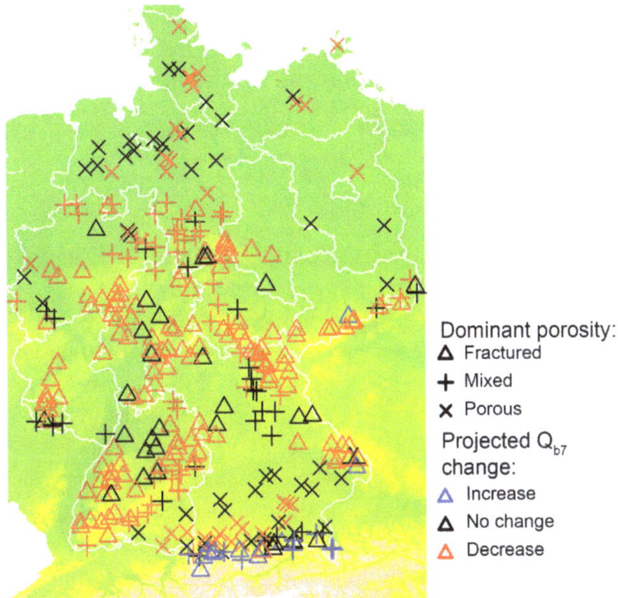

Figure 7. Future changes in Q_{b7} according to the test scenario.

may be shorter than 1 month and therefore not detectable on the monthly scale, whereas the attenuation of meteorological variability is clearly attributable to characteristic precipitation accumulation periods.

The large differences of baseflow response times for different porosity classes match the theoretical assumptions that baseflow strongly depends on hydrogeological conditions. For the entire streamflow, differences were found to be much smaller (not shown; compare e.g. Haslinger et al., 2014), since other processes like overland flow are also important. The patterns of T_R which are related to the catchments' hydrogeology support the assumption that baseflow can be used as a proxy of the groundwater situation on a catchment scale. More detailed information on hydrogeology has the potential to further improve the understanding of differences in baseflow response. However, this information is not yet detailed enough for the scale and distribution of headwater catchments. For example, two large-scale estimates on hydraulic conductivity available for Germany (GLHYMPS from Gleeson et al., 2014, and HÜK200 from BGR and SGD, 2016) were considered and found to differ up to 6 orders of magnitude. Moreover, information on superficial deposits that might be particularly important for streamflow generation could enhance results but is not available in detail yet.

Consistent with the work by Bloomfield et al. (2015), we found that hydrogeology is a highly relevant factor for the explanation of different groundwater baseflow response times. Kumar et al. (2016) did not find a relationship between the hydraulic conductivity and groundwater response time for boreholes. A possible reason for this different finding is that for point data even small local influences, which are hard to determine, are quite relevant (e.g. human influences), whereas baseflow reflects more the overall situation within the catchment. Small influences may be negligible at this scale, and the underlying influence of hydrogeology may be easier to detect.

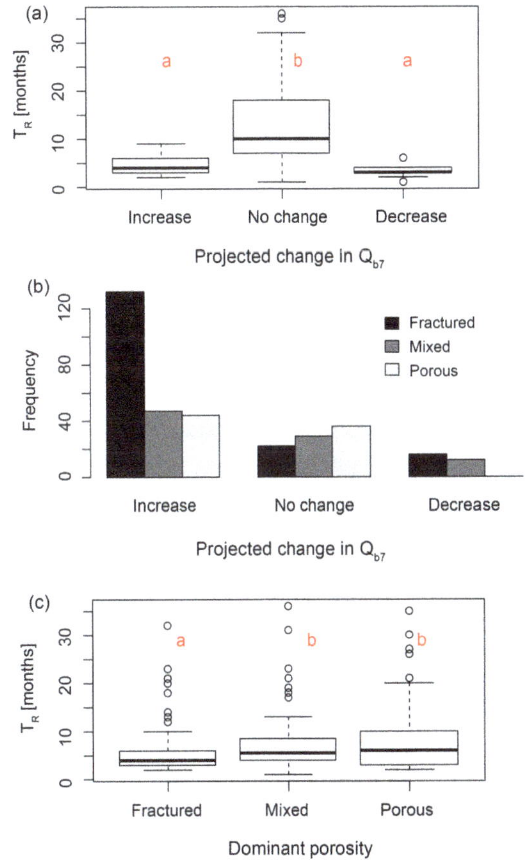

Figure 8. Relationships of projected changes, T_R and catchment's dominant porosity. **(a)** Distribution of T_R for projected changes, **(b)** frequency of catchments with a certain dominant porosity for different projected changes and **(c)** distribution of T_R for dominant porosities. Boxes represent the quartiles 0.25 and 0.75, and whiskers last to the most extreme value within a maximum distance of a 1.5-fold interquartile range. Red letters indicate results of the post hoc Tukey test to distinguish significantly different groups.

In general, the results indicate that groundwater storage – represented by baseflow – is vitally driven by precipitation on a catchment scale in the relevant recharge period. However, the season of low flow is also expected to have an influence: regimes with winter low flows in central Europe are governed by snow storage during that season. Thus, not only precipitation but also temperature is a major factor for that catchments. Moreover, it is impossible to distinguish snowmelt from groundwater outflow during baseflow separation. Therefore, a baseflow approach does not allow for conclusions on groundwater storage in snow-dominated catchments.

5.2 Past and potential future changes in low baseflow

The trend analysis revealed that the trend in baseflow minima is highly dependent on the period it is calculated for. The observation of a trend in Q_{b7} thus is a poor indicator of future developments. Despite the relatively narrow seasonality of low-baseflow timing (except for the alpine areas), trends are highly variable across the region – a result of the likewise variable response times. However, the correlation tests for attribution of the trends revealed that precipitation during the catchment-relevant recharge period is an important factor, which confirms previous model-based findings for part of the region (Stölzle et al., 2014). Thus, this catchment-relevant recharge period should be considered for projections.

The method employed in this study provides a straightforward projection of the probable future directions of changes in baseflow or groundwater drought under climate change. We accounted for the uncertainty in future climate projections by taking only concordant directions of precipitation change in our approach. Contrary to future climate projections, past trends in precipitation for the catchment-relevant recharge period were found to be small and mostly positive. Similarly, Kopp et al. (2018) found for southern Germany a high variability of annual groundwater recharge without distinct trends. In the past, precipitation trends have not been seasonally diverging, but climate models suggest that changes will become more relevant in the second half of this century (e.g. Jacob et al., 2012). As the magnitude of the trends differs for the climate models, we did not quantify our scenario and thus did not quantify the magnitude of future baseflow either.

Catchments' characteristic response times were assumed to remain constant. Under a more extreme climate change however, changes in catchments' responses (e.g. due to nonstationary response times) cannot be excluded. Based on assumed precipitation change in the catchments' respective recharge periods, decreasing Q_{b7} (i.e. increased groundwater drought hazard) were estimated for the majority of the catchments. This is because of the timing of $M_{Q_{b7}}$, which occurs predominantly in summer/autumn. Only those catchments with long response times can compensate for the decreases in summer precipitation, and only in the Alpine foothills, where $M_{Q_{b7}}$ is in winter, is a decrease in drought hazard predicted. However, in this region winter precipitation is mainly snow and therefore does not immediately contribute to groundwater recharge. Moreover, air temperature and its changes over time – which are not considered in the test scenario – are especially important for winter low flows. Hence, uncertainty is quite high for these projections. If response times are not assumed constant, further conclusions may be drawn. A decrease of response times might be caused by decreased storage, e.g. due to urbanization, or accelerated drying and increased evapotranspiration in spring. Such changes would increase the drought hazard for most parts of Germany due to the timing of $M_{Q_{b7}}$. An increase in response times by the opposite processes is possibly less likely but could compensate for the reduced precipitation during summer for the catchments that are currently estimated to have increased drought hazard.

6 Conclusions

Climate change is expected to alter the hydrological drought hazard. However, uncertainty of climate projections and even contrasting seasonal changes impede a straightforward assessment of the prospective changes in central Europe and elsewhere. Here we presented a statistical approach to estimating the potential direction of future changes in hydrological drought hazard. Past trends were found to be too variable to provide a consistent regional picture of past and expected changes because of their high dependency on the trend calculation period. But they did allow the attribution of trends in baseflow to precipitation changes in catchment-specific recharge periods. Based on that information, a more process-oriented approach was developed, using the catchments' characteristic response times to precipitation and the relevant recharge periods for projections valid for all common emission scenarios and climate models. These projections are efficient alternatives to ensemble projections and target the most important information for management. Especially for regions where directions of climate change are seasonally varying, they can provide valuable insights into the basic changes of the system.

Catchments with short response times were found to have a high probability for a decrease in baseflow minima and hence an increase in the groundwater drought hazard, as seasonal changes cannot compensate for each other. However, there is no homogeneous pattern of response times across central Europe, and so predicted changes of groundwater drought hazard also vary regionally. This urges for a regionally adapted groundwater management based on the local catchment response times. As past events like the 2015 central European drought have already caused groundwater-related drought impacts in headwater regions, there is an urgent need for adaptation in catchments facing even higher drought hazard in the future.

The diversity of response times, the dearth of long-time data on groundwater storage, and the absence of distinct past trends in precipitation and hydrological variables limit the potential to generalize the results. On the path to extensive predictions of future groundwater drought hazard across central Europe, further model-based work will be needed. Reproducing the catchment-relevant response times with high-resolution large-scale models may be key for an assessment of future changes and related implications for groundwater management under various scenarios and for ungauged catchments.

Appendix A: Mathematical relationship between baseflow and groundwater storage

In groundwater-dominated catchments Q_b is an integrated signal of groundwater conditions in the entire area and mainly depends as a function g on the hydraulic head H of the groundwater:

$$Q_b = g(H) + r \quad \text{with} \quad \frac{dQ_b}{dH} > 0 \, \forall H, \tag{A1}$$

where r is the additional flow from other stored sources. H is a monotonic function f of groundwater storage S:

$$H = f(S) \quad \text{with} \quad \frac{dH}{dS} > 0 \, \forall S. \tag{A2}$$

Combining Eqs. (A1) and (A2) leads to

$$Q_b = g(f(S)) + r = k(S) + r \quad \text{with} \quad \frac{dQ_b}{dS} > 0 \, \forall S. \tag{A3}$$

Because of the monotonic behaviour, function k is reversible:

$$S = k^{-1}(Q_b - r) \quad \text{with} \quad \frac{dS}{d(Q_b - r)} > 0 \, \forall Q_b - r. \tag{A4}$$

Potential sources for r include snowmelt, interflow, lake outflow, sewage discharge or other anthropogenic sources of discharge. Anthropogenic influences can be assumed negligible, as catchments in this study were selected as near-natural. Moreover, there are no big lakes or other surface water storages in the catchments. Snowmelt remains an important factor that might blur the true signal of groundwater storage. This is particularly important for the catchments at higher elevations, e.g. in the Alps, and must be considered for the interpretation of the results. However, in catchments where these factors are of minor relevance compared to groundwater outflow, Eq. (A4) simplifies to

$$S \simeq k^{-1}(Q_b) \quad \text{with} \quad \frac{dS}{d(Q_b)} > 0 \, \forall Q_b; \tag{A5}$$

i.e. higher baseflow indicates higher groundwater storage, and vice versa. However, without further knowledge about k^{-1}, the observation of Q_b allows for conclusions about S solely on an ordinal scale.

Appendix B: The Mann–Kendall trend test with pre-whitening

To calculate trends, the non-parametric Mann–Kendall (MK) test was applied. However, the results are affected by serial correlation which increases the type I error (i.e. rejecting the no-trend hypothesis although there is no trend). To test for serial correlation in the data, we fitted an autocorrelation AR(1) model to each Q_{b7} time series. The calculated autocorrelation is significant on a level of $\alpha = 0.05$ if absolute autocorrelation is higher than $1.96/\sqrt{n}$ (Douglas et al., 2000), where n is the length of the time series (in this case $n = 40$). One hundred thirty (130) out of 338 time series showed significant serial correlation, thus requiring a pre-processing before using MK. Kulkarni and Von Storch (1995) recommended pre-whitening time series to allow for the MK. Since the sample size is relatively small here and trend magnitude not too large, pre-whitening is not expected to reduce the MK test power much in this case, but rather to reduce the type I error (Bayazit and Önöz, 2007). Pre-whitening was done according to other studies (Kulkarni and Von Storch, 1995; Douglas et al., 2000) as in Eq. (A6):

$$Y_t = X_t - r_1 X_{t-1}, \tag{B1}$$

where Y_t is the pre-whitened time series at time step t, X_t is the original time series at time step t and r_1 is the serial correlation determined with the AR(1) model. Pre-whitening reduced the serial correlation in all time series to close to zero.

The MK test compares the number of concordant pairs in the data with the number of discordant pairs. This gives the Kendall score S_K (Kendall, 1948):

$$S_K = \sum\nolimits_{j>i} \text{sign}(Y_j - Y_i)\text{sign}(Z_j - Z_i), \tag{B2}$$

where sign() is a function that returns the algebraic sign. For the Mann–Kendall trend test that was applied here, Z is time. Therefore, the second part of Eq. (A7) always equals +1, and it simplifies to

$$S_K = \sum_{j>i} \text{sign}(Y_j - Y_i). \tag{B3}$$

In the case of no real trend, S_K has a mean value of zero; in the case of no ties in the data, it has a standard deviation σ_S of (Kendall, 1948)

$$\sigma_S = \sqrt{\frac{n(n-1)(2n+5)}{18}}. \tag{B4}$$

Thus, S_K can be transformed into the standard normal distributed z score by dividing by the standard deviation of S_K:

$$z = \begin{cases} \dfrac{S_K - 1}{\sigma_S} & \text{if } S_K > 0 \\ 0 & \text{if } S_K = 0 \\ \dfrac{S_K + 1}{\sigma_S} & \text{if } S_K < 0 \end{cases}. \tag{B5}$$

z was assumed to be significant if it was not within the interval ±1.96, i.e. a significance level of $\alpha = 0.05$.

Appendix C: ANOVA and Tukey's test

To detect the influence of a categorical variable with multiple levels on a numerical variable, an ANOVA was used in this work. The ANOVA compares S_e, the sum of squares explained by the categorical variable's levels, with the sum of squares of the residuals (S_r). The mean sums (M_{Se}, M_{Sr}) are tested for significance using the F test:

$$F = \frac{M_{Se}}{M_{Sr}} = \frac{S_e/d_{fe}}{S_r/d_{fr}}, \quad (C1)$$

where d_f denotes the degrees of freedom (d_{fe}: number of levels minus 1; d_{fr}: amount of data minus d_{fe} minus 1). The test statistic F follows the F distribution and was assumed to be significant for $p < 0.05$.

The ANOVA gives information about the general significance of the categorical variable. An equally important piece of information is which of the categorical variable's levels differ significantly regarding the target variable. This information was obtained using the post hoc Tukey test. For this analysis a pairwise t test between all levels is carried out:

$$t = \frac{\overline{x}_1 - \overline{x}_2}{\sqrt{\frac{s_1^2}{n_1} - \frac{s_2^2}{n_2}}}, \quad (C2)$$

where $\overline{x}_i$ and s_i^2 are the mean and the variance for the two levels, respectively, and n_i is the amount of data per level. Tukey's test accounts for the multiple-comparison problem (i.e. a higher probability of getting significant results by chance due to a higher number of comparisons) by using the studentized range distribution instead of the t distribution. Again, the results were assumed to be significant for $p < 0.05$.

Author contributions. JH and KS designed the study. JH performed all analyses and prepared the manuscript. Both authors revised the manuscript and approved the final version.

Competing interests. The authors declare that they have no conflict of interest.

Acknowledgements. We acknowledge the data providers at the federal states' agencies for the streamflow data, the E-OBS dataset from the EU-FP6 project ENSEMBLES (https://ensembles-eu.metoffice.com, last access: 28 November 2018) and the data providers in the ECA&D project (https://www.ecad.eu, last access: 28 November 2018). The authors were funded by the DFG (Jost Hellwig: project TrenDHy DFG STA632/4-1; Kerstin Stahl: Heisenberg programme STA632/3-1). Benedikt Heudorfer provided helpful comments on the manuscript. We are grateful for the comments of John Bloomfield and Spencer Sawaske, which helped to improve the quality of the manuscript. The article processing charge was funded by the German Research Foundation (DFG) and the University of Freiburg in the funding programme Open Access Publishing.

Edited by: Stacey Archfield

References

Apurv, T., Sivapalan, M., and Cai, X: Understanding the Role of Climate Characteristics in Drought Propagation, Water Resour. Res., 53, 9304–9329, https://doi.org/10.1002/2017WR021445, 2017.

Asarian, J. E. and Walker, J. D.: Long term trends in streamflow and precipitation in northwest California and southwest Oregon, 1953–2012, J. Am. Water. Resour. As., 52, 241–261, https://doi.org/10.1111/1752-1688.12381, 2016.

Barker, L. J., Hannaford, J., Chiverton, A., and Svensson, C.: From meteorological to hydrological drought using standardised indicators, Hydrol. Earth Syst. Sci., 20, 2483–2505, https://doi.org/10.5194/hess-20-2483-2016, 2016.

Bayazit, M. and Önöz, B.: To prewhiten or not to prewhiten in trend analysis? Hydrolog. Sci. J., 52, 611–624, https://doi.org/10.1623/hysj.52.4.611, 2007.

BGR and SGD: Bundesanstalt für Geowissenschaften und Rohstoffe and Staatliche Geologische Dienste: Hydrogeologische Übersichtskarte von Deutschland 1:200.000, Oberer Grundwasserleiter (HÜK200 OGWL), Digitaler Datenbestand, Version 3.0. – Hannover, 2016.

Bloomfield, J. P. and Marchant, B. P.: Analysis of groundwater drought building on the standardised precipitation index approach, Hydrol. Earth Syst. Sci., 17, 4769–4787, https://doi.org/10.5194/hess-17-4769-2013, 2013.

Bloomfield, J. P., Marchant, B. P., Bricker, S. H., and Morgan, R. B.: Regional analysis of groundwater droughts using hydrograph classification, Hydrol. Earth Syst. Sci., 19, 4327–4344, https://doi.org/10.5194/hess-19-4327-2015, 2015.

Bundesanstalt für Geowissenschaften und Rohstoffe: HÜK200 – Hydrogeologische Übersichtskarte von Deutschland 1 : 200000, available at: https://www.bgr.bund.de/DE/Themen/Wasser/Projekte/laufend/Beratung/Huek200/huek200_projektbeschr.html, last access: 28 November 2018.

Brunner, P., Cook, P. G., and Simmons, C. T.: Disconnected Surface Water and Groundwater: From Theory to Practice, Ground Water, 49, 460–467, https://doi.org/10.1111/j.1745-6584.2010.00752.x, 2011.

Brutsaert, W.: Long-term groundwater storage trends estimated from streamflow records: Climatic perspective, Water Resour. Res., 44, W02409, https://doi.org/10.1029/2007wr006518, 2008.

Brutsaert, W.: Annual drought flow and groundwater storage trends in the eastern half of the United States during the past two-third century, Theor. Appl. Climatol., 100, 93–103, https://doi.org/10.1007/s00704-009-0180-3, 2010.

Douglas, E. M., Vogel, R. M., and Kroll, C. N.: Trends in floods and low flows in the United States: impact of spatial correlation, J. Hydrol., 240, 90–105, https://doi.org/10.1016/s0022-1694(00)00336-x, 2000.

EC: Communication from the Commission to the European Parliament and the Council addressing the challenge of water scarcity and droughts in the European Union, Commission of the European Communities, COM(2007), 414 final, Brussels, 2007.

ECA and D: European Climate Assessment & Dataset, available at: https://www.ecad.eu/, last access: 28 November 2018.

Eckhardt, K.: How to construct recursive digital filters for baseflow separation, Hydrol. Process., 19, 507–515, https://doi.org/10.1002/hyp.5675, 2005.

Eltahir, E. A. B. and Yeh, P.: On the asymmetric response of aquifer water level to floods and droughts in Illinois, Water Resour. Res., 35, 1199–1217, https://doi.org/10.1029/1998wr900071, 1999.

European Environment Agency: Digital Elevation Model over Europe (EU-DEM), available at: https://www.eea.europa.eu/data-and-maps/data/eu-dem (last access: 28 November 2018), 2013.

Fendeková, M. and Fendek, M.: Groundwater drought in the Nitra river basin-identification and classification, J. Hydrol. Hydromech., 60, 185–193, 2012.

Fiorillo, F. and Guadagno, F. M.: Long karst spring discharge time series and droughts occurrence in Southern Italy, Environ. Earth Sci., 65, 2273–2283, 2012.

Forzieri, G., Feyen, L., Rojas, R., Flörke, M., Wimmer, F., and Bianchi, A.: Ensemble projections of future streamflow droughts in Europe, Hydrol. Earth Syst. Sci., 18, 85–108, https://doi.org/10.5194/hess-18-85-2014, 2014.

Giuntoli, I., Renard, B., Vidal, J. P., and Bard, A.: Low flows in France and their relationship to large-scale climate indices, J. Hydrol., 482, 105–118, https://doi.org/10.1016/j.jhydrol.2012.12.038, 2013.

Gleeson, T., Moosdorf, N. Hartmann, J., and van Beek L. P. H.: A glimpse beneath earth's surface: Global HYdrogeology MaPS (GLHYMPS) of permeability and porosity, Geophys. Res. Lett., 41, 3891–3898, https://doi.org/10.1002/2014GL059856, 2014.

Gosling, S. N., Zaherpour, J., Mount, N. J., Hattermann, F. F., Dankers, R., Arheimer, B., Breuer, L., Ding, J., Haddeland, I., Kumar, R., Kundu, D., Liu, J., Van Griensven, A., Veldkamp, T. I. E., Vetter, T., Wang, X., and Zhang, X. X.: A comparison of changes in river runoff from multiple global and catchment-scale hydrological models under global warming scenarios of 1 degrees C, 2 degrees C and 3 degrees C, Clim. Change, 141, 577–595, https://doi.org/10.1007/s10584-016-1773-3, 2017.

Green, T. R., Taniguchi, M., Kooi, H., Gurdak, J. J., Allen, D. M., Hiscock, K. M., Treidel, H., and Aureli, A.: Beneath the surface of global change: Impacts of climate change on groundwater, J. Hydrol., 405, 532–560, https://doi.org/10.1016/j.jhydrol.2011.05.002, 2011.

Haas, J. C. and Birk, S.: Characterizing the spatiotemporal variability of groundwater levels of alluvial aquifers in different settings using drought indices, Hydrol. Earth Syst. Sci., 21, 2421–2448, https://doi.org/10.5194/hess-21-2421-2017, 2017.

Hannaford, J. and Buys, G.: Trends in seasonal river flow regimes in the UK, J. Hydrol., 475, 158–174, https://doi.org/10.1016/j.jhydrol.2012.09.044, 2012.

Hannaford, J., Buys, G., Stahl, K., and Tallaksen, L. M.: The influence of decadal-scale variability on trends in long European streamflow records, Hydrol. Earth Syst. Sci., 17, 2717–2733, https://doi.org/10.5194/hess-17-2717-2013, 2013.

Haslinger, K., Koffler, D., Schoner, W., and Laaha, G.: Exploring the link between meteorological drought and streamflow: Effects of climate- catchment interaction, Water Resour. Res., 50, 2468–2487, https://doi.org/10.1002/2013wr015051, 2014.

Haylock, M. R., Hofstra, N., Tank, A., Klok, E. J., Jones, P. D., and New, M.: A European daily high-resolution gridded data set of surface temperature and precipitation for 1950–2006, J. Geophys. Res.-Atmos., 113, D20119, https://doi.org/10.1029/2008jd010201, 2008.

Hellwig, J., Stahl, K., Ziese, M., and Becker, A.: The impact of the resolution of meteorological data sets on catchment-scale precipitation and drought studies, Int. J. Climatol., 38, 3069–3081, https://doi.org/10.1002/joc.5483, 2018.

Heudorfer, B. and Stahl, K: Comparison of different threshold level methods for drought propagation analysis in Germany, Hydrol. Res., 48, 1311–1326, https://doi.org/10.2166/nh.2016.258, 2017.

Hübener, H., Bülow, K., Fooken, C., Früh, B, Hoffmann, P, Höpp, S., Keuler, K., Menz, C., Mohr, V., Radtke K., Ramthun, H., Spekat, A., Steger, C., Toussaint, F., Warrach-Sagi, K., and Woldt, M. ReKliEs-De, Ergebnisbericht, HLNUG, Hessisches Landesamt für Naturschutz, Umwelt und Geologie, 2017.

Jacob, D., Bülow, K., Kotova, L., Moseley, C., Petersen, J., and Rechid, D.: Regionale Klimaprojektionen für Europa und Deutschland: Ensemble Simulationen für die Klimafolgenforschung, MPI für Meteorologie, Climate Service Center, 2012.

Jacob, D., Petersen, J., Eggert, B., Alias, A., Christensen, O. B., Bouwer, L. M., Braun, A., Colette, A., Déqué, M., Georgievski, G., Georgopoulou, E., Gobiet, A., Menut, L., Nikulin, G., Haensler, A., Hempelmann, N., Jones, C., Keuler, K., Kovats, S., Kröner, N., Kotlarski, S., Kriegsmann, A., Martin, E., Van Meijgaard, E., Moseley, C., Pfeifer, S., Preuschmann, S., Radermacher, C., Radtke, K., Rechid, D., Rounsevell, M., Samuelsson, P., Somot, S., Soussana, J.-F., Teichmann, C., Valentini, R., Vautard, R., Weber, B., and Yiou, P.: EURO-CORDEX: new high-resolution climate change projections for European impact research, Reg. Environ. Change, 14, 563–578, https://doi.org/10.1007/s10113-013-0499-2, 2014.

Kendall, M. G.: Rank correlation methods, London: Griffin, 160 pp., 1948.

Haylock, M. R., Hofstra, N., Tank, A., Klok, E. J., Jones, P. D., and New, M.: A European daily high-resolution gridded data set of surface temperature and precipitation for 1950–2006, J. Geophys. Res.-Atmos., 113, D20119, https://doi.org/10.1029/2008jd010201, 2008.

Hellwig, J., Stahl, K., Ziese, M., and Becker, A.: The impact of the resolution of meteorological data sets on catchment-scale precipitation and drought studies, Int. J. Climatol., 38, 3069–3081, https://doi.org/10.1002/joc.5483, 2018.

Heudorfer, B. and Stahl, K: Comparison of different threshold level methods for drought propagation analysis in Germany, Hydrol. Res., 48, 1311–1326, https://doi.org/10.2166/nh.2016.258, 2017.

Hübener, H., Bülow, K., Fooken, C., Früh, B, Hoffmann, P, Höpp, S., Keuler, K., Menz, C., Mohr, V., Radtke K., Ramthun, H., Spekat, A., Steger, C., Toussaint, F., Warrach-Sagi, K., and Woldt, M. ReKliEs-De, Ergebnisbericht, HLNUG, Hessisches Landesamt für Naturschutz, Umwelt und Geologie, 2017.

Jacob, D., Bülow, K., Kotova, L., Moseley, C., Petersen, J., and Rechid, D.: Regionale Klimaprojektionen für Europa und Deutschland: Ensemble Simulationen für die Klimafolgenforschung, MPI für Meteorologie, Climate Service Center, 2012.

Jacob, D., Petersen, J., Eggert, B., Alias, A., Christensen, O. B., Bouwer, L. M., Braun, A., Colette, A., Déqué, M., Georgievski, G., Georgopoulou, E., Gobiet, A., Menut, L., Nikulin, G., Haensler, A., Hempelmann, N., Jones, C., Keuler, K., Kovats, S., Kröner, N., Kotlarski, S., Kriegsmann, A., Martin, E., Van Meijgaard, E., Moseley, C., Pfeifer, S., Preuschmann, S., Radermacher, C., Radtke, K., Rechid, D., Rounsevell, M.,

Samuelsson, P., Somot, S., Soussana, J.-F., Teichmann, C., Valentini, R., Vautard, R., Weber, B., and Yiou, P.: EURO-CORDEX: new high-resolution climate change projections for European impact research, Reg. Environ. Change, 14, 563–578, https://doi.org/10.1007/s10113-013-0499-2, 2014.

Kendall, M. G.: Rank correlation methods, London: Griffin, 160 pp., 1948.

Koffler, D., Gauster, T., and Laaha, G.: lfstat: Calculation of Low Flow Statistics for Daily Stream Flow Data. R package version 0.9.4. available at: https://CRAN.R-project.org/package=lfstat (last access: 28 November 2018), 2016.

Kulkarni, A. and von Storch, H.: Monte Carlo experiments on the effect of serial correlation on the Mann-Kendall test of trend, Meteorol. Z., 4, 82–85, 1995.

Kopp, B., Baumeister, C., Gudera, T., Hergesell, J., Morhard, A., and Neumann, J.: Entwicklung von Bodenwasserhaushalt und Grundwasserneubildung in Baden-Württemberg, Bayern, Rheinland-Pfalz und Hessen von 1951 bis 2015, Hydrol. Wasserbewirts., 62, 62–76, https://doi.org/10.5675/HyWa_2018,2_1, 2018.

Kumar, R., Musuuza, J. L., Van Loon, A. F., Teuling, A. J., Barthel, R., Ten Broek, J., Mai, J., Samaniego, L., and Attinger, S.: Multiscale evaluation of the Standardized Precipitation Index as a groundwater drought indicator, Hydrol. Earth Syst. Sci., 20, 1117–1131, https://doi.org/10.5194/hess-20-1117-2016, 2016.

Laaha, G., Gauster, T., Tallaksen, L. M., Vidal, J.-P., Stahl, K., Prudhomme, C., Heudorfer, B., Vlnas, R., Ionita, M., Van Lanen, H. A. J., Adler, M.-J., Caillouet, L., Delus, C., Fendekova, M., Gailliez, S., Hannaford, J., Kingston, D., Van Loon, A. F., Mediero, L., Osuch, M., Romanowicz, R., Sauquet, E., Stagge, J. H., and Wong, W. K.: The European 2015 drought from a hydrological perspective, Hydrol. Earth Syst. Sci., 21, 3001–3024, https://doi.org/10.5194/hess-21-3001-2017, 2017.

Leelaruban, N., Padmanabhan, G., and Oduor, P.: Examining the Relationship between Drought Indices and Groundwater Levels, Water, 9, 82, https://doi.org/10.3390/w9020082, 2017.

Li, B. and Rodell, M.: Evaluation of a model-based groundwater drought indicator in the conterminous US, J. Hydrol., 526, 78–88, https://doi.org/10.1016/j.jhydrol.2014.09.027 2015.

Lins, H. F. and Slack, J. R.: Streamflow trends in the United States, Geophys. Res. Lett., 26, 227–230, https://doi.org/10.1029/1998gl900291, 1999.

Lorenzo-Lacruz, J., Vicente-Serrano, S. M., López-Moreno, J. I., Morán-Tejeda, E., and Zabalza, J.: Recent trends in Iberian streamflows (1945–2005), J. Hydrol., 414, 463–475, https://doi.org/10.1016/j.jhydrol.2011.11.023, 2012.

Lyne, V. and Hollick, M.: Stochastic time-variable rainfall-runoff modelling, Paper presented at the Institute of Engineers Australia National Conference, Perth, Australia, 1979.

Marx, A., Kumar, R., Thober, S., Rakovec, O., Wanders, N., Zink, M., Wood, E. F., Pan, M., Sheffield, J., and Samaniego, L.: Climate change alters low flows in Europe under global warming of 1.5, 2, and 3 °C, Hydrol. Earth Syst. Sci., 22, 1017–1032, https://doi.org/10.5194/hess-22-1017-2018, 2018.

McKee, T. B., Doesken, N., J., and Kleist, J.: The relationship of drought frequency and duration to time scales. Eighth Conference on Applied Climatology, Anaheim CA, 6, 1993.

Mei, Y. and Anagnostou, E. N.: A hydrograph separation method based on information from rainfall and runoff records, J. Hydrol., 523, 636–649, https://doi.org/10.1016/j.jhydrol.2015.01.083, 2015.

Mishra, A. K. and Singh, V. P.: A review of drought concepts, J. Hydrol., 391, 204–216, https://doi.org/10.1016/j.jhydrol.2010.07.012, 2010.

Nathan, R. J. and McMahon, T. A.: Evaluation of automated techniques for base flow and recession analyses, Water Resour. Res., 26, 1465–1473, 1990.

Peters, E., Torfs, P., van Lanen, H. A. J., and Bier, G.: Propagation of drought through groundwater – a new approach using linear reservoir theory, Hydrol. Process., 17, 3023–3040, https://doi.org/10.1002/hyp.1274, 2003.

Peters, E., Bier, G., van Lanen, H. A. J., and Torfs, P.: Propagation and spatial distribution of drought in a groundwater catchment, J. Hydrol., 321, 257–275, https://doi.org/10.1016/j.jhydrol.2005.08.004, 2006.

R Core Team: R: A language and environment for statistical computing. R Foundation for Statistical Computing, Vienna, Austria, available at: https://www.R-project.org/ (last access: 28 November 2018), 2016.

Rennermalm, A. K., Wood, E. F., and Troy, T. J.: Observed changes in pan-arctic cold-season minimum monthly river discharge, Clim. Dynam., 35, 923–939, https://doi.org/10.1007/s00382-009-0730-5, 2010.

Sheffield, J., Wood, E. F., and Roderick, M. L.: Little change in global drought over the past 60 years, Nature, 491, 435–438, 2012.

Spinoni, J., Naumann, G., and Vogt, J. V.: Pan-European seasonal trends and recent changes of drought frequency and severity, Global Planet. Change, 148, 113–130, https://doi.org/10.1016/j.gloplacha.2016.11.013, 2017.

Stahl, K., Hisdal, H., Hannaford, J., Tallaksen, L. M., van Lanen, H. A. J., Sauquet, E., Demuth, S., Fendekova, M., and Jódar, J.: Streamflow trends in Europe: evidence from a dataset of near-natural catchments, Hydrol. Earth Syst. Sci., 14, 2367–2382, https://doi.org/10.5194/hess-14-2367-2010, 2010.

Stahl, K., Tallaksen, L. M., Hannaford, J., and van Lanen, H. A. J.: Filling the white space on maps of European runoff trends: estimates from a multi-model ensemble, Hydrol. Earth Syst. Sci., 16, 2035–2047, https://doi.org/10.5194/hess-16-2035-2012, 2012.

Staudinger, M., Stoelzle, M., Seeger, S., Seibert, J., Weiler, M., and Stahl, K.: Catchment water storage variation with elevation, Hydrol. Process., 31, 2000–2015, https://doi.org/10.1002/hyp.11158, 2017.

Stoelzle, M., Stahl, K., Morhard, A., and Weiler, M.: Streamflow sensitivity to drought scenarios in catchments with different geology, Geophys. Res. Lett., 41, 6174–6183, https://doi.org/10.1002/2014gl061344, 2014.

Tallaksen, L. M. and Stahl, K.: Spatial and temporal patterns of large-scale droughts in Europe: Model dispersion and performance, Geophys. Res. Lett., 41, 429–434, https://doi.org/10.1002/2013gl058573, 2014.

Tallaksen, L. M., Hisdal, H., and Van Lanen, H. A. J.: Space-time modelling of catchment scale drought characteristics, J. Hydrol., 375, 363–372, https://doi.org/10.1016/j.jhydrol.2009.06.032, 2009.

Van Lanen, H. A. J., Laaha, G., Kingston, D. G., Gauster, T., Ionita, M., Vidal, J. P., Vlnas, R., Tallaksen, L. M., Stahl, K., Hannaford, J., Delus, C., Fendekova, M., Mediero, L., Prudhomme, C., Rets, E., Romanowicz, R., Gailliez, S., Wong, W. K., Adler, M.-J., Blauhut, V., Caillouet, L., Chelcea, S., Frolova, N., Gudmundsson, L., Hanel, M., Haslinger,

K., Kireeva, M., Osuch, M., Sauquet, E., Stagge, J. H., and Van Loon, A. F.: Hydrology needed to manage droughts: the 2015 European case, Hydrol. Process., 30, 3097–3104, https://doi.org/10.1002/hyp.10838, 2016.

Van Loon, A. F. and Van Lanen, H. A. J.: A process-based typology of hydrological drought, Hydrol. Earth Syst. Sci., 16, 1915–1946, https://doi.org/10.5194/hess-16-1915-2012, 2012.

Van Loon, A. F., Kumar, R., and Mishra, V.: Testing the use of standardised indices and GRACE satellite data to estimate the European 2015 groundwater drought in near-real time, Hydrol. Earth Syst. Sci., 21, 1947–1971, https://doi.org/10.5194/hess-21-1947-2017, 2017.

Van Vliet, M. T. H., Donnelly, C., Stromback, L., Capell, R., and Ludwig, F.: European scale climate information services for water use sectors, J. Hydrol., 528, 503–513, https://doi.org/10.1016/j.jhydrol.2015.06.060, 2015.

WMO: World Meteorological Organization: Manual on Low-flow Estimation and Prediction, Operational Hydrology Report No. 50, WMO-NO. 1029, 2008.

Yevjevich, V. M.: An objective approach to definitions and investigations of continental hydrologic droughts, Hydrology Papers, 23, 1967.

Zebisch, M., Grothmann, T., Schröter, D., Hasse, C., Fritsch, U., and Cramer, W.: Climate Change in Germany, Vulnerability and Adaptation of climate sensitive Sectors, Environmental research of the federal ministry of the environment, nature conservation and nuclear safety, Research Report 20141253, 2005.

8

Contributions to uncertainty related to hydrostratigraphic modeling using multiple-point statistics

Adrian A. S. Barfod[1,2], **Troels N. Vilhelmsen**[2], **Flemming Jørgensen**[1], **Anders V. Christiansen**[2], **Anne-Sophie Høyer**[1], **Julien Straubhaar**[3], **and Ingelise Møller**[1]

[1]Department of Groundwater and Quaternary Geological Mapping, Geological Survey of Denmark & Greenland (GEUS), C.F. Møllers Allé 8, 8000 Aarhus C, Denmark

[2]Hydrogeophysics Group, Department of Geoscience, Aarhus University, C.F. Møllers Allé 4, 8000 Aarhus C, Denmark

[3]Centre d'Hydrogéologie et de Géothermie (CHYN), Université de Neuchâtel, Neuchâtel, Switzerland

Correspondence: Adrian A. S. Barfod (adrian.s.barfod@gmail.com)

Abstract. Forecasting the flow of groundwater requires a hydrostratigraphic model, which describes the architecture of the subsurface. State-of-the-art multiple-point statistical (MPS) tools are readily available for creating models depicting subsurface geology. We present a study of the impact of key parameters related to stochastic MPS simulation of a real-world hydrogeophysical dataset from Kasted, Denmark, using the snesim algorithm. The goal is to study how changes to the underlying datasets propagate into the hydrostratigraphic realizations when using MPS for stochastic modeling. This study focuses on the sensitivity of the MPS realizations to the geophysical soft data, borehole lithology logs, and the training image (TI). The modeling approach used in this paper utilizes a cognitive geological model as a TI to simulate ensemble hydrostratigraphic models. The target model contains three overall hydrostratigraphic categories, and the MPS realizations are compared visually as well as quantitatively using mathematical measures of similarity. The quantitative similarity analysis is carried out exhaustively, and realizations are compared with each other as well as with the cognitive geological model.

The results underline the importance of geophysical data for constraining MPS simulations. Relying only on borehole data and the conceptual geology, or TI, results in a significant increase in realization uncertainty. The airborne transient electromagnetic SkyTEM data used in this study cover a large portion of the Kasted model area and are essential to the hydrostratigraphic architecture. On the other hand, the borehole lithology logs are sparser, and 410 boreholes were present in this study. The borehole lithology logs infer local changes in the immediate vicinity of the boreholes, thus, in areas with a high degree of geological heterogeneity, boreholes only provide limited large-scale structural information. Lithological information is, however, important for the interpretation of the geophysical responses. The importance of the TI was also studied. An example was presented where an alternative geological model from a neighboring area was used to simulate hydrostratigraphic models. It was shown that as long as the geological settings are similar in nature, the realizations, although different, still reflect the hydrostratigraphic architecture. If a TI containing a biased geological conceptualization is used, the resulting realizations will resemble the TI and contain less structure in particular areas, where the soft data show almost even probability to two or all three of the hydrostratigraphic units.

1 Introduction

Geological models are important from both a societal and economic perspective, since they are used to locate essential natural resources, such as freshwater, oil, metals, and rare earth minerals. Additionally, they are used in risk assessment related to natural hazards, such as earthquakes, sinkholes, volcanic eruptions, and landslides. Building 3-D models depicting real-world subsurface geology is no trivial task. Information from multiple sources is required, i.e., conceptual geological understanding, geological information, lithology logs, and geophysical data. Such data are sparse, uncertain, and redundant. Dataset gaps force geoscientists to make uncertain predictions or estimates, which carries over into the resulting geological model. During the modeling procedure, such problems are dealt with as best as possible. Gaps in knowledge will render the resulting model uncertain, and quantifying such uncertainty is essential to making better use of the models and to making better predictions.

A common approach for building geological models is cognitive modeling (e.g., Jørgensen et al., 2013; Royse, 2010). Here, the datasets containing borehole lithology logs and geophysical models are co-interpreted by a professional with experience in the fields of geoscience, geophysics, and geological modeling, with a relevant regional conceptual model in mind. This modeling approach is deterministic and results in a single model realization. These specialists are trained in assessing the uncertainty of the underlying structures, and qualitative uncertainty estimates are often made on the structural model (e.g., indicating different levels of uncertainty in different subparts of the model domain). However, qualitative uncertainty estimates are difficult to carry over into the subsequent analysis, and the effect of the uncertainty of the geological model can therefore not be quantified in the resulting forecasts. If the forecasts are based on a single geologic model, the prediction does not encase the full complexity of the problem. Alternatively, if the model uncertainty can be quantified, it enables the option to include it in the forecast. However, quantifying the uncertainty in a cognitive modeling approach is difficult and tedious (Seifert et al., 2012). Another approach is stochastic modeling using multiple-point statistics (MPS) methodologies, where ensembles of models are produced, e.g., Comunian et al. (2012), Ferré (2017), He et al. (2016), Okabe and Blunt (2005), and Pirot (2017). MPS provides a framework which can integrate geophysical and borehole information, as well as conceptual geological information via a so-called training image (TI). Multiple model realizations are created from the dataset, and the resulting model ensemble reflects the uncertainty related to the underlying datasets and overall modeling procedure.

Studying model uncertainty using MPS results in numerous sets of model ensembles. Visual comparison of such a large number of realization ensembles is tedious and subjective but offers an overall understanding of the geological realism of the models (Barfod et al., 2018). Therefore, quantitative measures of similarity are desirable. Barfod et al. (2018) present a comparison of 3-D hydrostratigraphic models using the modified Hausdorff distance (d_{MH}). However, the d_{MH} was proven to be computationally expensive. Therefore, an alternative computationally feasible distance measure based on Euclidean distance transforms (EDT) (Maurer et al., 2003) is used in this paper. Generally, numerous mathematical methods for comparing images exist in the computer vision literature, e.g., image Euclidean distance (IMED) (Wang et al., 2005; Xiaofeng and Wei, 2008) and scale-invariant feature transform (SIFT) (Lowe, 2004). However, these alternative distance measures are, to our knowledge, an unexplored research avenue within comparison and uncertainty analysis of ensembles of 3-D hydrostratigraphic models.

Some types of forecasts are related to groundwater flow and accurate geological models are crucial to accurate predictions of hydraulic flow, since subsurface hydraulic flow is largely controlled by geological heterogeneity (e.g., Feyen and Caers, 2006; Fleckenstein et al., 2006; Fogg et al., 1998; Gelhar, 1984; LaBolle and Fogg, 2001; Zhao and Illman, 2017). Geological units, however, contain additional complexities not related to hydrologic units; therefore, the concept of hydrostratigraphic units is very useful. A detailed definition of hydrostratigraphic classification is given by Maxey (1964).

Stochastic geological modeling procedures like the MPS methodology apply two overall types of data, i.e., geophysical data and borehole data. Associated with these data types are the definitions of hard and soft data. Typically, hard data are considered certain information without an associated uncertainty, while soft data are uncertain information, which can be associated with an uncertainty. Geophysical data are typically considered soft data (Strebelle, 2002). Geophysical data are spatially dense and provide a smeared image of the overall subsurface geology. Resolution decreases with depth, especially beyond a specific depth, which is dependent on the geophysical method. Geophysical instruments portray bulk physical properties of the subsurface. Although geophysical data provide spatially dense information, it is not possible to exhaustively sample the subsurface. The density of the geophysical data will affect the final uncertainty. The raw geophysical data goes through a processing and modeling step, where the raw data are translated into geophysical models. During this step incorrect measurements, due to instrument error or interference, are identified and removed, further decreasing the geophysical information density. Such incomplete data can either be reconstructed or used as is during modeling. Another consideration in regards to modeling of geophysical data is the choice of inversion scheme and thereby the choice of a priori information (e.g., Ellis and Oldenburg, 1994; Tarantola and Valette, 1982). Here, several approaches can be taken, which yield different geophysical models. A common inversion scheme for airborne electromagnetic (AEM) data, such as SkyTEM data, is the spatially constrained inversion (SCI) (Viezzoli et al., 2008). However, this inversion approach does not represent the subsurface properly, e.g., layer boundaries are smeared and extreme values are not represented properly. A so-called sharp inversion scheme, suggested by Vignoli et al. (2015), tackles such issues. Therefore, the choice of inversion scheme influences the hydrostratigraphic model and should be considered as an integral step in the hydrostratigraphic modeling process.

The other overall type of data source is borehole lithology logs, which are commonly considered to be ground truth or hard data (e.g., Gunnink and Siemon, 2015; Tahmasebi et al., 2012). However, lithology logs are also uncertain. Generally, the uncertainty of borehole lithology logs relates to a number of parameters, such as drilling methods, the frequency with which sediment samples are collected precision with which the location is measured, the purpose of the borehole, and the choice of drilling contractor – see Barfod et al. (2016) and He et al. (2014) for more details. The resolution of borehole lithology logs is especially dependent on the sampling method. If a core is extracted for the entirety of the borehole, the resolution is, in principal, unlimited. However, this is expensive. It is more common to use either an auger drill, a rotary drill, or a cable tool, which yields a relatively limited resolution, compared to core drilling, depending on how samples are collected and handled.

We present a study of the uncertainty related to stochastic MPS modeling of hydrogeophysical data. The goal is (i) to understand the consequences of modifying the MPS setup to reflect some of the biases related to real-world hydrogeophysical datasets and (ii) to study the propagation of uncertainty in the resulting ensembles of stochastic hydrostratigraphic models. An essential choice during MPS modeling is, for example, the choice of TI, or how to incorporate borehole and/or geophysical data. A hydrogeophysical dataset, consisting of an airborne transient electromagnetic survey, numerous lithological borehole logs, and 3-D categorical TIs, is used to assess how different choices during MPS modeling influence the resulting uncertainty of the hydrostratigraphic model ensemble.

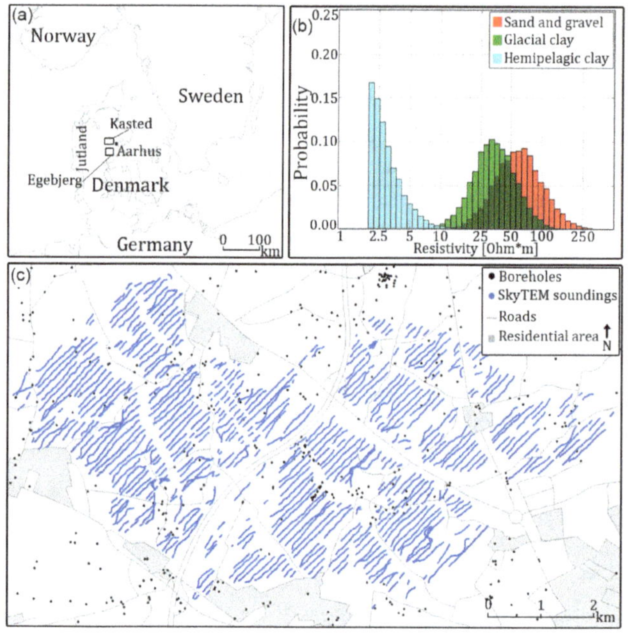

Figure 1. The Kasted survey area and resistivity–hydrostratigraphic relationship histograms. **(a)** shows the geographical location of the Kasted survey area and the Egebjerg model used as a secondary TI. **(b)** shows the reconstructed resistivity–hydrostratigraphic relationship histograms for the three main hydrostratigraphic unit categories based on SkyTEM resistivity models and borehole lithology logs. **(c)** shows the Kasted survey with the SkyTEM sounding and borehole locations. Note that the SkyTEM soundings are sampled so densely that the dots marking each individual sounding merge into blue lines.

2 The Kasted study area

The Kasted area is located northwest of Aarhus, Denmark (Fig. 1a), and has also been presented by Barfod et al. (2018), Marker et al. (2017), and Høyer et al. (2015). The regional geology of the Kasted area is dominated by a Quaternary buried valley complex with complex abutting relationships between the individual valleys. The buried valleys are infilled with a combination of till and glacial meltwater deposits. The valleys are incised into the substratum, which consists of hemipelagic clay. The regional geology has been described in detail by Høyer et al. (2015), who created a detailed cognitive geological model of the area.

An important geological feature of the Danish subsurface is the buried tunnel valleys (e.g., Jørgensen and Sandersen, 2006; Sandersen et al., 2009). The geological heterogeneity varies considerably across Denmark and can in some places be quite complicated, such as in the Egebjerg area (Jørgensen et al., 2010). In the Kasted area, the main buried valleys are clearly outlined in the geophysical dataset thanks to the significant resistivity contrasts between the infill of the buried valleys and the underlying Paleogene clay (Høyer et al., 2015). Therefore, the Kasted survey is ideal for studying the uncertainty related to stochastic hydrostratigraphic modeling using MPS methods.

The survey covers an area of 45 km^2 and is composed of a spatially dense SkyTEM survey with a total of 333 line km (Sørensen and Auken, 2004), with a line spacing of 100 m. The resulting SkyTEM soundings have been processed according to the description by Auken et al. (2009). Finally two sets of geophysical models were produced using either the smooth spatially constrained inversion models (Constable et al., 1987; Viezzoli et al., 2008) or the sharp SCI (sSCI) models (Vignoli et al., 2015). Furthermore, there are 948 boreholes scattered throughout the Kasted survey area, each with a corresponding lithology log of a varying quality. The quality assessment presented by He et al. (2014) and Barfod et al. (2016) is used to divide the boreholes into quality groups, ranking between 1 and 5. Only 410 of the boreholes are above the selected quality threshold, i.e., within quality groups 1–3, and contain lithological information relevant to this study. An overview of the dataset is found in Fig. 1c and described in further detail in Barfod et al. (2018) and Høyer et al. (2015).

3 Methods

3.1 Multiple-point statistics (MPS)

The multiple-point statistics framework stems from the general geostatistics framework. Here, multiple-point (MP) information from a training image is used to condition simulations to probable geological patterns (Journel and Zhang, 2007). The TI thus provides a conceptual geological understanding of a given area and can be viewed as a database containing probable geological patterns, which are used to condition the MPS simulation. The choice of TI is an important step in any MPS setup and influences the realization results, as will be illustrated. The TI does not need to carry locally accurate information – i.e., the TI does not need to spatially or geographically overlap with real-world geological units – and can be purely conceptual in nature. Together with the TI, it is also possible to use geophysical datasets for constraining MPS simulations, resulting in realizations that reflect real-world regional geology. Today, MPS is a widely used tool, which is used in a variety of geoscience fields, including, but not limited to, reservoir modeling (e.g., Okabe and Blunt, 2004; Strebelle and Journel, 2001), hydrology (e.g., Le Coz et al., 2011; Hermans et al., 2015; Høyer et al., 2017), and geological modeling (e.g., de Iaco and Maggio, 2011).

3.1.1 Single normal equation simulation (snesim)

The MPS method used in this paper is known as the single normal equation simulation (snesim) framework (Strebelle, 2002), and is implemented in the Stanford Geostatistical Modeling Software, or SGeMS. The snesim framework allows for simulating a real-world categorical geological model using a TI, constrained using soft geophysical data and hard borehole data. The snesim algorithm scans the entire TI, ahead of simulation, and stores the MP information contained in the TI in a search-tree database. The MP information can then be retrieved from the database during simulation. The integration of soft geophysical data for constraining the simulations is achieved by utilizing the tau model, which will be described in detail in Sect. 3.2 (Journel, 2002; Krishnan, 2004). Here, the continuous soft data variable needs to be translated into a probability grid, describing the probability of finding a given geological unit based on the geophysical data; see Barfod et al. (2018). In order to guarantee the reproduction of geological patterns at all scales, snesim uses the multiple-grid formulation, presented by Tran (1994).

3.1.2 Reconstructing incomplete datasets using direct sampling

In the field of geoscience, we are always dealing with incomplete datasets, since we cannot sample the subsurface exhaustively. Several approaches exist for dealing with incomplete datasets, of which two general approaches can be defined. A common approach is to reconstruct incomplete datasets using geostatistical tools (e.g., Goovaerts, 1997; Mariethoz and Renard, 2010), which means that during the hydrostratigraphic modeling process no information is present in the dataset gaps. However, it is important to emphasize that the reconstructed information is not as valuable as the actual measured geophysical information. The other common approach is to just use the incomplete dataset as is. This means that no information is present in the dataset gaps during the hydrostratigraphic modeling process, which, depending on the modeling method, might result in large uncertainties.

In this study the MPS method called direct sampling (DS) is used for stochastic reconstruction of incomplete datasets (Mariethoz and Renard, 2010). The DS method uses the dataset we wish to reconstruct both as a simulation grid and a TI. This means that the patterns that are present in the incomplete dataset are inserted into the simulation grid before reconstruction. It is, according to Mariethoz and Renard (2010), important that the patterns we wish to reconstruct are actually present in the incomplete dataset, since we are borrowing the patterns from the TI, or incomplete dataset, to stochastically reconstruct the dataset. If the patterns are not present in the incomplete dataset they will, simply put, not be inferred in the reconstructed dataset. Provided that enough information on the overall patterns is available in the incomplete dataset, the DS method is a straightforward approach for reconstructing incomplete datasets.

3.2 The tau model: combining conditional probabilities

Combining information from different sources is a frequent challenge in subsurface modeling. A fundamental challenge of the research conducted in this paper was to combine conditional probabilities from different sources. In this paper we used the common tau model approach (Journel, 2002). The tau model generally combines the probability values from different sources using Bayes' theorem and a set of τ values, or τ weights, for determining how to weigh the probabilities. The choice of τ weights is subjective, and assigning these is not a trivial task. It is recommended to run a series of exhaustive tests when assigning the τ weights.

We will now briefly introduce the tau model; for more details see Journel (2002). Firstly, we define a data event, D_i, as a location vector, $\boldsymbol{u}$, and a data value. Suppose we have a set of such data events, $D_i i = 1, \ldots, n$, and the goal is to estimate the probability that a hydrostratigraphic unit (A) is present provided all data events:

$$P(A|\mathbf{D}_1, \ldots, \mathbf{D}_n). \tag{1}$$

The first step is then to define the prior probability distribution, $P(A)$. Generally the tau model can be applied to as many different probability grids as desired, but for the purpose of simplification two probability distributions are defined as follows: $P(A|\mathbf{D}_1)$ and $P(A|\mathbf{D}_2)$. In this study we will consider $\mathbf{D}_1$ and $\mathbf{D}_2$ as 2-D or 3-D probability grids from different sources. As an example $\mathbf{D}_1$ could be a probability grid from geophysical data and $\mathbf{D}_2$ a probability grid from borehole lithology logs. The probability grids are translated into distance grids by applying the "probability-into-distance" transform:

$$x_o = \frac{1-P(A)}{P(A)}, x_1 = \frac{1-P(A\,|\mathbf{D}_1)}{P(A\,|\mathbf{D}_1)},$$
$$\text{and } x_2 = \frac{1-P(A\,|\mathbf{D}_2)}{P(A\,|\mathbf{D}_2)}. \tag{2}$$

Then the following distance ratio is computed using the tau model expression:

$$\frac{x}{x_0} = \prod_{i=1}^{n=2}\left(\frac{x_i}{x_0}\right)^{\tau_i}, \tau\epsilon[-\infty;+\infty], \tag{3}$$

where the tau values are $[\tau_1, \tau_2]$. The final conditional probability is computed as follows:

$$P(A|\mathbf{D}_1,\mathbf{D}_2) = \frac{1}{1+x}, \tag{4}$$

where the value of x is computed from Eq. (2), as follows:

$$x = x_0 \cdot \left(\frac{x_1}{x_0}\right)^{\tau_1} \cdot \left(\frac{x_2}{x_0}\right)^{\tau_2}. \tag{5}$$

3.3 Comparing simulation results

Comparing a large set of extensive 3-D models is a common problem encountered in stochastic MPS modeling. A common approach is visual comparison, which is not an objective or quantitative comparison method. Each equiprobable hydrostratigraphic model in this study contains 1 187 823 cells. Furthermore, a total of 400 MPS realizations were computed, Table 1, which makes it difficult to visually compare modeling results. This, along with advances in stochastic modeling tools such as MPS, motivated Tan et al. (2014) to develop a framework in which multiple 2-D or 3-D realizations can be compared quantitatively. The idea is to use a distance measure, which measures the distance between two realizations. Realizations which are geometrically similar have small distance values, while dissimilar realizations have a large distance value. The comparison techniques in this study are based on the principles presented by Tan et al. (2014). In this study the distances between individual realizations are based on the Euclidean distance transforms (Maurer et al., 2003). The usage of EDT as a measure for similarity will be described in more detail below. A full distance matrix is computed containing distances between each individual realization for all the different cases. The resulting 400 by 400 distance matrix is then interpreted by itself.

3.3.1 Ensemble mode ratio maps (EMR maps)

The visual comparison can be helped by creating so-called ensemble mode ratio maps, or EMR maps. The idea is to create a summary map portraying the mode ratio of a given ensemble of models, ranging between $1/K$ and 1, where K is the number of hydrostratigraphic categories. The EMR maps describe the certainty of the simulation based on the resulting realization ensemble. If the EMR map shows a value of 1, then every single realization in the present ensemble has simulated the same category or, in this case, hydrostratigraphic unit. On the other hand, if the EMR map shows a ratio of $1/K$ the ensemble of realizations shows equal probability for each of the K categories. Each realization is equiprobable, and the EMR values of the categorical variables are computed from the probability distribution of a given cell with location, u. The probability that the attribute S is equal to s_k, $P_k(\boldsymbol{u})$, is computed as follows:

$$P_k(\boldsymbol{u}) = \frac{1}{N}\sum\nolimits_{i=1}^{N}\left(s_{k,i}(\boldsymbol{u}) = s_k\right), \tag{6}$$

where N is the number of realizations, s_k is the state of attribute S for which we are currently computing the probability, and $s_{k,i}(\boldsymbol{u})$ is the state of the attribute at location $\boldsymbol{u}$ and for the ith realization. The EMR values for a given cell, $\boldsymbol{u}$, can then be computed as follows:

$$r_{\text{EMR}}(\mathbf{u}) = (P_k(\mathbf{u})), \tag{7}$$

where $P_k(\boldsymbol{u})$ denotes the probability for category k at location $\boldsymbol{u}$ computed using Eq. (6).

The EMR values are then computed for each grid cell using Eqs. (6) and (7), which, simply put, is the occurrence ratio of the mode category of a given ensemble containing a given number of realizations, N_{reals}. In other words, at a given location, $\boldsymbol{u}$, if 45 out of 50 realizations yield the same category, then the EMR value is 0.9, and the ensemble certainty for the given cell is high. On the other hand, with three possible lithological categories, i.e., $K = 3$, the lowest possible certainty is $1/K = 1/3$, which means there is an equal probability of occurrence for each lithological category. This means that $P(s_1) = P(s_2) = P(s_3) = 1/3$, and therefore at the given location, $\boldsymbol{u}$, the $r_{\text{EMR}} = 1/3$ and the simulation is uncertain.

3.3.2 Euclidean distance transforms (EDT) – measuring similarity between 3-D hydrostratigraphic realizations

The hydrostratigraphic realizations are categorical and contain three hydrostratigraphic units. Comparing two realization grids, they first need to be transformed from a categorical grid into continuous Euclidean distance grids by using EDT (Maurer et al., 2003). The two 3-D EDT grids are then compared by calculating the average difference in the respective grids. Similar images have a small average EDT distance, and dissimilar images have a large average EDT distance. The EDT computes the Euclidean distances for all locations of a binary grid, i.e., a grid containing only two states (codes 0 and 1). The EDT map is simply the Euclidean distance of

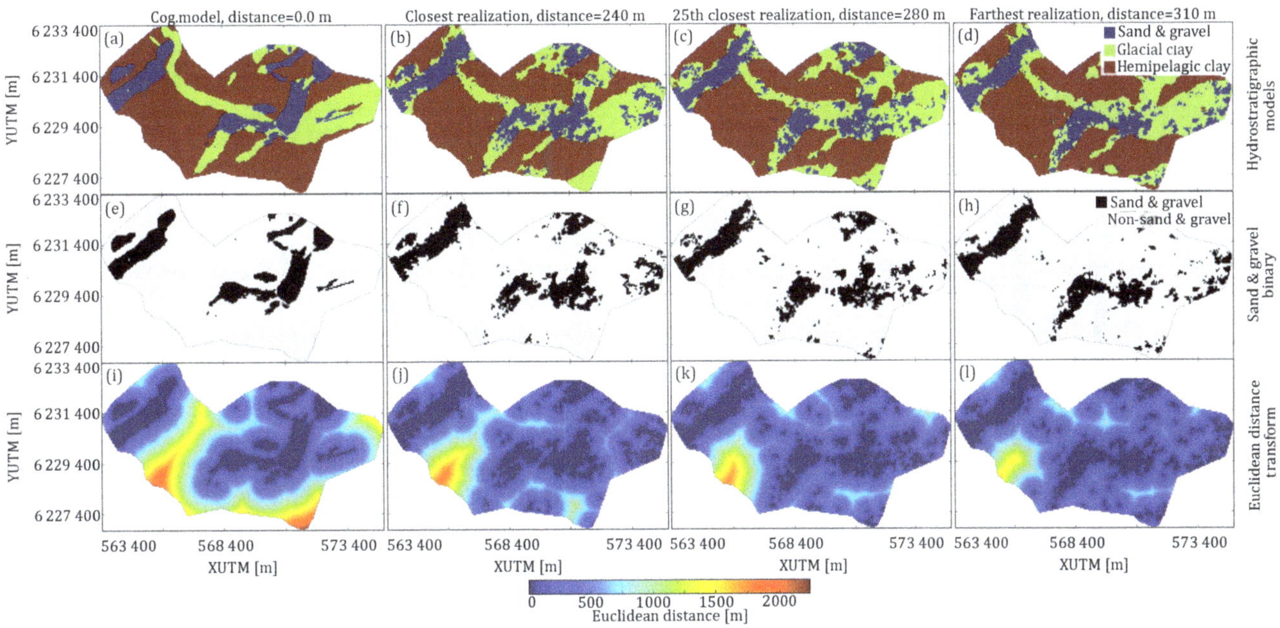

Figure 2. A 2-D example of the Euclidean distance transforms (EDT) as a measure for the similarity between categorical MPS realizations. In this example a set of 50 realizations, from the basic modeling setup, are compared based on the differences in EDT for sand and gravel units. **(a–d)** show the hydrostratigraphic models for the TI, closest realization, 25th closest realizations, and farthest realization, respectively. **(e–h)** show, in the same order as above, the binary images of the sand and gravel units of the 2-D hydrostratigraphic model layers. **(i–l)** show, in the same order as above, the Euclidean distances layers computed from the 2-D sand and gravel binary layers.

the medium depicted by the state code 1, i.e., for grid cell at location $\boldsymbol{u}$:

$$d_{\mathrm{EDT}}(\boldsymbol{u}) = (\|\boldsymbol{u} - \boldsymbol{v}\|_2), \tag{8}$$

where $\boldsymbol{v}$ is a set of grid cells with a state code equal to 1.

The d_{EDT} implementation presented by Maurer et al. (2003) uses a computationally favorable method for computing the exhaustive EDT at all locations in a binary grid.

To illustrate the d_{EDT} approach for comparing realizations a 2-D example case is presented. The basic modeling setup contains 50 realizations, i.e., $N_{\mathrm{realizations}} = 50$, which are going to be compared to the cognitive model, which in this case also happens to be the TI. The 2-D example is created by selecting the horizontal cross section at 20 m b.s.l., for each of the 50 basic modeling setup realizations and the single cognitive geological model (Fig. 2a–d). Each of the 2-D layers are transformed into 2-D binary layers, portraying sand and gravel as the main variable, and glacial clay and hemipelagic clay as a background variable (Fig. 2e–h). The 2-D binary layers are then translated into 2-D d_{EDT} layers by using Eq. (8) to exhaustively compute the d_{EDT} at each grid cell for all of the 50 realizations. The resulting d_{EDT} layers, of which three are seen in Fig. 2i–l, are used to compute an average Euclidean distance between each realization, $m_{r,i}$, and the cognitive geological model, m_{cog}:

$$\Delta d_{\mathrm{EDT}}\left(m_{\mathrm{cog}}, m_{r,i}\right) = \frac{1}{M}\sum\nolimits_{j=1}^{M}\left[d_{\mathrm{EDT}}^{m_{\mathrm{cog}}}(u_j) - d_{\mathrm{EDT}}^{m_{r,i}}(u_j)\right], \tag{9}$$

where $i \in \{1, \ldots, N\}$, with N being the number of realizations, which in this case is $N = 50$, and M being the number of cells in the simulation grid, or in this case, the 2-D layer. The Δd_{EDT}, Eq. (9), then describes the average difference of the distance to the nearest active cell in the binary grid. The 50 realizations are then ranked by the average Euclidean distance differences, Δd_{EDT}, as seen in Fig. 2, where the realization which is closest to the cognitive geological model (Fig. 2b, f and j) has a Δd_{EDT} value of 240 m, the realization which was ranked 25th closest (Fig. 2c, g and k) has a Δd_{EDT} value of 280 m, and lastly the realization which was farthest (Fig. 2d, h and l) has a Δd_{EDT} value of 310 m. It should be noted that the Δd_{EDT} computation, described by Eq. (9), is not limited to comparing a realization to a cognitive model and can in fact be used to compare any pair of 3-D categorical models. In fact, a generalized version of Eq. (9) can be defined as follows:

$$\Delta d_{\mathrm{EDT}}(m_{\mathrm{A}}, m_{\mathrm{B}}) = \frac{1}{M}\sum\nolimits_{j=1}^{M}\left[d_{\mathrm{EDT}}^{m_{\mathrm{A}}}(u_j) - d_{\mathrm{EDT}}^{m_{\mathrm{B}}}(u_j)\right], \tag{10}$$

where the number of cells in model-A, m_{A}, must be equal to the number of cells in model-B, m_B, i.e., $M_{m_{\mathrm{A}}} = M_{m_{\mathrm{B}}} = M$.

From this point forward we leave the 2-D example behind and will from here on only consider Δd_{EDT} computations on 3-D hydrostratigraphic grids. Furthermore, the Δd_{EDT} computations are carried out on a set of three binary grids, one for each of the three hydrostratigraphic categories. The distance value between two hydrostratigraphic grids is the summed

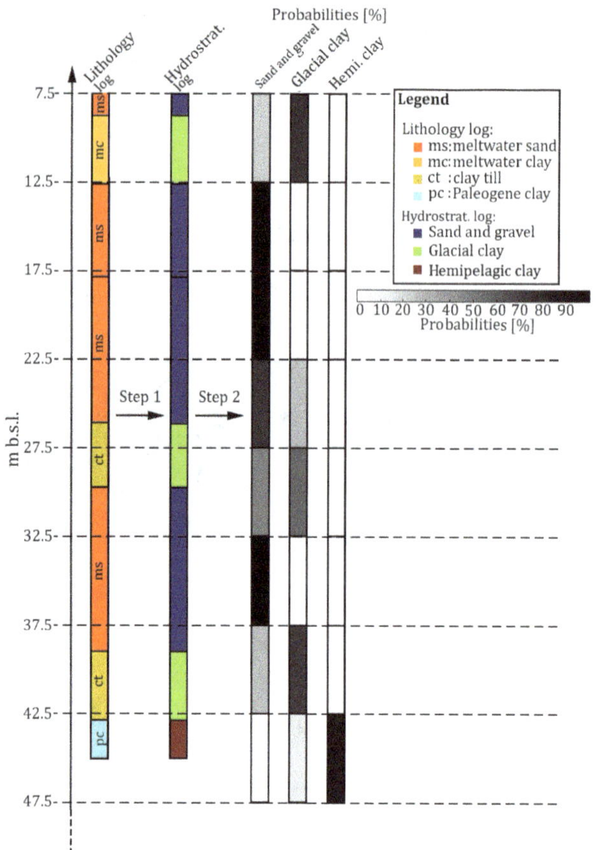

Figure 3. A schematic diagram presenting the conversion of the lithological logs into probability logs for the three hydrostratigraphic units: sand and gravel, glacial clay, and hemipelagic clay. Step 1: the lithology log is translated into a hydrostratigraphic log. Step 2: the hydrostratigraphic logs are resampled according to the vertical modeling grid intervals and an interval probability is calculated for each of the hydrostratigraphic units.

distance for each of the three hydrostratigraphic categories, ensuring that the distance values reflect the complexities related to each of the hydrostratigraphic categories.

3.3.3 Evaluating the distance matrix

The average Euclidean distance difference, Δd_{EDT}, from here on referred to as the distance between two realizations, is exhaustively computed between all realizations and compiled into an exhaustive 400 by 400 distance matrix. The distance matrix, **D**, contains all distance values between all hydrostratigraphic realizations computed using Eq. (9) and is defined as follows:

$$D_{i,j} = d_{\mathrm{EDT}}\left(m_{r,i}, m_{r,j}\right), \tag{11}$$

where $ij = \{1, \ldots, N\}$, with N being the number of realizations. The distance matrix, **D**, can be evaluated directly by comparing the distances between individual realizations to each other. Another option is to summarize the distance matrix in a table representing the distances between the different cases. This is achieved by organizing the distance matrix according to which case they belong to. In this study the distance matrix is sorted according to the order of the individual cases, as in Table 1. The distance matrix can then be summarized by computing the average distance for each group of realizations pertaining to a specific case. The concepts of distance variability and distance to cognitive model were presented by Barfod et al. (2018) and are also used here. The concept is that the variability pertaining to a specific case can be determined by computing the average of the distances of the 50 realizations for a given case ensemble. Another measure is the distance to the cognitive model. The distances between all realizations and the cognitive model are computed, and this provides a reference point to which the realizations are compared.

4 MPS modeling setup

The Kasted dataset used in this study is comprised of a dense geophysical dataset acquired using the SkyTEM system (Sørensen and Auken, 2004), borehole lithology logs and a cognitive geological model (Høyer et al., 2015). We show how uncertainty related to resistivity data, measured with the SkyTEM system, and borehole lithology logs influences the hydrostratigraphic modeling realizations. Two readily available MPS tools are showcased. The first tool is the direct sampling method for reconstruction of incomplete datasets (Mariethoz and Renard, 2010). The other MPS tool is the single normal equation simulation, which is used for stochastic hydrostratigraphic modeling (Strebelle, 2002). The MPS modeling setup is similar to the one presented by Barfod et al. (2018). However, the goal of this study is different. The study is divided into a total of six cases, or eight sub-cases, which are designed to study how perturbations of the underlying MPS setup affect the hydrostratigraphic realizations using snesim and study the propagation of uncertainties into the hydrostratigraphic models. First the basic case (case 0), from which the other cases are perturbed, uses a hydrostratigraphic simplification of the 3-D cognitive model of the Kasted area as a TI, the gap-filled SkyTEM data present as smooth inversion models as soft data, and the borehole lithological logs as hard data. Then, in case 1, the TI is substituted by two other TIs. The cases 2 and 3 are related to the SkyTEM data, where the incomplete SkyTEM data and sSCI inversion models are used, respectively. The influence of the boreholes is studied in case 4 either leaving out boreholes as hard data or changing them into soft data. Finally, in case 5 the SkyTEM data are not used. In other words, one of the overall goals of this study is to improve the Kasted model by using stochastic ensemble modeling to quantify the uncertainty of the model, such as suggested by Ferré (2017) and Pirot (2017). Further details on the individual cases follow in the coming sections and Table 1 summarizes each case. First of all, the model discretization and parameterization as well as construction of hard and soft data grids are described.

The Kasted model covers an area of 12 km by 7 km, discretized on a modeling grid with 229 by 133 by 39 cells, containing a total of 1 187 823 cells. Each cell has a size of 50 m by 50 m by 5 m. It is parameterized into three hydrostratigraphic units, listed as follows.

1. Sand and gravel: a combination of coarse lithological units, including sand till, meltwater sand, gravel and pebbles of glacial origin, late glacial freshwater sand, and postglacial freshwater sand.

2. Glacial clay: this category contains silty and sandy clays, including clay till and meltwater clay of glacial origin.

3. Hemipelagic clay: a combination of fine grained conductive clays, containing the extensive and homogeneous hemipelagic Paleogene and Oligocene clays found in Denmark.

These three categories serve the purpose of simplifying the geology of the Kasted area. The Kasted survey lithology logs reveal a combination of 59 geological categories, which are translated into a set of hydrostratigraphic logs (Step 1, Fig. 3) using these hydrostratigraphic categories. Similarly, the 42 geological units in the cognitive geological model are divided into the three abovementioned categories (Fig. 4a). The vertical proportions of the three-category hydrostratigraphic Kasted model can be viewed in Fig. 5a.

The borehole lithology logs need to be assigned to a 3-D grid, which is carried out in three overall steps. The first step is to translate the borehole lithology logs into hydrostratigraphic logs using the above mentioned three categories; Step 1, Fig. 3. The second step is then to divide the hydrostratigraphic logs into intervals identical to the vertical intervals of the model grid. At each resampled interval a probability value is directly calculated for each hydrostratigraphic unit; Step 2, Fig. 3. The probability is simply the percentage of the given unit, which is present within the interval. Finally, the last step is to assign the hydrostratigraphic probabilities to a grid. The probability values are assigned to the grid cell in which the given hydrostratigraphic log is present. On the rare occasion that multiple logs are present within a given cell, the probabilities are combined accordingly to one representative probability value. The end result is a grid containing the borehole probability values of each hydrostratigraphic unit: sand and gravel, glacial clay, and hemipelagic clay. It is common to view borehole lithology logs as hard information, or ground truth. The borehole probability grid can therefore be translated into a hard data grid, by assigning the most probable hydrostratigraphic unit in each grid cell.

The 1-D SkyTEM resistivity models are assigned to a 3-D grid, identical to the modeling grid. The first step is to fill all

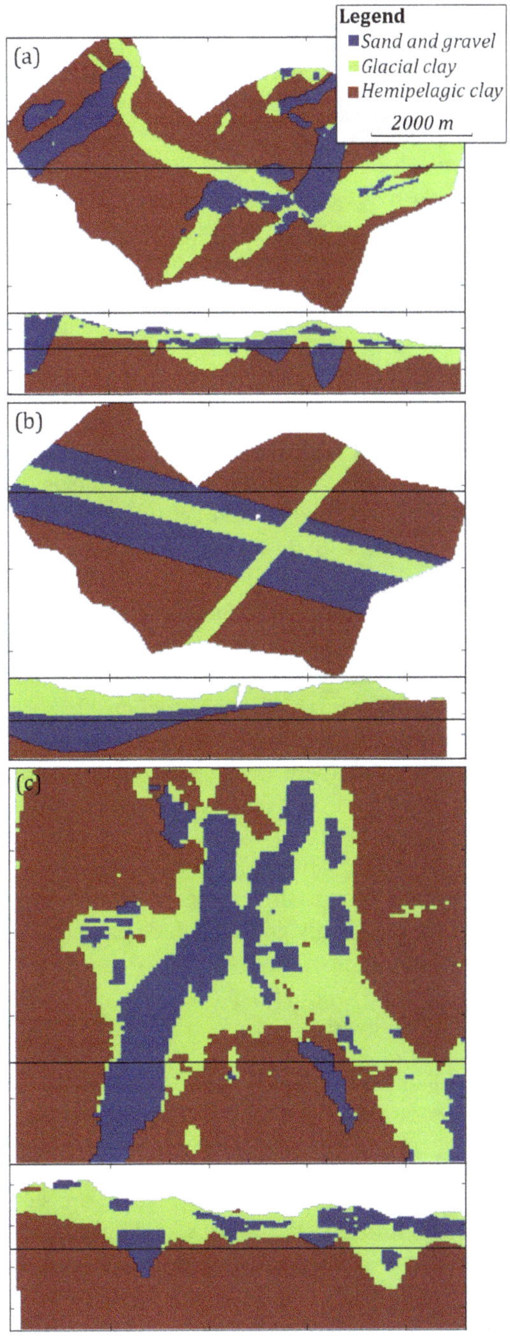

Figure 4. An overview of the training images (TIs) which are used during MPS simulation. A horizontal slice and vertical cross section is presented for each TI, portraying the hydrostratigraphic architecture; **(a)** shows the Kasted TI, **(b)** shows the conceptual TI, and **(c)** shows the Egebjerg TI, which is generally larger than the Kasted model.

grid cells containing a resistivity model. This is carried out using block kriging and results in an incomplete resistivity grid of block average resistivities (Fig. 6a). The second and final step is to stochastically reconstruct the incomplete resistivity grid using DS stochastic reconstruction (Mariethoz and Renard, 2010) (Fig. 7). The reconstruction procedure was originally presented by Barfod et al. (2018); however, we have made some improvements for this study. Originally, a simple kriging estimation approach was used to assign the resistivity models to a 3-D modeling grid. This resulted in an incomplete resistivity grid, which contained resistivity information not only pertaining to grid cells containing a SkyTEM sounding; i.e., the resistivity grid had already been partly reconstructed in the proximity of the geophysical soundings. To avoid this, block kriging estimation was used instead. The block kriging method is also a variogram-based estimation method, which estimates the average value of a rectangular block (Goovaerts, 1997). For more details on reconstructing incomplete resistivity grids see Barfod et al. (2018).

The SGeMS snesim framework utilizes the tau model for soft data conditioning (Journel, 2002), which requires the translation of resistivity grids into probability grids. This requires information on the regional resistivity–hydrostratigraphic relationship. Such knowledge is not always available, but if enough boreholes and electromagnetic geophysical data are available, the framework for studying the resistivity–hydrostratigraphic relationship, presented by Barfod et al. (2016), can be used to create a set of histograms. The resistivity–hydrostratigraphic histograms, Fig. 1b, are compiled from available hydrostratigraphic logs and SkyTEM resistivity models and are presented in more detail in Barfod et al. (2016, 2018). The estimated histograms (Fig. 1b) are then used to directly translate each resistivity value, in a given resistivity grid, into three probabilities, one for each hydrostratigraphic unit.

The general MPS workflow can be summarized in seven overall steps as follows:

1. Using block kriging, the SkyTEM resistivity models are assigned to a 3-D grid identical to the Kasted model grid.
2. The incomplete resistivity grids (Fig. 6a) are stochastically reconstructed using direct sampling (Fig. 7a). The result is an ensemble of 50 equiprobable reconstructed resistivity grids.
3. The reconstructed resistivity grids are translated into probability grids using the resistivity–hydrostratigraphic relationship histograms (Fig. 1b).
4. The borehole lithology logs are translated into hydrostratigraphic logs; Step 1, Fig. 3.
5. The hydrostratigraphic logs are resampled and three probability values, one for each hydrostratigraphic unit, are directly computed at each resampled interval; Step 2, Fig. 3.
6. The borehole probabilities are assigned to a grid identical to the cognitive Kasted model grid.
7. The borehole probability grid is translated into a hard data grid, by assigning the most likely hydrostratigraphic unit to each grid cell.

A total of 400 realizations are created, with 50 realizations per sub-case (Table 1). In snesim a random number seed needs to be manually selected for each realization to initialize the random number generator and in particular define a random path through the modeling grid. The random seed convention chosen in this paper was to apply the same random seed vector to each sub-case. The vector contains 50 linearly increasing random seed numbers, ensuring consistency when comparing realizations from the individual sub-cases.

4.1 Basic modeling setup

The basic modeling setup is designed to act as the base from which all other cases are built. The different sub-cases are simply modified versions of the basic modeling setup, each designed to study how modification to the base setup relates to hydrostratigraphic MPS modeling. The basic modeling setup uses the borehole data as hard information, SCI models with smooth inversion constraints as soft data, and the cognitive hydrostratigraphic Kasted model as a TI (Fig. 4a), for which the global proportions are listed in Table 2 and vertical proportions are displayed in Fig. 5a.

4.2 Case 1 – conceptual geological understanding

The basic modeling setup uses the actual cognitive geological model of the Kasted survey area as a TI (Høyer et al., 2015). In Denmark, it is common practice to build 3-D cognitive geological models of the near-subsurface. Many cognitive models exist and are publicly available. Such models can easily be adapted and used as 3-D TIs to simulate new survey areas, provided the geological settings are similar. Case 1 is divided into two sub-cases. The first sub-case, case 1a, uses the basic setup, but, in place of the cognitive Kasted model, the cognitive geological model of the Egebjerg area (Fig. 1a) is used as a TI (Fig. 4c). The geologic setting in Egebjerg is relevant since it is partly dominated by a buried valley complex (Jørgensen et al., 2010).

The Egebjerg model consists of a total of 72 geological units which are categorized accordingly to reflect the three hydrostratigraphic units of the Kasted hydrostratigraphic model. Egebjerg additionally contains undesired features, such as local Miocene complexes. Two such local geological environments, which do not reflect the geological setting of the Kasted area, are present. One is found south of the buried valley complex and the other to the west. By cropping the model and rotating it 90° counterclockwise, a relevant TI without undesired geological architecture is produced (Fig. 4c); this is referred to as case 1a. It is clearly seen by comparing Fig. 4a and c that the Kasted and Egebjerg TIs are different. The Kasted TI is smaller and contains smooth geological features, while the Egebjerg model is larger and

Table 1. An overview table showing information on the MPS cases along with information on the number of realizations for each case or sub-case, and a brief description of each case.

Case name	Sub-case names	Number of realizations	Total number of realizations	Case description
Basic setup	Basic modeling setup	50	50	The basic setup uses boreholes as hard data, smooth resistivity models as soft data, and the cognitive Kasted model as a TI.
Case 1a	(a) Egebjerg TI	50	100	Two different TIs are used to study the uncertainty related to the choice of TI, which reflects the conceptual geological understanding.
Case 1b	(b) Conceptual TI	50		
Case 2	Incomplete soft data grid	50	50	The uncertainty related to the reconstruction of the resistivity grid is studied by running simulations with an incomplete resistivity grid.
Case 3	Sharp resistivity models	50	50	The sharp resistivity models are used for simulations instead of the smooth models to study how the choice of resistivity model influences the hydrostratigraphic models.
Case 4a	(a) No borehole data	50	100	Simulations are run without hard data to see how much the hard data influences the results.
Case 4b	(b) Soft borehole data	50		The borehole data are used as soft information instead of hard by combining the borehole probability grid with the SkyTEM probability grid using the tau model.
Case 5	No soft resistivity data	50	50	Simulations are run using only the hard data and the cognitive Kasted TI.
Total	–	–	400	–

Table 2. The global proportions related to each of the three TIs presented in Fig. 4.

	Sand and gravel	Glacial clay	Hemipelagic clay
Kasted TI	0.17	0.21	0.62
Conceptual TI	0.17	0.22	0.61
Egebjerg TI	0.10	0.22	0.68

contains coarse, block-like geological features. The important features, in relation to hydrostratigraphic modeling, are the buried valley complexes, which are present in the Egebjerg model (Fig. 4c). The global proportions of the Egebjerg TI (Table 2) are similar to the ones found in the Kasted TI. However, the vertical proportions of the Egebjerg TI (Fig. 5c) are different, especially in the upper part of the TI where glacial clay units dominate.

The second sub-case, case 1b, utilizes a purely conceptual TI. The conceptual TI is created by using a set of hyperbolic secant functions to populate a 3-D matrix and is purely mathematical in nature. The conceptual TI can be seen in Fig. 4b and is designed to have three overall buried valleys eroded into a hemipelagic clay substratum. There are two narrow and shallow glacial clay valleys, and a broad and deep sand and gravel valley. One of the glacial clay valleys is a younger valley that is eroded into the older sand and gravel valley and they run roughly parallel to each other. The last glacial clay valley is almost orthogonal to the other valleys, and also erodes into the sand and gravel valley. The upper part of the TI contains a cover layer of glacial clay (Fig. 5b). The simple conceptual TI is designed to contain the main geological architecture of the Kasted area, namely the buried valley complexes. The sand and gravel valley, trending west–northwest to east–southeast, was chosen on purpose to study what happens when oversimplified and smooth MP information is added to a TI. The global proportions of the conceptual TI are consistent with the other TIs, while the vertical proportions for sand and gravel and glacial clay units show a significantly different pattern (Fig. 5b).

4.3 Case 2 – incomplete soft data

During reconstruction of the resistivity grid, it is assumed that the patterns in the incomplete dataset contain information regarding the content of the dataset gaps. This is true only when the incomplete grid contains a sufficient amount of data. Sufficient, in this case, means that the parameter space is sampled densely enough to reflect the patterns we wish to reconstruct (Mariethoz and Renard, 2010). If the grid is too sparse, then limited or no information which can help

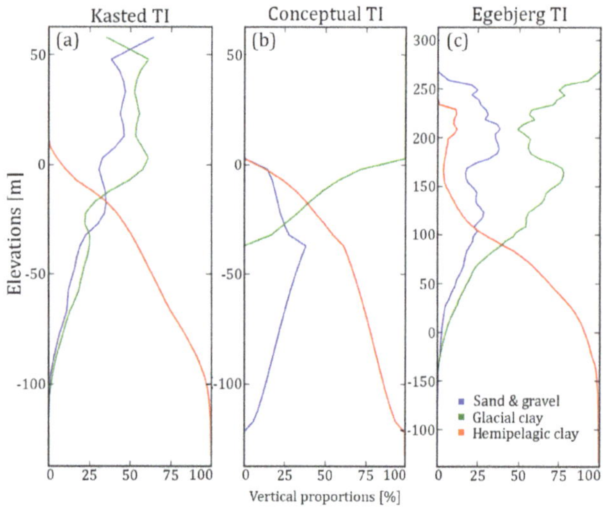

Figure 5. The vertical proportions of the three training images for each of the hydrostratigraphic categories, where **(a)** portrays the Kasted TI, **(b)** the conceptual TI, and **(c)** the Egebjerg TI.

reconstruct missing patterns is present. Signs of mediocre data density are seen in the incomplete grids (Fig. 6a). Artifacts from the DS reconstruction are present in the completed resistivity grids. The resistive valley to the west in the horizontal slices and vertical cross sections in Fig. 7a and b reveal a striated pattern. An alternative to reconstructing the resistivity grid beforehand is to use the incomplete resistivity grids for simulation, meaning no information is present in the resistivity dataset gaps. Grid cells containing a resistivity model are translated into three probability values using the resistivity–hydrostratigraphic relationship histograms (Fig. 1b, 6b–d). Areas without soft resistivity data rely on the TI during simulation, emphasizing the fact that no actual information is present between soundings. The overall setup is identical to the basic setup; the only difference is the reconstructed soft data grids are interchanged for the incomplete soft data grid (Fig. 6b–d).

4.4 Case 3 – choice of resistivity model

The choice of inversion algorithm results in different SkyTEM resistivity models. The purpose of this case is to study how using sSCI (Vignoli et al., 2015) models influences the modeling results. A common inversion approach is SCI, where a smooth regularization is used (Constable et al., 1987). Such resistivity models have a smooth transition from resistive to conductive features, and vice versa. Geological layer boundaries are rarely smooth in nature, meaning such soft transitions in resistivities seldom reflect reality. Furthermore, extreme resistivity values are not presented correctly in the smooth model inversions. Vignoli et al. (2015) propose an alternative SCI approach, employing a minimum gradient support regularization term instead. Such sSCI models produce resistivity models with sharp layer boundaries and a better representation of extreme values. The setup in case 3 is identical to the basic setup, except that the SCI models are interchanged for sSCI models. The DS grid reconstruction is then conducted on the sSCI models, which are then translated into probability grids. Finally, these grids are used as soft data for simulation using the snesim method.

The sharp resistivity models are different from the smooth models, but no particularly sharp layer boundaries are reflected in the reconstructed resistivity grid (Fig. 7b).

One of the obvious differences is found in the resistivity patterns of the sand and gravel valley to the far west of the survey area. The valley itself is not significantly different; however, the small resistive patch, west of the large valley, is more pronounced in the sharp model and has an overall more pronounced fingerprint (Fig. 7b). The sharp resistivity models better estimate the true bulk resistivity values of specific geological units, such as the resistive patch accentuated here. The ensemble standard deviation grid, Fig. 7c and d, shows a general reduction in the ambiguity of the reconstructed sharp resistivity models. This is clear from the reduction areas with large standard deviation, shown in red colors, which are overall reduced in size.

4.5 Case 4 – borehole lithology logs

This case is dedicated towards how the borehole data are handled and how they influence the hydrostratigraphic modeling results. The hard borehole data are normally sparse, relative to geophysical data. Boreholes are commonly considered ground truth since they directly sample the subsurface sediments or petrological units. This case is divided into two sub-cases. The first sub-case, case 4a, portrays what happens when hard data are not included in the snesim simulation. The model setup is therefore identical to the basic MPS setup, but without including the borehole data.

The second sub-case, case 4b, incorporates the borehole lithology logs as soft data. The certainty of a lithological log varies depending on a range of factors, e.g., drilling method, the purpose of the borehole, and sampling frequency (e.g., Barfod et al., 2016; He et al., 2014). The hydrostratigraphic probability logs, introduced in the basic modeling setup (Step 2, Fig. 3), are utilized in place of the hard borehole grid. The boreholes are assigned a lateral footprint, so the information is not only found at the borehole locations. The borehole footprint is assigned by creating a grid where the borehole probability values have been estimated in a radius of 200 m around each borehole using simple kriging with a search radius of 200 m and a mean of $1/K = 1/3$, where K is the number of unique hydrostratigraphic units (Fig. 8d–f) (for additional information see Sect. A1 in the Appendix). The tau model is then used to combine the SkyTEM (Fig. 8a–c) and borehole (Fig. 8d–f) probability grids (e.g., Journel, 2002; Krishnan, 2004; Remy et al., 2014). The first step is then to define the prior probability distribution, $P(A)$, which

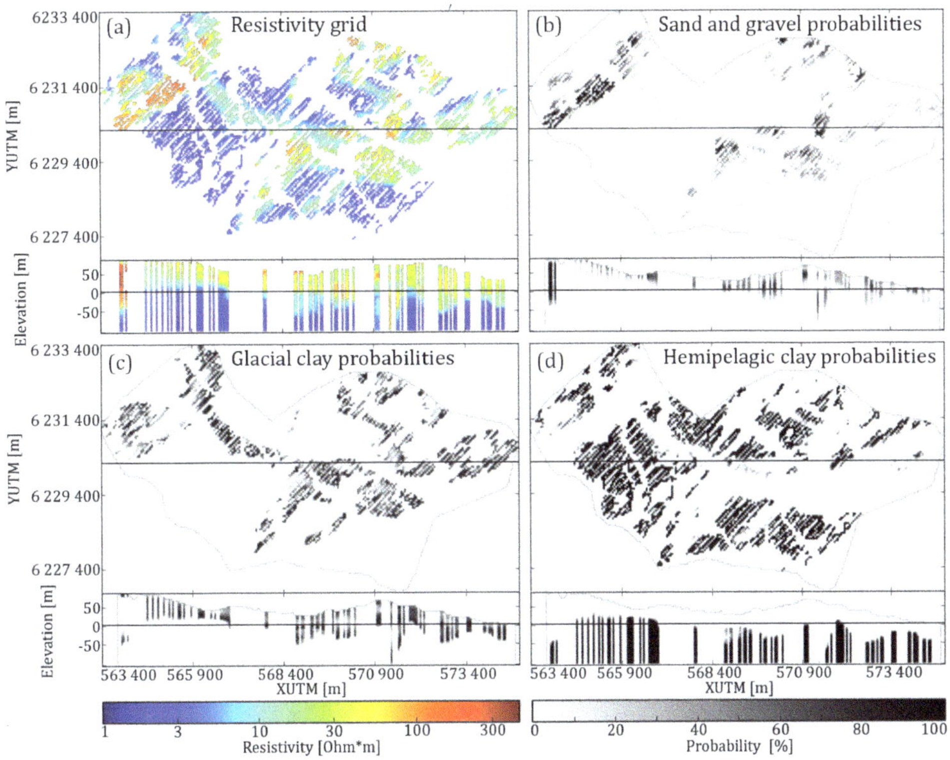

Figure 6. A presentation of the incomplete resistivity grid. Each grid is portrayed as a horizontal slice at 20 m b.s.l. and a vertical cross section intersecting at YUTM 6 230 100 m. **(a)** shows the resistivity grid which is translated into three probability grids using the resistivity–hydrostratigraphic relationship histograms (Fig. 1b). Grid cells without SkyTEM soundings are not assigned a probability value. **(b–d)** show the sand and gravel, glacial clay, and hemipelagic clay probability values, respectively.

in this case is the vertical proportion taken from each layer of the cognitive Kasted TI (Fig. 5a). Then the probability distributions are defined as follows: $P(A|D_r)$ and $P(A|D_b)$, where D_r is the resistivity probability grid and D_b the borehole probability grid. The 3-D probability grids are translated into distance grids by applying the probability-into-distance transform and computing the distance ratio using Eqs. (2) and (3), where the tau values were assigned based on a series of exhaustive tests. The final tau values were selected based on the criteria that the transitions in areas where both borehole and resistivity information is available should be as smooth as possible in the resulting combined probability grid, as seen in Fig. 8. Based on the tests, the resulting tau values were $[\tau_r, \tau_b] = [2.0, 1.0]$. The final conditional probability was computed using Eq. (4) and resulted in the three hydrostratigraphic probability grids, as seen in Fig. 8g–i. The combined probability grids replace the smooth probability grids used in the basic setup.

4.6 Case 5 – excluding the soft resistivity data

The final case, case 5, illustrates the consequences of not including the soft SkyTEM resistivity information in the MPS simulation routine. The basic setup is simply run without the inclusion of soft data; i.e., the setup only uses the cognitive Kasted TI and hard borehole information.

5 Results

5.1 Visual comparison of hydrostratigraphic realizations and ensemble mode ratio maps

For each of the presented cases two hydrostratigraphic realizations are presented (Fig. 9), along with an EMR map (Fig. 10). The EMR maps show the occurrence ratio of the most likely simulated category for each grid cell based on 50 realizations. The two realizations and EMR map of the basic modeling setup, Figs. 9a and 10a, reveal the same overall trends as the cognitive geological model, Fig. 4a: namely the western sand and gravel valley striking $\sim$N40° E, the glacial clay valley striking $\sim$E30° S, the large mixed sand and gravel and glacial clay valley striking $\sim$N20° E to the south, and the small subsidiary glacial clay valley striking $\sim$N50° E to the south. However, even though the main hydrostratigraphic architecture of the cognitive geological model is similar, there are still differences between the snesim realizations and the cognitive geological model. The cognitive model shows clear-cut, smooth, and ordered hy-

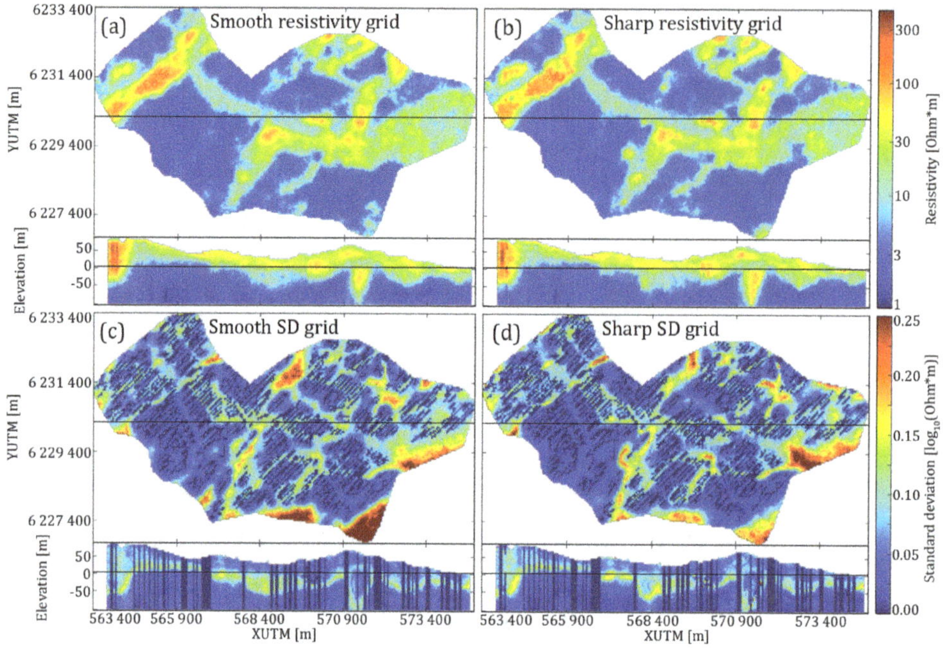

Figure 7. An overview of the key differences between reconstructing the resistivity grid using smooth and sharp inversion resistivity models. Each grid is portrayed as a horizontal slice at 20 m b.s.l. and a vertical cross section intersecting at YUTM 6 230 100 m. **(a)** shows the smooth reconstructed resistivity grid. **(b)** portrays the sharp reconstructed resistivity grid. **(c)** shows the standard deviation calculated from 50 stochastic reconstructions of the smooth resistivity grid. **(d)** shows the standard deviation calculated from 50 stochastic reconstructions of the sharp resistivity grid.

drostratigraphic units. The basic modeling setup realizations reveal sporadic and random patterns. The sand and gravel units are placed in small lumps throughout the glacial clay units, but are not present within the homogenous hemipelagic clay. Patches of uncertain glacial clay units are, however, found in the homogeneous hemipelagic clay, especially in the southeast corner of the Kasted survey area (Figs. 9a, 10a). The same sporadic picture is seen in the vertical slices of the realizations (Fig. 9a), although here an additional trend is revealed. The sand and gravel valley to the far west and at XUTM 570 900 m is not consistently filled with sand and gravel (Fig. 9a) as in the cognitive geological model (Fig. 4a). Furthermore, the EMR map reveals that the valley margins are subject to a larger degree of ambiguity (Fig. 10a); in fact at some locations the r_{EMR} value is close to 1/3, which means that for the model ensemble the occurrence of either hydrostratigraphic unit is possible.

The case 1a realizations (Fig. 9b, 10b), which use the Egebjerg TI (Fig. 4c), show the same overall trends as in the basic modeling setup. The subset of buried valleys mentioned above is present; however, an obvious difference is the coarse and block-like appearance of the case 1a realization ensemble. This appearance is similar to the block-like appearance of the Egebjerg TI (Fig. 4c). Furthermore, the horizontal slice of the realizations and EMR map reveals that the glacial-clay-dominated area to the east has a generally larger occurrence ratio and is thus more certain. The realizations are clearly influenced by the choice of TI, especially when case 1b is also considered (Fig. 9c). The hydrostratigraphic realizations of case 1b (Fig. 9c, 10c) clearly depict the same overall buried valley trends, but the valleys in the central part of the model are largely filled with the opposite of the valley filling hydrostratigraphic units. Furthermore, the occurrence ratio seems quite low in certain areas, such as to the south of the model, which means the ambiguity has increased. Finally, the realizations also reveal an absence of small-scale patterns, which corresponds to the conceptual TI that only contains homogenous hydrostratigraphic units.

The importance of reconstructing the incomplete resistivity grid is seen in case 2 (Fig. 9d, 10d). The two realizations in Fig. 9d show the main buried valley features, e.g., the western sand and gravel valley. However, the hydrostratigraphic units are sporadic, especially in areas with no data. Patches of sand and gravel and glacial clay are randomly spread throughout the presented horizontal slice and vertical cross section (Fig. 9d). The EMR map also reveals an increase in low occurrence ratios in areas without soft data (Fig. 10d).

The uncertainty related to the choice of geophysical modeling procedure is portrayed by case 3. Here, snesim realizations are constrained to sharp resistivity models. Generally, the realizations (Fig. 9e) are quite similar to the basic mod-

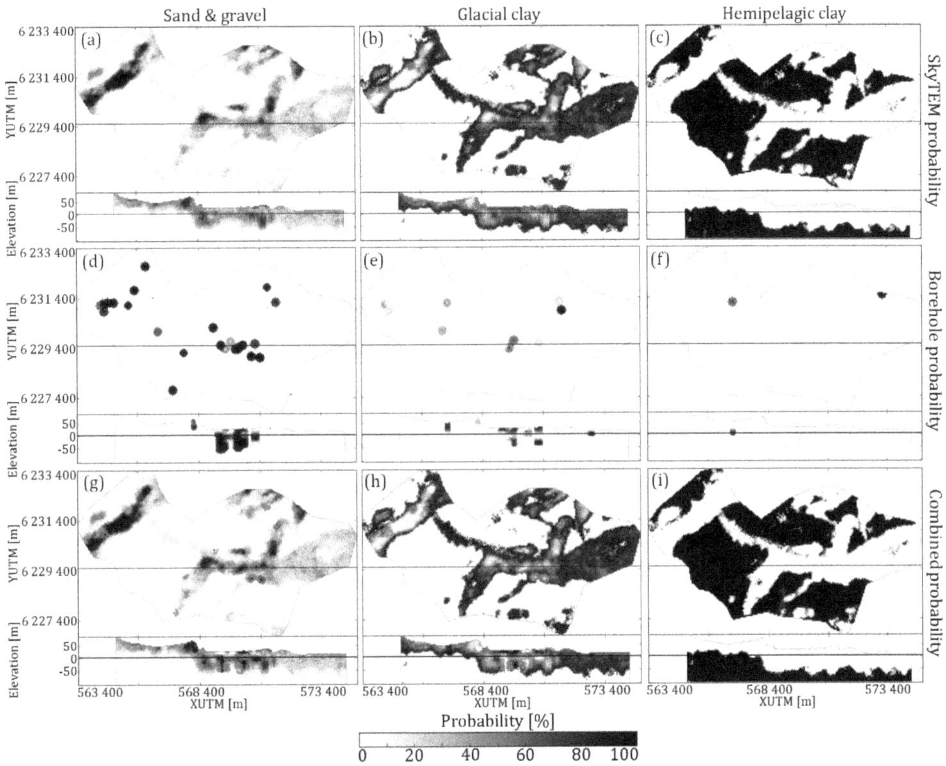

Figure 8. A visual representation of the tau model procedure for combining the soft resistivity and borehole grids. Each grid is portrayed as a horizontal slice at 20 m b.s.l. and a vertical cross section intersecting at YUTM 6 230 100 m. **(a–c)** show the sand and gravel, glacial clay, and hemipelagic clay probability maps, respectively, for one DS reconstructed resistivity grid; **d–f** show the 200 m radius kriged borehole probability; and **(g–i)** show the combined resistivity grid, which has been combined using a tau model with the values $[\tau_r, \tau_b] = [2, 1]$.

eling setup realizations (Fig. 9a). However, a key difference is the significant reduction or absence of patches of glacial clay in the homogeneous hemipelagic clay. In fact only one patch is found in the first realization (Fig. 9e) in the southwest corner, while it is not present in the second realization, and the EMR map further reveals a reduction of the occurrence ratios generally, especially along the southern margin of the realizations (Fig. 10e).

Case 4 shows the influence that the hard data have on the hydrostratigraphic realizations in two sub-cases: case 4a, where snesim simulations are run without hard data; and case 4b, where the borehole data are treated as soft information. Fig. 9f and g shows two hydrostratigraphic realizations without hard data and with soft borehole data, respectively. These realizations do not differ significantly from the basic modeling setup realizations and in fact are quite similar. One key difference is the central glacial clay valley striking ~E30° S, which does not contain any sand and gravel to the west (Fig. 9f and g). The EMR maps reveal that without boreholes (Fig. 10f) the occurrence ratios generally decrease, making the realizations more ambiguous. The usage of the borehole data as soft information also seems to reduce the occurrence ratios compared to the basic modeling setup. Generally, leaving out the borehole data, or treating it as soft data, results in local changes in areas with a high density of boreholes.

The final case, case 5, illustrates the importance of the SkyTEM soft data. The snesim simulations are run using only hard data and the cognitive geological model as a TI. The output realizations (Fig. 9h) portray smooth and large-scale hydrostratigraphic units. The hydrostratigraphic architecture of the buried valleys is not reflected. However, the sand and gravel valley, to the west, does seem to protrude slightly in the realizations (Fig. 9h) and the EMR map reveals a significant decrease in the occurrence ratio and thus an increase in the ambiguity of the model ensemble (Fig. 10h).

5.2 Quantitative comparison using differences in object-based Euclidean distances as a measure for similarity

The distances between each of the 400 realizations have been computed using Eqs. (8) and (10). The full distance matrix is presented in Fig. 11a. The distances between each realization and the cognitive geological models have also been computed and plotted in Fig. 11b. To aid the interpretation of the distance matrix and distances to the cognitive model a summary table, Table 3, has been compiled.

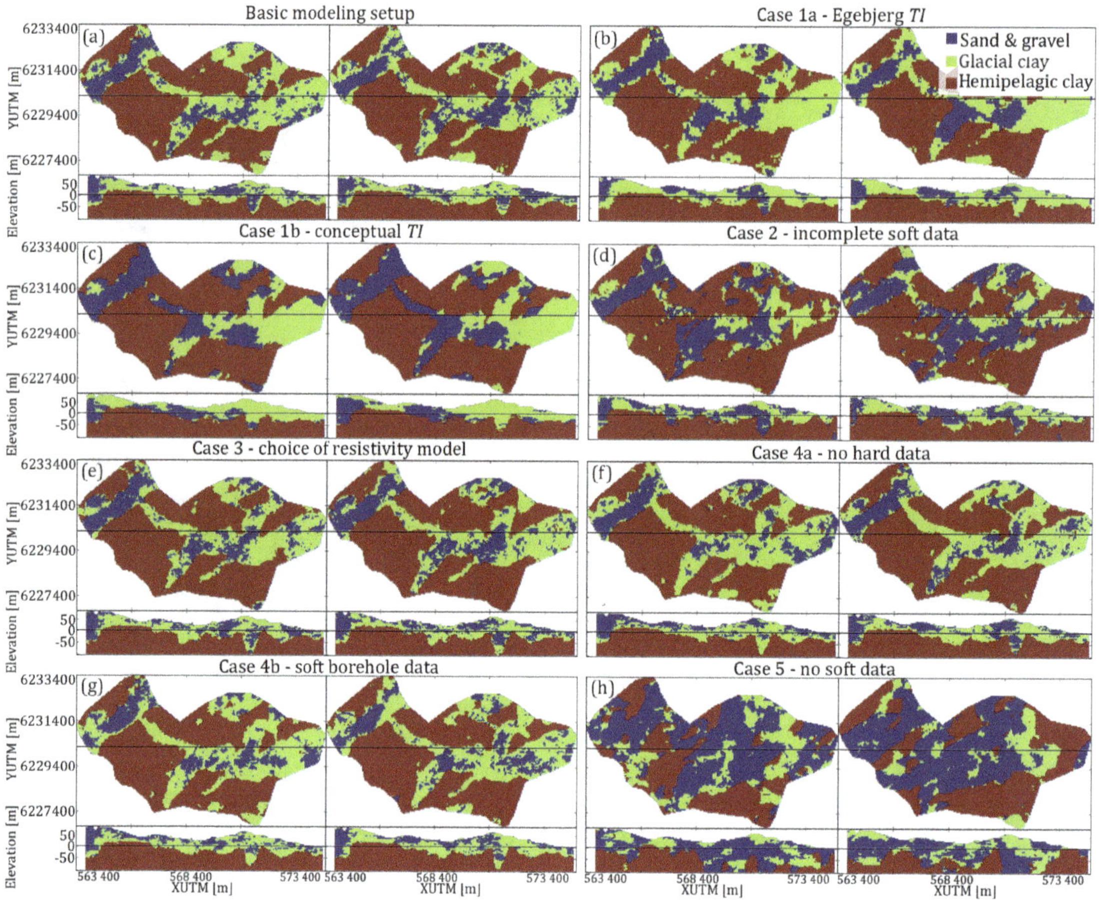

Figure 9. Each case is displayed by two realizations: realization no. 1 of 50 and realization no. 30 of 50. Each realization is portrayed as a horizontal slice at 20 m b.s.l. and a vertical cross section intersecting at YUTM 6230100 m. **(a)** shows the realization results for the basic modeling setup, **(b)** shows the realization results for case 1a, **(c)** shows the realization results for case 1b, **(d)** shows the results for case 2, **(e)** shows the results for case 3, **(f)** shows the results for case 4a, **(g)** shows the results for case 4b, and **(h)** shows the results for case 5. For more details on individual cases the reader is referred to Table 1.

The basic modeling setup constitutes a common snesim setup, with the geophysical data as soft data, boreholes as hard data, and a 3-D geological conceptualization encased in a TI. The ensemble average variability is computed according to the equations presented by Barfod et al. (2018), and the resulting ensemble average variability is 10.1 m, with an average distance to the cognitive model of 24.3 m. This means that the Euclidean distance mismatch between the individual realizations related to basic modeling setup is 10.1 m, and the average difference in Euclidean distance to the nearest active cell between the realizations and the cognitive model was 24.3 m.

The 3-D geological conceptualization contained in the TI influences the final hydrostratigraphic realizations as illustrated in case 1, which is divided into two sub-cases: case 1a and case 1b. In case 1a, using a 3-D cognitive geological model from the Egebjerg area as a TI for hydrostratigraphic simulation increases the average distance to the cognitive model to 24.9 m (Fig. 11a, Table 3). Furthermore, the average variability has increased to 13.6 m (Fig. 11b, Table 3). The other sub-case revolves around using an entirely conceptual geological model as a TI. The conceptual TI was designed to reflect the overall geology, yet still contains some bias. The results reflect the bias, with increased distances to the cognitive geological model, which are now centered on 25.6 m (Fig. 11a, Table 3). The ensemble variability has increased to 14.8 m (Fig. 11b, Table 3).

The importance of proper reconstruction of the incomplete resistivity grid is illustrated in case 2, where the incomplete resistivity grid was used for simulation. The resulting realization ensemble has a large ensemble variability centered on 24.1 m (Fig. 11a, Table 3). The distance to the cognitive geological model is also large, with an average value of 33.1 m (Fig. 11b, Table 3).

In case 3 the sharp SCI models were used for simulation in place of the smooth SCI models. The realizations related to case 3 were the closest to the cognitive model with an average value of 21 m (Fig. 11b, Table 3). The variability of the case 3

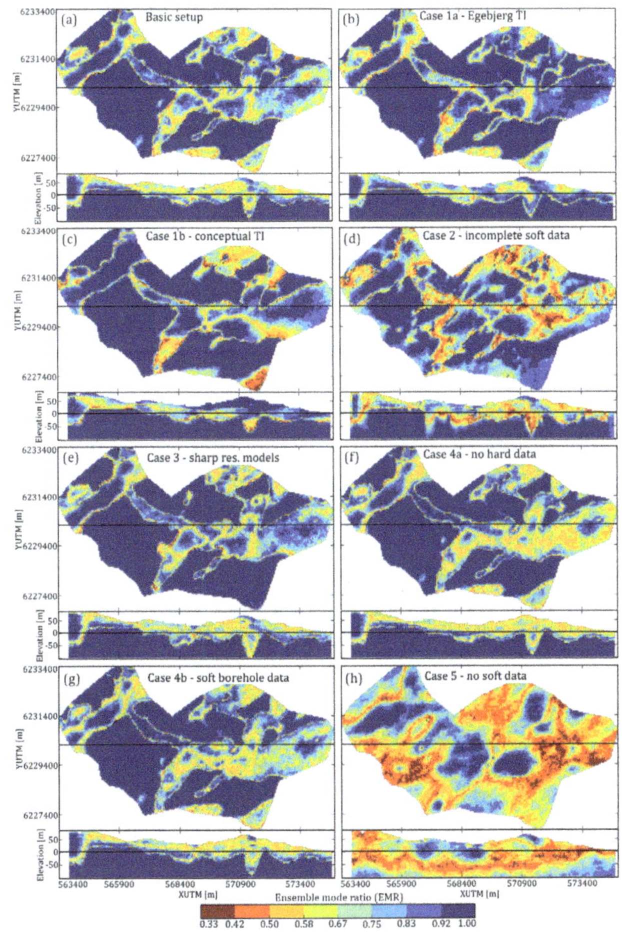

Figure 10. A presentation of the ensemble mode ratio (EMR) maps, computed for the different case ensembles of hydrostratigraphic models. Each EMR map is presented as a horizontal slice centered on 20 m b.s.l. and a vertical cross section intersecting at YUTM 6 230 150 m. **(a–h)** present the EMR-type uncertainty map for each of the different cases, which are summarized in Table 1. The EMR values portray how certain the ensemble of MPS realizations are; i.e., if $r_{\text{EMR}} = 1/3$ then the realization is uncertain, and we have equal probability of finding either hydrostratigraphic unit since $P(s_1) = P(s_2) = P(s_3) = 33\,\%$. On the other hand, if $r_{\text{EMR}} = 1$, then each realization of the given ensemble contains the same hydrostratigraphic unit at the given grid cell.

realization ensemble, i.e., the distances between the realizations pertaining to case 3, is small, with an average value of 9.4 m – see Table 3. Recalling the raw hydrostratigraphic realizations (Fig. 9e) and the EMR map (Fig. 10e), the large reduction in distances could partly be related to the removal of non-hemipelagic clay units along the southern border of the model and an overall increase in confidence along the southern and southeastern border of the model.

The influence of the borehole lithology logs on the hydrostratigraphic realizations is reflected in case 4, which is divided into two sub-cases. In the first sub-case, case 4a, the borehole information is not used as hard data, and the realizations are created only using soft geophysical data and the Kasted TI. However, the borehole data are still used for creating the resistivity–hydrostratigraphic histograms (Fig. 1b), which are used for creating the probability grids. The ensemble average variability is 10.7 m (Fig. 11a, Table 3) and the average distance to the cognitive model is 24.3 m (Fig. 11b, Table 3). In the second sub-case, case 4b, the boreholes are used as soft information to reflect the uncertainty of the borehole information. The ensemble average variability is 10.9 m, and the average distance to the cognitive model is 24.3 m. This illustrates how the snesim realizations are not particularly sensitive towards the sparse borehole hard data.

Not including the geophysical soft data in the snesim simulations, case 5, resulted in the largest ensemble average variability of 40.0 m (Fig. 11a, Table 3). The average distance between case 5 realizations and the cognitive model was 59.3 m. This means that the realizations of case 5 are the most different from the rest of the realizations. The snesim realizations are sensitive towards not including the geophysical data, or using the incomplete resistivity grid. This underlines the importance of the geophysical soft data in relation to hydrostratigraphic modeling using the snesim methodology.

6 Discussion

The cognitive geological model was created based on smooth SkyTEM resistivity models and lithological logs (Høyer et al., 2015) as well as the conceptual geological understanding of the area. The model was simplified from a full 3-D geological model containing a total of 42 unique geological units to a hydrostratigraphic model containing only 3 hydrostratigraphic units. The cognitive geological model, although detailed and extensive, is not the true geological model. The ensemble realizations should not directly reflect the cognitive model, yet the cognitive model can be thought of as a reference point in modeling space, which we would prefer our models to resemble.

The results revealed the importance of the SkyTEM dataset. Not including the resistivity models in the MPS simulations, case 5, yielded realizations which were both the least similar to the cognitive geological model and had the largest variability between the individual realizations. Including the incomplete resistivity grid, case 2, improved the realization results compared to not including them at all. Yet, the ensemble variability was large and resulting realizations were ranked second least similar to the cognitive geological model. The realization ensemble which was closest to the cognitive geological model belongs to case 3. Here, the resistivity grid was reconstructed from the sharp SCI models that, in this case, increased the fingerprint of resistive extreme values, which in turn results in less ambiguous reconstructed resistivity grids; compare Fig. 7c and d. It should be noted that the usage of block kriging for assigning the sharp resistivity

Table 3. A summary table showing the average distance value for each 50 by 50 square representing a given case in the distance matrix (Fig. 11a). The final column, labeled Distance$_{cog}$, summarizes the distances to the cognitive geological model, presented as the average of each colored point cloud in Fig. 11b. The distances in parenthesis represent ensemble variabilities, and the remaining values represent average distances between different ensembles. The unit of the average distances is meter.

Distance (m)	Basic setup	Case 1a	Case 1b	Case 2	Case 3	Case 4a	Case 4b	Case 5	Distance$_{cog}$
Basic setup	(10.1)	12.9	16.9	24.0	12.7	11.1	11.2	49.6	24.3
Case 1a	12.9	(13.6)	18.3	26.0	15.1	14.0	13.9	51.7	24.9
Case 1b	16.9	18.3	(14.8)	27.9	17.8	18.1	18.1	52.5	25.6
Case 2	24.0	26.0	27.9	(24.1)	23.5	25.0	24.9	45.2	33.1
Case 3	12.7	15.1	17.8	23.5	(9.4)	13.6	13.6	49.6	21.6
Case 4a	11.1	14.0	18.1	25.0	13.6	(10.7)	11.1	50.7	24.3
Case 4b	11.2	13.9	18.1	24.9	13.6	11.1	(10.9)	50.6	24.3
Case 5	49.6	51.7	52.5	45.2	49.6	50.7	50.6	(40.0)	59.3

models to the modeling grid resulted in smoothing of sharp vertical boundaries otherwise found in sSCI models. These three cases together reveal the importance of the geophysical soft data when using the snesim setup presented in this study.

In relation to case 5, it can be argued that, even though the SkyTEM resistivity models are not used as soft data, they are still included indirectly since the TI, or cognitive geological model, was created using smooth SkyTEM resistivity models. However, the realizations related to case 5 revealed an ensemble of realizations, which did not replicate the overall geological architecture, implying the importance of using the SkyTEM models as soft data.

On the other hand, the cases related to studying the sensitivity towards borehole information, case 4a and case 4b, revealed that the large-scale hydrostratigraphic architecture was not changed significantly. The distance measure used in this study observes similarities or dissimilarities of large-scale hydrostratigraphic architecture and is not sensitive towards local changes in small-scale patterns. The amount of geophysical information is relatively large, meaning the relative influence of (few) borehole data becomes less significant. This does not mean that the borehole data are not important; they both contain locally accurate information and are used to estimate the regional resistivity–hydrostratigraphic relationship (Fig. 1b). In other surveys, where the contrast between geophysical and lithological information is smaller, the importance of the borehole data will likely increase. In relation to this study, such small-scale changes are insignificant. Yet if the realizations are to be used for flow simulations or predictions on a smaller scale, such smaller scales might suddenly have an important impact on prediction accuracy. Additionally, if such small-scale patterns are important, the size of the model grid cells should be smaller to accommodate simulations of these variations. Discretizing hydrostratigraphic and groundwater models with relatively small grid cells can be CPU demanding, depending on the total number of grid cells.

In case 4b, the borehole data were used as soft information as in the study by Høyer et al. (2017). This was done since boreholes are associated with uncertainty related to a number of factors, as described above. Therefore, the soft borehole probability values derived during the assigning of the boreholes to the modeling grid are combined with the SkyTEM-based probability grids using the tau model. This approach enables the borehole probability to alter the final probability grid, while still conditioning the SkyTEM data. Combining the information rather than letting the borehole data count as ground truth, i.e., hard data, allows the borehole data to influence the realizations, especially if the soft borehole information disagrees with the soft geophysical data (Fig. 8).

The conceptual geological understanding has always been considered an integral part of geological modeling. In this case the conceptual geological understanding is implemented via the TI, which makes it easy to change the underlying conceptual geological understanding of a given model. A total of three different TIs were used for simulation in this study: the Kasted, Egebjerg, and conceptual TIs. The results showed that models simulated using the Egebjerg TI, case 1a, portrayed the same overall hydrostratigraphic architecture. This opens for the possibility of using 3-D cognitive geological models as TIs for new survey areas, as long as the geological settings are similar. One key difference between the models, however, was the more block-like and coarse nature of the realizations using the Egebjerg TI, due to the coarseness of the Egebjerg TI. An important observation is that, when a spatially dense and extensive geophysical dataset, such as SkyTEM, is present, the snesim realizations are not as sensitive towards the choice of TI when the TI is relatively similar to the expected scenario. However, as illustrated by case 1b, a TI which has significantly differing vertical proportions in comparison with the actual model, which in this case is assumed to be the cognitive model (Fig. 5a, b), causes a conflict between the TI and the soft data. In such a case the TI dominates the realizations and simulates improbable hydrostratigraphic units in places where the soft data reveal low probability for those specific units. In Figs. 9c and 10c, it can be seen that the glacial clay valley, both present in the soft data variable (Fig. 8b) and the cognitive Kasted geological

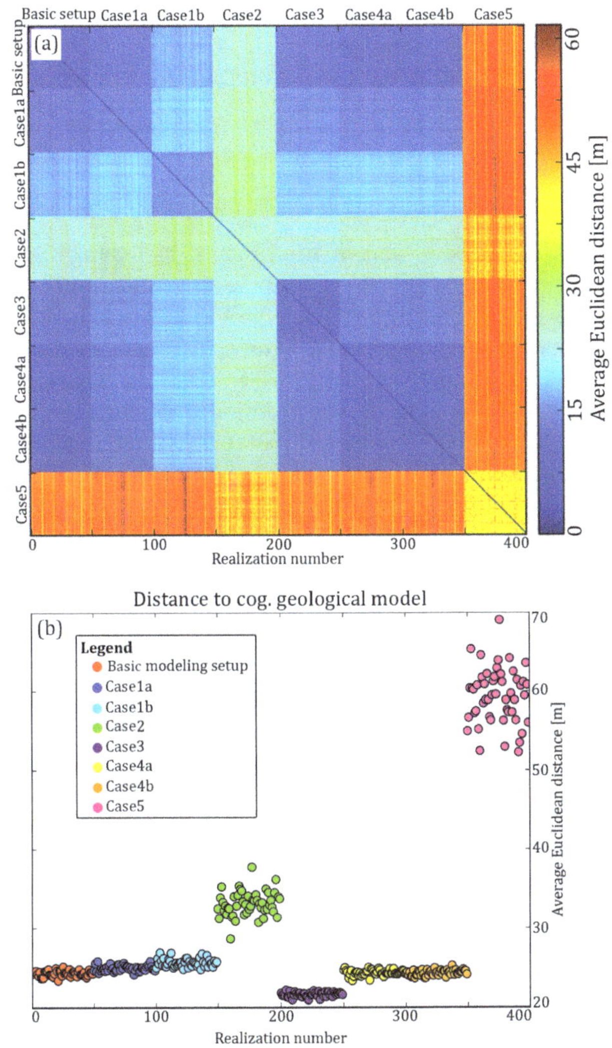

Figure 11. A presentation of the average Euclidean distance calculations. **(a)** shows the full distance matrix; **(b)** shows the average Euclidean distances between each individual hydrostratigraphic realization and the cognitive geological model.

model (Fig. 4a), is represented as sand and gravel. This leads to the conclusion that one needs to pay attention to the construction of the TI, as also witnessed in the study by Høyer et al. (2017). Furthermore, the large-scale and homogenous nature of the hydrostratigraphic architecture in the conceptual TI results in realizations which reflects the homogeneity. In comparison with the realizations based on the TIs derived from cognitive models, the realizations contain much fewer small-scale patterns.

Small-scale patterns are present in the realizations although they are not at the same degree present in the TIs. This can be explained by the fact that the simulations are probabilistic in nature and are based on random processes. At the beginning of the simulation a random path is drawn so that the simulation grid is filled by visiting each grid cell only once. The small-scale patterns are partly due to the hard data that are inserted into the simulation grid before the simulation commences, which is excluded from the simulation path. As the grid is filled out, the hard borehole data might suggest a certain category but the soft data suggest another category. As the grid is filled out the overall category from the soft data dominates and, if the random path visits the grid cells near the hard data point towards the end of the random path, then we are left with a small intercalation (Hansen et al., 2018). The intercalations are also inherent in the simulations without hard data, which is mainly due to process randomness related to how the snesim algorithm draws from a cumulative density function (Strebelle, 2002).

The Kasted model and TI are influenced by non-stationarity, which has not been dealt with in the MPS setup. Even though the models are influenced by non-stationarity, the simulations result in models which overall resemble the cognitive model, e.g., cases 1–4. However, once the geophysical data are removed in case 5, the resulting MPS models are increasingly random and are heavily influenced by non-stationarity. It is important to note that the increasing amounts of soft geophysical data generally decrease the effects of non-stationarity, due to the increased conditioning of the soft data.

The reconstruction of the resistivity grid is an important step of the MPS setup presented in this study. This was illustrated in case 2, where the incomplete resistivity grid was used instead of the reconstructed resistivity grid, resulting in larger realization variability and distance to the cognitive geological model. These realizations could have been improved by increasing the prior knowledge provided to snesim before simulation. One such option is to provide so-called vertical proportions, in place of solely the target global proportions. The global proportions simply give a percentage fraction of the different hydrostratigraphic units in the outcome realizations. The vertical proportions are defined for each simulation grid layer and determine the proportions as a function of depth. This makes sense if the different units in the realizations are clearly linked to geological units, which in turn have clear stratigraphic layering. In our case, this would have impacted the realizations by not allowing the presence of hemipelagic clay at the top of the model. Furthermore, sand and gravel and glacial clay would not be allowed at the bottom of the model. However, vertical proportions were not used in any other cases and were therefore not used in case 2. The usage of vertical proportions for conditioning could also improve the results of case 5.

Part of the considerations of this study was to utilize the DeeSse code for direct sampling simulation. Whereas Chugunova and Hu (2008) present an MPS method for constraining a categorical simulation using continuous auxiliary data, DS is much more flexible and allows multivariate simulations reproducing spatial statistics within and between variables, which can be categorical or continuous. DS requires the construction of a multivariate TI, and it is important that every variable reflects the spatial relationship to be modeled. In our context, one could envision creating a bivariate TI consisting of a hydrostratigraphic model and a continuous auxiliary variable reflecting an AEM dataset. It is no trivial task, and, to our knowledge, no studies have been presented where an auxiliary variable is created for a 3-D AEM dataset. However, Lochbühler et al. (2014) presented a generalized example on creating auxiliary variables for tomographic images, i.e., 2-D images. Generally, the requirements for the geophysical modeling procedure are twofold. Firstly, the categorical TI needs to be populated with resistivity values, e.g., as in Christensen et al. (2017) where a Bayesian Markov chain Monte Carlo (MCMC) algorithm is used to create 1-D resistivity models drawn from a posterior probability distribution. This is no straightforward task. Secondly, the populated resistivity model then ideally needs to be forward modeled using full 3-D forward modeling code, which is computationally expensive. Alternatively, approximate 1-D forward modeling is also an option. The correct system parameters of the AEM instrument and data processing parameters have to be taken into account. Thirdly, the synthetic data obtained by forward modeling must be inverted using the same procedure as the field dataset. To our knowledge, such usage of an auxiliary variable for constraining soft geophysical models is not widespread within the domain of AEM geophysical methods. In this study, DS was only used to reconstruct the incomplete AEM dataset (a univariate case with the dataset as a TI, and hard data; see Sect. 3.2) and the snesim method was used for hydrostratigraphic modeling, due to its usage of the τ model (Journel, 2002). The τ model proved a more straightforward approach when combined with the method for creating the resistivity–hydrostratigraphic histograms presented by Barfod et al. (2016).

The study presented by Barfod et al. (2018) used the alternative modified Hausdorff distance measure for comparing realizations. Due to the computational burden of the method, it was difficult to create exhaustive distance computations, i.e., where all information from individual realizations is used. The usage of differences in EDT of binary translations of the categorical realizations for comparing the individual realizations proved to be a more computationally feasible approach. In this paper an efficient algorithm for computing the EDT was used (Maurer et al., 2003). This computationally advantageous approach for computing the distance between two realizations allows for a full analysis of the realizations. Each realization is then compared based on each of the hydrostratigraphic categories and on the entire 3-D objects, resulting in a detailed comparison. The resulting distance matrix (Fig. 11a) was able to differentiate between the realizations pertaining to the different cases. The random number seed between cases was chosen so the first realization of each case has the same random seed and the second realization has the same seed. This can be seen in the distance matrix (Fig. 11a), where off-diagonal cases have smaller distance values along the diagonal within the given 50 by 50 sub-matrix. An example is the 50 by 50 sub-matrix between the basic setup and case 1a, where the diagonal is clearly marked by lower distances relative to the remaining sub-matrix.

7 Conclusion

A hydrogeophysical dataset from Kasted in Denmark was used for stochastic hydrostratigraphic simulation using the snesim algorithm. The main goal of this study was to improve our understanding of ensemble hydrostratigraphic modeling variability related to stochastic MPS modeling. The study was divided into eight sub-cases designed to reflect the impact related to key components of the hydrostratigraphic modeling setup, i.e., the TI, borehole lithology logs, and SkyTEM resistivity models. The results revealed that the hydrostratigraphic realizations were sensitive first and foremost to the geophysical dataset due to its extensive nature. Not including the geophysical data in the realizations resulted in an average Euclidean distance variability of 40 m and a distance to the cognitive model of 59 m, which was, by far, the largest distance of all realizations. Furthermore, the geophysical modeling procedure influences the resulting realizations. It was shown that choosing so-called sharp inversion models (sSCI), in place of smooth inversion models (SCI), resulted in a realization ensemble which had similar distance-based variabilities, 9.4 and 10.1 m, respectively. However, using sSCI models decreased the distance to the cognitive geological model from 24.3 to 21.6 m. The choice of a TI containing a relevant geological conceptualization is important. The cognitive Egebjerg model was used as a TI to simulate the hydrostratigraphic Kasted model, which yielded similar realizations to the case where the cognitive Kasted model was used as a TI. The Egebjerg TI contained relevant geological architecture, but if a conceptual TI is introduced containing significantly different vertical proportions, the resulting realizations will reflect these differing vertical proportions. Finally, it was seen that the borehole lithology logs did not significantly influence the realizations. The lithology logs only carry information in the immediate vicinity of the borehole and are sparse in comparison to the resistivity data. The boreholes therefore only have a minor influence on the realizations. The comparison measures used here mainly compare the overall large-scale architecture components of the realizations and do not reflect small-scale changes. In relation to this study the usage of the lithology logs as hard data does not show a significant impact on the MPS realizations. However, if the hydrostratigraphic models are used for predicting groundwater flow, the boreholes might be important. However, it should be mentioned that the resistivity–hydrostratigraphic histograms, which are used extensively in this research, are created from the borehole information.

Appendix A

Table A1. Simple kriging parameters for creating borehole probability grids (case 4)

Variogram model type exponential			
$mean_{SK} = 1/3$			
Search ellipsoid			
	Max	Med	Min
Ranges	200	200	10
Angles	0	0	0
Variogram			
Contribution = 1			
	Max	Med	Min
Ranges	1000	1000	50
Angles	0	0	0

Table A2. General SGeMS parameters used for the snesim realizations.

Property name	Value/count
algorithm name	snesim_std
use_pre_simulated_gridded_data	0
Use_ProbField	1
ProbField_properties	count=3, value="sg_0;ct1;pc2"
TauModelObject	[1 1]
use_vertical_proportion	0
Cmin	5
Constraint_Marginal_ADVANCED	0
resimulation_criterion	−1
resimulation_iteration_nb	1
Nb_Multigrids_ADVANCED	5
Debug_Level	0
Subgrid_choice	0
expand_isotropic	1
expand_anisotropic	0
aniso_factor	NA
Use_Affinity	0
Use_Rotation	0
Nb_Facies	3
Marginal_Cdf	0.19 0.24 0.57
Max_Cond	100
Search_Ellipsoid	[750 750 0 0 0 0]

Marginal cdf			
	Sand and gravel	Glacial clay	Hemipelagic clay
Value	0.19	0.24	0.57

Author contributions. The main author, AASB, developed the concepts and ideas for the paper during a 6-month sabbatical at Stanford University and was the main contributor in regards to writing and compiling the paper. IM helped refine and develop the concepts presented in the paper and aided in the writing process with iterative internal paper reviews. Co-author TNV provided valuable insights into MPS modeling and helped with the practical process of MPS modeling, as well as internal co-author reviews. Co-author FJ helped to guide the research topics that are addressed in this paper. The remaining co-authors, AVC, ASH, and JS, all aided in the co-author review process and contributed with ideas and insights.

Competing interests. The authors declare that they have no conflict of interest.

Acknowledgements. We would like to thank Senior Research Engineer Celine Scheidt of Stanford University for pointing us in the direction of using Euclidean distance transforms for comparing MPS realizations. Three anonymous reviewers and the editor are thanked for valuable comments that improved the clarity of the manuscript. This study is supported by HyGEM (Integrating geophysics, geology, and hydrology for improved groundwater and environmental management, project no. 11-116763). The funding for HyGEM is provided by the Danish Council for Strategic Research.

Edited by: Jesús Carrera

References

Auken, E., Christansen, A. V., Westergaard, J. H., Kirkegaard, C., Foged, N., and Viezzoli, A.: An integrated processing scheme for high-resolution airborne electromagnetic surveys, the SkyTEM system, Explor. Geophys., 40, 184–192, https://doi.org/10.1071/EG08128, 2009.

Barfod, A. A. S., Møller, I., and Christansen, A. V.: Compiling a national resistivity atlas of Denmark based on airborne and ground-based transient electromagnetic data, J. Appl. Geophys., 134, 199–209, https://doi.org/10.1016/j.jappgeo.2016.09.017, 2016.

Barfod, A. A. S., Møller, I., Christiansen, A. V., Høyer, A.-S., Hoffimann, J., Straubhaar, J., and Caers, J.: Hydrostratigraphic modeling using multiple-point statistics and airborne transient electromagnetic methods, Hydrol. Earth Syst. Sci., 22, 3351–3373, https://doi.org/10.5194/hess-22-3351-2018, 2018.

Christensen, N. K., Minsley, B. J., and Christensen, S.: Generation of 3-D hydrostratigraphic zones from dense airborne electromagnetic data to assess groundwater model prediction error, Water Resour. Res., 53, 1019–1038, https://doi.org/10.1002/2016WR019141, 2017.

Chugunova, T. and Hu, L. Y.: Multiple-Point Simulations Constrained by Continuous Auxiliary Data, Math. Geosci., 40, 133–146, https://doi.org/10.1007/s11004-007-9142-4, 2008.

Chugunova, T. and Hu, L. Y.: Multiple-Point Simulations Constrained by Continuous Auxiliary Data, Math. Geosci., 40, 133–146, https://doi.org/10.1007/s11004-007-9142-4, 2008.

Comunian, A., Renard, P., and Straubhaar, J.: 3D multiple-point statistics simulation using 2D training images, Comput. Geosci., 40, 49–65, https://doi.org/10.1016/j.cageo.2011.07.009, 2012.

Constable, S. C., Parker, R. L., and Constable, C. G.: Occam's inversion: A practical algorithm for generating smooth models from electromagnetic sounding data, Geophysics, 52, 289–300, https://doi.org/10.1190/1.1442303, 1987.

de Iaco, S. and Maggio, S.: Validation techniques for geological patterns simulations based on variogram and multiple-point statistics, Math. Geosci., 43, 483–500, https://doi.org/10.1007/s11004-011-9326-9, 2011.

Ellis, R. G. and Oldenburg, D. W.: Applied geophysical inversion, Geophys. J. Int., 116, 5–11, https://doi.org/10.1111/j.1365-246X.1994.tb02122.x, 1994.

Ferré, T. P. A.: Revisiting the Relationship Between Data, Models, and Decision-Making, Groundwater, 55, 604–614, https://doi.org/10.1111/gwat.12574, 2017.

Feyen, L. and Caers, J.: Quantifying geological uncertainty for flow and transport modeling in multi-modal heterogeneous formations, Adv. Water Resour., 29, 912–929, https://doi.org/10.1016/j.advwatres.2005.08.002, 2006.

Fleckenstein, J. H., Niswonger, R. G., and Fogg, G. E.: River-aquifer interactions, geologic heterogeneity, and low-flow management, Ground Water, 44, 837–852, https://doi.org/10.1111/j.1745-6584.2006.00190.x, 2006.

Fogg, G. E., Noyes, C. D., and Carle, S. F.: Geologically based model of heterogeneous hydraulic conductivity in an alluvial setting, Hydrogeol. J., 6, 131–143, https://doi.org/10.1007/s100400050139, 1998.

Gelhar, L. W.: Stochastic analysis of flow in heterogeneous porous media, in Fundamental of Transport Phenomena in Porous Media, edited by: Bear, J. and Corapcioglu, Y. M., Springer, the Netherlands, 673–717, 1984.

Goovaerts, P.: Geostatistics for Natural Resource Evaluation, 1st ed., edited by: Journel, A. G., Oxford University Press, New York, 1997.

Gunnink, J. L. and Siemon, B.: Applying airborne electromagnetics in 3D stochastic geohydrological modelling for determining groundwater protection, Near Surf. Geophys., 13, 45–60, https://doi.org/10.3997/1873-0604.2014044, 2015.

Hansen, T. M., Vu, L. T., Mosegaard, K., and Cordua, K. S.: Multiple point statistical simulation using uncertain (soft) conditional data, Comput. Geosci., 114, 1–10, https://doi.org/10.1016/j.cageo.2018.01.017, 2018.

He, X., Koch, J., Sonnenborg, T. O., Jørgensen, F., Schamper, C., and Christian Refsgaard, J.: Transition probability-based stochastic geological modeling using airborne geophysical data and borehole data, Water Resour. Res., 50, 3147–3169, https://doi.org/10.1002/2013WR014593, 2014.

He, X. L., Sonnenborg, T. O., Jørgensen, F., and Jensen, H. J.: Modelling a real-world buried valley system with vertical non-stationarity using multiple-point statistics, Hydrogeol. J., 25, 359–370, https://doi.org/10.1007/s10040-016-1486-8, 2016.

Hermans, T., Nguyen, F., and Caers, J.: Uncertainty in training image-based inversion of hydraulic head data constrained to ERT data: Workflow and case study, Water Resour. Res., 51, 5332–5352, https://doi.org/10.1002/2014WR016460, 2015.

Høyer, A.-S., Jørgensen, F., Sandersen, P. B. E., Viezzoli, A., and Møller, I.: 3D geological modelling of a complex buried-valley network delineated from borehole and AEM data, J. Appl. Geophys., 122, 94–102, https://doi.org/10.1016/j.jappgeo.2015.09.004, 2015.

Høyer, A.-S., Vignoli, G., Hansen, T. M., Vu, L. T., Keefer, D. A., and Jørgensen, F.: Multiple-point statistical simulation for hydrogeological models: 3-D training image development and conditioning strategies, Hydrol. Earth Syst. Sci., 21, 6069–6089, https://doi.org/10.5194/hess-21-6069-2017, 2017.

Jørgensen, F. and Sandersen, P. B. E.: Buried and open tunnel valleys in Denmark—erosion beneath multiple ice sheets, Quaternary Sci. Rev., 25, 1339–1363, https://doi.org/10.1016/j.quascirev.2005.11.006, 2006.

Jørgensen, F., Møller, R. R., Sandersen, P. B. E., and Nebel, L.: 3-D geological modelling of the Egebjerg area , Denmark, based on hydrogeophysical data, Geol. Surv. Den. Green., 20, 27–30, 2010.

Jørgensen, F., Møller, R. R., Nebel, L., Jensen, N. P., Christansen, A. V., and Sandersen, P. B. E.: A method for cognitive 3D geological voxel modelling of AEM data, B. Eng. Geol. Environ., 72, 421–432, https://doi.org/10.1007/s10064-013-0487-2, 2013.

Journel, A. G.: Combining Knowledge From Diverse Sources: An Alternative to Traditional Data, Math. Geol., 34, 573–596, 2002.

Journel, A. G. and Zhang, T.: The Necessity of a Multiple-Point Prior Model, Math. Geol., 38, 591–610, https://doi.org/10.1007/s11004-006-9031-2, 2007.

Krishnan, S.: Combining individual data information: A review and the tau model, Stanford Center for Reservoir Forecasting (SCRF) annual report 17, 1–54, 2004.

LaBolle, E. M. and Fogg, G. E.: Role of Molecular Diffusion in Contaminant Migration and Recovery in an Alluvial Aquifer System, Transport Porous Med., 42, 155–179, https://doi.org/10.1023/A:1006772716244, 2001.

Le Coz, M., Genthon, P., and Adler, P. M.: Multiple-Point Statistics for Modeling Facies Heterogeneities in a Porous Medium: The Komadugu-Yobe Alluvium, Lake Chad Basin, Math. Geosci., 43, 861–878, https://doi.org/10.1007/s11004-011-9353-6, 2011.

Lochbühler, T., Pirot, G., Straubhaar, J., and Linde, N.: Conditioning of Multiple-Point Statistics Facies Simulations to Tomographic Images, Math. Geosci., 46, 625–645, https://doi.org/10.1007/s11004-013-9484-z, 2014.

Lowe, D. G.: Distinctive Image Features from Scale-Invariant Keypoints, Int. J. Comput. Vision, 60, 91–110, https://doi.org/10.1023/B:VISI.0000029664.99615.94, 2004.

Mariethoz, G. and Renard, P.: Reconstruction of Incomplete Data Sets or Images Using Direct Sampling, Math. Geosci., 42, 245–268, https://doi.org/10.1007/s11004-010-9270-0, 2010.

Marker, P. A., Vilhelmsen, T. N., Foged, N., Wernberg, T., Auken, E., and Bauer-Gottwein, P.: Probabilistic predictions using a groundwater model informed with airborne EM data, Adv. Water Resour., 103, 86–98, https://doi.org/10.1016/j.advwatres.2017.03.002, 2017.

Maurer, C. R., Qi, R., Raghavan, V., and Member, S.: A linear time algorithm for computing exact Euclidean distance transforms of binary images in arbitrary dimensions, IEEE T. Pattern Anal., 25, 265–270, https://doi.org/10.1109/TPAMI.2003.1177156, 2003.

Maxey, G. B.: Hydrostratigraphic units, J. Hydrol., 2, 124–129, 1964.

Okabe, H. and Blunt, M. J.: Prediction of permeability for porous media reconstructed using multiple-point statistics, Phys. Rev. E, 70, 1–2, https://doi.org/10.1103/PhysRevE.70.066135, 2004.

Okabe, H. and Blunt, M. J.: Pore space reconstruction using multiple-point statistics, J. Pet. Sci. Eng., 46, 121–137, https://doi.org/10.1016/j.petrol.2004.08.002, 2005.

Pirot, G.: Using Training Images to Build Model Ensembles with Structural Variability, Groundwater, 55, 656–659, https://doi.org/10.1111/gwat.12556, 2017.

Remy, N., Boucher, A., and Wu, J.: Applied geostatistics with SGeMS: a user's guide, Cambridge University Press, 2014.

Royse, K. R.: Combining numerical and cognitive 3D modelling approaches in order to determine the structure of the Chalk in the London Basin, Comput. Geosci., 36, 500–511, https://doi.org/10.1016/j.cageo.2009.10.001, 2010.

Sandersen, P. B. E., Jørgensen, F., Larsen, N. K., Westergaard, J. H., and Auken, E.: Rapid tunnel-valley formation beneath the receding Late Weichselian ice sheet in Vendsyssel, Denmark, Boreas, 38, 834–851, https://doi.org/10.1111/j.1502-3885.2009.00105.x, 2009.

Seifert, D., Sonnenborg, T. O., Refsgaard, J. C., Højberg, A. L., and Troldborg, L.: Assessment of hydrological model predictive ability given multiple conceptual geological models, Water Resour. Res., 48, 1–16, https://doi.org/10.1029/2011WR011149, 2012.

Sørensen, K. I. and Auken, E.: SkyTEM – a new high-resolution transient electromagnetic system, Explor. Geophys., 35, 191–199, https://doi.org/10.1071/EG04194, 2004.

Strebelle, S.: Conditional Simulation of Complex Geological Structures Using Multiple-Point Statistics, Math. Geol., 34, 1–21, https://doi.org/10.1023/A:1014009426274, 2002.

Strebelle, S. and Journel, A. G.: Reservoir modeling using multiple-point statistics, Proc. SPE Annu. Tech. Conf. Exhib., 11, https://doi.org/10.2523/71324-MS, 2001.

Tahmasebi, P., Hezarkhani, A., and Sahimi, M.: Multiple-point geostatistical modeling based on the cross-correlation functions, Comput. Geosci., 16, 779–797, https://doi.org/10.1007/s10596-012-9287-1, 2012.

Tan, X., Tahmasebi, P., and Caers, J.: Comparing training-image based algorithms using an analysis of distance, Math. Geosci., 46, 149–169, https://doi.org/10.1007/s11004-013-9482-1, 2014.

Tarantola, A. and Valette, B.: Generalized nonlinear inverse problems solved using the least squares criterion, Rev. Geophys., 20, 219–232, https://doi.org/10.1029/RG020i002p00219, 1982.

Tran, T.: Improving variogram reproduction on dense simulation grids, Comput. Geosci., 20, 1161–1168, https://doi.org/10.1016/0098-3004(94)90069-8, 1994.

Viezzoli, A., Christansen, A. V., Auken, E., and Sørensen, K. I.: Quasi-3D modeling of airborne TEM data by spatially constrained inversion, Soundings, 73, F1015–F113, https://doi.org/10.1190/1.2895521, 2008.

Vignoli, G., Fiandaca, G., Christansen, A. V., Kirkegaard, C., and Auken, E.: Sharp spatially constrained inversion with applications to transient electromagnetic data, Geophys. Prospect., 63, 243–255, https://doi.org/10.1111/1365-2478.12185, 2015.

Wang, L., Zhang, Y., and Feng, J.: On the Euclidean distance of images, IEEE T. Pattern Anal., 27, 1334–1339, https://doi.org/10.1109/TPAMI.2005.165, 2005.

Xiaofeng, F. and Wei, W.: Centralized binary patterns embedded with image euclidean distance for facial expression recognition, Proc. – 4th Int. Conf. Nat. Comput. ICNC 2008, 4, 115–119, https://doi.org/10.1109/ICNC.2008.94, 2008.

Zhao, Z. and Illman, W. A.: On the importance of geological data for three-dimensional steady-state hydraulic tomography analysis at a highly heterogeneous aquifer-aquitard system, J. Hydrol., 544, 640–657, https://doi.org/10.1016/j.jhydrol.2016.12.004, 2017.

9

Redressing the balance: quantifying net intercatchment groundwater flows

Laurène Bouaziz[1,2], Albrecht Weerts[2,3], Jaap Schellekens[4], Eric Sprokkereef[5], Jasper Stam[5], Hubert Savenije[1], and Markus Hrachowitz[1]

[1]Water Resources Section, Faculty of Civil Engineering and Geosciences, Delft University of Technology, P.O. Box 5048, 2600 GA Delft, the Netherlands
[2]Department Catchment and Urban Hydrology, Deltares, Boussinesqweg 1, 2629 HV Delft, the Netherlands
[3]Hydrology and Quantitative Water Management Group, Department of Environmental Sciences, Wageningen University, Wageningen, the Netherlands
[4]VanderSat, Wilhelminastraat 43A, 2011 VK Haarlem, the Netherlands
[5]Ministry of Infrastructure and Water Management, Zuiderwagenplein 2, 8224 AD Lelystad, the Netherlands

Correspondence: Laurène Bouaziz (laurene.bouaziz@deltares.nl)

Abstract. Intercatchment groundwater flows (IGFs), defined as groundwater flows across topographic divides, can occur as regional groundwater flows that bypass headwater streams and only drain into the channel further downstream or directly to the sea. However, groundwater flows can also be diverted to adjacent river basins due to geological features (e.g., faults, dipping beds and highly permeable conduits). Even though intercatchment groundwater flows can be a significant part of the water balance, they are often not considered in hydrological studies. Yet, assuming this process to be negligible may introduce misrepresentation of the natural system in hydrological models, for example in regions with complex geological features. The presence of limestone formations in France and Belgium potentially further exacerbates the importance of intercatchment groundwater flows, and thus brings into question the validity of neglecting intercatchment groundwater flows in the Meuse basin. To isolate and quantify the potential relevance of net intercatchment groundwater flows in this study, we propose a three-step approach that relies on the comparison and analysis of (1) observed water balance data within the Budyko framework, (2) results from a suite of different conceptual hydrological models and (3) remote-sensing-based estimates of actual evaporation. The data of 58 catchments in the Meuse basin provide evidence of the likely presence of significant net intercatchment groundwater flows occurring mainly in small headwater catchments underlain by fractured aquifers. The data suggest that the relative importance of net intercatchment groundwater flows is reduced at the scale of the Meuse basin, as regional groundwater flows are mostly expected to be self-contained in large basins. The analysis further suggests that net intercatchment groundwater flow processes vary over the year and that at the scale of the headwaters, net intercatchment groundwater flows can make up a relatively large proportion of the water balance (on average 10 % of mean annual precipitation) and should be accounted for to prevent overestimating actual evaporation rates.

1 Introduction

Intercatchment groundwater flows are defined as groundwater fluxes crossing topographic divides, implying that precipitation falling in one catchment affects the streamflow in another catchment. A theoretical framework to describe groundwater flows was introduced by Tóth (1963). He classified different systems of groundwater flows, starting from local flow paths, nested in larger intermediate systems, which in turn are nested in regional flow systems. The theory describes that regional groundwater flow paths transport water from small headwaters to the larger and lower elevation basin, meaning that small basins tend to export or import water and large basins are likely self-contained (Schaller and Fan, 2009). This is based on the assumption that regional flow paths occur within surface drainage boundaries at the largest scale; however, systems with dipping sedimentary beds can divert groundwater away from the basin, leading to complications of the above-described theories and to intercatchment groundwater flows between adjacent basins (Schaller and Fan, 2009; Frisbee et al., 2016). Regional flow paths within a basin and between adjacent basins are the subject of this study as they characterize intercatchment groundwater flows.

Large-scale studies and theoretical models can help to understand the link between intercatchment groundwater flows and physical catchment characteristics. Schaller and Fan (2009) assessed the role of topography, aquifer properties, climate and geology on intercatchment groundwater flows. On the continental scale, they found that arid climates favor intercatchment groundwater flows. However, on the regional and basin scale, geology exerts the strongest control on intercatchment groundwater flows. The particularities of the geological systems (e.g., faults, connectivity between faults, subsurface flow conduits) can inhibit expected correlations between the magnitude of intercatchment groundwater flows and physical catchment characteristics (e.g., lithology), as was also pointed out by Le Moine et al. (2007). This highlights the difficulty to generalize the presence of intercatchment groundwater flows based on similarities in climate and topography between catchments.

Intercatchment groundwater flows cannot be directly measured and are therefore difficult to quantify, which can explain why they are often neglected in small catchment studies (Genereux et al., 2002). However, Schaller and Fan (2009) showed that intercatchment groundwater flows can be a significant portion of a basin's water balance across the continental United States; with up to 90 % of flow leaving catchments as groundwater export and up to 50 % of river flow originating from groundwater imported from other basins. Methods to identify and quantify intercatchment groundwater flows in real-world basins either rely on stream chemistry and isotope analyses (Genereux et al., 2002; Genereux and Jordan, 2006; Ajami et al., 2011; Frisbee et al., 2011, 2012, 2016), numerical groundwater flow and transport modeling (Gleeson and Manning, 2008; Welch and Allen, 2012; Ameli et al., 2018), or on water budget analyses (within the Budyko framework) (Genereux et al., 2005; Le Moine et al., 2007, 2008; Schaller and Fan, 2009; Hrachowitz et al., 2014). Depending on the type of solute (Ameli et al., 2017), higher solute concentrations in regional groundwater flows (due to longer residence time) compared to local flow paths can provide evidence for groundwater gains through intercatchment groundwater flows. Water budget analyses, using observed discharges in real-world catchments, can in contrast show net gains or losses due to intercatchment groundwater inflow or outflow (Genereux et al., 2002).

Intercatchment groundwater flows impact water quality in higher order streams, the alteration of nonpoint source agricultural pollution, water replenishment in aquifers, the generation and migration of petroleum and mineral deposits and the ecological functioning of a catchment (Ameli et al., 2018), and it is therefore essential to understand intercatchment groundwater flows in spite of the difficulties in quantifying them.

Most conceptual hydrological models, including HBV (Bergström, 1992), TOPMODEL (Beven and Kirkby, 1979), HyMOD (Wagener et al., 2001), SUPERFLEX (Fenicia et al., 2013), VHM (Willems, 2014) and NAM (Nielsen and Hansen, 1973), solely rely on closing the water balance and neglect the possible presence of intercatchment groundwater flows by relating the change in storage over time to the difference between precipitation and the sum of actual evaporation and discharge. These models assume watertight catchment boundaries derived from surface elevation, an impermeable substratum and no deep subsurface flow bypassing the stream. These assumptions imply the absence of intercatchment groundwater flows. Adding a loss or gain term to represent such intercatchment groundwater flows is often not warranted in models due to limited data availability for calibration (often only streamflow) and the difficulties involved in determining potential and actual evaporation (Beven, 2001; Mouelhi et al., 2006). Conceptual models have several possibilities to adjust the water balance and a "correction" factor on climatic input data has often been favored over an explicit representation of intercatchment groundwater flows. Yet, this common practice may introduce misrepresentation of the natural system in hydrological models, for example in regions with complex geological features (Zhang et al., 2005; Zhang and Savenije, 2005; Reggiani and Rientjes, 2010). In the absence of robust quantitative evidence on the magnitude and temporal variability of intercatchment groundwater flow, the errors introduced by an omission of this process in models is typically compensated for by the actual evaporation term. Examples of conceptual (or empirical) models that explicitly account for net intercatchment groundwater flows include the GR4J empirical model (Perrin et al., 2003) often applied in French catchments, HYDROLOG (Chiew and McMahon, 1990), SMAR (Goswami et al., 2007; Goswami and O'Connor, 2010), mHM (Samaniego et al., 2011) and the flexible model structure used in Hrachowitz et al. (2014).

Including intercatchment groundwater flows in conceptual models has been studied in a large set of French catchments (Le Moine et al., 2007) and results in a more plausible partitioning between evaporation, streamflow and underground fluxes than methods correcting for potential errors in climatic input data or catchment area instead. Isotopic and chemical analyses indicate an intra-annual variability of intercatchment groundwater flow processes (Ajami et al., 2011; Frisbee et al., 2012).

While several studies used extensive tracer and geochemical data or developed detailed flow and transport models to quantify intercatchment groundwater flows, we propose a framework that uses widely available hydrometric observations. Previous research also using water balance data shows that different methods for estimating intercatchment groundwater flows are characterized by different uncertainties. The novelty of this study is that, here, we aim to limit these uncertainties and to detect and quantify net intercatchment groundwater flows (i.e., $Q_{\mathrm{IGF,\,in}} - Q_{\mathrm{IGF,\,out}}$) in a complementary three-step approach through (1) water budget accounting, (2) testing a set of model concepts and (3) evaluating the results against remote sensing estimates of actual evaporation. In a proof-of-concept study in the Meuse basin, we test the following hypotheses:

1. Observed water balance data in combination with the Budyko framework can provide robust evidence of the likelihood and spatial variability of net intercatchment groundwater flows.
2. Simple hydrological conceptual models enable the magnitude and intra-annual variability of net intercatchment groundwater flows over mesoscale catchments to be quantified and to assess the likelihood that intercatchment groundwater flows occur within a basin or between neighboring basins.
3. Actual evaporation estimates from remote sensing provide additional evidence to support the presence of net intercatchment groundwater flows.

2 Study areas and data

2.1 Meuse basin

This study uses data from 58 catchments within the Meuse basin upstream of Eijsden (where the Meuse flows into the Netherlands), which includes the French and Belgian part of the basin, with an area of approximately 21 000 km^2; see Fig. 1. The 58 catchments have areas varying from 50 to 16 500 km^2, with a median value of 370 km^2 and mean annual precipitation between 750 and 1200 mm yr^{-1}. Median annual runoff and potential evaporation in these catchments is approximately 420 and 620 mm yr^{-1}, respectively. Elevation in the basin ranges from 50 to 700 m. The Meuse is a typically rain-dominated river with large variations in seasonal runoff. Snow occurs relatively frequently, but is not a major factor in the discharge regime. The discharge seasonality is mainly caused by high summer and low winter evaporation, as mean precipitation displays little seasonal variation (de Wit et al., 2001).

The Meuse basin is underlain by a complex geology that combines limestone from the Middle and Late Jurassic in the southern part of the basin (mainly in the French part) with relatively impermeable metamorphic Cambrian rock and Early Devonian sandstone in the Ardennes Massif and Plateau.

From the 58 available stations, five stations are available in the Semois River catchment (Fig. 2 and Table 1) and are studied in more detail along with five additional stations (Fig. 1a and Table 1).

The Semois catchment upstream of Membre-Pont is interesting because it combines both the Jurassic and Early Devonian geological horizons: only the upstream catchment of Sainte-Marie consists of marl (and limestone), while further downstream the basin is underlain by relatively impermeable sandstone and schist. In addition, several discharge stations along the Semois river are available and allow us to detect how net intercatchment groundwater flows ($\mathrm{IGF_{net}}$) evolve as we move further downstream along the same river. Characteristics of the Semois catchments are included in Table 1 and a map is provided in Fig. 2.

In the French part of the Meuse basin, the tributary of the Aroffe River at Vannes-le-Châtel (198 km^2; see Fig. 1a) flows underground through limestone deposits towards the Moselle catchment (Fister, 2012). The Aroffe is a typical example of an overflow spring that is activated when the capacity of the conduit is exceeded, while it flows underground to the Moselle the rest of the time. The Aroffe is one of the additional five catchments where $\mathrm{IGF_{net}}$ is quantified (see Sect. 4.2.3).

2.2 Meteorological and hydrological data

For each catchment, areal averages of precipitation, potential evaporation and observed discharges (available between 2006 and 2016) are required for the analyses.

Hourly precipitation measurements are interpolated using climatological monthly background grids, using a combination of the HYRAS (Rauthe et al., 2013) and E-OBS (Haylock et al., 2008) datasets and following the method described in (van Osnabrugge et al., 2017). Precipitation measurements in Belgium were provided by the Service Public de Wallonie (2017); in France data were retrieved from the Dutch operational forecasting system. Potential evaporation estimates are based on the Makkink formula (Hooghart and Lablans, 1988) and rely on hourly interpolated temperature station data (using a lapse rate of 6.6×10^{-3} °C m^{-1}) and hourly radiation data from Maastricht (Royal Netherlands Meteorological Institute). Mean hourly values of precipitation and potential evaporation are derived from the 1200 m resolution gridded data for each catchment.

Observed discharge data are available at the hourly time step for the stations in Belgium from the Service Public de Wallonie (2017) and at the daily timestep for the stations in France from Banque Hydro. In the Semois catchments, discharges between March and mid-June 2013 were set to missing because of high observed discharges with too-limited precipitation amounts.

Table 1. Catchment characteristics. Meteorological and hydrological data are based on data between October 2006 and September 2016.

Station River	Straimont Vierre	Ste-Marie Semois	Tintigny Semois	Chiny Semois	Membre P Semois	Huccorgne Mehaigne	Yvoir Bocq	Belval Sormonne	Pierrepont Crusnes	V-le-C Aroffe
Area (km^2)	182	143	381	738	1226	305	230	369	207	198
Mean elev. (m)	440	366	405	407	390	158	268	254	340	367
Mean slope (–)	0.067	0.044	0.055	0.060	0.083	0.026	0.064	0.066	0.054	0.060
Forest (%)	34	38	50	47	56	3	16	28	23	48
Pasture (%)	29	26	22	24	18	1.7	14	48	13	20
Urban (%)	8	11	6	6	5	15	10	4	4	1
Crop (%)	29	26	22	22	21	80	60	20	60	30
Hillslopes (%)	7.4	1.5	4.6	6.0	15	0.9	7.5	9.1	6.6	8.4
Fissured* (%)	0	63	27	16	9	16	71	48	94	72
P ($mm\,y^{-1}$)	1176	1041	1110	1152	1183	753	867	1114	939	833
Q_{obs} ($mm\,y^{-1}$)	665	455	570	600	665	206	297	422	337	88
E_p ($mm\,y^{-1}$)	608	615	611	614	611	627	618	620	618	621
Q_{obs}/P (–)	0.57	0.44	0.51	0.52	0.56	0.27	0.34	0.38	0.36	0.11
E_p/P (–)	0.52	0.59	0.55	0.53	0.52	0.83	0.71	0.56	0.66	0.75

* Fissured denotes the percentage of highly productive fissured aquifers based on the International Hydrogeological Map of Europe, IHME.

2.3 Remote-sensing-based actual evaporation estimates

Two products of remote-sensing-based actual evaporation estimates are used for comparison with modeled actual evaporation: the Global Land Evaporation Amsterdam Model (GLEAM, Miralles et al., 2011; Martens et al., 2017) and Satellite Application Facility on Land Surface Analysis daily MSG actual evaporation (LSA SAF, Trigo et al., 2011).

GLEAM v3a calculates actual evaporation based on satellite-observed soil moisture, vegetation optical depth and snow water equivalent, reanalysis air temperature and radiation, and a multi-source precipitation product. GLEAM provides actual evaporation estimates at a spatial resolution of 0.25°and accounts for subgrid heterogeneity by considering three land surface types (bare soil, short vegetation and vegetation with a tall canopy). GLEAM estimates are available for the entire studied period between 2006 and 2016.

LSA SAF daily MSG (Meteosat Second Generation) actual evaporation (hereafter referred to as LSA SAF) includes soil evaporation, interception and transpiration and is calculated by solving the energy balance by combining radiative, land surface, vegetation and meteorological data. Each pixel (3 km × 3 km resolution at nadir) is split into four tiles to represent main land cover types (bare soil, grassland, crops and forests) and the surface energy balance is solved for each tile separately and results in an actual evaporation value per pixel based on the weighted average of the tiles (https://landsaf.ipma.pt/en/products/evapotranspiration/dailymet/, 29 November 2018). LSA SAF estimates are only available for the validation period (2012–2016).

3 Methods

This study consists of three parts aimed to identify, quantify and test for the presence of net intercatchment groundwater flows (IGF_{net}) in the Meuse basin. First, we use long-term observed water balance data in combination with the Budyko framework (Budyko, 1961) to identify catchments with evidence of water losses or gains through IGF_{net}. Second, we use conceptual hydrological models to assess the magnitude and temporal variability of potential IGF_{net} in the Meuse basin and we assume that they are the main cause of water balance discrepancies and thereby neglect uncertainties in forcing data. We model IGF_{net} as independent losses or gains in alternative model concepts and evaluate their magnitude in several catchments of the Meuse basin. To assess if part of the groundwater flow bypasses the headwater stream to reach the river further downstream, we model the losses or gains in increasingly large catchments along the same tributary. Thirdly, we use actual evaporation from remote sensing estimates to provide additional evidence for the likelihood and magnitude of IGF_{net}.

3.1 Identification: how to detect net intercatchment groundwater flows from observed data signals?

The water balance of a catchment reads as follows:

$$\frac{dS}{dt} = p - q_{obs} - e_a - q_{IGF}, \quad (1)$$

where S is the storage in the catchment, p is the precipitation, q_{obs} is the observed discharge at the catchment outlet, e_a is the actual evaporation and q_{IGF} is the groundwater net loss (if q_{IGF} is positive, meaning that the groundwater flow out of the catchment is larger than the flow into the catchment)

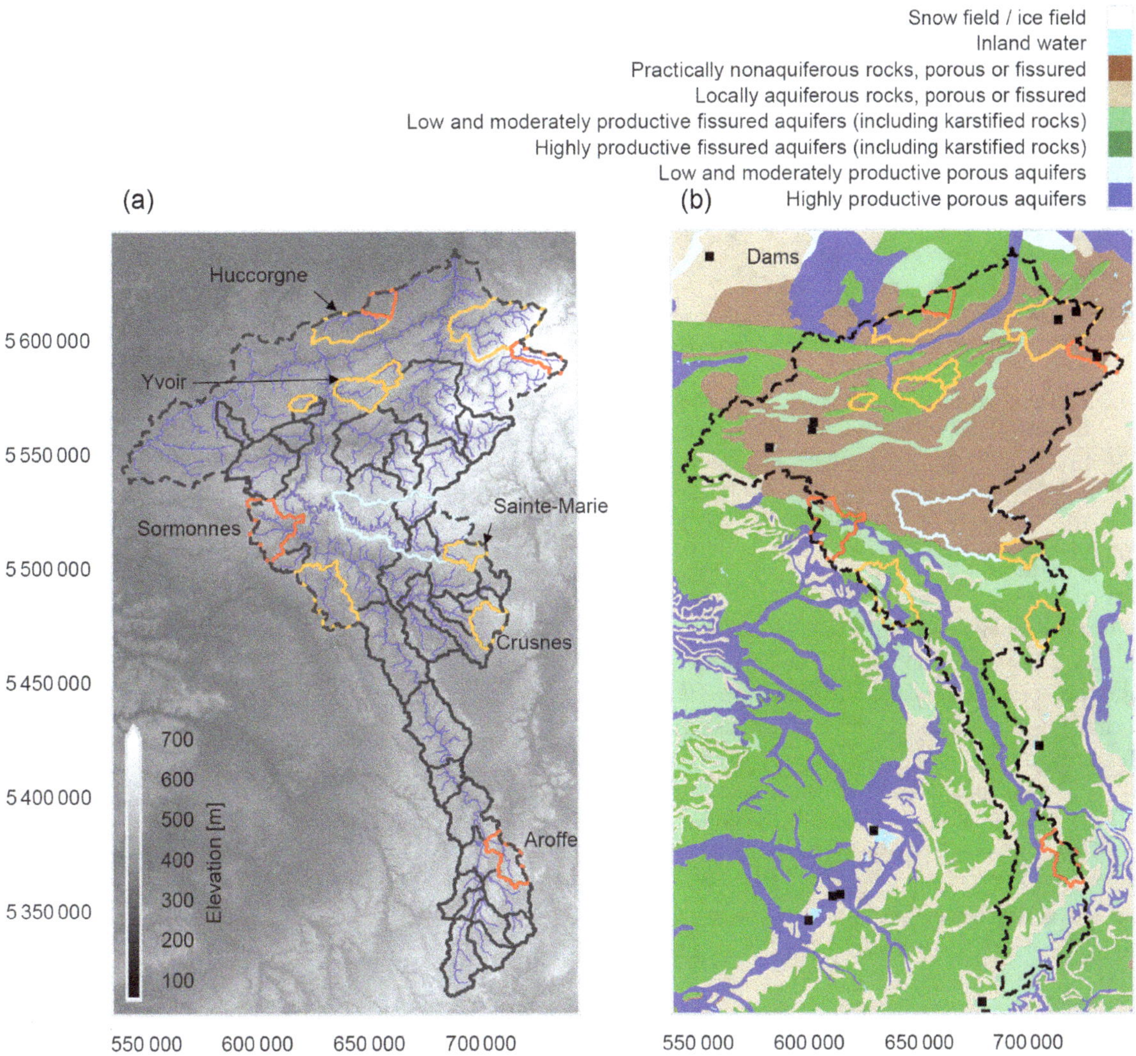

Figure 1. (a): Digital elevation model and outline of the Meuse basin with all catchments (black), catchments plotting beyond the energy limit (red), catchments very close to the energy limit (orange). The location of the Semois catchment at Membre-Pont is indicated in pale turquoise. **(b)**: International Hydrogeological Map of Europe (IHME), location of main dams (black squares, FAO database) and catchments close to (orange) and beyond (red) the energy limit.

or net gain (if q_{IGF} is negative) to the catchment, where all variables represent instantaneous fluxes (in $\mathrm{mm\,h^{-1}}$).

Intercatchment groundwater flows are often not considered and over a long period (several years), the change in storage is assumed to be zero, and long-term mean precipitation P, actual evaporation E_{a} and observed discharge Q_{obs} (in $\mathrm{mm\,yr^{-1}}$) can be reduced to the following:

$$P = Q_{\mathrm{obs}} + E_{\mathrm{a}}. \tag{2}$$

The Budyko framework (Budyko, 1961) describes the empirical global relation between the long-term evaporative index (E_{a}/P) and the dryness index (E_{p}/P, with E_{p} the long-term mean potential evaporation) and shows that natural catchments show a tendency to plot along the Budyko curve in the theoretical range located in between the energy and water limits. The water limit implies that a catchment cannot evaporate (or discharge) more water than it receives from precipitation; this implies that catchments with higher runoff than precipitation plot beyond the water limit (gaining catchments) in the Budyko framework. The energy limit implies that catchments cannot evaporate (E_{a}) more than the energy available for evaporation (E_{p}), and therefore catchments where the difference between precipitation and runoff is larger than potential evaporation are beyond the energy limit (leaky catchments), as shown in Fig. 3a. Assuming negligible observation errors, they are likely affected by net intercatchment groundwater inflows (gaining catchments) or outflows (leaky catchments). Andréassian and Perrin (2012) suggest replacing the axis of the evaporative index ($E_{\mathrm{a}}/P = 1 - Q_{\mathrm{obs}}/P$) with the runoff coefficient (Q_{obs}/P)

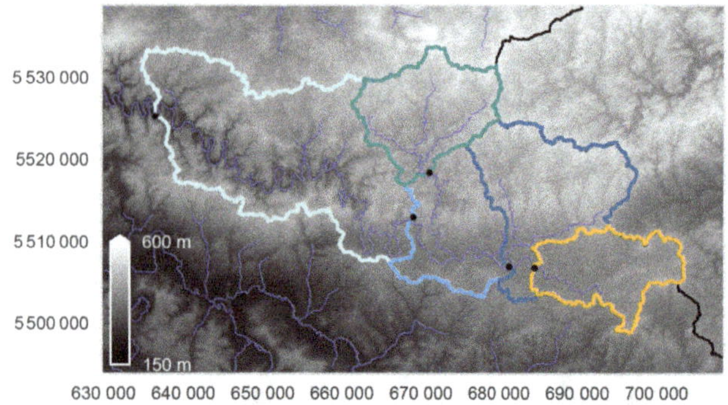

Figure 2. Digital elevation model of the Semois catchment and location of the stations from upstream to downstream: Sainte-Marie (orange), Tintigny (dark blue), Chiny (light blue), Membre-Pont (pale turquoise) and additional tributary in the north is the Vierre at Straimont (green). The catchment of Sainte-Marie plots very close to the energy limit as shown in Fig. 3.

in the Budyko framework because gaining catchments would otherwise have a negative evaporative index and because E_a itself is not measured at the catchment scale. We therefore plot each catchment in the nondimensional representation of the runoff coefficient (Q_{obs}/P) as a function of the dryness index (E_p/P), hereinafter referred to as the Budyko framework for the sake of convenience, using hydrological years between October 2006 and September 2016 (10 years) with more than 350 days of streamflow data per year.

Catchments show a tendency to plot close to the Budyko curve or other alternative expressions. The Turc–Mezentsev formula (Turc, 1954; Mezentsev, 1955) plots very close to the Budyko curve (Fig. 3) and has often been used in studies of French catchments (Le Moine et al., 2007) and was therefore applied in our analysis. The Turc–Mezentsev formula is the most general function that fulfills the two conditions $Q_{obs} \sim 0$ when $P \ll E_p$ (in very dry, moisture-constrained catchments) and $Q_{obs} \sim P - E_p$ when $P \gg E_p$ (in very wet, energy-constrained catchments) (Turc, 1954; Lebecherel et al., 2013), and transposed to streamflow, it reads as follows:

$$\frac{Q_{obs}}{P} = 1 - \frac{1}{(1+(\frac{P}{E_p})^n)^{\frac{1}{n}}}, \tag{3}$$

in which n is an exponent to estimate. Depending on the value of the parameter n, the Turc–Mezentsev relation occupies the domain from the energy limit to the water limit; Turc (1954) retained a value of $n = 2$. Here we define catchments plotting more than 5 % away from this curve (which implies a narrower range than in Gentine et al., 2012, but wider than in Li et al., 2014) and close to the limits as likely to be affected by IGF_{net}. More specifically, catchments plotting beyond the energy limit and between the energy limit and the lower boundary of the Turc–Mezentsev uncertainty range (see Fig. 3) potentially indicate the presence of net subsurface losses. Indeed, catchments that plot very close to the energy limit imply that the difference between precipitation and discharge approximates the total energy available for evaporation ($P - Q_{obs} \approx E_p$). During dry and/or very warm periods, however, evaporation is constrained by water availability and mean annual actual evaporation is therefore expected to be considerably lower than potential; this in turn means that water must be leaving the catchment through another route to comply with the observed long-term water balance. We hypothesize that water is leaving the catchment through underground pathways.

We consider the shortest distance between each catchment and the energy limit in the Budyko framework as a proxy for the presence of IGF_{net}. The closer a catchment is to the energy limit, the higher the probability of IGF_{net}. We adjust this distance by the shortest distance of the point on the Turc–Mezentsev curve at the catchment's E_p/P to the energy limit (see Fig. 3a) because arid catchments have lower runoff coefficients and are therefore expected to be further away from the energy limit. Negative distances imply that catchments plot beyond the energy limit.

We then assess if the adjusted distance to the energy limit is correlated with several physical catchment characteristics that may influence the formation of IGF_{net}, including the percentage of highly productive fissured aquifers (including karstified rocks) as provided by the International Hydrogeological Map of Europe (IHME, https://www.bgr.bund.de/ihme1500, last access: 29 November 2018) and International Geological Map of Europe (IGME), catchment area, and percentage of hillslopes (slopes steeper than 13 %, Gharari et al., 2011).

3.2 Quantification: how to quantify the variation of net intercatchment groundwater flow processes over the Meuse basin using conceptual models?

3.2.1 Models description

A reference conceptual model is developed including interception, soil moisture, fast and slow reservoirs, but no IGF_{net} (see Fig. 4a). This conceptual model is similar to the model used by Fovet et al. (2015) and has 10 calibration parameters. The characteristic timescale of the recession of the slow reservoir is determined with a master recession curve analysis.

Two options are investigated to incorporate IGF_{net} in the reference model. The first one involves a continuous constant groundwater exchange flux (loss or gain) from or to the slow reservoir ($q_{IGF}(t) = c_{IGF}$), assuming a slowly draining, homogeneous, low-permeability aquifer (Fig. 4b). The second relies on preferential permeable pathways, activated above a certain threshold, to lose or gain water (see Fig. 4c and Sects. S1 and S2 in the Supplement). In the preferential model, part of the recharge is lost or gained (before entering the slow reservoir) when the recharge exceeds a cer-

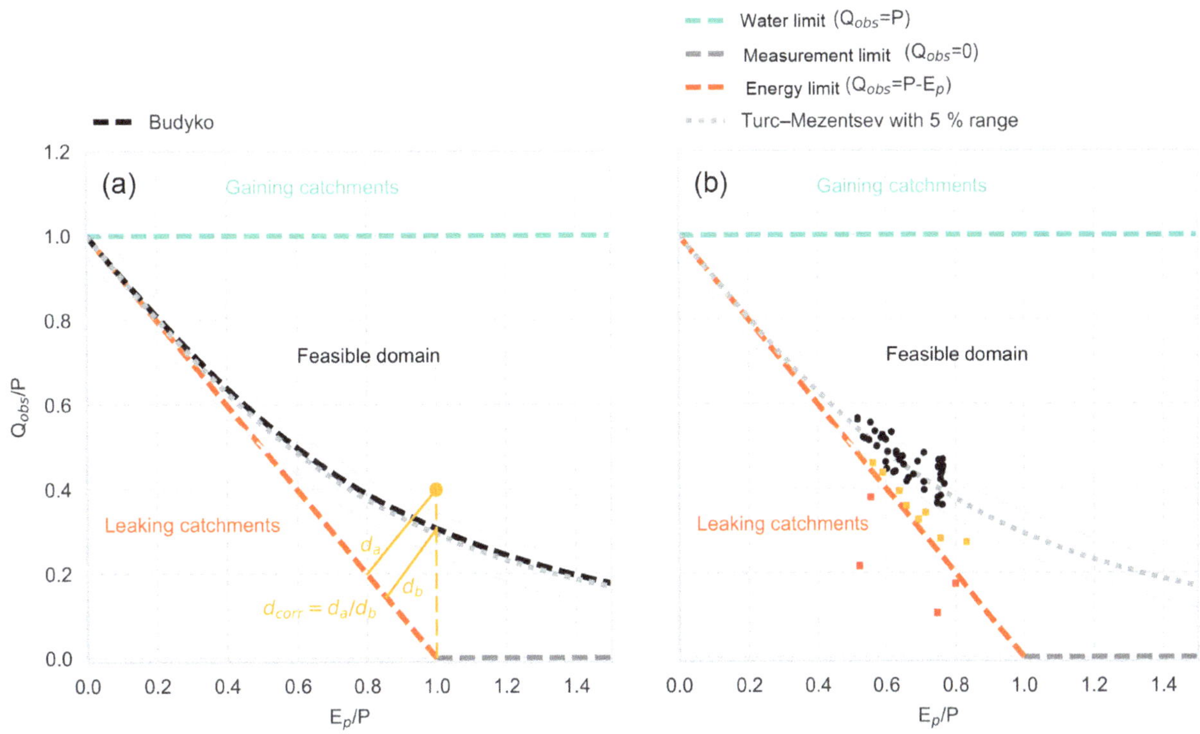

Figure 3. (a) Dimensionless representation of the runoff coefficient (Q_{obs}/P) as a function of the dryness index (E_{p}/P), referred to as the Budyko framework. The red line is the energy limit ($Q_{\mathrm{obs}} = P - E_{\mathrm{p}}$) beyond which catchments are leaking water; the blue line is the water limit ($Q = P$) above which catchments are gaining water; the dark grey line is the measurement limit ($Q = 0$). The domain within these three limits is the theoretical feasible domain. The Turc–Mezentsev and the Budyko curves plot very close to each other. The 5 % uncertainty bound around the Turc–Mezentsev curve is also shown. For each catchment, the ratio of the distance to the energy limit (d_a) over the distance of Turc–Mezentsev to the energy limit (d_b) is used as a proxy for the presence of net intercatchment groundwater flows. **(b)** The catchments of the Meuse basin are located around the Turc–Mezentsev curve (black circles). However, four catchments plot beyond the energy limit (red squares) and eight catchments plot very close to the energy limit and are beyond the lower 5 % range of Turc–Mezentsev (orange squares). In these catchments, we expect net intercatchment groundwater flow losses to occur.

tain threshold. An error function is used to simulate this behavior: $q_{\mathrm{IGF}}(t) = \mathrm{erf}(r_{\mathrm{US}}(t),\ \mu,\ m_3) \cdot p_{\mathrm{erc}} \cdot r_{\mathrm{US}}(t)$, with $r_{\mathrm{US}}(t)$ the recharge from the root zone storage to the slow reservoir, μ the threshold parameter of the recharge above which $\mathrm{IGF_{net}}$ occurs, p_{erc} the maximum fraction of the recharge to $\mathrm{IGF_{net}}$ and m_3 a shape parameter of the error function (not calibrated). The constant loss–gain model resembles the one in Hrachowitz et al. (2014) and counts one extra parameter, while the preferential $\mathrm{IGF_{net}}$ model has two additional parameters compared to the reference model.

In the catchment of the Aroffe River, water sinks in the karstified limestone after traveling through sandstone and marl deposits and emerges again in the neighboring catchment of the Moselle (which is a tributary of the Rhine River). During peak flows, the conduit capacity is exceeded and water flows in the river bed of the Aroffe (Fister, 2012). To simulate the hydrological functioning of the Aroffe river, an overflow type of model is developed to quantify the losses of this catchment to the neighboring Moselle basin, according to $q_{\mathrm{IGF}}(t) = k_{\mathrm{IGF}}^{-1} \cdot S_{\mathrm{S}}(t)$, with k_{IGF} the characteristic timescale of the underground stores (S_{S}), as shown in Fig. 4d and in Sect. S1.

Parameters, water balance and constitutive equations of all models are provided in Sects. S1 and S2 and model schematizations are shown in Fig. 4. All models are programmed in Python and an implicit Euler time stepping scheme is used to solve the model equations.

3.2.2 Model experiments – general procedure

The model was run between 1 January 2006 and 31 December 2011, using 2006 as a warm-up year, to explore the parameter space with a Monte Carlo strategy and sampling from uniformed prior parameter distributions (10^5 realizations). This was done at an hourly time step because of the fast processes occurring in the Meuse River basin. Feasible parameter sets are retained based on their simultaneous ability to reproduce high- and low-flow metrics during calibration with Nash–Sutcliffe efficiencies of at least 0.7 for different indicators (Nash–Sutcliffe efficiency of the flows $E_{\mathrm{NS},\,Q}$ and of the log of the flows $E_{\mathrm{NS},\,\log Q}$, Nash–

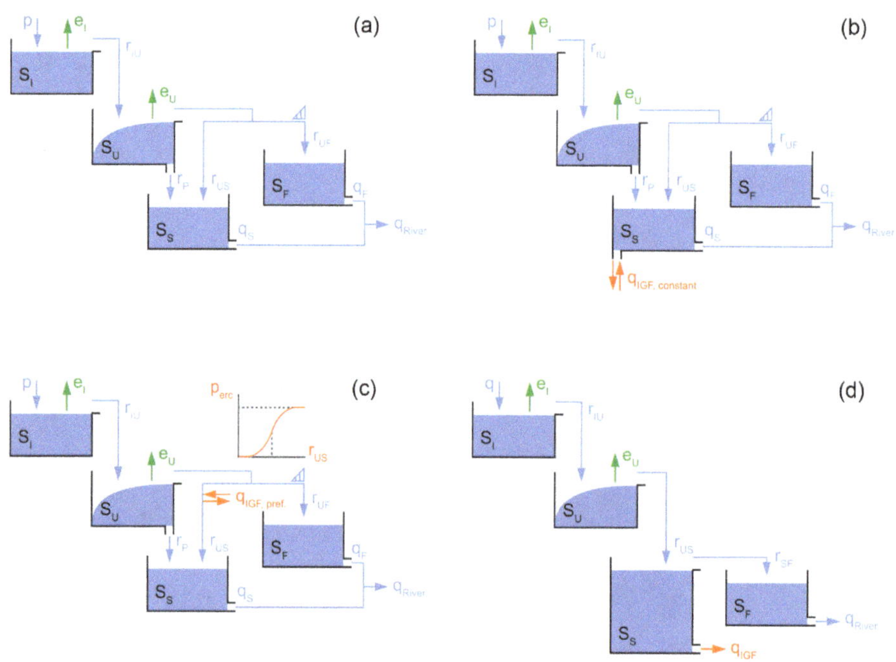

Figure 4. Conceptual model schematizations. **(a)** Reference model without net intercatchment groundwater flows. **(b)** Reference model with net constant intercatchment groundwater flows from the slow reservoir. **(c)** Reference model with net preferential intercatchment groundwater flows retrieved from or added to the recharge to the slow reservoir (using an error function that relates the percentage or recharge lost or gained to net intercatchment groundwater flows as a function of the recharge rate). **(d)** Overflow model used for the Aroffe catchment at Vannes-le-Châtel that simulates river flows in the Aroffe only when the underground storage capacity is exceeded. The rest of the time, flows occur underground towards neighboring basins. Here, we define p as precipitation, e as evaporation, r as an internal flux and q as surface or subsurface discharge (all in mm h^{-1}) and S as storage (in mm). For the subscripts, we define I as interception, U as root zone, S as slow response, F as fast response and P as percolation. The parameter p_{erc} defines the maximum percentage of recharge as net intercatchment groundwater flow.

Sutcliffe efficiency of the flow duration curve of the log of the flows $E_{\mathrm{NS,\,FDC,\,log}\,Q}$), and to reproduce discharge volumes at different temporal scales (relative volume error E_{RVE}, Nash–Sutcliffe efficiency of runoff coefficients for 6-monthly $E_{\mathrm{NS,RC,\,6\,m}}$, monthly $E_{\mathrm{NS,RC,\,m}}$ and weekly $E_{\mathrm{NS,RC,\,w}}$ periods). The tested models are evaluated in an independent validation period running from 1 January 2012 to 31 December 2016.

Prior and posterior parameter ranges are provided in Sects. S2. The characteristic timescale of the recession of the slow reservoir is estimated with a master recession curve analysis for each catchment (Fenicia et al., 2006). A range of 10 days around the derived value is used as a calibration range to account for nonlinear recession when a constant loss or gain is added to the slow reservoir.

The experiments designed to test the hypotheses of this paper are described in the following sections.

3.2.3 Representation: how to represent net intercatchment groundwater flows – zero, constant or preferential flows?

The stations on the Semois River and its tributary (Vierre at Straimont and Semois at Sainte-Marie, Tintigny, Chiny, Membre-Pont, shown in Fig. 2) are used to assess three alternative model concepts: the reference model without $\mathrm{IGF_{net}}$, constant $\mathrm{IGF_{net}}$ from or to the slow reservoir and preferential $\mathrm{IGF_{net}}$ from or to the recharge to the slow reservoir. These stations are selected because they also allow us to quantify how $\mathrm{IGF_{net}}$ evolve from upstream to downstream along the same river (see Sect. 3.2.4). The five stations are calibrated independently using the three models to quantify the magnitude of $\mathrm{IGF_{net}}$ in the subsequent catchments. The most suitable model structure is determined based on a visual inspection of hydrographs and modeled discharge regime, a comparison of performance indicators in the validation period, and a comparison between the magnitude of the loss and the distance to the energy limit (long-term mean and annual variability). Additionally, modeled mean annual actual evaporation are compared to Turc–Mezentsev estimates and we assess the shift of the modeled water balance in the Budyko framework when IGF_{net} are considered versus neglected.

3.2.4 Direction: where do intercatchment groundwater flows go?

To test if part of the groundwater flow bypasses the headwater stream to reach the river only further downstream, we model the Semois River catchments (using the experiments described in Sect. 3.2.3) to quantify how the loss–gain term varies as catchment size increases along the same river. Additionally, we looked for examples in the literature located in the Meuse basin to highlight the possible difference between IGF_{net} that is internal to a river basin and IGF_{net} to neighboring river basins.

3.2.5 Magnitude: what is the magnitude of net intercatchment groundwater flows at the scale of the Meuse basin?

Several catchments plotting close to or beyond the energy limit (from the analysis described in Sect. 3.1) are modeled to quantify the magnitude of potential IGF_{net} at several locations in the Meuse basin. Additional catchments where the magnitude of IGF_{net} is evaluated using the preferential model (because it performed better for the Semois at Sainte-Marie; see the results in Sect. 4.2.1) include the Sormonne at Belval, the Mehaigne at Huccorgne, the Bocq at Yvoir and the Crusnes at Pierrepont (Fig. 1a). For the Aroffe at Vannes-le-Châtel, the overflow type of model (Fig. 4d) is used to model the loss towards the Moselle basin, based on findings from the literature (Fister, 2012).

3.3 Evaluation: is the presence of net intercatchment groundwater flows supported by remotely sensed actual evaporation estimates?

We test for the presence of IGF_{net} using independent additional data sources. Actual evaporation is a major component of the water balance at the catchment scale, but it is also a great unknown. Reliable estimates of actual evaporation at the catchment scale would allow us to attribute the gap in the water balance to IGF_{net}, assuming minor anthropogenic activities. Global evaporation products are, however, not derived directly from earth observations, but rely on remotely sensed data in combination with models to derive actual evaporation. In this study, we compare two sources of remotely sensed actual evaporation estimates (LSA SAF and GLEAM) with our modeled actual evaporation to test the hypothesis of IGF_{net}.

4 Results and discussion

4.1 Identification: observed data and Budyko framework to detect net intercatchment groundwater flows and link with physical catchment characteristics

The analysis of observed water balances in the Budyko framework shows that relatively small headwater catchments of the Meuse basin (50–700 km^2; see Fig. 1) plot closest to or beyond the energy limit (Fig. 3b); this suggests that these catchments exhibit the highest potential for the presence of net intercatchment groundwater flows (IGF_{net}). Amongst them is the headwater catchment of the Semois at Sainte-Marie (Fig. 2), which plots close to the energy limit, suggesting underground losses towards other catchments. The water balance of two catchments in the northeast (Fig. 1) might be affected by the presence of dams (FAO, 2016) and the two catchments are therefore left out of further analyses. The net losses calculated with long-term observed runoff, precipitation and Turc–Mezentsev estimates of actual evaporation in these headwater catchments range between 70 mm yr^{-1} (for the Semois at Sainte-Marie, which corresponds to 7 % of mean annual precipitation), and 260 mm yr^{-1} (for the Aroffe catchment at Vannes-le-Châtel, which is 31 % of annual precipitation), with a median of 100 mm yr^{-1} (or 12 % of median annual precipitation). The distance of the Aroffe catchment to the energy limit is negative (the catchment plots beyond the energy limit) and approximately three times larger than the (positive) distance of the Semois at Sainte-Marie.

The catchments of the Meuse basin show a significant trend ($p = 0.001$ and $R^2 = 0.22$) indicating more losses from the catchment (negative or shorter distance to the energy limit) as the percentage of highly productive fissured aquifers increases, as shown in Fig. 5a. Intercatchment groundwater flows in the Meuse basin are therefore likely to occur in catchments with highly productive fissured aquifers, including karstified rocks (see the IHME hydrogeological map in Fig. 1b). These productive aquifers are characterized by limestone, marl or chalk lithologies (IGME). Karstification processes may cause "piracy" routes to develop (Hartmann et al., 2014) and therefore be at the origin of IGF_{net}.

We use the percentage of hillslopes in a catchment (defined as areas with a slope steeper than 13 %, Gharari et al., 2011) as a proxy for how well the drainage network is defined from the surface and relate it to the potential presence of IGF_{net} (through the distance to the energy limit) as shown in Fig. 5b. The data show a significant trend ($p = 0.001$ and $R^2 = 0.22$), indicating fewer losses from the catchment (larger distance towards the energy limit) as the percentage hillslope increases. The underlying idea is that surface topography displays the result of a competition between surface and subsurface flows. Catchments dominated by steep valleys, as encountered in the Ardennes, clearly show their drainage network at the surface. The steeper the catchment, the higher the

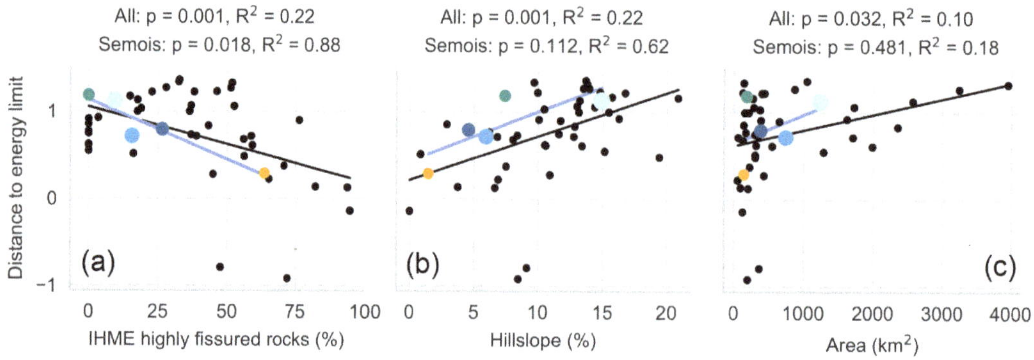

Figure 5. Adjusted distance of each catchment to the energy limit in the Budyko framework (as explained in Fig. 3) is plotted as a function of several catchment characteristics. This distance is used as a proxy for the presence of net intercatchment groundwater flows. The black line and dots show the correlation for all stations of the Meuse basin and the colored dots (with sizes scaled to catchment area) and blue line display the catchments of the Semois River only. **(a)** Distance to the energy limit as a function of the percentage highly fissured aquifers including karstified rocks based on the International Hydrogeological Map of Europe (IHME), indicating larger net losses as the percentage of highly fissured aquifers increases because of lower (or negative) values of the distance to the energy limit. **(b)** Distance to the energy limit as a function of percentage of hillslopes defined as slopes above 13 % and representative for the competition between surface and subsurface drainage. **(c)** Distance to the energy limit as a function of catchment areas of the main tributaries (up to 4000 km^2).

relative importance of lateral flow through a subsurface preferential path network to the channel or stream. On the other hand, catchments lying on permeable lithologies as chalk and middle Jurassic limestones may be dominated by rivers cutting through relatively flat plateaus and may hide an underground network of subsurface flow paths from the surface (Le Moine, 2008). The flatter the catchment, the higher the potential importance of an underground flow network and therefore of subsurface losses or gains. In the Meuse basin, IGF_{net} is therefore likely to occur in catchments dominated by a relatively flat topography.

We also tested the hypothesis that part of the groundwater flow bypasses the channel to reach the river only further downstream by correlating the distance of each catchment to the energy limit (as a proxy for the presence of IGF_{net}) to the catchment area, for the main tributaries of the Meuse basin (Fig. 5c). We expected the presence of IGF_{net} to be reduced as catchment size increases, and although this trend is significant ($p = 0.032$ and $R^2 = 0.10$), the correlation is weak. The data shown in Fig. 5c, however, suggest that evidence for IGF_{net} is highest in small catchments (with areas less than 500 km^2) and much less pronounced in larger downstream catchments, although there are also small catchments with little evidence of it. This is likely related to the variability of local geological features underlying these small catchments.

4.2 Quantification: variation of net intercatchment groundwater flow processes across the Meuse basin

4.2.1 Representation: a preferential model to represent net intercatchment groundwater flows

The reference (without IGF_{net}), constant and preferential IGF_{net} models are calibrated on subsequent catchments along the Semois river. In the following sections, the models are evaluated based on (1) performance indicators during the validation period and visual inspection of the hydrographs and seasonal behavior, (2) the magnitude of modeled IGF_{net}, and (3) modeled actual evaporation.

Performance indicators and visual inspection of the hydrographs

Performance indicators of the feasible realizations of the three models in the Semois catchments during the calibration and validation period are shown in Fig. 6. The preferential model shows an improvement in high- and low-flow indicators, and in modeled runoff coefficients in the Semois catchment at Sainte-Marie compared to the constant and zero IGF_{net} models, whereas in the other catchments of the Semois River, performance indicators are similar for the three models. Nash–Sutcliffe efficiencies of daily flows ($E_{NS,\,Q}$) and log of the flows ($E_{NS,\,\log Q}$) increase when the reference model (no IGF_{net}) is extended with a constant IGF_{net} term and increase even more when a preferential IGF_{net} term is included in the catchment upstream of Sainte-Marie. This also applies for the Nash–Sutcliffe efficiency applied on monthly and weekly runoff coefficients ($E_{NS,\,RC,\,m}$ and $E_{NS,\,RC,\,w}$). On the other hand, all performance indicators for

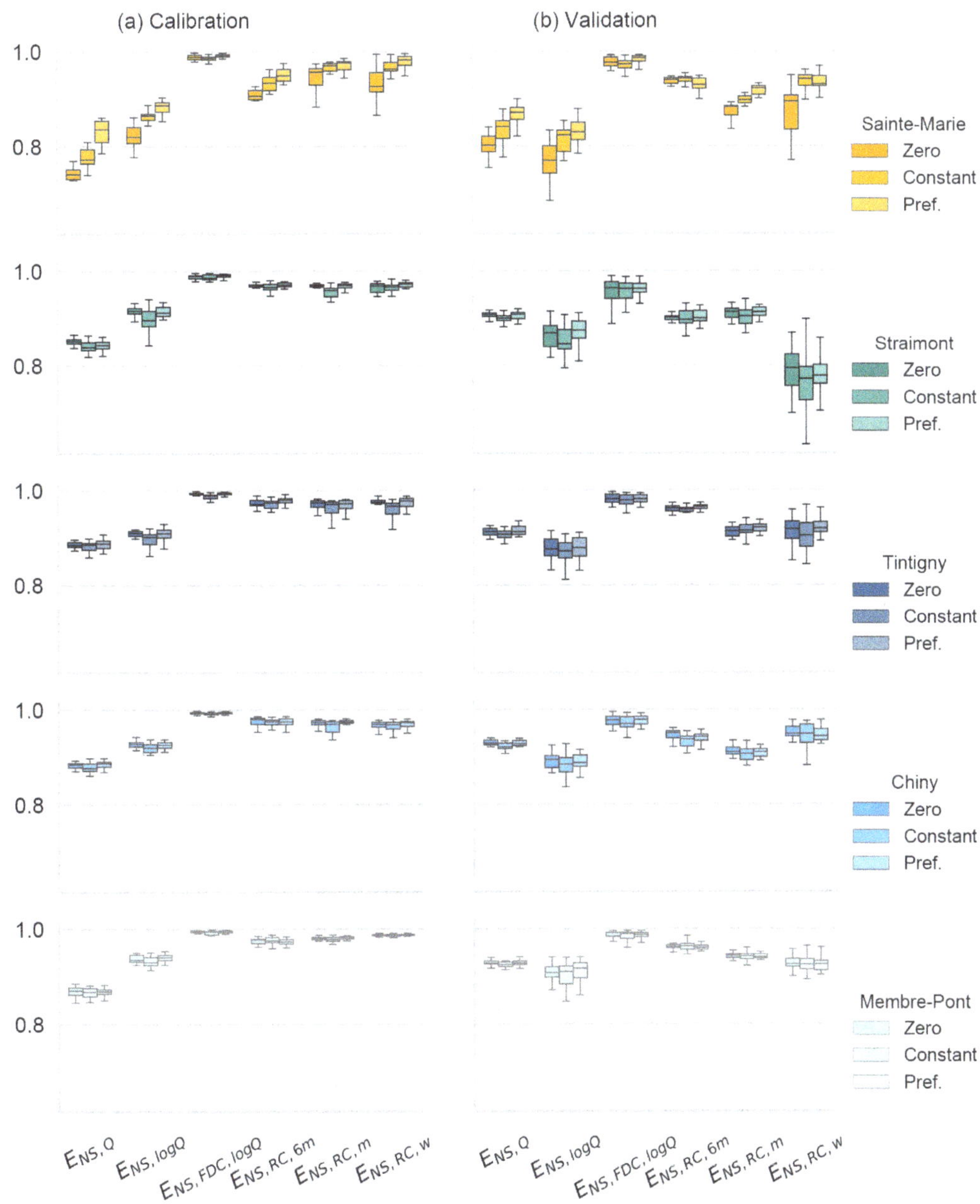

Figure 6. Performance indicators during the calibration (2007–2011, **a**) and the validation period (2012–2016, **b**) for the zero, constant and preferential models for the Semois at Sainte-Marie, the Vierre at Straimont, the Semois at Tintigny, the Semois at Chiny and the Semois at Membre-Pont. Including net intercatchment groundwater flows leads to an improved performance in the catchment of Sainte-Marie but not in the other catchments of the Semois.

the Vierre at Straimont (Fig. 6) show similar results for the three models. Adding an exchange term in this sandstone-dominated catchment (constant or preferential) does not lead to an improved performance. This behavior also characterizes the other catchments at Tintigny, Chiny and Membre-Pont (Fig. 6).

A visual inspection of the in 2014 modeled and observed hydrographs at Sainte-Marie (Fig. 7) shows a decrease in modeled winter peak flows at the beginning of the year and an increase of modeled peak flows after the dry season (October) for the preferential model compared to the zero IGF_{net} model, which better approximates observed behavior. Although this behavior might vary throughout the years, a higher performance of the preferential model in reproducing the observed discharge regime is also visible in Fig. 8. Including preferential IGF_{net} in the model reduces the mean overestimation of 9 mm month^{-1} at the beginning of the year and the underestimation of 11 mm month^{-1} in October and

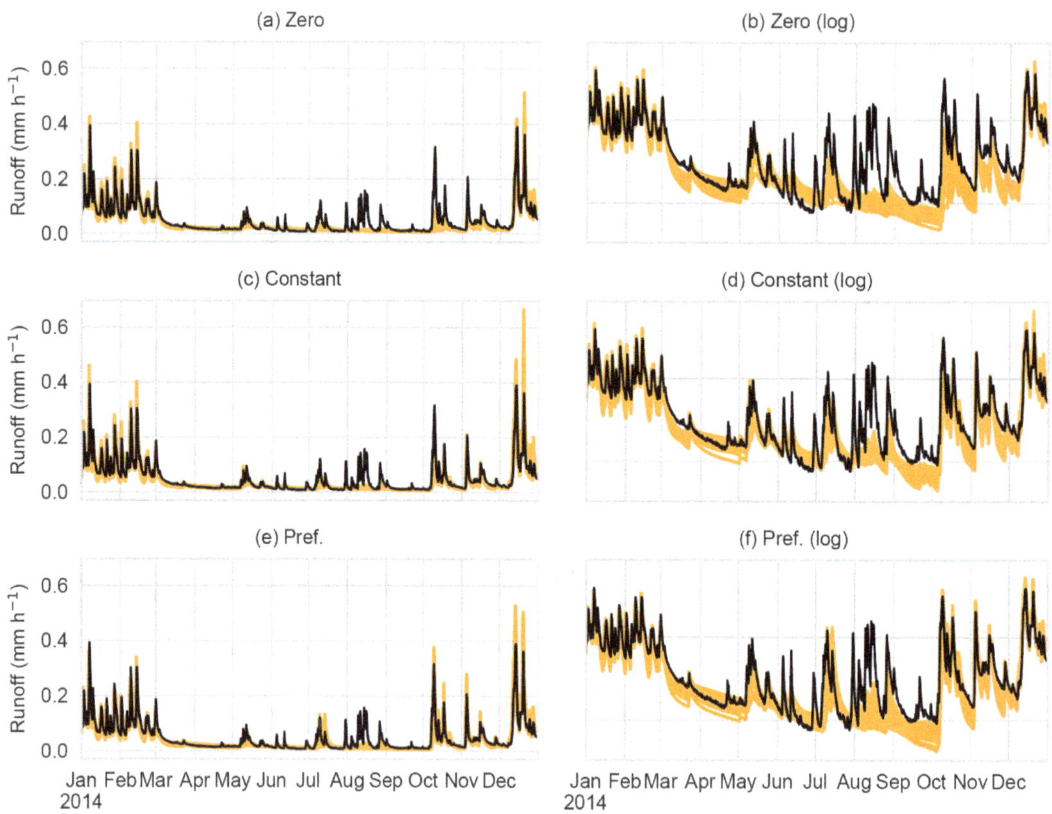

Figure 7. Observed (black line) and feasible realizations of modeled hydrographs (orange) in the catchment of the Semois at Sainte-Marie in 2014 for the three models (**a, b**: zero, **c, d**: constant and **e, f**: preferential model) on a normal **(a, c, e)** and log **(b, d, f)** scale. Including net intercatchment groundwater flows leads to lower simulated winter runoff (January–March) and higher runoff in the wetting-up period (October–November).

November, simulated by the zero IGF_{net} model, to respectively 0.5 and 3 mm month^{-1} on average. This implies that the error is reduced by 94 % at the beginning of the year and by 73 % in October and November. The improved simulation of the seasonal behavior indicates a better representation of the underlying processes and the resulting partitioning of water fluxes.

An analysis of the interannual variability of modeled IGF_{net} (see Sect. S3) also shows better performances achieved with the preferential IGF_{net} model.

Groundwater net losses or gains in the Semois catchment

In the catchment upstream of Sainte-Marie, a median annual loss term of 17 % and 20 % of observed discharge (corresponding to 77 and 90 mm yr^{-1}) is modeled by the feasible realizations of the preferential and constant IGF_{net} model, respectively, as shown in Fig. 9a, b. The magnitude of IGF_{net} decreases in the catchments further downstream on the Semois River. At the catchment outlet (Membre-Pont) and in the Vierre tributary, the magnitude of IGF_{net} is centered around zero. The range of IGF_{net} is larger for the constant model compared to the preferential model. For the preferential model, IGF_{net} approximates a value of zero for all other catchments than Sainte-Marie. In the constant model, median values of IGF_{net} are positive (losses), but some realizations imply a slight gain. Additionally, Fig. 9c, d show that the magnitude of IGF_{net} decreases as the distance to the energy limit increases. This means that as catchments plot closer to the Budyko curve (and further away from the energy limit), we see the relative importance of IGF_{net} decreasing, which is in line with expectations.

Effect on actual evaporation

Turc–Mezentsev estimates of actual evaporation are compared with modeled mean annual actual evaporation of the feasible realizations of the three models in all Semois stations in Fig. 10. Including (constant or preferential) IGF_{net} in the catchment of Sainte-Marie leads to median annual actual evaporation rates close to Turc–Mezentsev estimates, whereas the reference model leads to 10 % higher actual evaporation rates (535 mm yr^{-1} for the preferential model versus 590 mm yr^{-1} for the zero IGF_{net} model). The refer-

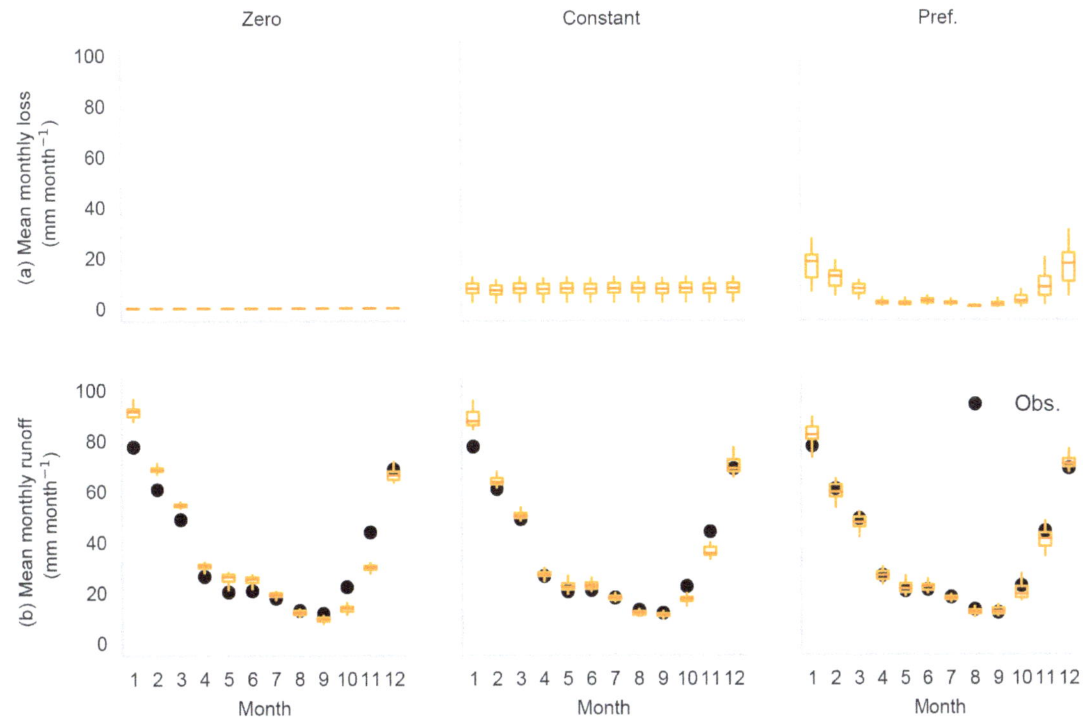

Figure 8. (a) Mean monthly loss between 2007 and 2016 for the feasible model realizations in the Semois catchment at Sainte-Marie. **(b)** Mean monthly discharge between 2007 and 2016 for the feasible model realizations (orange) and observations (black dots) for the three models at Sainte-Marie. The preferential model leads to better performances with lower simulated runoff in the first half-year and higher runoff in the wetting-up period (October–November).

ence model compensates for the absence of an intercatchment groundwater flow term by increasing actual evaporation rates to reproduce observed flow volumes. For the majority of the other catchments, the effect of adding IGF_{net} on modeled actual evaporation rates is less pronounced, but still visible.

When using observed river discharges and neglecting IGF_{net}, the catchment of Sainte-Marie plots close to the energy limit in the Budyko framework; however, when IGF_{net} is modeled and added to river flows, the catchment of Sainte-Marie plots close to the Turc–Mezentsev curve, as shown in Fig. 11. This shift in the Budyko framework occurs because we acknowledge that part of the discharge produced from the catchment bypasses the measuring gauge in the river. Including IGF_{net} in the representation of the system results in a higher degree of plausibility, based on the Budyko framework. The shift is most obvious for the catchment of Sainte-Marie, although it also occurs in the other catchments.

4.2.2 Direction: groundwater bypass routes versus intercatchment groundwater flows to external basins

The magnitude of modeled IGF_{net} decreases from nested upstream to downstream catchments along the Semois River (see Fig. 9a, b), which is an indication that "losses" modeled at Sainte-Marie are internal to the catchment of the Semois River. Losses in the upstream catchment of Sainte-Marie reappear as additional groundwater inflows in the downstream parts of the Semois, thereby reducing the IGF_{net} from upstream to downstream.

In contrast, experiments previously conducted in the Aroffe River catchment (Fister, 2012; Martin and Zany, 2012a) revealed the presence of groundwater flows, leaving the Meuse basin towards the Moselle catchment (which is part of the Rhine basin). Losses from the Meuse basin also occur along the northern boundary of the tributary of the Geer River catchment (Reggiani and Rientjes, 2010). Additionally, downstream of the village of Bazoilles, the Meuse flows underground during a large part of the year, leaving its surface bed empty, before emerging again at Noncourt, just upstream of Neufchâteau (in the upstream part of the Meuse basin); this is referred to as "la Perte de la Meuse" (translation: the Loss of the Meuse) (Newman, 1949; Martin and Zany, 2012b). This variety of processes highlights the contrast between stations that are losing water to neighboring catchments (Aroffe to the Rhine) and catchments that are losing water to themselves further downstream.

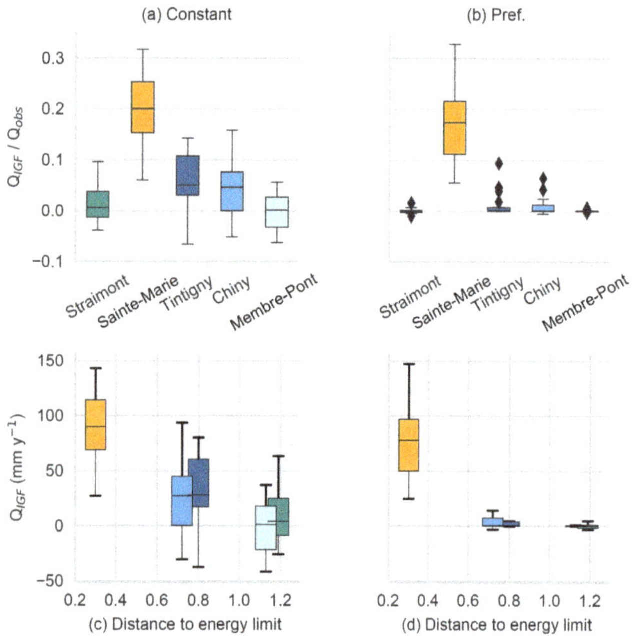

Figure 9. (a, b) Ratio of modeled net loss over observed discharge for the Semois stations for the period 2007–2016 (positive values indicate a net loss, whereas negative values indicate a net gain) for the constant model **(a)** and the preferential model **(b)**. **(c, d)** Mean annual net intercatchment groundwater flow rates (for the feasible model realizations) as a function of the observed distance to the energy limit for the catchments of the Semois river (same color code as **a, b**) for the constant model **(c)** and preferential model **(d)**. Both models show a decrease in net intercatchment groundwater flows as the distance to the energy limit increases. The discharge observations of all Semois stations are provided by the Service Public de Wallonie.

4.2.3 Magnitude: quantification of net intercatchment groundwater flows at the scale of the Meuse basin

The magnitude of IGF_{net} is assessed in several other catchments of the Meuse basin that plot close to or beyond the energy limit (annotated catchments in Fig. 1a). The preferential model is used to assess the magnitude of IGF_{net} because it performed better in the catchment of Sainte-Marie. In the Aroffe catchment, an overflow type of model is applied to represent the functioning of the system based on a priori available knowledge (see Sect. 3.2.1). Modeled mean annual flows between 2007 and 2016 overlap well with observations, as shown in Fig. 12a. The ratio of mean annual net intercatchment groundwater flows to observed discharges is always positive (indicating a loss). Modeled losses can be substantial compared to observed discharges, as shown in Fig. 12b. In the Aroffe, the median loss rate (of 208 mm yr^{-1}) is approximately 2.5 times higher than observed river flows (85 mm yr^{-1}). Median values of annual loss rates over observed discharges range from 0.1 % to 32 % (0.3 to 130 mm yr^{-1}) in the other catchments. Modeled actual evaporation is close to or slightly overestimates Turc–Mezentsev estimates (Fig. 12c), showing that the models are able to reproduce the observed long-term water balance in a meaningful way.

At the scale of the Meuse basin, intercatchment groundwater flow processes play only a small role because they occur in relatively small catchments and because part of these losses may be internal to the Meuse basin. However, IGF_{net} occurring at the scale of headwater catchments make up a considerable part of the water balance (on average 10 % and up to 25 % of mean annual precipitation), which in many current models is wrongly attributed to actual evaporation.

4.3 Evaluation: comparison with actual evaporation from remote sensing

GLEAM estimates of mean annual actual evaporation approximate or slightly overestimate (< 5 %) modeled and Turc–Mezentsev estimates of actual evaporation, as shown in Figs. 10 and 12c, whereas estimates from a land surface modeling approach, such as LSA SAF data, are considerably lower (between 400 and 470 mm yr^{-1}, Fig. 12c) in the studied catchments. While the difference in both products highlights the uncertainty in remote-sensing-based estimates of actual evaporation, it also shows that actual evaporation might even be less than that resulting from our models, which might imply even larger magnitudes of losses due to IGF_{net}. The simple conceptualization of soil-moisture-constrained evaporation used in our models, which does not account for a temperature-based stress function, might lead to an overestimation of transpiration. Thus, being arguably conservative modeled estimates, the low estimates of LSA SAF evaporation lend further credibility to evidence suggesting the presence of considerable IGF_{net}.

5 Limitations and advances

5.1 Limitations

In this work, we rely on the empirical organizing principle provided by the Turc–Mezentsev or Budyko curves (Turc, 1954; Mezentsev, 1955; Budyko, 1961) and assume that catchments of the Meuse basin plotting close to or beyond the energy limit (Fig. 3) may be subject to losses due to net intercatchment groundwater flows. Changing vegetation, climate and human interactions might, however, also be at the origin of catchments deviating from the Budyko curve (Velde et al., 2014; Berghuijs et al., 2014). The location of each catchment within the Budyko framework is also subject to uncertainties in the data used to calculate long-term mean precipitation, discharge and potential evaporation. Data uncertainties can originate from the spatial interpolation of the precipitation, the choice of a potential evaporation formula, errors in discharge measurements or in catchment delineation, or the presence of unknown anthropogenic activities affecting the

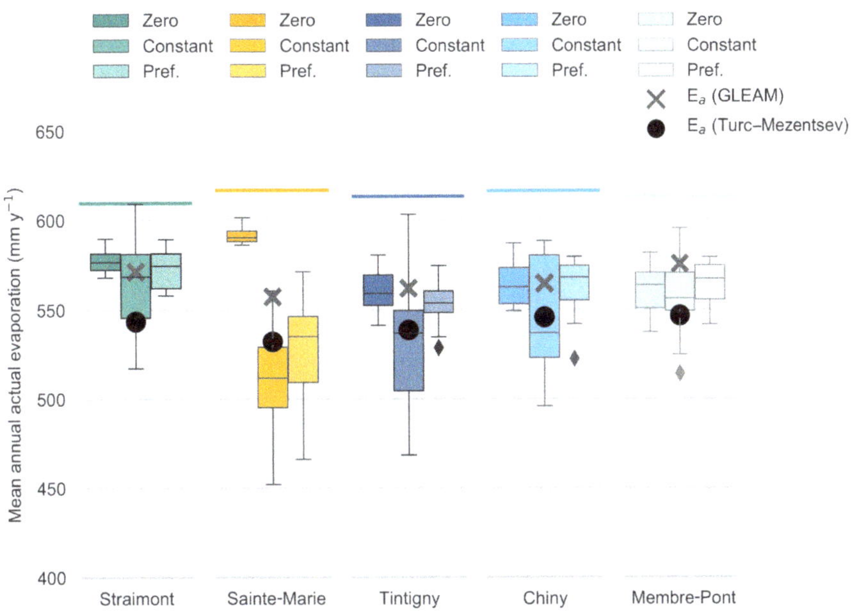

Figure 10. Best realizations of modeled mean annual actual evaporation in all stations in the Semois catchment for the three models (zero, constant and preferential models) during 2007–2016. Colored horizontal lines indicate mean annual potential evaporation used as forcing. Estimates of actual evaporation from the Turc–Mezentsev curve are shown as black dots and GLEAM estimates are shown as grey crosses. In the catchment of Sainte-Marie, the reference model without net intercatchment groundwater flows overestimates actual evaporation compared to the other two models and Turc–Mezentsev estimates.

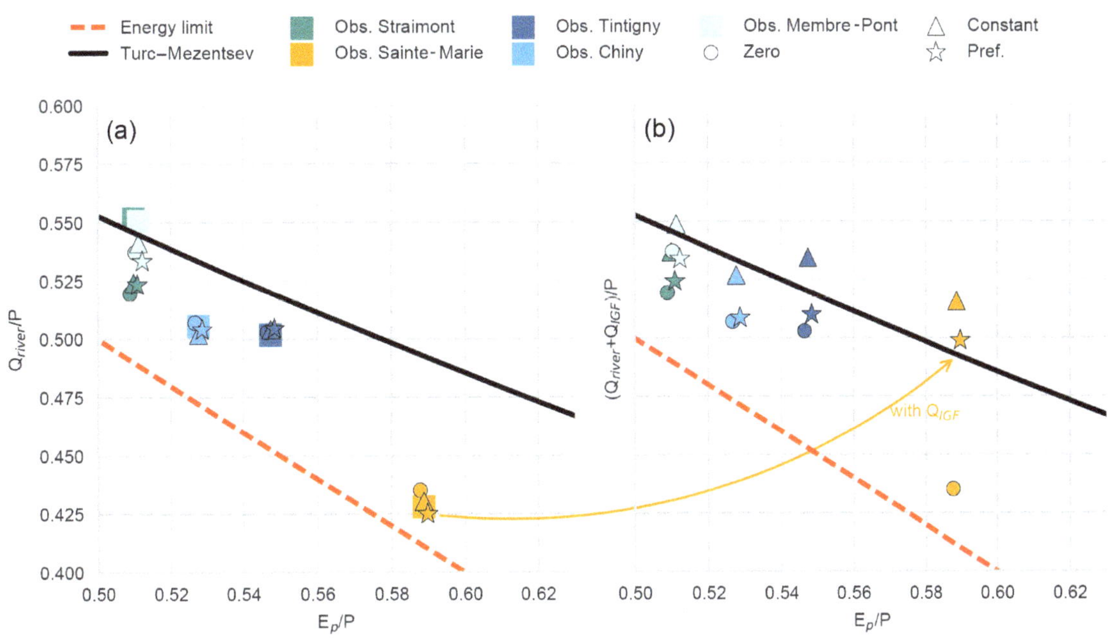

Figure 11. **(a)** Dimensionless representation of Q_{river}/P as a function of E_{p}/P. Long-term observed values between 2007 and 2016 are shown together with modeled river flows (runoff from fast and slow reservoirs) using the three models for all stations of the Semois River. **(b)** Dimensionless representation of $(Q_{\text{river}} + Q_{\text{IGF}})/P$ as a function of E_{p}/P. In this plot, we acknowledge that part of the groundwater bypasses the gauging station and we consider this flow in addition to the river flow. For the catchment of Sainte-Marie, we see a shift towards the Turc–Mezentsev curve when net intercatchment groundwater flows are acknowledged.

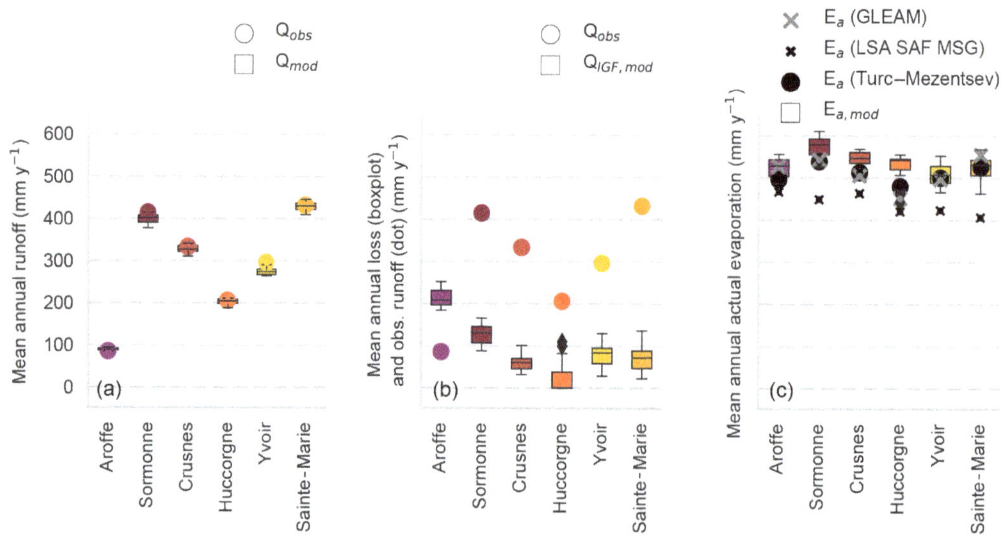

Figure 12. Water balance components of additional modeled catchments over the period 2007–2016. **(a)** Modeled (box plot) and observed (dot) mean annual discharge overlap well. **(b)** Modeled mean annual net loss (box plot) and observed mean annual discharge (dot), showing the large proportion of net intercatchment groundwater flows especially in the Aroffe catchment. **(c)** Modeled actual evaporation (box plot), GLEAM actual evaporation (grey cross), LSA SAF actual evaporation (black cross) and Turc–Mezentsev estimates (black dot). It should be noted that LSA SAF estimates are only available during the validation period (2012–2016). Model results overlap relatively well with GLEAM and Turc–Mezentsev estimates of actual evaporation, but LSA SAF estimates are lower.

water balance. The Budyko framework relies on long-term mean precipitation, potential evaporation and discharge data as well as the underlying assumption that changes in total catchment water storage (for example in lakes, wetlands or groundwater) can be treated as negligible, i.e $dS \sim 0$ [L T^{-1}]. However, the assumption of $dS \sim 0$ [L T^{-1}] is unlikely to hold over shorter time periods, which implies that the framework cannot be used in a meaningful way to estimate short-term differences (e.g., annual) in intercatchment groundwater flows. In spite of these shortcomings, the three-step approach of this study, which combines different perspectives and data to estimate net intercatchment groundwater flows, allowed us to plausibly attribute deficits in the observed water balance to the potential presence of net intercatchment groundwater flows.

We treated intercatchment groundwater flows as independent net losses or gains in lumped conceptual catchment models, without explicitly connecting the loss of one catchment to the gain of another. By modeling several stations along the same tributary (the Semois), we hypothesized that the loss in the headwater catchment at Sainte-Marie might bypass the channel to reach the river only further downstream, implying an "internal" loss within the river system; but other configurations of groundwater flows in this area might lead to similar results. Additionally, we found evidence in literature (Fister, 2012) that the Aroffe catchment flows underground to the Moselle catchment (a tributary of the Rhine), but we could not relate the flow out of the relatively small Aroffe catchment (198 km^2) to its emergence in the much larger Moselle catchment near Toul (3338 km^2) due to the difference in catchment area. Interestingly, in the recent geological past (250 000 years ago), the upstream catchment of the Moselle at Toul was flowing through the Meuse valley before it changed course to join the Rhine basin (de Wit, 2008). Subsurface flow paths connecting both catchments may therefore still remain from these earlier geological times.

5.2 Advances

In this study, we question in three steps the validity of neglecting intercatchment groundwater flows in catchment-scale hydrological studies. In the Meuse basin, the potential presence of net intercatchment groundwater flows is detected from observed water balance data in relatively small headwater catchments (< 500 km^2) and is much less pronounced in larger downstream catchments (Fig. 1). In the theory advanced by Tóth (1963), regional groundwater flows occur from the headwaters to the bottom of the basin. This implies that headwater catchments may export water through groundwater flow paths into the river further downstream, thereby increasing the groundwater contributions in larger downstream catchments; this suggests a variability of dominant hydrological (subsurface) processes across spatial scales, as also demonstrated by Frisbee et al. (2011). Schaller and Fan (2009) found that the largest magnitudes of intercatchment groundwater flow occur at catchment size near 100 km^2, which also results from our analysis (Fig. 5c). Catchment size might not be the most important control (as compared to geology or topography), but it is to some extent a proxy of the position of a catchment in the landscape. We could show that the largest evidence for intercatchment groundwater flows occurs in small headwater catchments, whereas in the lowlands, where the proportion of larger catchments is also higher, net intercatchment flow is of less relevance: the further downstream the catchment is situated, the more the losses that occurred upstream become accounted for. Schaller and Fan (2009) also report that efficient aquifers favor intercatchment groundwater flows. In the Meuse basin, the identified headwater catchments are relatively flat and underlain by highly productive and fissured aquifers (Fig 5b and c), where karstification processes might be at the origin of underground exchange flow paths between catchments. While previous research (Gleeson and Manning, 2008; Ameli et al., 2018) showed that more intercatchment groundwater flows are to be expected with increasing catchment slope, these studies assume a homogeneous subsurface, which is not the case in the Meuse basin. The relatively weak correlations between physical catchment characteristics and intercatchment groundwater flows shown in Fig. 5 can be explained by the high spatial variability of intercatchment groundwater flows due to local geological features that overrule theoretical relations at the basin scale, as also argued by Genereux et al. (2002), Schaller and Fan (2009) and Frisbee et al. (2016).

We make one of the first steps to bridge the gap between regional groundwater models in which topographic catchment boundaries are not considered and lumped conceptual hydrological models that treat catchments as well-defined impermeable entities, by adding an additional flux in conceptual models to represent net intercatchment groundwater flows. We model net intercatchment groundwater flows as preferential fluxes, occurring when recharge exceeds a threshold, to represent the filling of underground stores before intercatchment flow paths are activated (Fig. 4c), rather than as constant matrix flow. Interestingly, we show that accounting for net preferential intercatchment groundwater flows improves not only low-flow performance indicators, but also high-flow simulations (Fig. 6). The increased performance achieved with the preferential model during both high and low flows suggests that the relative importance of intercatchment groundwater flow processes change throughout the year, as also found by Frisbee et al. (2012) based on a chemical and isotopic analysis. Ajami et al. (2011) also suggest that local, intermediate and regional groundwater flow paths are active during winter, while mainly local groundwater flow paths are active during summer. The ratio of net intercatchment groundwater exports over total discharge ($Q_{\mathrm{IGF}}/(Q_{\mathrm{IGF}}+Q_{\mathrm{river}})$) is about 70 % in the Aroffe catchment (where the flow is diverted into the neighboring Moselle river) and is on average 17 % in the other catchments; these values are within the range provided by Schaller and Fan (2009).

We use independent data sources of remotely sensed actual evaporation estimates to quantify the overestimation of actual evaporation modeled when intercatchment groundwater flows are neglected. Both global actual evaporation products (GLEAM and LSA SAF) rely on different models and remotely sensed data and provide relatively large differences in mean annual values (up to 150 mm yr^{-1}), highlighting the large uncertainty in estimating actual evaporation. While GLEAM actual evaporation estimates approximate our model results and Turc–Mezentsev estimates, LSA SAF estimates indicate lower evaporation rates (Figs. 10, 12c), potentially indicating an underestimation of actual evaporation in this area, or the even larger importance of losses due to net intercatchment groundwater flows in the studied catchments.

6 Conclusions

This proof-of-concept study in the Meuse basin shows strong evidence that we can identify net intercatchment groundwater flow processes from analyzing the long-term observed water balance of a catchment. The results suggest that intercatchment groundwater flows mainly play a role in headwater catchments (< 500 km^2) with productive aquifers. In these catchments, we then use simple conceptual models to show that a net groundwater loss occurs when recharge exceeds a threshold. This preferential net loss term represents the filling of underground stores before intercatchment flow paths are activated, and ranges between 0 and 208 mm yr^{-1} (0 % and 25 % of annual precipitation) with an average of 100 mm yr^{-1} (10 % of mean annual precipitation) in the studied catchments (Fig. 12b). Some of these underground flow paths may lead to downstream catchments along the same river (regional groundwater flow paths), while others may lead to neighboring river basins (diverted groundwater flows due to the presence of geological features), which explains why these net losses can be considerable at the headwater catchment scale and negligible at the scale of larger catchments (modeled net intercatchment groundwater flows reduced to zero at the most downstream station of the Semois tributary). These findings therefore highlight that dominant streamflow generation processes vary across spatial scales. Additionally, errors in simulating the seasonal behavior are reduced by more than 70 % with the preferential model (Fig. 8b), this suggests a pronounced intra-annual variability of the magnitude of net intercatchment groundwater flow processes. Neglecting net intercatchment groundwater flows in conceptual models may still result in high performances of streamflow simulation; however, it comes at the cost of overestimating actual evaporation rates to compensate for this lack (Fig. 10). Including net intercatchment groundwater flow processes in models can considerably increase the correspondence between modeled actual evaporation and remote sensing estimates, and this provides additional evidence for the presence and magnitude of net intercatchment groundwater flows.

Data availability. The following sources of data were used for this study. Discharge data in France were retrieved from the Banque Hydro portal (http://www.hydro.eaufrance.fr/, Banque Hydro, 2015). Discharge and precipitation data in Belgium were provided by the Service Public de Wallonie in Belgium (Direction générale opérationnelle de la Mobilité et des Voies hydrauliques, Département des Etudes et de l'Appui à la Gestion, Direction de la Gestion hydrologique intégrée (Bld du Nord 8-5000 Namur, Belgium)). Hourly radiation data were retrieved from the portal of the Royal Netherlands Meteorological Institute (2017; http://www.knmi.nl/nederland-nu/klimatologie/uurgegevens). Remote-sensing-based actual evaporation (ET and DMET) were provided by the EUMETSAT Satellite Application Facility on Land Surface Analysis (LSA SAF; Trigo et al., 2011; http://lsa-saf.eumetsat.int). Actual evaporation estimates from the Global Land Evaporation Amsterdam Model (GLEAM) are available through the SFTP server of GLEAM, https://www.gleam.eu/ (Miralles et al., 2011 and Martens et al., 2017). The International Hydrogeological map of Europe (IHME1500) is available through https://www.bgr.bund.de/ihme1500 (BGR, 2008). Locations of dams in Belgium and France were provided by FAO (http://www.fao.org/nr/water/aquastat/dams/, FAO, 2016).

Author contributions. LB and MH designed the study. LB conducted all the analyses and wrote the paper. All authors discussed the design and results, and contributed to the final manuscript.

Competing interests. The authors declare that they have no conflict of interest.

Acknowledgements. This work has been conducted with funding from Rijkswaterstaat and Deltares, the Netherlands. We thank Rijkswaterstaat and Deltares for their support. The authors would like to thank the Service Public de Wallonie, Direction générale opérationnelle de la Mobilité et des Voies hydrauliques, Département des Etudes et de l'Appui à la Gestion, and Direction de la Gestion hydrologique intégrée (Bld du Nord 8-5000 Namur, Belgium) for providing the precipitation and discharge data. We would also like to thank the editor and two anonymous referees for their valuable comments, which helped us improve the paper.

Edited by: Martijn Westhoff

References

Ajami, H., Troch, P. A., Maddock, T., Meixner, T., and Eastoe, C.: Quantifying mountain block recharge by means of catchment-scale storage-discharge relationships, Water Resour. Res., 47, W04504, https://doi.org/10.1029/2010WR009598., 2011.

Ameli, A., Gabrielli, C., Morgenstern, U., and McDonnell, J.: Groundwater subsidy from headwaters to their parent water watershed: A combined field-modeling approach, Water Resour. Res., 54, 5110–5125, https://doi.org/10.1029/2017WR022356, 2018.

Ameli, A. A., Beven, K., Erlandsson, M., Creed, I. F., McDonnell, J. J., and Bishop, K.: Primary weathering rates, water transit

Ameli, A., Gabrielli, C., Morgenstern, U., and McDonnell, J.: Groundwater subsidy from headwaters to their parent water watershed: A combined field-modeling approach, Water Resour. Res., 54, 5110–5125, https://doi.org/10.1029/2017WR022356, 2018.

Ameli, A. A., Beven, K., Erlandsson, M., Creed, I. F., McDonnell, J. J., and Bishop, K.: Primary weathering rates, water transit

times, and concentration-discharge relations: A theoretical analysis for the critical zone, Water Resour. Res., 53, 942–960, 2017.

Andréassian, V. and Perrin, C.: On the ambiguous interpretation of the Turc-Budyko nondimensional graph, Water Resour. Res., 48, W10601, https://doi.org/10.1029/2012WR012532, 2012.

Banque Hydro: ©Ministère de l'Ecologie, du Développement Durable et de l'Energie, available at: http://hydro.eaufrance.fr/ (last access: 15 May 2017), 2015.

Berghuijs, W., Woods, R., and Hrachowitz, M.: A precipitation shift from snow towards rain leads to a decrease in streamflow, Nat. Clim. Change, 4, 583–586, https://doi.org/10.1038/nclimate2246, 2014.

Bergström, S.: The HBV model: Its structure and applications, Swedish Meteorological and Hydrological Institute, Norrkoping, Sweden, 1992.

Beven, K.: On hypothesis testing in hydrology, Hydrol. Process., 15, 1655–1657, 2001.

Beven, K. J. and Kirkby, M. J.: A physically based, variable contributing area model of basin hydrology, Hydrolog. Sci. J., 24, 43–69, 1979.

Budyko, M. I.: The heat balance of the earth's surface, Sov. Geogr., 2, 3–13, 1961.

Chiew, F. and McMahon, T.: Estimating groundwater recharge using a surface watershed modelling approach, J. Hydrol., 114, 285–304, https://doi.org/10.1016/0022-1694(90)90062-3, 1990.

de Wit, M.: Van regen tot Maas, chap. De ondergrond, p. 54, Uitgeverij Veen Magazines, Diemen, the Netherlands, 2008.

de Wit, M., Warmerdam, P., Torfs, P., Uijlenhoet, R., Roulin, E., Cheymol, A., Van Deursen, W., Van Walsum, P., Ververs, M., Kwadijk, J., and Buiteveld, H: Effect of climate change on the hydrology of the river Meuse, Wageningen UR, the Netherlands, 2001.

FAO: AQUASTAT website, Food and Agriculture Organization of the United Nations (FAO), available at: http://www.fao.org/nr/water/aquastat/dams/ (last access: 4 December 2018), 2016.

Federal Institute for Geosciences and Natural Resources (BGR): EuroGeoServeys (EGS) and United Nations Educational, Scientific and Cultural Organization (UNESCO), International Hydrogeological Map of Europe 1 : 5 000 000 (IHME1500), Hannover, Germany, available at: https://www.bgr.bund.de/ihme1500 (last access: 29 November 2018), 2008.

Fenicia, F., Savenije, H. H. G., Matgen, P., and Pfister, L.: Is the groundwater reservoir linear? Learning from data in hydrological modelling, Hydrol. Earth Syst. Sci., 10, 139–150, https://doi.org/10.5194/hess-10-139-2006, 2006.

Fenicia, F., Kavetski, D., Savenije, H. H. G., Clark, M. P., Schoups, G., Pfister, L., and Freer, J.: Catchment properties, function, and conceptual model representation: is there a correspondence?, Hydrol. Process., 28, 2451–2467, https://doi.org/10.1002/hyp.9726, 2013.

Fister, V.: Dynamique des écoulements dans les aquifères calcaires de bas plateaux: de l'identification à la quantification des types de circulation. Exemple des formations triasiques et jurassiques dans le Nord-Est de la France, PhD thesis, Université de Lorraine, Lorraine, France, 2012.

Fovet, O., Ruiz, L., Hrachowitz, M., Faucheux, M., and Gascuel-Odoux, C.: Hydrological hysteresis and its value for assessing process consistency in catchment conceptual models, Hydrol. Earth Syst. Sci., 19, 105–123, https://doi.org/10.5194/hess-19-105-2015, 2015.

Frisbee, M. D., Phillips, F. M., Campbell, A. R., Liu, F., and Sanchez, S. A.: Streamflow generation in a large, alpine watershed in the southern Rocky Mountains of Colorado: Is streamflow generation simply the aggregation of hillslope runoff responses?, Water Resour. Res., 47, W06512, https://doi.org/10.1029/2010WR009391, 2011.

Frisbee, M. D., Phillips, F. M., Weissmann, G. S., Brooks, P. D., Wilson, J. L., Campbell, A. R., and Liu, F.: Unraveling the mysteries of the large watershed black box: Implications for the streamflow response to climate and landscape perturbations, Geophys. Res. Lett., 39, L01404, https://doi.org/10.1029/2011GL050416, 2012.

Frisbee, M. D., Tysor, E. H., Stewart-Maddox, N. S., Tsinnajinnie, L. M., Wilson, J. L., Granger, D. E., and Newman, B. D.: Is there a geomorphic expression of interbasin groundwater flow in watersheds? Interactions between interbasin groundwater flow, springs, streams, and geomorphology, Geophys. Res. Lett., 43, 1158–1165, 2016.

Genereux, D. P. and Jordan, M.: Interbasin groundwater flow and groundwater interaction with surface water in a lowland rainforest, Costa Rica: a review, J. Hydrol., 320, 385–399, 2006.

Genereux, D. P., Wood, S. J., and Pringle, C. M.: Chemical tracing of interbasin groundwater transfer in the lowland rainforest of Costa Rica, J. Hydrol., 258, 163–178, 2002.

Genereux, D. P., Jordan, M. T., and Carbonell, D.: A paired-watershed budget study to quantify interbasin groundwater flow in a lowland rain forest, Costa Rica, Water Resour. Res., 41, W04011, https://doi.org/10.1029/2004WR003635, 2005.

Gentine, P., D'Odorico, P., Lintner, B. R., Sivandran, G., and Salvucci, G.: Interdependence of climate, soil, and vegetation as constrained by the Budyko curve, Geophys. Res. Lett., 39, L19404, https://doi.org/10.1029/2012GL053492, 2012.

Gharari, S., Hrachowitz, M., Fenicia, F., and Savenije, H. H. G.: Hydrological landscape classification: investigating the performance of HAND based landscape classifications in a central European meso-scale catchment, Hydrol. Earth Syst. Sci., 15, 3275–3291, https://doi.org/10.5194/hess-15-3275-2011, 2011.

Gleeson, T. and Manning, A. H.: Regional groundwater flow in mountainous terrain: Three-dimensional simulations of topographic and hydrogeologic controls, Water Resour. Res., 44, W10403, https://doi.org/10.1029/2008WR006848, 2008.

Goswami, M. and O'Connor, K. M.: A "monster" that made the SMAR conceptual model "right for the wrong reasons", Hydrolog. Sci. J., 55, 913–927, https://doi.org/10.1080/02626667.2010.505170, 2010.

Goswami, M., O'Connor, K., and Bhattarai, K.: Development of regionalisation procedures using a multi-model approach for flow simulation in an ungauged catchment, J. Hydrol., 333, 517–531, https://doi.org/10.1016/j.jhydrol.2006.09.018, 2007.

Hartmann, A., Goldscheider, N., Wagener, T., Lange, J., and Weiler, M.: Karst water resources in a changing world: Review of hydrological modeling approaches, Rev. Geophys., 52, 218–242, 2014.

Haylock, M. R., Hofstra, N., Klein Tank, A. M. G., Klok, E. J., Jones, P. D., and New, M.: A European daily high-resolution gridded data set of surface temperature and precipitation for 1950–2006, J. Geophys. Res., 113, D20119, https://doi.org/10.1029/2008JD010201, 2008.

Hooghart, J. and Lablans, W.: Van Penman naar Makkink: een nieuwe berekeningswijze voor de klimatologische verdampingsgetallen, De Bilt, Royal Netherlands Meteorological Institute (KNMI), De Bilt, the Netherlands, 1988.

Hrachowitz, M., Fovet, O., Ruiz, L., Euser, T., Gharari, S., Nijzink, R., Freer, J., Savenije, H. H. G., and Gascuel-Odoux, C.: Process consistency in models: The importance of system signatures, expert knowledge, and process complexity, Water Resour. Res., 50, 7445–7469, https://doi.org/10.1002/2014WR015484, 2014.

Lebecherel, L., Andréassian, V., and Perrin, C.: On regionalizing the Turc-Mezentsev water balance formula, Water Resour. Res., 49, 7508–7517, 2013.

Le Moine, N.: Le bassin versant de surface vu par le souterrain: une voie d'amélioration des performances et du réalisme des modèles pluie-débit ?, PhD thesis, CEMAGREF, UR HBAN, Antony, France, 2008.

Le Moine, N., Andréassian, V., Perrin, C., and Michel, C.: How can rainfall-runoff models handle intercatchment groundwater flows? Theoretical study based on 1040 French catchments, Water Resour. Res., 43, W06428, https://doi.org/10.1029/2006WR005608, 2007.

Le Moine, N., Andréassian, V., and Mathevet, T.: Confronting surface-and groundwater balances on the La Rochefoucauld-Touvre karstic system (Charente, France), Water Resour. Res., 44, W03403, https://doi.org/10.1029/2007WR005984, 2008.

Li, H.-Y., Sivapalan, M., Tian, F., and Harman, C.: Functional approach to exploring climatic and landscape controls of runoff generation: 1. Behavioral constraints on runoff volume, Water Resour. Res., 50, 9300–9322, 2014.

Martens, B., Miralles, D. G., Lievens, H., van der Schalie, R., de Jeu, R. A. M., Fernández-Prieto, D., Beck, H. E., Dorigo, W. A., and Verhoest, N. E. C.: GLEAM v3: satellite-based land evaporation and root-zone soil moisture, Geosci. Model Dev., 10, 1903–1925, https://doi.org/10.5194/gmd-10-1903-2017, 2017.

Martin, P. and Zany, D.: Géologie de la Lorraine, Les pertes de l'Aroffe, available at: http://www4.ac-nancy-metz.fr/base-geol/fiche.php?dossier=167&p=3descrip (last access: 29 November 2018), 2012a.

Martin, P. and Zany, D.: Géologie de la Lorraine. Pertes de la Meuse, available at: http://www4.ac-nancy-metz.fr/base-geol/fiche.php?dossier=157&p=3descrip (last access: 29 November 2018), 2012b.

Mezentsev, V.: Back to the computation of total evaporation, Meteorologia i Gidrologia, 5, 24–26, 1955.

Miralles, D. G., Holmes, T. R. H., De Jeu, R. A. M., Gash, J. H., Meesters, A. G. C. A., and Dolman, A. J.: Global land-surface evaporation estimated from satellite-based observations, Hydrol. Earth Syst. Sci., 15, 453–469, https://doi.org/10.5194/hess-15-453-2011, 2011.

Mouelhi, S., Michel, C., Perrin, C., and Andréassian, V.: Stepwise development of a two-parameter monthly water balance model, J. Hydrol., 318, 200–214, 2006.

Newman, B.: The lazy Meuse, Jenkins, London, 1949.

Nielsen, S. and Hansen, E.: Numerical simulation of the rainfall runoff process on a daily basis, Nord. Hydrol., 4, 171–190, 1973.

Perrin, C., Michel, C., and Andréassian, V.: Improvement of a parsimonious model for streamflow simulation, J Hydrol., 279, 275–289, https://doi.org/10.1016/S0022-1694(03)00225-7, 2003.

Rauthe, M., Steiner, H., Riediger, U., Mazurkiewicz, A., and Gratzki, A.: A Central European precipitation climatology – Part I: Generation and validation of a high-resolution gridded daily data set (HYRAS), Meteorol. Z., 22, 235–256, 2013.

Reggiani, P. and Rientjes, T.: Closing horizontal groundwater fluxes with pipe network analysis: An application of the REW approach to an aquifer, Environ. Modell. Softw., 25, 1702–1712, 2010.

Royal Netherlands Meteorological Institute: http://www.knmi.nl/nederland-nu/klimatologie/uurgegevens, last access: 13 July 2017.

Samaniego, L., Kumar, R., and Jackisch, C.: Predictions in a data-sparse region using a regionalized grid-based hydrologic model driven by remotely sensed data, Hydrol. Res., 42, 338–355, 2011.

Schaller, M. F. and Fan, Y.: River basins as groundwater exporters and importers: Implications for water cycle and climate modeling, J. Geophys. Res.-Atmos., 114, D04103, https://doi.org/10.1029/2008JD010636., 2009.

Service Public de Wallonie: Direction générale opérationnelle de la Mobilité et des Voies hydrauliques, Département des Etudes et de l'Appui à la Gestion, Direction de la Gestion hydrologique intégrée, Bld du Nord 8-5000 Namur, Belgium, 2017.

Tóth, J.: A theoretical analysis of groundwater flow in small drainage basins, J. Geophys. Res., 68, 4795–4812, 1963.

Trigo, I. F., Dacamara, C. C., Viterbo, P., Roujean, J.-L., Olesen, F., Barroso, C., Camacho-de Coca, F., Carrer, D., Freitas, S. C., García-Haro, J., Geiger, B., Gellens-Meulenberghs, F., Ghilain, N., Meliá, J., Pessanha, L., Siljamo, N., and Arboleda, A.: The satellite application facility for land surface analysis, Int. J. Remote Sens., 32, 2725–2744, https://doi.org/10.1080/01431161003743199, 2011.

Turc, L.: Le bilan d'eau des sols. Relations entre les précipitations, l'évaporation et l'écoulement, Ann. Agron., 5, 491–596, 1954.

van Osnabrugge, B., Weerts, A., and Uijlenhoet, R.: genRE: A Method to Extend Gridded Precipitation Climatology Data Sets in Near Real-Time for Hydrological Forecasting Purposes, Water Resour. Res., 53, 9284–9303, 2017.

Velde, Y., Vercauteren, N., Jaramillo, F., Dekker, S. C., Destouni, G., and Lyon, S. W.: Exploring hydroclimatic change disparity via the Budyko framework, Hydrol. Process., 28, 4110–4118, 2014.

Wagener, T., Boyle, D. P., Lees, M. J., Wheater, H. S., Gupta, H. V., and Sorooshian, S.: A framework for development and application of hydrological models, Hydrol. Earth Syst. Sci., 5, 13–26, https://doi.org/10.5194/hess-5-13-2001, 2001.

Welch, L. and Allen, D.: Consistency of groundwater flow patterns in mountainous topography: Implications for valley bottom water replenishment and for defining groundwater flow boundaries, Water Resour. Res., 48, W05526, https://doi.org/10.1029/2011WR010901, 2012.

Willems, P.: Parsimonious rainfall-runoff model construction supported by time series processing and validation of hydrological extremes – Part 1: Step-wise model-structure identification and calibration approach, J. Hydrol., 510, 578–590, https://doi.org/10.1016/j.jhydrol.2014.01.017, 2014.

Zhang, G. P. and Savenije, H. H. G.: Rainfall-runoff modelling in a catchment with a complex groundwater flow system: application of the Representative Elementary Watershed (REW) approach, Hydrol. Earth Syst. Sci., 9, 243–261, https://doi.org/10.5194/hess-9-243-2005, 2005.

Zhang, G., Fenicia, F., Rientjes, T., Reggiani, P., and Savenije, H.: Modeling runoff generation in the Geer river basin with improved model parameterizations to the REW approach, Phys. Chem. Earth Pt. A/B/C, 30, 285–296, 2005.

10

Integrated assessment of future potential global change scenarios and their hydrological impacts in coastal aquifers – a new tool to analyse management alternatives in the Plana Oropesa-Torreblanca aquifer

David Pulido-Velazquez[1,2], Arianna Renau-Pruñonosa[3], Carlos Llopis-Albert[4], Ignacio Morell[3], Antonio-Juan Collados-Lara[1], Javier Senent-Aparicio[2], and Leticia Baena-Ruiz[1]

[1]Instituto Geológico y Minero de España, Granada, Spain
[2]Universidad Católica de Murcia, Murcia, Spain
[3]Jaume I University, Castellón, Spain
[4]Universitat Politècnica de València, Valencia, Spain

Correspondence: Antonio-Juan Collados-Lara (ajcollados@gmail.com)

Abstract. Any change in the components of the water balance in a coastal aquifer, whether natural or anthropogenic, can alter the freshwater–salt water equilibrium. In this sense climate change (CC) and land use and land cover (LULC) change might significantly influence the availability of groundwater resources in the future. These coastal systems demand an integrated analysis of quantity and quality issues to obtain an appropriate assessment of hydrological impacts using density-dependent flow solutions. The aim of this work is to perform an integrated analysis of future potential global change (GC) scenarios and their hydrological impacts in a coastal aquifer, the Plana Oropesa-Torreblanca aquifer. It is a Mediterranean aquifer that extends over 75 km^2 in which important historical LULC changes have been produced and are planned for the future. Future CC scenarios will be defined by using an equi-feasible and non-feasible ensemble of projections based on the results of a multi-criteria analysis of the series generated from several regional climatic models with different downscaling approaches. The hydrological impacts of these CC scenarios combined with future LULC scenarios will be assessed with a chain of models defined by a sequential coupling of rainfall-recharge models, crop irrigation requirements and irrigation return models (for the aquifer and its neighbours that feed it), and a density-dependent aquifer approach. This chain of models, calibrated using the available historical data, allow testing of the conceptual approximation of the aquifer behaviour. They are also fed with series representatives of potential global change scenarios in order to perform a sensitivity analysis regarding future scenarios of rainfall recharge, lateral flows coming from the hydraulically connected neighbouring aquifer, agricultural recharge (taking into account expected future LULC changes) and sea level rise (SLR). The proposed analysis is valuable for improving our knowledge about the aquifer, and so comprises a tool to design sustainable adaptation management strategies taking into account the uncertainty in future GC conditions and their impacts. The results show that GC scenarios produce significant increases in the variability of flow budget components and in the salinity.

1 Introduction

Certain coastal regions simultaneously suffer scarce surface water resources and significant water demand. As a result, the reliability of supplying the demand depends on groundwater resources, which therefore play an important role in the management of these systems (Sola et al., 2013; Renau-

Pruñonosa et al., 2016). The analysis of coastal aquifer management problems is an important and complex issue in which water quantity and quality have to be considered together to predict the salinization process, which depend on aquifer stratigraphy and other hydrodynamic factors (precipitation regime, tides, wave set-up and storm surges, etc.) (Vallejos et al., 2015). Due to the interaction between freshwater and seawater, coastal aquifers have important hydrodynamic and hydrogeochemical peculiarities (Custodio, 2010). Any change in the components of the water balance can modify the freshwater–salt water equilibrium that defines the seawater intrusion processes (Yechieli and Sivan, 2011; Arslan and Demir, 2013). In the future, the difficulty of meeting demand in coastal systems will increase due to the impact of GC, which will reduce freshwater recharge, raise sea level and increase irrigation demand (Fujinawa, 2011; Unsal et al., 2014). Therefore, GC impacts will challenge the current water supply management of coastal aquifers (Rasmussen et al., 2013).

In recent years the number of studies of CC impacts focusing on aquifers has grown quickly (Green et al., 2011; Molina et al., 2013). A few of these studies consider quality impacts, though groundwater quality can be affected by GC in many different ways (Pulido-Velazquez et al., 2014; Dragoni and Sukhija, 2008). Only some of them are focused on the impacts of GC on coastal aquifers (Yechieli et al., 2010). Though coastal aquifers warrant greater attention since they are more vulnerable to GC (due to their connection with the sea and the interaction between freshwater and seawater), they have not been extensively studied (Rasmussen et al., 2013). To assess the salinization process and possible adaptation strategies properly, coastal aquifers require that water quantity and quality issues are analysed in an integrated way.

In order to analyse the potential impacts of future scenarios of CC on any hydrological system, we need to generate time series of climate variables and to use them as inputs of previously calibrated hydrological models with them.

In coastal aquifers there is another important issue that needs to be considered when analysing CC impacts, namely any change in sea level (Ketabchi et al., 2016). A number of authors have studied the impacts of SLR in various coastal aquifers. Chang et al. (2011) showed that in a synthetic confined coastal aquifer in which recharge is steady there is no long-term impact on seawater intrusion. Werner and Simmons (2009) showed that in unconfined aquifers the influence of the inland boundary condition can be significant to its sensitivity to SLR. Rasmussen et al. (2013) analysed an inland coastal aquifer and found that minor SLRs did not seem to affect seawater intrusion as much. There are also reports by the IPCC (Church et al., 2013) and the European Environment Agency (EEA, 2014), which focus on the analysis of historical and potential future SLR scenarios.

In addition, urban and agricultural development forces appropriate management rules to be applied if groundwater resources under different LULC scenarios are to be exploited sustainably (Robins et al., 1999; Grundmann et al., 2012). A certain degree of overpumping usually occurs in Mediterranean coastal aquifers, particularly in summer, which encourages salinization processes (Rosenthal et al., 1992). This could be exacerbated by the future LULC scenarios.

Few studies have been published that analyse GC in an integrated way – along with its impacts on future LULC change scenarios – to produce an overall analysis of GC (e.g. Pulido-Velazquez et al., 2014; Guo et al., 2015); some address cases of coastal aquifers (Benini et al., 2016; Gorelick and Zheng, 2015).

From a methodological point of view, the study of the impacts of potential GC on groundwater using an integrated and holistic climatic–agronomical–hydrological model that includes water quantity and quality continues to be a big challenge. From an operational point of view, research aimed at solving these problems has focussed on sequential coupling of models. The assessment of these impacts requires models that can predict the evolution of the freshwater–salt water interface. In order to obtain an accurate representation of the physical process involved in seawater intrusion, flow and transport need to be coupled and solved simultaneously for each time step. Such an approach gives density-dependent solutions (e.g. Shammas and Thunvik, 2009; Doulgeris and Zissis, 2014) that take the salinization process into account. Simplified sharp interface models would provide a less accurate approach with a lower number of parameters and fewer computational requirements (e.g. Llopis-Albert and Pulido-Velazquez, 2014, 2015).

The objective of this work is to perform an integrated analysis of future potential GC scenarios (including CC, LULC change and SLR) and their hydrological impacts in a coastal aquifer, the Plana Oropesa-Torreblanca aquifer. We simultaneously consider water quantity and quality in order to approach the salinization process. Section 2 describes the aquifer and the available data, while Sect. 3 presents the methodology. We propose a method to analyse and generate potential future GC scenarios involving different sources of uncertainty (Sect. 3.1). A modelling framework is defined to assess hydrological impacts on the coastal aquifer based on a sequential coupling of rainfall-recharge models, crop irrigation requirements and irrigation-return models, and a density-dependent solution that couples the resolution of flow and transport calculations for each time step (Sect. 3.2). Section 4 presents the results and their discussion. It includes an analysis of the sensitivity to potential future changes in rainfall recharge, LULC and sea level. We consider the limitations of this study and propose future research. Lastly, Sect. 5 presents the main conclusions of this research.

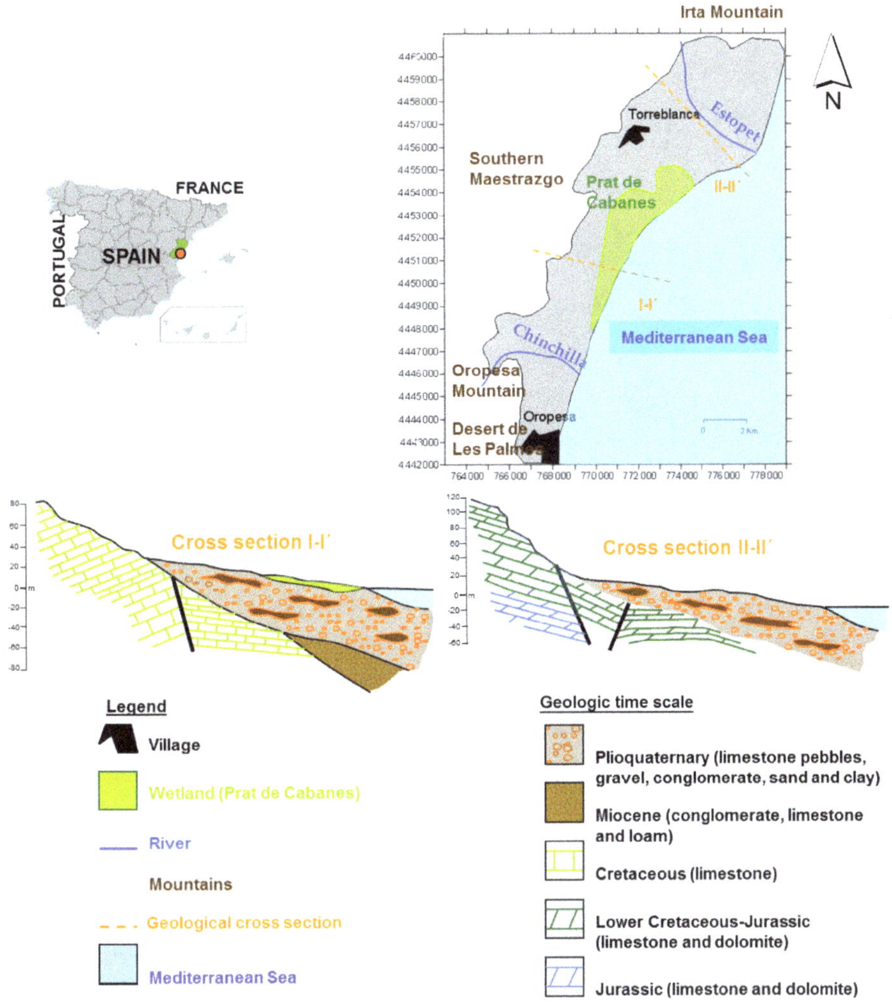

Figure 1. Location map area and cross sections of the study.

2 Materials: description of the aquifer and the data available

2.1 Location and hydrogeology

The Plana Oropesa-Torreblanca is a shallow heterogeneous detrital aquifer that extends over approximately 75 km^2. It is oriented NE–SW, parallel to the Mediterranean coast along 21 km, its inland dimension being in a range between 2.5 and 6 km. The ground surface over the aquifer comprises a gentle relief, steepening towards the surrounding limestone massifs (Fig. 1).

Geographically, the Plana Oropesa-Torreblanca aquifer borders the Sierra de Irta to the north (Cretaceous-Jurassic limestone), which is in hydraulic connection with the Plana. To its south lie the Oropesa Mountains (Cretaceous limestone). The western border (southern Maestrazgo) is formed by the Aptian and Gargasian limestone massif, which is in hydraulic connection with the Plana Oropesa-Torreblanca aquifer (except in the immediate vicinity of the Chinchilla and Estopet rivers where the impermeable Miocene base appears; Morell and Giménez, 1997; Renau-Pruñonosa et al., 2016).

The Plana Oropesa-Torreblanca is composed of Plioquaternary detrital materials comprising limestone pebbles, gravel and conglomerates derived from the adjacent mountain ranges, with abundant lenses of coarse sand, silt and clays. There are frequent lateral and vertical changes in facies and the overall distribution is irregular. The aquifer is overlain by more recent alluvial fans, colluviums, dunes and peatlands. Its geometry is lenticular – it is thinnest in the interior and thickest near the coast, exceeding 80 m at the mouths of the Estopet and Chinchilla rivers. Several studies have demonstrated how the transmissivity of the aquifer varies over a wide range: from 5000 to 100 $m^2\,day^{-1}$; the calculated effective porosity varies from 1 to 13 %, with the highest porosity nearest to the coastline. Inflows to the system consist of lateral groundwater transfers coming from adjacent aquifers, infiltration from precipitation and irrigation returns. Outflows comprise the pumped abstractions, together with

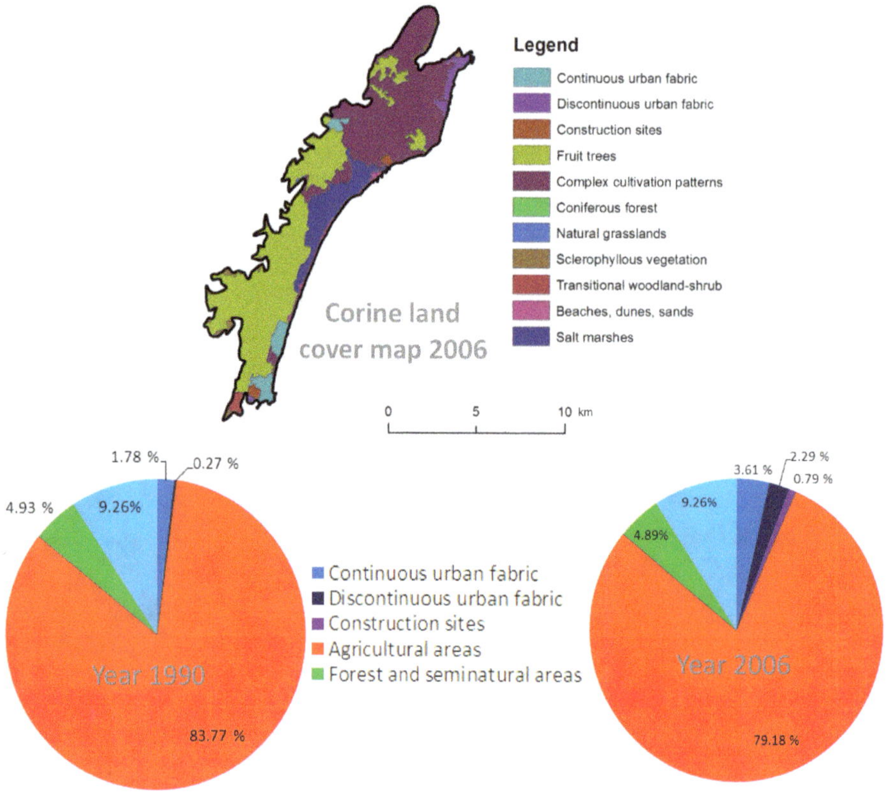

Figure 2. Land use change (comparing the CORINE Land Cover databases for 1990 and 2006) over the Plana Oropesa-Torreblanca and Maestrazgo aquifers.

groundwater discharges to the sea and seeps/springs in the Prat de Cabanes wetland. Under natural conditions, groundwater flows NW–SE, perpendicular to the coastline (Morell and Giménez, 1997; Renau-Pruñonosa et al., 2016).

The carbonate basement receives the contributions coming from the bordering aquifer (Maestrazgo aquifer) and feeds the detrital aquifer. The direction of the groundwater flow is from the carbonate basement to the detrital aquifer.

The Prat de Cabanes is a wetland located in the centre of the Plana. It extends approximately 9 km^2, parallel to the coastline with an elongated shape. It is separated from the sea by a coastal bar some 8 km long, 20 m wide and 3 m high, consisting of sorted pebbles. It is composed of brown and black silt and loam, with a recognized peat level some 3 to 4 m thick, which is commercially exploited.

2.2 Historical LULC changes and climatic–hydrological data

In the 1960s and early 1970s the Oropesa-Torreblanca Plain was sparsely populated and land was dedicated mostly to non-irrigated cropping. From 1975 to 1995 there was a significant transformation from dry to irrigated lands, especially in the period 1985–1995. From 1995 to 2010, marked changes occurred in LULC, with a generalized overhaul of the irrigation systems used and a conversion from agricultural to residential LULC, particularly along the coastal belt (Oropesa del Mar, Amplaries and Torre la Sal).

The following historical data for the period 1973–2010 were used to define the inputs of the coupled modelling framework described in Sect. 3.

1. Changes in LULC (Fig. 2), obtained from both fieldwork undertaken in the area and from the European CORINE Land Cover database (Feranec et al., 2010). These data were used to estimate the irrigation returns, following the procedure described in Sect. 3.2.2.

   A comparison of the land cover maps for 1990 and 2006 shows that the main change in LULC classes was an increase (+227 %; from 182 ha in 1996 to 592 ha in 2006) in artificial surfaces (transport infrastructure, urban sprawl, tourism and recreation facilities), mainly at the expense of agricultural areas (−5.5 %; from 7420 ha in 1996 to 7015 ha in 2006). According to the Júcar River Basin Management Plan (CHJ, 2015), the percentage LULC for major crops is citrus fruits 72.4 %, vines 12.9 %, outdoor vegetables 7.9 % and others 6.8 %. As for irrigation techniques, drip irrigation supplies 65.5 % of the total irrigated area, flood irrigation provides 34 % and spray irrigation provides 0.5 %, with an overall irrigation efficiency of 60.1 %.

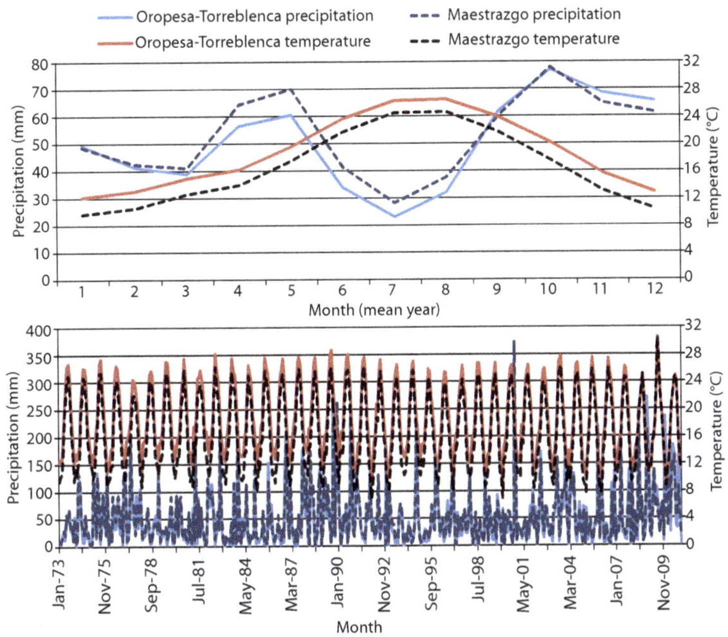

Figure 3. Historical rainfall and temperature in the Plana de Oropesa-Torreblanca and Maestrazgo aquifers.

2. Historical rainfall and temperature (T) for the Plana Oropesa-Torreblanca and Maestrazgo aquifers (see Fig. 3) were taken from the Spain02 project dataset (Herrera et al., 2016). They were used to estimate rainfall recharge (see Sect. 3.2.2).

3. Historical evolution of total pumping in the Plana Oropesa-Torreblanca aquifer was deduced from historical data.

   The transformation from dry to irrigated croplands led to an increase in pumped abstractions that extended over 2 decades (1975–1995, especially in the period 1985–1995), provoking a drop in groundwater level and seawater intrusion problems. From 1995 to 2010 there was a progressive reduction in pumping due to the abandonment of certain crops and irrigated areas.

A graphical representation of these series is included in Sect. 3.2.3 (density-dependent flow model).

Other hydrological information used to calibrate the models is the following.

1. An infiltration rate coefficient of 14 % for the historical period, which was obtained from previous lysimiter readings from a neighbouring aquifer with similar characteristic (Plana de Castellón; Tuñon, 2000). It has been used to generate historical series of rainfall recharge by applying an infiltration coefficient directly to rainfall data, which is a simple approximation commonly applied (Kirn et al., 2017). The mean historical recharge (85 mm yr^{-1}) obtained from this infiltration rate coefficient is quite similar to the mean (89 mm yr^{-1}) estimated by other authors who applied an atmospheric chloride mass balance (Alcalá and Custodio, 2014).

2. Irrigation return coefficients for the crops in the area were taken from a previous study performed by Tuñon (2000). They have been used to assess return from irrigation demands.

3. Hydraulic heads and salinity in different observation wells.

The available observation network (21 points for hydraulic head and 31 points for salinity) yielded good distributed information on aquifer state from both the quantitative and qualitative points of view. The location of the observation wells and the evolution of the variables at some of these points are described in Sect. 3.2.3 (density-dependent flow model). We have used the information available for the period 1973–2010.

3 Method – application to the case study

The flowchart of the method has been represented in Fig. 4. It summaries the steps that we propose to follow in order to perform an integrated assessment of potential GC scenarios. First of all, we propose and approach to generate future potential GC scenarios (Sect. 3.1) considering, LULC, CC and SLR. Then, a modelling framework (Sect. 3.2) was defined to assess hydrological impacts on the coastal aquifer based on a density-dependent simulation whose inputs are defined by sequential coupling of different models. Finally,

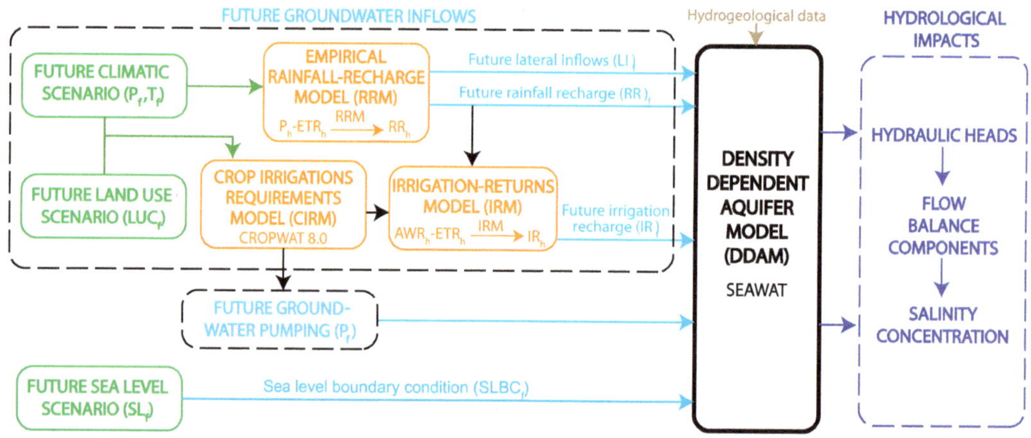

Figure 4. Flowchart of the modelling framework.

it is used to propagate the generated potential future GC scenarios (Sect. 3.3).

3.1 Generation of future GC scenarios

They have been defined by combining the future LULC scenario approved in the General Town Plans in the area and the generated CC and SLR scenarios.

3.1.1 Future LULC change scenarios

The predicted future changes in LULC over the Plana Oropesa-Torreblanca are of greater magnitude than the historical ones and could drastically modify the rural and urban landscape. The already-approved tourist developments (the public urbanization work – PAI – for the Marina d'Or Golf in Oropesa and Cabanes, and the General Town Plan – PGOU – for Torreblanca) anticipate an increase in population of more than 130 000 inhabitants, as well as the disappearance of most of the agricultural activity in the area. These significant changes in LULC will produce significant impacts on water demands, and, therefore in pumping and recharge and so to the hydrodynamics of the aquifer. In contrast, there are no significant changes to LULC anticipated in the area belonging to the municipality of Alcalà de Xivert, also situated on the Plana.

The General Town Plan for Torreblanca (PGOU Torreblanca, 2009) approves the conversion of 70 % of the municipality's area included in the General Town Plan, which is currently used for citrus agriculture, into land classified as buildable residential or industrial. Even along the coast, to the north of Prat de Cabanes, is the projected urbanization – included in the so-called Integrated Activity Plan (PAI) – of Doña Blanca Golf (Fig. 5).

The municipalities of Cabanes and Oropesa in the southern inland part of the Plana de Oropesa-Torreblanca have approved PAIs for the Marina d'Or Golf, which will include three golf courses, private residential developments, hotel complexes and associated garden areas. Once all these planned constructions are completed, they will cover some 16 km^2 (Fig. 5). Two of the conditions that the Valencian Government imposed before it approved the PAI Marina D'Or Golf are that the whole of the area that requires irrigation (both golf course and garden areas) must use recycled residential water from the wastewater treatment works; and that all water destined for urban supply must be sourced from a desalination plant. Thus, as the various individual projects are built, the groundwater abstractions in the area are falling, until they will cease completely at the end of the PAI development period. Over the neighbouring Maestrazgo aquifer, we will assume that there are no changes in LULC.

3.1.2 Generation of potential future climate scenarios for the system

We propose a method to generate consistent potential future climate scenarios for a short-term horizon (2011–2035) from the historical (1973–2010) data (see Sect. 2.2) and the climate model simulations performed in the framework of the EU CORDEX project (2013). It requires an analysis of the results obtained by applying different downscaling techniques. A multi-criteria analysis of some statistics of these series was performed to identify the best simulations of the historical data. Different ensemble hypotheses have been adopted to define more representative potential future climate scenarios to be employed in the groundwater impact study. Our main target is to provide an estimate of the most representative plausible future climate scenarios. For this reason we propose generating and propagating four plausible representative climate scenarios defined by an ensemble of different climate models, which provide a better approach of future climate scenarios than taking directly a scenario defined by a single model. The "ensembles" coalesce and consolidate the results of individual climate projections, thus allowing for more robust climate projections that are more representative than those based on a single model (Spanish Meteorologi-

Table 1. RCMs and GCMs considered.

GCMs\RCMs	CNRM-CM5	EC-EARTH	MPI-ESM-LR	IPSL-CM5A-MR
CCLM4-8-17	X	X	X	
RCA4	X	X	X	
HIRHAM5		X		
RACMO22E		X		
WRF331F				X

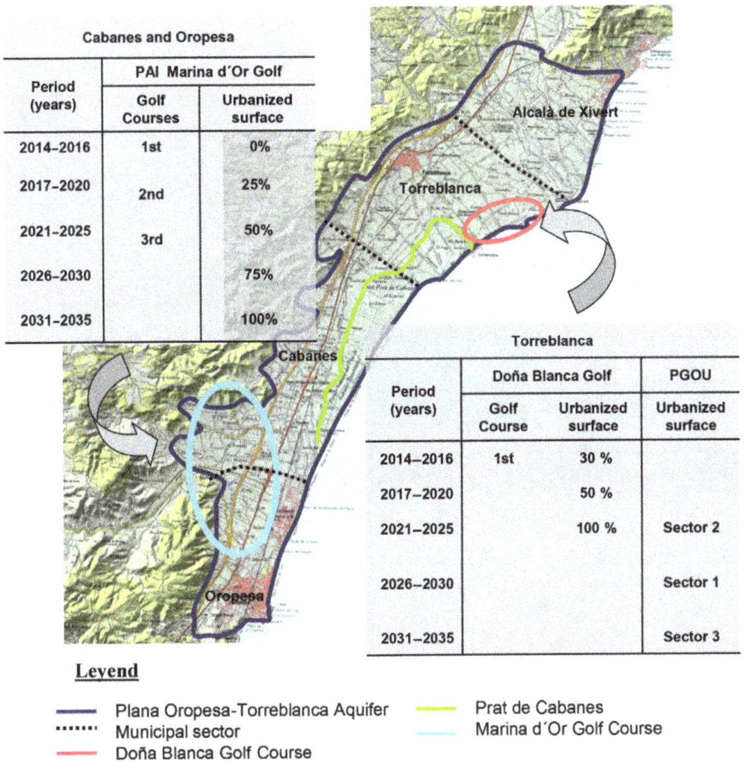

Figure 5. Future land use scenarios in 2035.

cal Agency, AEMET, 2009). In this paper we do not intend to perform a detailed analysis of hydrological uncertainty. In this case, in order to assess the uncertainty on hydrological impacts it would be more appropriate to obtain results from each individual climate model.

Climate model simulation data – control and future scenarios

In this work we have focused on the information available for the most pessimistic emission scenario (RCP8.5). We analyse the information coming from the EU CORDEX project (2013), where we find nine climate-change scenarios (see Table 1) defined with the simulations (control and future series) obtained with five regional climate models (CCLM4-8-17, RCA4, HIRHAM5, RACMO22E and WRF331F) nested with some GCMs (four GCMs were available).

We have obtained representative lumped series of these simulations for our system, by weighting the values in each CORDEX cell according to its surface in the domain. Figure 6 shows a significant bias between the historical data and the control simulation that will force us to apply a correction technique to generate future scenarios.

We also observe important differences between the statistics of future (2011–2035) and control (1976–2000) series (rainfall and T) for each RCM (Fig. 7).

Application of different downscaling techniques (bias correction and delta change techniques)

In accordance with the hypotheses assumed to define future climate series in a water resource system (starting from the climate model simulations) we can consider two different kinds of downscaling techniques: bias correction techniques and delta change techniques (Räisänen and Räty, 2012). In

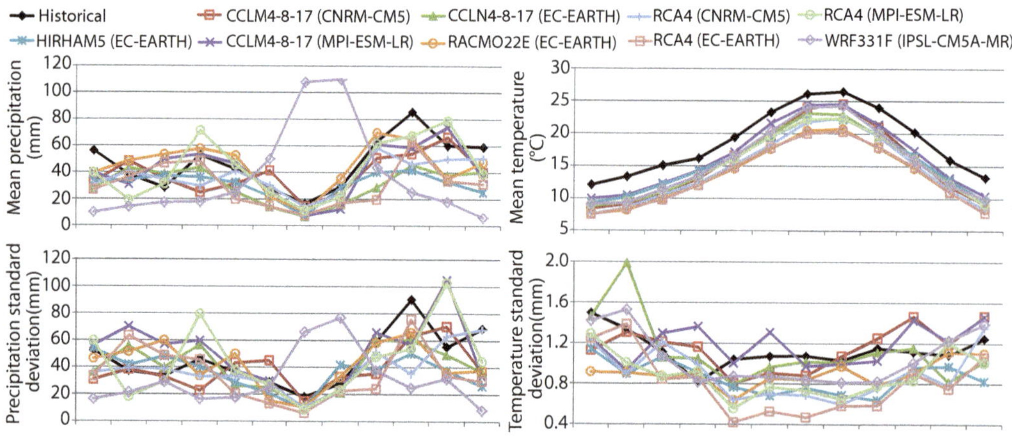

Figure 6. Monthly mean and standard deviation of the historical and RCMs control series (rainfall and temperature) for the mean year in the period 1976–2000. RCMs data obtained from CORDEX project.

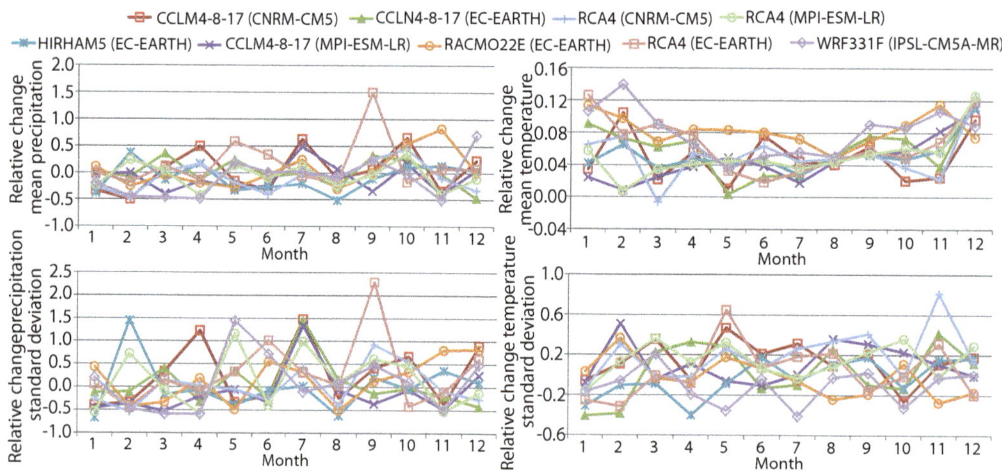

Figure 7. Relative monthly change in mean and standard deviation of the future series (2011–2035) with respect to the control series (1976–2000) for the considered RCMs under the RCP8.5 emission scenario.

the present study we apply both conceptual approaches (bias correction and delta change techniques).

The bias correction techniques are based on the analysis of the statistical difference between the climatic variables in the historical data and the control simulations produced by the climate models for the same period. They aim to define a transformation function to correct the control series to obtain a better approximation of the historical statistic. They assume that in the future the bias between model and data will be the same as observed in the historical period (e.g. Watanabe et al., 2012; Haerter et al., 2011). The delta change approaches assume that the model can obtain a good approximation of the relative changes in climate variables' statistics, but do not provide a good prediction of the absolute values. Accordingly, they try to characterize the "delta change" produced in the main statistics of the climatic variables by analysing the relative difference between the future and control scenario simulations. The future series will be obtained by perturbation of the historical series in accordance with the estimated "delta change" (e.g. Pulido-Velazquez et al., 2011a, 2014; Räisänen and Räty, 2012).

When applying both correction techniques (bias and delta change) the spatial resolution of the historical data available for our systems is usually more detailed than those adopted by the climate model and, therefore, these transformations indirectly produce solutions with higher spatial resolution than the RCM one. Therefore, they are commonly known as downscaling transformations. We have applied two downscaling techniques (correction of first and second order moments) for both conceptual approaches (bias correction and delta change techniques). The correction of the first and second moments for the delta change approach was defined at

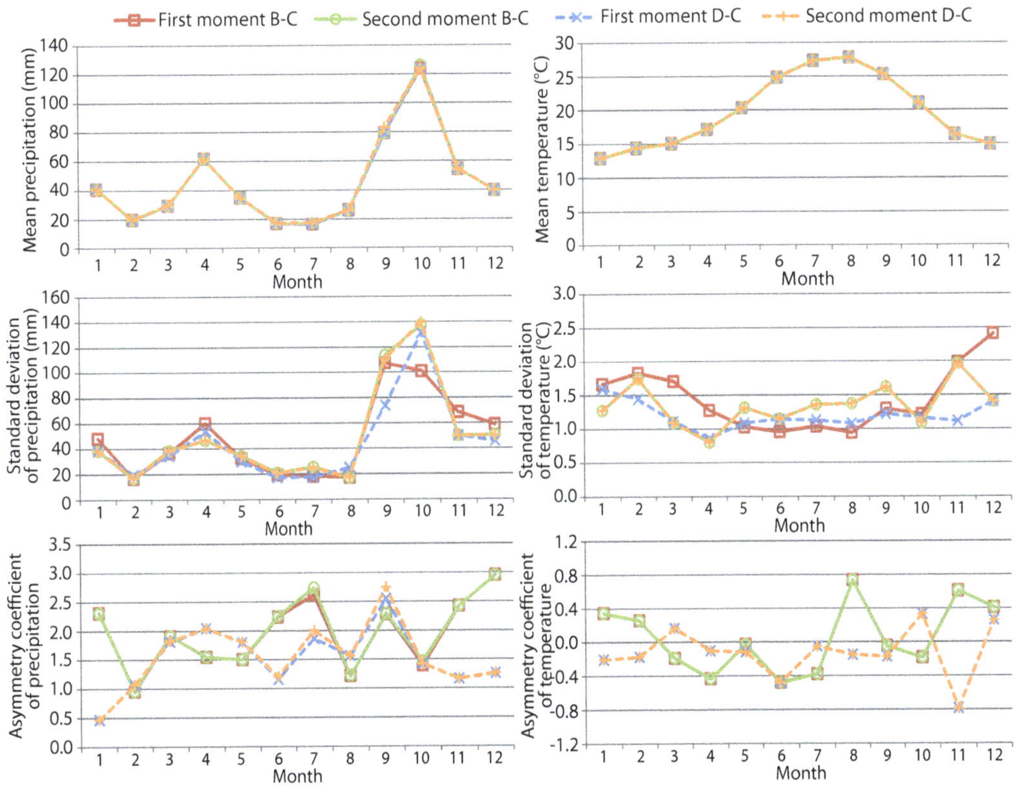

Figure 8. Mean, standard deviation and asymmetry coefficients of future precipitation series (A) and future temperature series (B) for the average year for the RCM RCA4 model linked to the CNRM-CM5 GCM.

monthly scale as described in Pulido-Velazquez et al. (2014). The bias correction case was also defined in an analogous way. The difference is that the perturbation is defined by modifying the statistics (first and second moments) of the control series to approximate the historical ones. It will be applied assuming to correct the future assuming that will be invariant. As example, the first, second and third moments of the series obtained with both approaches for one of the RCM (RCA4 linked to CNRM-CM5) are represented in Fig. 8.

Multi-criteria analysis of the main statistic

Two multi-objective analyses are proposed: one related to the bias approaches and another to the delta ones.

In the delta change approaches we applied the multi-criteria analysis proposed by Pulido-Velazquez et al. (2014). It intends to identify the models that provided the best approximations to the main historical statistics (mean, standard deviation and asymmetry coefficients) based on the analysis of their control simulation of the historical period. The dominated solution or "inferior" models approaching the historical statistic were identified and eliminated. A model is eliminated (see Table 2) if any other model's prediction provides better approximation to the cited statistics.

Table 2. Eliminated and non-eliminated models in the multiple-criteria analysis.

Eliminated?	RCM	GCM
NO	CCLM4-8-17	CNRM-CM5
NO	CCLM4-8-17	EC-EARTH
NO	CCLM4-8-17	MPI-ESM-LR
NO	HIRHAM5	EC-EARTH
NO	RACMO22E	EC-EARTH
NO	RCA4	CNRM-CM5
NO	RCA4	EC-EARTH
YES	RCA4	MPI-ESM-LR
NO	WRF331F	IPSL-CM5A-MR

In this study we also propose a multiple-criteria analysis for the bias correction approaches. This allows us to identify the best combination of model and bias correction techniques (see Table 3) to approximate the main statistics of the historical series. Since most of the combinations of model and bias correction techniques provide very good approximations to the first moment, a relative error threshold was defined to consider a corrected control to approach better a statistic when significant differences (higher than the threshold) are obtained.

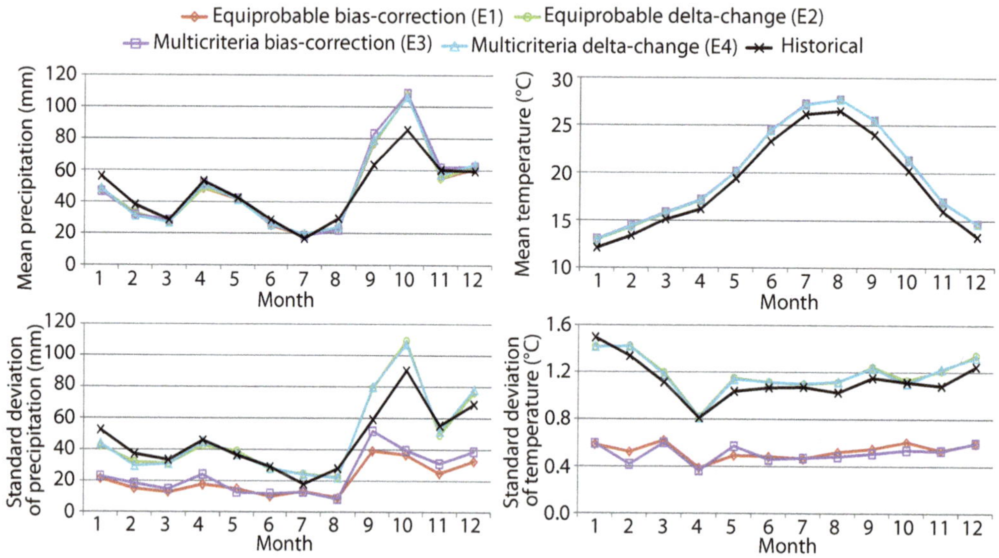

Figure 9. Monthly mean and standard deviation of future precipitation and temperature series obtained by the four ensemble options.

Table 3. Eliminated and non-eliminated combinations of model and bias-correction techniques.

Threshold 1 % eliminated?	RCM	GCM	Technique
NO	CCLM4-8-17	CNRM-CM5	First moment
NO	CCLM4-8-17	CNRM-CM5	Second moment
YES	CCLM4-8-17	EC-EARTH	First moment
NO	CCLM4-8-17	EC-EARTH	Second moment
YES	CCLM4-8-17	MPI-ESM-LR	First moment
YES	CCLM4-8-17	MPI-ESM-LR	Second moment
NO	HIRHAM5	EC-EARTH	First moment
NO	HIRHAM5	EC-EARTH	Second moment
NO	RACMO22E	EC-EARTH	First moment
NO	RACMO22E	EC-EARTH	Second moment
YES	RCA4	CNRM-CM5	First moment
NO	RCA4	CNRM-CM5	Second moment
NO	RCA4	EC-EARTH	First moment
NO	RCA4	EC-EARTH	Second moment
YES	RCA4	MPI-ESM-LR	First moment
NO	RCA4	MPI-ESM-LR	Second moment
YES	WRF331F	IPSL-CM5A-MR	First moment
NO	WRF331F	IPSL-CM5A-MR	Second moment

Ensembles of predictions to define more representative future climate scenarios

We considered four options to define representative future scenarios by applying different ensembles of corrected projections. Two ensemble scenarios (E_i) were generated by a linear combination of all the future series generated by delta change (E_1) or bias correction (E_2). Two other options were defined by combining only the non-eliminated models (E3, for the delta change approach) or combinations of models and correction techniques (E4, for the bias correction techniques). They do not consider the eliminated options because they provide inferior approximations to the historical series that make us mistrust their predictions.

All ensemble predictions show a very similar increase in mean T (see Fig. 9). The standard deviation estimated with the delta change approaches is quite similar to the historical one, but both ensembles defined by applying bias correction show smaller standard deviations.

A reduction in future mean rainfall is predicted by all the ensembles for every month except September and October, when relative increases in rainfall are predicted. Since these are the months with the highest historical rainfall, the overall effect is a higher total annual rainfall compared to the historical. During September and October in this Mediterranean area we have important storms related to the phenomenon known as the "cold drop" (Roth, 2003). It is related to the higher total rainfall observed in these months. In accordance with the obtained potential scenarios, in the future we would have an increment in these extreme rainfall events.

As we also observed for the T, the standard deviation of the future rainfall predicted with the delta change approaches is quite similar to the historical one, but both ensembles defined by applying bias correction show significant reductions.

3.1.3 Sea level rise (SLR) scenarios

Based on the European Environment Agency analysis (EEA, 2014) of historical and potential future SLRs, we propose a SLR scenario to study the sensitivity of our GC simulations to a potential SLR. The EAA report states that, over the last 2 decades, satellite measurements have indicated a mean rate of SLR of more than $3.2\,\mathrm{mm\,yr^{-1}}$. If we assume that this rate remains constant, then, by the end of a future horizon of 25 years, we would have a rise in sea level of 0.08 m. On the other hand, model simulations for the RCP8.5 emission scenario show a rise in sea level for 2081–2100 in the range 0.45–0.81 m. If we assume a constant rate of SLR from now until 2100, the sea level would rise a maximum of 0.19 m (more than double the observed rate over the last 2 decades). This value (a rise of 0.19 m by the end of the future horizon) was used to define a very pessimistic scenario of maximum SLR. A linear SLR was considered during the future horizon (2011–2035).

3.2 Definition of a coupled modelling framework

In order to assess quantity and quality impacts on groundwater systems, we needed to calibrate a density-dependent model that simulates flow and transport within the porous aquifer medium. We propose a sequential coupling of three "auxiliary models" (rainfall-recharge models, crop irrigation requirements and irrigation return models) with this density-dependent model, in which the outputs of the auxiliary models are used as inputs of the groundwater model (see Fig. 4).

The models were calibrated with the available historical data (1981–2010). Historical data for the period 1973–1981 were used to validate them. These models were then used to simulate the impacts of future LULC and CC scenarios.

3.2.1 Rainfall-recharge models

Based on the historical climate (rainfall and T) and recharge series for the period 1973–2010 described in Sect. 2.2, we applied the simple empirical rainfall-recharge proposed by Pulido-Velazquez et al. (2017) to generate yearly aquifer recharge series. The model assumes that rainfall and T are the most important climatic variables determining potential aquifer recharge, and the variability of both of them will determine the impacts of future potential climatic scenarios on the recharge.

It intends to define a correction function for the perturbation of the historical series defined as the difference in rainfall and evapotranspiration (ETR) (difference hereafter referred to as PE series), modifying its mean and standard deviation to make them equal to the statistic of the historical aquifer recharge previously deduced from the lysimeter measurements (see Sect. 2.2). The calibrated function is employed to obtain the precipitation recharge series (PR) also for the future period. The steps required are the following (Pulido-Velazquez et al., 2017).

1. Estimation of the change in the mean and standard deviation of PE and PR series for the historical period:

$$\Delta\mu = \frac{\mu(\mathrm{PR}) - \mu(\mathrm{PE})}{\mu(\mathrm{PR})} \text{ and } \Delta\sigma = \frac{\sigma(\mathrm{PR}) - \sigma(\mathrm{PE})}{\sigma(\mathrm{PR})} \quad (1)$$

2. Standardization of the PE series (historical and future):

$$\mathrm{PEn}_i = \frac{\mathrm{PE}_i - \overline{\mathrm{PE}}}{\sigma_{\mathrm{PE}}} \quad (2)$$

3. Generation of PR series from PE series

   The future R series will be obtained from the future PR series assuming that the bias ($\Delta\mu$ and $\Delta\sigma$) remains invariant in the future.

$$\mathrm{PR}_i = \sigma_c \cdot \mathrm{PEn}_i + \mu_c \quad (3)$$

   where

$$\mu_c = \mu(\mathrm{PE}) \cdot (1 + \Delta\mu)\mu_c \text{ and } \sigma_c = \sigma(\mathrm{PE}) \cdot (1 + \Delta\sigma) \quad (4)$$

Taking the positive relationship of T and E into account (Arora, 2002; Gerrits et al., 2009), changes in T will determine the available non-evaporative fraction of rainfall available for aquifer recharge. Different non-global empirical models could be applied to assess the historical E from T series (e.g. Turc, 1954, 1961; Coutagne, 1954; Budyko, 1974; amongst others) as described in Arora (2002), Gerrits et al. (2009), and España et al. (2013). In this study, we applied Turc's model (1954, 1961), in which the results depend on mean annual T and solar irradiation over the latitude.

In order to approach the impact of seasonal variability, the annual rainfall-recharge values obtained using the simplified model were distributed between the 12 months maintaining the pattern of the historical rainfall-recharge series. We assume that recharge from a rainfall event will reach the aquifer in less than 1 month, so working with stress periods of 1 month means there is no delay between the rainfall and the aquifer recharge.

Two simplified rainfall-recharge models were developed, one for the Plana de Oropesa-Torreblanca aquifer and the other for the Maestrazgo aquifer.

The Maestrazgo aquifer recharge model is used to assess inflows to the carbonate basement that produces lateral inflows (LI) to the Plana Oropesa-Torreblanca aquifer under various potential GC scenarios. The Maestrazgo aquifer has an important storage capacity which is almost in a natural regime and does not present significant changes in hydraulic head. The Maestrazgo rainfall-recharge model is employed to assess future recharge, the LI to the carbonate basement being obtained by assuming a constant ratio between Maestrazgo rainfall recharge and these LI.

3.2.2 Modelling crop irrigation demands and irrigation returns

The LULC information was used to estimate agricultural water requirements following a procedure to compute crop water requirements based on the FAO Irrigation and Drainage Paper (Allen et al., 1998). This approximation was applied in previous CC impact research studies (e.g. Escriva-Bou et al., 2016). The irrigation values added to the rainfall constitute the total inflows coming from the surface system. A Turc (1954) model was employed to estimate the total ETR in the area considering not only rainfall, but also irrigation water. The difference between the total ETR and the ETR for the rainfall allows us to determine the ETR related to irrigation, taking into account the climate conditions (rainfall and T). The estimated irrigation demands have also been employed to assess pumping taking into account information about the origin of the water that supplies each demand. The irrigation demands are multiplied by the irrigation return coefficients obtained for the crops in this area in previous studies (Tuñon, 2000) to assess recharge from irrigation.

3.2.3 Density-dependent flow model (flow and transport)

Based on the hydrological description in Sect. 2.1, a conceptual model was defined to approach the aquifer as an unconfined heterogeneous detritic aquifer. The inflows to the aquifer include rainfall recharge, lateral groundwater inflows from the bordering aquifers (LGI) and irrigation returns. The outflows include natural springs that feed the wetland, pumping wells and outflows to the sea. The historical evolution of the inflows and outflows is represented in Fig. 10.

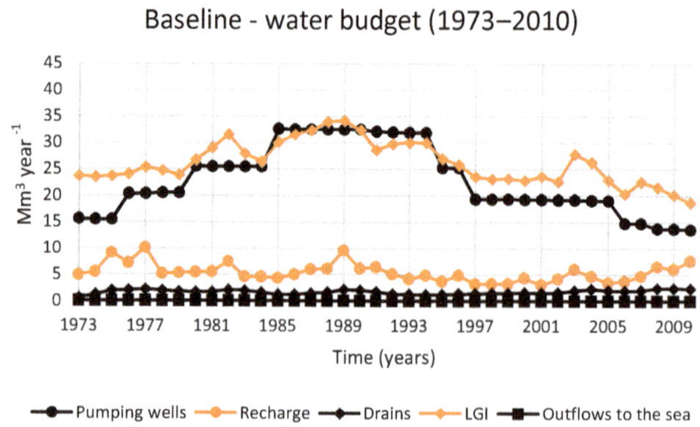

Figure 10. Historical evolution of inflows and outflows in the aquifer.

The SEAWAT 3-D finite-difference numerical code (Guo and Langevin, 2002) was used to solve the coupled partial differential equations for variable-density flow and transport. It combines MODFLOW (McDonald and Harbough, 1988) and MT3DMS (Zheng and Wang, 1999) into a single code that conserves fluid mass, rather than fluid volume, and uses the equivalent freshwater head as the primary dependent variable.

The domain has been divided into 32 columns and 90 rows in the horizontal plane, with a cell size of 250 m × 250 m. Therefore, the groundwater flow and transport domain extends over a size of 8000 m × 22 500 m. We adopted a vertical discretization consist of 11 layers in order to avoid a high computational cost (e.g. Sreekanth and Datta, 2010) while maintaining reasonable levels of numerical dispersion and representing complex flow patterns near areas of high concentration gradients (Guo and Langevin, 2002). These 11 layers, were defined as confined/unconfined layers where the transmissivity varies. On the one hand, for the flow boundary conditions a prescribed head of 0 m has been assigned to all cells belonging to the sea frontier of the model. For the transport boundary conditions a prescribed concentration of 35 g L^{-1} has been assigned to all cells belonging to the sea frontier of the model.

We assumed a density of freshwater of 1000 and 1025 kg m^{-3} for the seawater with 0.7143 as the slope of the linear equation of state that relates fluid density to solute concentration.

The groundwater model covers the Plio-Quaternary and Pre-Quaternary formations. The spatial heterogeneity is tackled using the concept of multiple statistical populations (Llopis-Albert and Capilla, 2010), in which the rock matrix and each fracture are represented as independent statistical populations. The random function for each structure (i.e. the aquifer matrix and fractures) is modelled based on a geostatistical analysis conditional on its own statistical distribution (i.e. hydraulic conductivity data as well as geological information and expert judgment). The random function is supposed to be multi-Gaussian for the rock matrix, while the fractures are considered non-multi-Gaussian. In this way, the rock matrix is generated by sequential Gaussian simulation using the GCOSIM3D code (Gómez-Hernández and Srivastava, 1990), while the fractures are generated by sequential indicator simulation using the ISIM3D code (Gómez-Hernández and Journel, 1993). The latter code makes use of local conditional cumulative density functions (ccdfs) defined by conductivity measurements and the corresponding indicator variograms. Therefore, the spatial heterogeneity is modelled as an equivalent porous medium (e.g. Llopis-Albert and Capilla, 2010). On the one hand, the hydraulic conductivity data for fractures present high values, greater than 1000 m day^{-1}. This allows the reproduction of strings of extreme values of hydraulic conductivity that often take place in nature and can be crucial in order to obtain realistic and safe estimations of mass transport predictions. That is, it allows reproduction of preferential flow channels in strongly heterogeneous aquifers or fractured formations. On the other hand, the hydraulic conductivity data for the aquifer matrix cover a wide range of values, i.e. from 5 to 200 m day^{-1}. In addition, for each cell we have defined a vertical hydraulic conductivity equal to a tenth of the horizontal hydraulic conductivity. The position of fractures is deterministically incorporated into the model based on geological information and expert judgment, thus allowing us to classify the cell models. Those cells of the model intersected by a fracture are assigned conductivities according to the intersecting fracture, and those that are not are considered to be cells belonging to the rock matrix.

The model was calibrated with the available historical data in the period (1981–2010), while the data in the period 1973–1981 were used to validate it. This was carried out by applying a trial and error procedure simultaneously considering quantity and quality. We have not used an inverse model

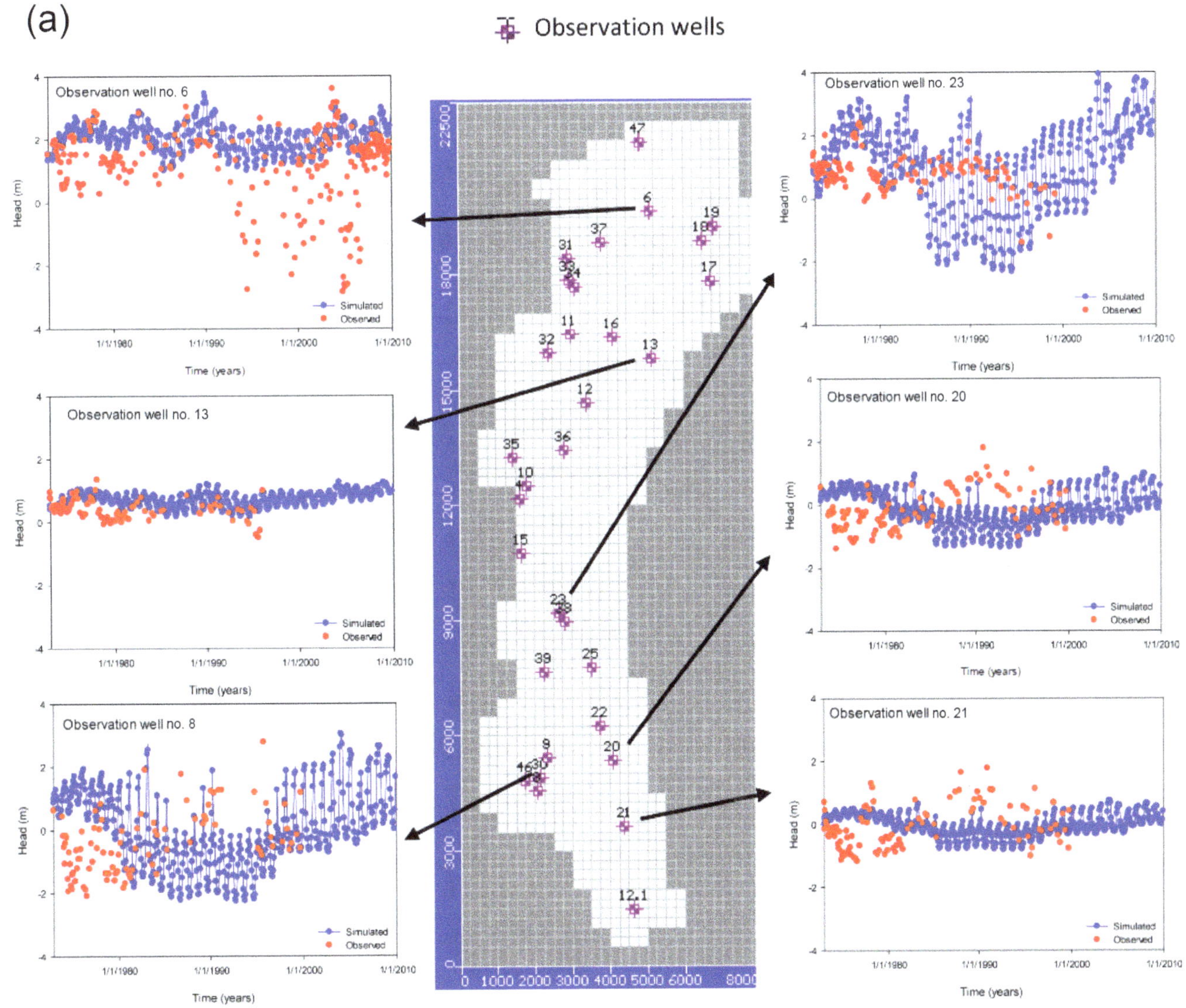

Figure 11.

(e.g. Llopis-Albert et al., 2016) because its computational cost would be important in order to deal with such a quantity of parameters (hydraulic conductivity, storativity, porosity, dispersion coefficients, etc.) and variables (hydraulic head and salinity) within a SEAWAT model over a long period of time (from 1973 to 2010), so that we have decided to apply a trial and error procedure.

The calibrated model parameters encompass different zones of horizontal hydraulic conductivities ranging from 5 to 200 m day^{-1}, while the vertical hydraulic conductivities are between 0.5 and 20 m day^{-1}. There are also different zones of specific storage values, which range from 10^{-5} to 5×10^{-4} 1/m; specific yield, with values from 0.01 to 0.15; effective porosity values from 0.01 to 0.13; and dispersion coefficients from 50 to 100 m. We have replaced a highly heterogeneous porous medium with an upscaled "equivalent" homogeneous porous medium to represent the hydrogeological parameters since the cell size of the discretization is 250 m × 250 m (e.g. Llopis-Albert and Capilla, 2010). Then, we used a value of the effective porosity based on available data, which were subsequently calibrated and upscaled using expert judgment. The results of the calibration process prove the worth of this approach.

Reasonably good results in terms of goodness of fit to the historical hydraulic head and salinity were obtained, as shown in Fig. 11 for various head and salinity observation wells.

The root mean square value (rms) of the departures between observed and simulated values for both piezometric heads and salt concentrations is presented for the whole domain and temporal discretization due to the large number of boreholes (21 boreholes for piezometric heads

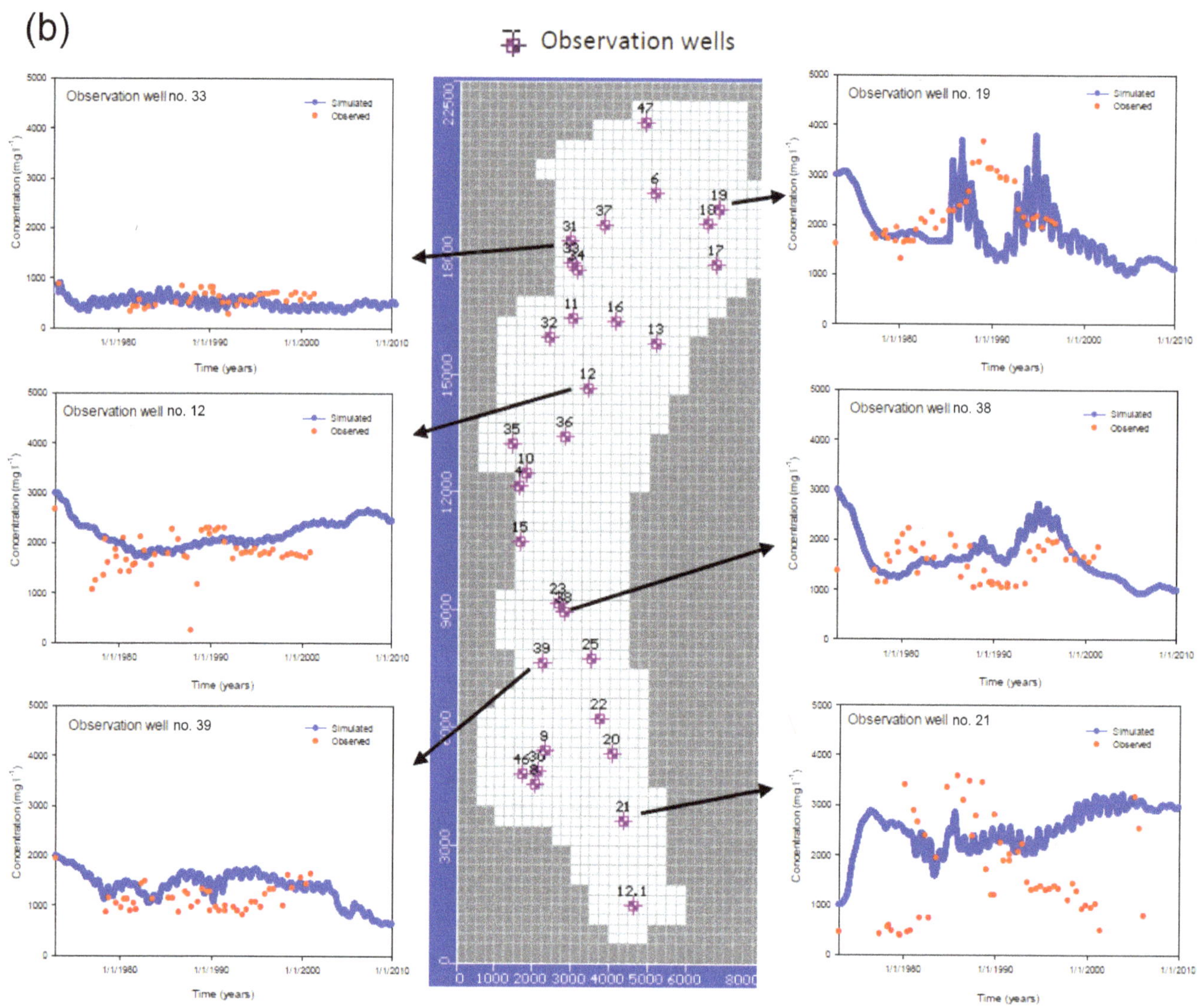

Figure 11. **(a)** Hydraulic head obtained with the models vs. data at some observation points. **(b)** Salinity obtained with the models vs. data at some observation points.

and 31 for salinity). These values are $\eta_h = 0.7$ m and $\eta_c = 391.8$ mg L^{-1}.

Note that this $\eta_h = 0.7$ m could seems to be a little high but we should take into account that we have observation wells where the historical hydraulic head measurement fluctuates sometimes more than 6 m during the same month (see for example observation well 6). Note that the high value for $\eta_c = 391.8$ mg L^{-1} could be also explained by the scale of the salt concentration, which range from 0 to 35 000 mg L^{-1}, and the measurement fluctuations, which in some wells is even higher than 4000 mg L^{-1} during some months.

Finally, the calibrated model is used to propagate impacts of future LULC and CC scenarios.

3.3 Hydrological impacts: propagation of future climatic scenarios

Analysis of GC impacts will require a sequential simulation of the LULC change scenarios for the different climate scenarios (E_i) in the auxiliary models (rainfall-recharge models, the crop irrigation requirement and irrigation return models) in order to define the inputs for the groundwater density-dependent model.

In order to assess the potential impacts of the future LULC scenario (LULC scenario) and different GCs (CC and LULC change scenarios), we have simulated the following scenarios using the density-dependent flow model.

1. Baseline (BL) scenario: no LULC change and no CC. We simulate a future scenario for horizon 2011–2035 assuming that from 2011 we would have the same

LULC that we observed in 2010. We also assume that the hydrological characteristic does not change, and we have simulated assuming that the rainfall recharge and the LI from the neighbouring aquifer are equal to those estimated in the last 5 years of the historical periods (2006–2010). In this period of 5 years (2006–2010) there was no significant change in LULC, and so this period could be adopted as being representative of the mean recent climatic–hydrological conditions. This scenario was defined in order to compare against the others to analyse the sensitivity to GC.

2. LULC scenario: it considers the described future LULC scenario and assumes that there is not CC.

3. Global change scenarios (GC_i) assuming constant sea level: we consider four GC scenarios that simultaneously consider the potential impacts of the described future LULC scenario under the four generated CC scenarios (Eisee Sect. 3.1.2.4). The comparison of these scenarios with the BL provides information about the GC impacts.

4. SLR scenarios: we have also simulated the four GC scenarios assuming that the sea level grows in a linear way to 0.19 m during the future horizon (2011–2035).

The results obtained are summarized and discussed in Sect. 4.

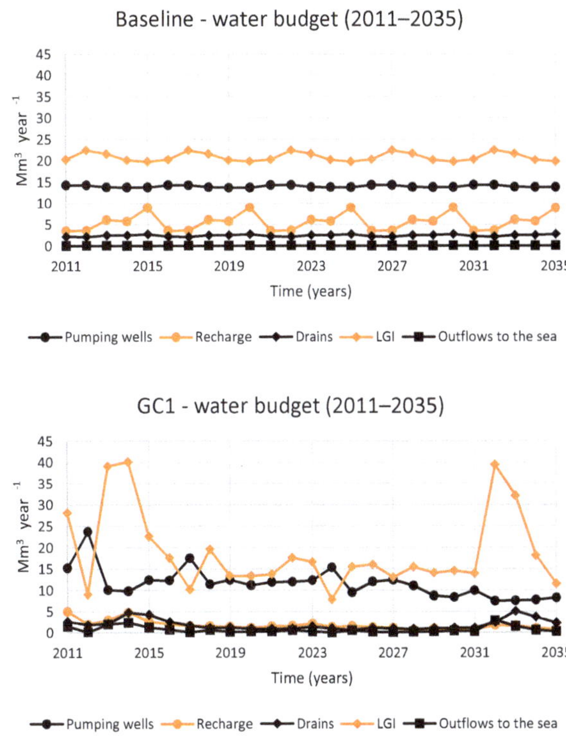

Figure 12. Components of the flow budget evolution for the BL scenario and the GC1 scenario (horizon 2011–2035).

4 Results and discussion

The proposed approach has similarities to the one described by Pulido-Velazquez et al. (2015), in which an integrated analysis of GC is performed including CC and LULC changes. The most important differences are related to the fact that a coastal aquifer is studied and we consider quality issues simulating with a variable density flow and transport model. The model was calibrated with the historical data. The approach allows us to propagate impacts of different CC and LULC change scenarios in terms of global flow balance, as well as a distributed approximation of the hydraulic head and salinity.

The historical simulation in accordance with the data shows that, in most of the aquifer area (more than 80 %), the salinity is above the natural background of the aquifer ($1100\,\mathrm{mg\,L^{-1}}$), and, therefore, it is in general affected by intrusion.

The GC scenarios have been analysed in terms of impacts on global flow balance components and salinity at specific points.

The four GC scenarios show higher variability in all the inflows and outflows in the aquifer (see GC1 in Fig. 12). If we focus on the components of the global balance, in global terms, this analysis does not show an increment in the intrusion, and it would even be slightly reduced with respect to the baseline scenario due to the outflows to the sea not decreasing (see Fig. 13). These results show a lower sensitivity to the LULC than to the GC scenarios in which the impacts also include climate change.

Figure 14 shows the evolution in terms of salinity at four specific observation points roughly equi-spaced. They were selected to cover the extension of the aquifer from north to south (starting with the more northerly one and moving towards the south, we have observation wells 33, 12, 39 and 21 respectively).

From the results in terms of salinity at specific observation points (Fig. 14), we can observe the heterogeneity of the impacts of the LULC scenario and CC scenarios. The area around observation point 33 is not affected by LULC changes and we only observe sensitivity to CC scenarios that produce a higher variability in the salinity evolution, but the mean trend of the concentration does not change. Around observation point 12, in the Torreblanca area, the LULC change scenario would produce a reduction in the recharge whose impacts can be observed as an increment in the salinity during the first decade of the future horizon. Nevertheless, in the last 10 years the reduction in pumping produced by the successive transformations of irrigated areas to residential land would reduce the salinity significantly. In this Torreblanca area CC scenarios would impact the salt concentration significantly during the first years due to the increment in pumping requirements produced by the higher water irrigation requirements obtained for these CC scenarios. Even considering the impacts of CC scenarios, during the last 10 years of the horizon it starts to recover due to the reduction in pumping produced by the new transformations of irrigated areas to residential land defined in the LULC scenario, the concentration at the end of the horizon (2035) even being under the values obtained for the BL scenario. In observation point 39, located further away from the coast, the reduction in pumping due to LULC change would reduce the salinity during most of the year of the horizon. Nevertheless, the reduction in recharge in some years makes the salinity get close to the one obtained without LULC changes (BL scenario), being very similar at the end of the period. Again, CC scenarios would increase the variability of the salinity in the simulated period.

We can also identify in the southern part areas where the situation will clearly improve throughout the future horizon contemplated with the proposed scenarios. For example, at observation point 21, in the Oropesa area, the salinity would be reduced with the contemplated scenarios, which would be mainly related to the reduction of pumping in this area due to the transformation of irrigated areas to residential land defined in the LULC scenario.

As mentioned above, the expected results considering GC impacts are likely to be too pessimistic or optimistic, depending on the location. These results can be useful for the authorities in charge of implementing management policies in the Plana Oropesa-Torrablanca. We can use this coupled modelling framework to assess potential effects of adaptation measures on GC by modifying the inputs of the models. Participatory processes including the relevant stakeholders might be essential in the definition of scenarios and successful adaptation measures (Pulido-Velazquez et al., 2015). This modelling framework could be useful in the search for consensus ("shared vision" models) between different stakeholders.

Lastly, we analysed the sensitivity of the GC scenarios to a SLR scenario (Fig. 15). A rise of 0.19 m at the end of the future horizon was used to define a very pessimistic scenario of maximum SLR (see Sect. 3.1.3). A linear SLR was considered during the future horizon (2011–2035). Figure 15 shows the results obtained in terms of hydraulic head and salinity at a number of observation points. It shows that the sensitivity of hydraulic head is very low. The sensitivity of the salinity, although not very significant, is higher than that observed for hydraulic head. The low sensitivity of the results should be due to the maximum value of SLR considered, 0.19 m in 2035, quite low with respect to the level fluctuations experienced in most of the observation wells (see Fig. 15). For this reason, the sensitivity of the flow and transport solutions is low. We find in the literature other examples in which the sensitivity of seawater intrusion to the SLR would be low. Chang et al. (2011) obtained this conclusion in a synthetic confined coastal aquifer in which recharge is unchanged; Rasmussen et al. (2013) obtained the same con-

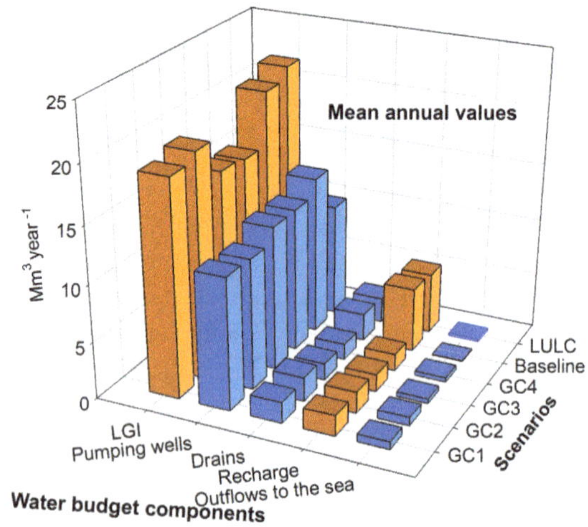

Figure 13. Mean inflows and outflows for various global scenarios (GC1–GC4).

clusion for an inland coastal aquifer with minor SLRs. Nevertheless, other authors, such as Werner and Simmons (2009), showed that in unconfined aquifers the influence of the inland boundary condition can be significant for its sensitivity to SLR.

The methodology proposed in this paper to perform an integrated analysis of future potential GC scenarios (considering LULC, CC and SLR) and their hydrological impacts is general. It can be used to assess the potential status of any coastal aquifer in terms of flow balance components, hydraulic head, and salinity. It includes the definition of future CC scenarios by using an equi-feasible and non-equi-feasible ensemble of projections based on the results of a multicriteria analysis of the series generated from several regional climatic models with different downscaling approaches. A modelling framework was proposed to assess hydrological impacts of future climatic scenarios on the coastal aquifer based on a density-dependent simulation whose inputs are defined by sequential coupling of rainfall-recharge models, crop irrigation requirements and irrigation return models (a chain of models). This chain of models, calibrated using the available historical data, allows testing of the conceptual approximation of the aquifer behaviour and the propagation of the generated potential future scenarios.

The application of the proposed method allows us to improve our knowledge about the case study and so comprise a tool to support decisions about sustainable adaptation management strategies. In the definition of scenarios and plausible adaptation measures to be analysed participatory processes including the relevant stakeholders might be essential (Pulido-Velazquez et al., 2015). This modelling framework could be useful in the search for consensus ("shared vision" models) between different stakeholders in the definition of the cited scenarios. The limitations and potential future ex-

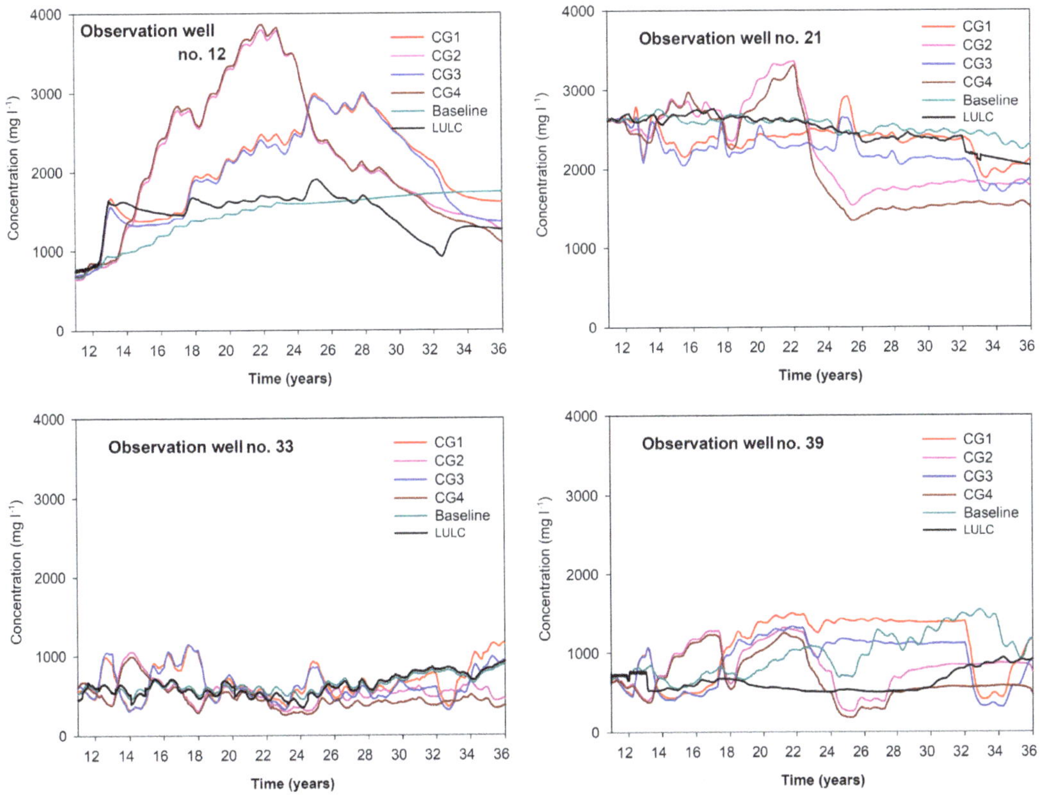

Figure 14. Evolution of future salinity concentrations at four observation points for the baseline (BL) scenario and the four global change scenarios.

tensions of this work, for example in terms of analyses of uncertainties, are discussed in Sect. 4.1.

4.1 Hypotheses and limitations – usefulness of the results

From the methodological point of view, the proposed approach is general, ambitious and valid scientifically. Nevertheless there are some assumptions and limitations in the application performed that we wanted to highlight and summarize in this section in order to clarify the accuracy and utility of the results obtained. We have grouped them into three categories.

1. Generation of future potential climatic scenarios
   - The research is focused on the analysis of a future short-term horizon (2011–2035). Other future horizons, such as mid-term and long-term scenarios, in which there should be even higher uncertainties about the impacts of CC because of being further away in time, are not considered.
   - Only the most severe IPCC scenario (RCP8.5) was analysed to assess the most pessimistic impacts in the future short-term scenario.
   - Potential plausible future climate scenarios are defined by combining information coming from different regional climatic models (RCMs) and general circulation models (GCMs).
   - Two downscaling approaches (correction of first- and second-order moments) under two different hypotheses (bias correction and delta change techniques) were applied to generate future climatic series in accordance with RCM simulations. Note that, depending on the problem and the target solution, several downscaling techniques of varying complexity and accuracy (correction of first- and second-order moments, regression approach, quantile mapping, etc.) can be applied by assuming different conceptual approaches, such as bias correction and delta change techniques (Räisänen and Räty, 2012).
   - Two different hypotheses, equi-feasible members or non-equi-feasible members, were applied to define ensembles of the obtained future series for each RCM. They may help to achieve more representative future potential climate scenarios for assessing impacts on aquifer recharge.
2. Hydrological propagation of the climatic impact on aquifer recharge

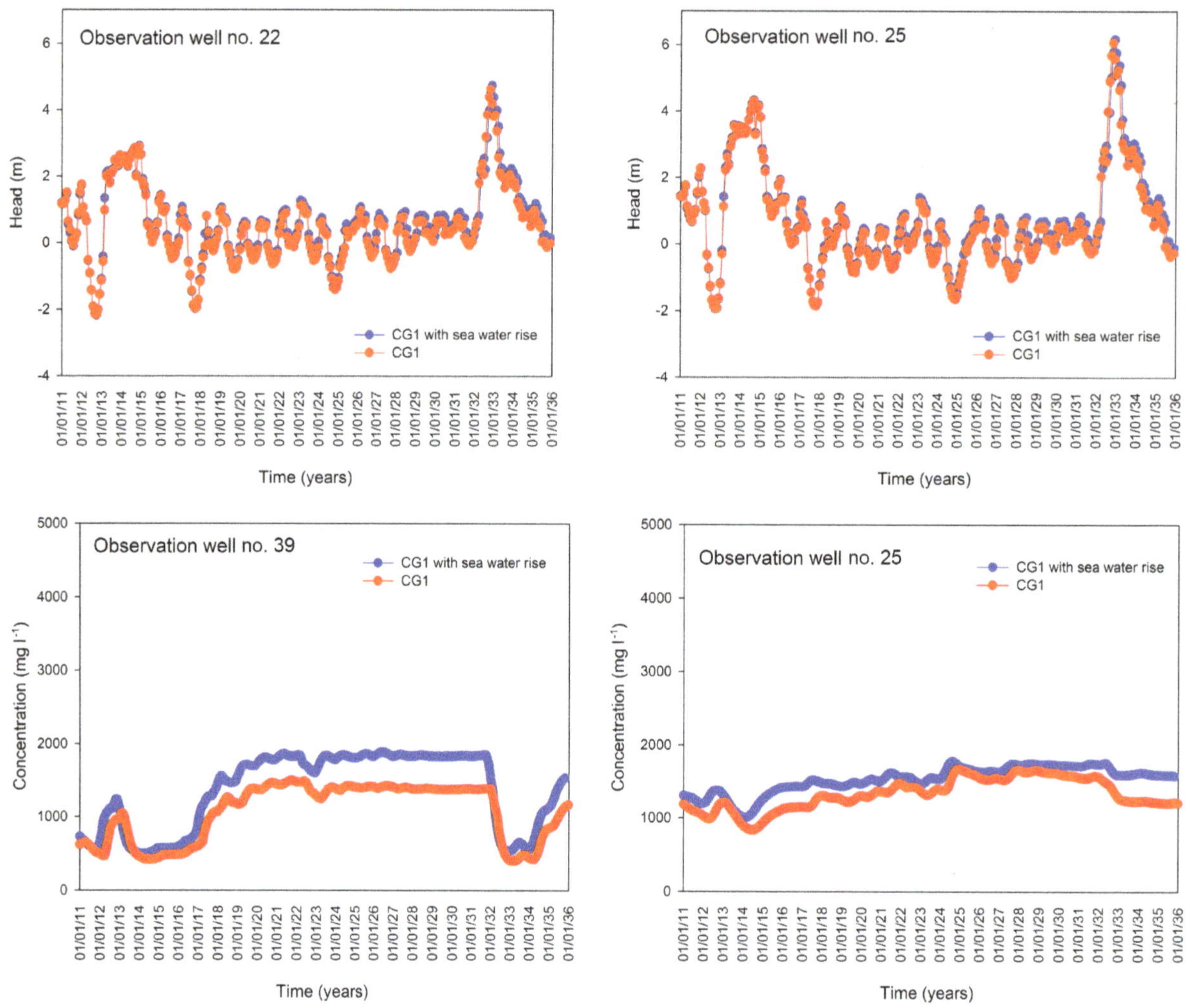

Figure 15. Sensitivity of head and salinity obtained for GC1 to a SLR (0.19 m) scenario.

- A simple empirical precipitation-recharge model has been adopted whose inputs are rainfall and E. We have not tested other hydrological models, for example, based on a more physically based or detailed representation of the processes involved in the hydrological balance and the geological structures. We assume that rainfall and T are the variables determining NAR, and their spatiotemporal variability determines the impacts of future potential climatic scenarios on renewable groundwater resources. We do not consider the changes in other variables affecting recharge, such as soil properties, vegetation patterns and land use. They are considered steady, despite their being expected to change according to global climate driving forces and new human actions on a local scale that will be induced by human adaptation to climate and water resource availability (Martínez-Valderrama et al., 2016).
- We assume that the climatic fields (rainfall and T) taken from the Spain02 project (Herrera et al., 2016) are good enough to approximate the historical climate.
- Turc's (1954, 1961) model was applied to estimate E. Its results depend on mean annual T and rainfall. Different non-global empirical models could be applied to assess the historical E from T and rainfall time series.

3. Hydrological impacts on aquifer status: seawater intrusion simulation: in this study we decided to employ a variable-density model instead of a sharp-interface solution, which has been extensively used to define management models because of its simplicity in terms of required parameters and computational burden (e.g. Mantoglou et al., 2004). This is because they better describe the dynamic of real complex coastal aquifers despite the limitations in the available data and the course discretization used for the Plana Oropesa-Torreblanca

aquifer. Better adjustments between observed and modelled hydraulic head and salinity would provide greater confidence in the model predictions. Nevertheless, although there are some differences between observed and simulated data, the uncertainty of the prediction coming from the model growing, and having to be cautious with the conclusions obtained, the fit is good enough to capture the general trend of the hydraulic head and salinity variables within a quite long calibration period (37 years, from 1973 to 2010), and, therefore, to assess general impacts of climate and LULC changes. Note that it is difficult to improve the calibration of the proposed approach in this work for Plana de Oropesa-Torreblanca due to the following facts.

- The hydrogeological complexity of the aquifer makes it difficult to define a better approach taking into account its spatial heterogeneity, with fractured formations and preferential flow channels existing in the aquifer (Morell and Giménez, 1997). The spatial heterogeneity is handled by means of sequential indicator simulation using the computer code ISIM3D (Gómez-Hernández and Srivastava, 1990).
- The quality of some observation data is poor. This problem is accentuated when considering the long simulated time period with historical data, which spans from 1973 to 2010. In this way, for a certain observation borehole there are data close in time with measurements quite disparate, which cannot be explained by any physical phenomenon. A statistical processing of data with expert judgment would be advisable to dismiss the wrong data. We have opted for using all available data (as a transparency measurement and in order to ease the reproduction of this exercise by other researchers).
- The lack of reliable estimates of dispersion coefficients (Naji et al., 1999) may prevent a proper adjustment for the Plana Oropesa-Torreblanca aquifer.
- A better fit might be achieved using a more refined spatial discretization, which would allow modelling of the preferential flow channels existing in the aquifer (Morell and Giménez, 1997). However, the high computational burden when solving the variable-density flow and transport equations with SEAWAT prevents the use of a fine grid (e.g. Sreekanth and Datta, 2010).
- As further research, we intended to couple the seawater intrusion model with a management simulation–optimization model for control and remediation that would further prevent the use of a fine grid.

4. Analysis of uncertainty in impacts: note that the information about seawater intrusion status provided by the analyses of flow balance is quite limited. We have not used indices based on the salinity distribution to assess the global intrusion status (see Baena-Ruiz et al., 2018). Finally, in this paper we do not intend to perform a detailed analysis of uncertainty in impacts, which could be deeply analysed as a further study. In order to assess the uncertainty in hydrological impacts it would be more appropriate to assess results from each individual climate model. Note that it would require us to deal with different sources of uncertainty (Matott et al., 2009). The complexity is even greater for the presented methodology, since it entails the coupling of several numerical codes and a large amount of data and a long simulation time period.

5 Conclusions

We have proposed a general method to perform an integrated analysis of the potential impacts of future CC, LULC change and SLR in a coastal aquifer. It is assessed in terms of flow balance components, hydraulic head, and salinity distribution. It includes the definition of future CC scenarios by using equi-feasible and non-equi-feasible ensembles of projections based on the results of a multi-criteria analysis of the series generated from several regional climatic models with different downscaling approaches. A modelling framework defined with a chain of models was proposed to assess hydrological impacts of these future climatic scenarios.

It has been applied in the Plana Oropesa-Torreblanca aquifer assuming some hypotheses or simplifications. Representative future CC scenarios are generated by applying the proposed method A future LULC scenario defined in accordance with the plan approved by the local government (PGOU Torreblance, 2009). Four GC scenarios were defined by combining the LULC scenario and the CC scenarios. These GC scenarios have been propagated to assess hydrological impact by simulating them within a coupled modelling framework based on a density-dependent model whose inputs are defined by a sequential coupling of different models (rainfall-recharge models, crop irrigation requirements and irrigation return models). The historical status shows that the majority of the aquifer is affected by seawater intrusion. The analysis of the future flow balance components does not show an increment in the global intrusion. These results show a lower sensitivity to the LULC than to the GC scenarios in which the impacts also include climate change. These global scenarios' simulations show a significant increase (with respect to a BL scenario with no GC) in the variability of the flow budget components and in the salinity. The impacts on the aquifer salinity will be heterogeneous. We observed specific areas where the situation gets worse and another where it will clearly improve. The proposed analysis is valuable for improving our knowledge about the aquifer and so comprising a tool to support decisions about sustainable adaptation management strategies. From an operational point of view, this modelling framework could also be useful in the search for a consensus ("shared vision" models) between different stakeholders involved in the management of the aquifer.

Competing interests. The authors declare that they have no conflict of interest.

Special issue statement. This article is part of the special issue "Assessing impacts and adaptation to global change in water resource systems depending on natural storage from groundwater and/or snowpacks". It is not associated with a conference.

Acknowledgements. This research work has been partially supported by the GESINHIMPADAPT project (CGL2013-48424-C2-2-R) with Spanish MINECO funds, the PMAFI/06/14 project with UCAM funds and the Plan de Garantía Juvenil from MINECO, co-financing by BEI and FSE. We would like to thank the Spain02 (AEMET and UC) and CORDEX projects and the Júcar Water Agency (CHJ) for the data provided for this study. We appreciate the valuable comments and suggestions provided by the editor and two anonymous referees.

Edited by: Jesús Carrera

References

Alcalá, F. J. and Custodio, E.: Spatial average aquifer recharge through atmospheric ride mass balance and its uncertainty in continental Spain, Hydrol. Process, 28, 218–236, 2014.

Allen, R. G., Pereira, L. S., Raes, D., and Smith, M.: Crop evapotranspiration – Guidelines for computing crop water requirements, FAO Irrigation and drainage paper 56, FAO, Rome, http://www.fao.org/docrep/X0490E/x0490e00.htm (last access: May 2018), 1998.

Arora, V. K.: The use of the aridity index to assess climate change effect on annual runoff, J. Hydrol., 265, 164–177, 2002.

Arslan, H. and Demir, Y.: Impacts of seawater intrusion on soil salinity and alkalinity in Bafra Plain, Turkey, Environ. Monit. Assess., 185, 1027–1040, 2013.

Baena-Ruiz, L., Pulido-Velazquez, D., Collados-Lara, A. J., Renau-Pruñonosa, A., and Morell, I.: Global assessment of seawater intrusion problems (status and vulnerability), Water Resour. Manage., 32, 2681–2700, https://doi.org/10.1007/s11269-018-1952-2, 2018.

Benini, L., Antonellini, M., Laghi, M., and Mollema, P. N.: Assessment of Water Resources Availability and Groundwater Salinization in Future Climate and Land use Change Scenarios: A Case Study from a Coastal Drainage Basin in Italy, Water Resour. Manage., 30, 731–745, 2016.

Brunet, M., Casado, M. J., de Castro, M., Galán, P., López, J. A., Martín, J. M., Pastor, A., Petisco, E., Ramos, P., Ribalaygua, J., Rodríguez, E., Sanz, I., and Torres, L.: Generación de escenarios regionalizados de cambio climático para España, Ministerio de Medio Ambiente y Medio Rural y Marino; Agencia Estatal de Meteorología, Madrid, 158 pp., 2009.

Budyko, M. I.: Climate and Life, Academic Press, New York, 508 pp., 1974.

Chang, S., Clement, T. P., Simpson, M., and Lee, K.: Does sea-level rise have an impact on saltwater intrusion?, Adv. Water Resour., 34, 1283–1291, https://doi.org/10.1016/j.advwatres.2011.06.006, 2011.

CHJ – Júcar Water Agency: Júcar River Basin Plan, Demarcación hidrográfica del Júcar, Confederación Hidrográfica del Júcar, Ministry of Agriculture, Food and Environment, Madrid, Spain, 2015.

Church, J. A., Clark, P. U., Cazenave, A., Gregory, J. M., Jevrejeva, S., Levermann, A., Merrifield, M. A., Milne, G. A., Nerem, R. S., Nunn, P. D., Payne, A. J., Pfeffer, W. T., Stammer, D., and Unnikrishnan, A. S.: Sea Level Change, in: Climate Change 2013: The Physical Science Basis, Contribution of Working Group I to the Fifth Assessment Report of the Intergovernmental Panel on Climate Change Cambridge University Press, Cambridge, UK and New York, NY, USA, 2013.

Control networks of the Júcar River Basin Authority: https://www.chj.es/es-es/medioambiente/redescontrol/Paginas/RedesdeControl.aspx/, last access: 28 May 2018.

CORDEX PROJECT: The Coordinated Regional Climate Downscaling Experiment CORDEX, Program sponsored by World Climate Research Program (WCRP), available at: http://wcrp-cordex.ipsl.jussieu.fr/ (last access: 4 May 2017), 2013.

CORDEX Regional Data Portals: http://www.cordex.org/data-access/regional-data-portals/, last access: 28 May 2018.

Coutagne, A.: Quelques considérations sur le pouvoir évaporant de l'atmosphere, le déficit d'écoulement effectif et le déficit d'écoulement maximum, La Houille Blanche, 3, 360–374, https://doi.org/10.1051/lhb/1954036, 1954.

Custodio, E.: Coastal Aquifers of Europe: an overview, Hydrogeol. J., 18, 269–280, https://doi.org/10.1007/s10040-009-0496-1 2010.

Doulgeris, C. and Zissis, T.: 3D variable density flow simulation to evaluate dumping schemes in coastal aquifers, Water Resour. Manage., 28, 4943–4956, 2014.

Dragoni, W. and Sukhija, B.S.: Climate change and groundwater: a short review, Geol. Soc. Lond. Spec. Publ., 288, 1–12, 2008.

EEA – European Environment Agency: Global and European sea-level rise, http://www.eea.europa.eu/data-and-maps/indicators/sea-level-rise-2/assessment (last access: 4 May 2017), 2014.

Escriva-Bou, A., Pulido-Velazquez, M., and Pulido-Velazquez, D.: The Economic Value of Adaptive Strategies to Global Change for Water Management in Spain's Jucar Basin, J. Water Resour. Pl. Manage., 143, 04017005, https://doi.org/10.1061/(ASCE)WR.1943-5452.0000735, 2016.

España, S., Alcalá, F. J., Vallejos, A., and Pulido-Bosch, A.: A GIS tool for modelling annual diffuse infiltration on a plot scale, Comput. Geosci., 54, 318–325, 2013.

Feranec, J., Hazeu, G., Soukup, T., and Jaffrain, G.: Determining changes and flows in European landscapes 1990–2000 using CORINE land cover data, Appl. Geogr., 30, 19–35, 2010.

Fujinawa, K.: Anthroscape of the Mediterranean Coastal Area in the Context of Hydrogeology: Projected Impacts of Climate Change, in: Sustainable Land Management, edited by: Kapur, S.,

Eswaran, H., and Blum, W., Springer, Berlin, Heidelberg, 311–332, https://doi.org/10.1007/978-3-642-14782-1_14, 2011.

Gerrits, A. M. J., Savenije, H. H. G., Veling, E. J. M., and Pfister, L.: Analytical derivation of the Budyko curve based on rainfall characteristics and a simple evaporation model, Water Resour. Res., 45, W04403, https://doi.org/10.1029/2008WR007308, 2009.

Gómez-Hernández, J. J., and Journel, A. G.: Joint simulation of MultiGaussian random variables, Geostatistics tróia~92, edited by: Soares, A., in: Vol. 1, Kluwer, Dordrecht, the Netherlands, 85–94, 1993.

Gómez-Hernández, J. J. and Srivastava, R. M.: ISIM3D: An ANSI-C three dimensional multiple indicator conditional simulation program, Comput. Geosci., 16, 395–440, 1990.

Gorelick, S. M. and Zheng, C.: Global change and the groundwater management challenge, Water Resour. Res., 51, 3031–3051, https://doi.org/10.1002/2014WR016825, 2015.

Green, T. R., Taniguchi, M., Kooi, H., Gurdak, J. J., Allen, D. M., Hiscock, K. M., Treidel, H., and Aureli, A.: Beneath the surface of global change: impacts of climate change on groundwater, J. Hydrol., 405, 532–560, https://doi.org/10.1016/j.jhydrol.2011.05.002, 2011.

Grundmann, J., Schutze, N., and Schmitz, G. H.: Towards an integrated arid zone water management using simulation-based optimisation, Environ. Earth Sci., 65, 1381–1394, https://doi.org/10.1007/s12665-011-1253-z, 2012.

Guo, F., Jiang, G., Polk, J. S., Huang, X. F., and Huang, S. Y.: Resilience of Groundwater Impacted by Land Use and Climate Change in a Karst Aquifer, South China, Water Environ. Res., 87, 1990–1998, 2015.

Guo, W. and Langevin, C. D.: User's guide to SEAWAT: a computer program for simulation of three-dimensional variable-density groundwater flow, Report No. US Geol. Surv. Open File 01-434, US Geological Survey, Tallahassee, Florida, 2002.

Haerter, J. O., Hagemann, S., Moseley, C., and Piani, C.: Climate model bias correction and the role of timescales, Hydrol. Earth Syst. Sci., 15, 1065–1079, https://doi.org/10.5194/hess-15-1065-2011, 2011.

Herrera, S., Fernández, J., and Gutiérrez, J. M.: Update of the Spain02 Gridded Observational Dataset for Euro-CORDEX evaluation: Assessing the Effect of the Interpolation Methodology, Int. J. Climatol., 36, 900–908, https://doi.org/10.1002/joc.4391, 2016.

Ketabchi, H., Mahmoodzadeh, D., Ataie-Ashtiani, B., and Simmons, C. T.: Sea-level rise impacts on seawater intrusion in coastal aquifers. Review and integration, J. Hydrol., 535, 235–255, 2016.

Kirn, L., Mudarra, M., Marín, A., Andreo, B., and Hartmann, A.: Improved Assessment of Groundwater Recharge in a Mediterranean Karst Region: Andalusia, Spain, in: Renard P. and Bertrand, C., in: EuroKarst 2016, Neuchâtel, Advances in Karst Science, Springer, Cham, 117–125, https://doi.org/10.1007/978-3-319-45465-8_13, 2017.

Llopis-Albert, C. and Capilla, J. E.: Stochastic inverse modelling of hydraulic conductivity fields taking into account independent stochastic structures: A 3D case study, J. Hydrol., 391, 277–288, https://doi.org/10.1016/j.jhydrol.2010.07.028, 2010.

Llopis-Albert, C. and Pulido-Velazquez, D.: Discussion about the validity of sharp-interface models to deal with seawater intrusion in coastal aquifers, Hydrol. Process., 28, 3642–3654, https://doi.org/10.1002/hyp.9908, 2014.

Llopis-Albert, C. and Pulido-Velazquez, D.: Using MODFLOW code to approach transient hydraulic head with a sharp-interface solution, Hydrol. Process., 29, 2052–2064, https://doi.org/10.1002/hyp.10354, 2015.

Llopis-Albert, C., Merigó, J. M., and Xu, Y.: A coupled stochastic inverse/sharp interface seawater intrusion approach for coastal aquifers under groundwater parameter uncertainty, J. Hydrol., 540, 774–783, https://doi.org/10.1016/j.jhydrol.2016.06.065, 2016.

Mantoglou, A., Papantoniou, M., and Giannoulopoulos, P.: Management of coastal aquifers based on nonlinear optimization and evolutionary algorithms, J. Hydrol., 297, 209–228, https://doi.org/10.1016/j.jhydrol.2004.04.011, 2004.

Martínez-Valderrama, J., Ibáñez, J., Del Barrio, G., Sanjuán, M. E., Alcalá, F. J., Martínez-Vicente, S., Ruiz, A., and Puigdefábregas, J.: Present and future of desertification in Spain: implementation of a surveillance system to prevent land degradation, Sci. Total Environ., 563–564, 169–178, 2016.

Matott, L. S., Babendreier, J. E., and Purucker, S. T.: Evaluating uncertainty in integrated environmental models: a review of concepts and tools, Water Resour. Res., 45, W06421, https://doi.org/10.1029/2008WR007301, 2009.

McDonald, M. G. and Harbough, A. W.: A Modular Three-Dimensional Finite-Difference Groundwater Flow Model, US Geological Survey Technical Manual of Water Resources Investigation, Book 6, US Geological Survey, Reston, VA, p. 586, 1988.

Molina, J. L., Pulido-Velázquez, D., García-Aróstegui, J. L., and Pulido-Velázquez, M.: Dynamic Bayesian Networks as a Decision Support tool for assessing Climate Change impacts on highly stressed groundwater systems, J. Hydrol., 479, 113–129, https://doi.org/10.1016/j.jhydrol.2012.11.038, 2013.

Morell, I. and Giménez, E.: Hydrogeochemical analysis of salinization processes in the coastal aquifer of Oropesa (Castellón, Spain), Environ. Geol., 29, 118–131, 1997.

Naji, A., Cheng, A. D., and Quazar, D.: BEM solution of stochastic seawater intrusion problems, Eng. Anal. Bound. Elements, 23, 529–537, https://doi.org/10.1016/S0955-7997(99)00012-0, 1999.

PGOU Torreblanca: Plan General de Ordenación Urbana de Torreblanca, Ayuntamiento de Torreblanca, Torreblanca, 2009.

Pulido-Velazquez, D., Garrote, L., Andreu, J., Martin-Carrasco, F. J., and Iglesias, A.: A methodology to diagnose the effect of climate change and to identify adaptive strategies to reduce its impacts in conjunctive-use systems at basin scale, J. Hydrol., 405, 110–122, https://doi.org/10.1016/j.jhydrol.2011.05.014, 2011.

Pulido-Velazquez, D., García-Aróstegui, J. L., Molina, J. L., and Pulido-Velázquez, M.: Assessment of future groundwater recharge in semi-arid regions under climate change scenarios (Serral-Salinas aquifer, SE Spain). Could increased rainfall variability increase the recharge rate?, Hydrol. Process., 29, 828–844, https://doi.org/10.1002/hyp.10191, 2014.

Pulido-Velazquez, D., Collados-Lara, A.-J., and Alcalá, F. J.: Assessing impacts of future potential climate change scenarios on aquifer recharge in continental Spain, J. Hydrol., https://doi.org/10.1016/j.jhydrol.2017.10.077, in press, 2017.

Pulido-Velazquez, M., Peña-Haro, S., García-Prats, A., Mocholi-Almudever, A. F., Henriquez-Dole, L., Macian-Sorribes, H., and Lopez-Nicolas, A.: Integrated assessment of the impact of climate and land use changes on groundwater quantity and quality in the Mancha Oriental system (Spain), Hydrol. Earth Syst. Sci., 19, 1677–1693, https://doi.org/10.5194/hess-19-1677-2015, 2015.

Räisänen, J. and Räty, O.: Projections of daily mean temperature variability in the future: cross-validation tests with ENSEMBLES regional climate simulations, Clim. Dynam., 41, 1553–1568, https://doi.org/10.1007/s00382-012-1515-9, 2013.

Rasmussen, P., Sonnenborg, T. O., Goncear, G., and Hinsby, K.: Assessing impacts of climate change, SLR, and drainage canals on saltwater intrusion to coastal aquifer, Hydrol. Earth Syst. Sci., 17, 421–443, https://doi.org/10.5194/hess-17-421-2013, 2013.

Renau-Pruñonosa, A., Morell, I., and Pulido-Velazquez, D.: A methodology to analyse and assess pumping management strategies in coastal aquifers to avoid degradation due to seawater intrusion problems, Water Resour. Manage., 30, 4823–4837, https://doi.org/10.1007/s11269-016-1455-y, 2016.

Robins, N. S., Jones, H. K., and Ellis, J.: An aquifer management case study – The Chalk of the English South Downs, Water Resour. Manage., 13, 205–218, 1999.

Rosenthal, E., Vinokurov, A., Ronen D., Magaritz M., and Moshkovitz, S.: Anthropogenically induced salinization of groundwater: A case study from the Coastal Plain aquifer of Israel, J. Contam. Hydrol., 11, 149–171, 1992.

Roth, G. D.: Meteorología, Formaciones nubosas y otros fenómenos meteorológicos, Situaciones meteorológicas generales, Pronósticos del tiempo Barcelona, Ediciones Omega, Barcelona, Spain, p. 301, 2003.

Shammas, M. I. and Thunvik, R.: Predictive simulation of flow and solute transport for managing the Salalah coastal aquifer, Oman, Water Resour. Manage., 23, 2941, 2009.

Sola, F., Vallejos, A., Moreno, L., López-Geta, J. A., and Pulido-Bosch, A.: Identification of hydrogeochemical process linked to marine intrusion induced by pumping of a semi-confined Mediterranean coastal aquifer, Int. J. Environ. Sci. Technol., 10, 63–76, 2013.

Spain02: A set of gridded precipitation and temperature datasets: http://www.meteo.unican.es/datasets/spain02/, last access: 28 May 2018.

Sreekanth, J. and Datta, B.: Multi-objective management of saltwater intrusion in coastal aquifers using genetic programming and modular neural network based surrogate models, J. Hydrol., 39, 245–256, 2010.

Tuñon, J.: Determinación experimental del balance hídrico del suelo y evaluación de la contaminación asociada a las prácticas agrícolas, PhD Thesis, Universitat Jaume I de Castellón, Castellón, Spain, 2000.

Turc, L.: Water balance of soils: relationship between precipitation, evapotranspiration and runoff, Ann. Agron., 5, 49–595 and 6, 5–131, 1954.

Turc, L.: Estimation of irrigation water requirements, potential evapotranspiration: A simple climatic formula evolved up to date, Ann. Agron. 12, 13—49, 1961.

Unsal, B., Yagbasan, O., and Yazicigil, H.: Assessing the impacts of climate change on sustainable management of coastal aquifers, Environ. Earth Sci., 72, 2183–2193, 2014.

Vallejos, A., Sola, F., and Pulido-Bosch, A.: Processes Influencing Groundwater Level and the Freshwater-Saltwater Interface in a Coastal Aquifer, Water Resour. Manage., 29, 679–697, https://doi.org/10.1007/s11269-014-0621-3, 2015.

Watanabe, S., Kanae, S., Seto, S., Yeh, P. J.-F., Hirabayashi, Y., and Oki, T.: Intercomparison of bias-correction methods for monthly temperature and precipitation simulated by multiple climate models, J. Geophys. Res., 117, D23114, https://doi.org/10.1029/2012JD018192, 2012.

Werner, A. D. and Simmons, C. T.: Impact of sea-level rise on sea water intrusion in coastal aquifers, Ground Water, 47, 197–204, 2009.

Yechieli, Y. and Sivan, O.: The distribution of saline groundwater and its relation to the hydraulic conditions of aquifers and aquitards: example from Israel, Hydrogeol. J., 19, 71–87, 2011.

Yechieli, Y., Shalev, E., Wollman, S., Kiro, Y., and Kafri, U.: Response of the Mediterranean and Dead Sea coastal aquifers to sea level variations, Water Resour. Res., 46, W12550, https://doi.org/10.1029/2009WR008708, 2010.

Zheng, C. and Wang, P.: MT3DMS: A Modular Three-Dimensional Multispecies Transport Model for Simulation of Advection, Dispersion and Chemical Reactions of Contaminants in Groundwater Systems, Documentation and User's Guide, Alabama University, Tuscaloosa, Alabama, 1999.

11

Impacts of changes in groundwater recharge on the isotopic composition and geochemistry of seasonally ice-covered lakes: insights for sustainable management

Marie Arnoux[1,2], **Florent Barbecot**[1], **Elisabeth Gibert-Brunet**[2], **John Gibson**[3], **and Aurélie Noret**[2]

[1]GEOTOP, Université du Québec à Montréal, Montréal, Québec, H3C 3P8, Canada
[2]GEOPS, UMR 8148, CRNS-Université Paris Saclay/Paris-Sud, Orsay, France
[3]Alberta Innovates Technology Futures, 3-4476 Markham Street, Victoria, BC V8Z 7X8, Canada

Correspondence to: Marie Arnoux (marie.arnoux@protonmail.com), Florent Barbecot (barbecot.florent@uqam.ca), Elisabeth Gibert-Brunet (elisabeth.gibert@u-psud.fr), John Gibson (jjgibson@uvic.ca), and Aurélie Noret (aurelie.noret@u-psud.fr)

Abstract. Lakes are under increasing pressure due to widespread anthropogenic impacts related to rapid development and population growth. Accordingly, many lakes are currently undergoing a systematic decline in water quality. Recent studies have highlighted that global warming and the subsequent changes in water use may further exacerbate eutrophication in lakes. Lake evolution depends strongly on hydrologic balance, and therefore on groundwater connectivity. Groundwater also influences the sensitivity of lacustrine ecosystems to climate and environmental changes, and governs their resilience. Improved characterization of groundwater exchange with lakes is needed today for lake preservation, lake restoration, and sustainable management of lake water quality into the future. In this context, the aim of the present paper is to determine if the future evolution of the climate, the population, and the recharge could modify the geochemistry of lakes (mainly isotopic signature and quality via phosphorous load) and if the isotopic monitoring of lakes could be an efficient tool to highlight the variability of the water budget and quality.

Small groundwater-connected lakes were chosen to simulate changes in water balance and water quality expected under future climate change scenarios, namely representative concentration pathways (RCPs) 4.5 and 8.5. Contemporary baseline conditions, including isotope mass balance and geochemical characteristics, were determined through an intensive field-based research program prior to the simulations. Results highlight that future lake geochemistry and isotopic composition trends will depend on four main parameters: location (and therefore climate conditions), lake catchment size (which impacts the intensity of the flux change), lake volume (which impacts the range of variation), and lake G index (i.e., the percentage of groundwater that makes up total lake inflows), the latter being the dominant control on water balance conditions, as revealed by the sensitivity of lake isotopic composition. Based on these model simulations, stable isotopes appear to be especially useful for detecting changes in recharge to lakes with a G index of between 50 and 80 %, but response is non-linear. Simulated monthly trends reveal that evolution of annual lake isotopic composition can be dampened by opposing monthly recharge fluctuations. It is also shown that changes in water quality in groundwater-connected lakes depend significantly on lake location and on the intensity of recharge change.

1 Introduction

For decades, climate change, combined with rapidly expanding urban, industrial, and agricultural water needs, has placed increasing stress on water resources and on groundwater resources in particular. Future pressure on these resources is likely to be even more pronounced, as groundwater is likely to be increasingly exploited to enhance water supply and to alleviate the worsening drought situation in some arid regions (Dragoni and Sukhija, 2008). Many studies have suggested that sustainable groundwater use has to be based on, among other things, a reliable assessment of recharge, which largely controls its evolution. Aquifer recharge refers to the quantity of water reaching the saturated zone of an aquifer, and therefore replenishing the water table. Unfortunately, in many parts of the world, recharge rates are often not well known at the regional scale (Rivard et al., 2013). While aquifer recharge is crucial to supporting sustainable management of regional groundwater resources, it is difficult to accurately estimate, owing mainly to limited data availability, as well as limitations inherent to estimation methods and field measurements (Rivard et al., 2013). Recharge rates are controlled by geology, soil characteristics, topography, land cover, land use, and climate (Rivard et al., 2014). Thorough literature reviews of the various techniques that exist to quantify groundwater recharge are provided in Scanlon et al. (2002) and Healy (2011). Many methods can be used to estimate groundwater recharge, such as water budget methods, modeling methods, tracer methods, and methods based on surface water interaction studies. The latter is based on the estimation of groundwater discharge to surface water, mainly by streambed seepage determination, streamflow duration curves, or streamflow hydrograph separation (Scanlon et al., 2002). The recharge amount (in $mm\,yr^{-1}$) is then typically obtained by dividing measured or estimated discharge flow by the surface drainage area at the measurement site. This procedure assumes that aquifer boundaries coincide with watershed boundaries, and consequently that the area of the aquifer that contributes to groundwater discharge is equal to the surface drainage area (Kuniansky, 1989; Rutledge, 1998, 2007). However, this assumption must be considered carefully, as groundwater basins and watershed boundaries can differ drastically (Tiedeman et al., 1997). Miscalculation of the aquifer contributing area leads to a proportional error in recharge estimate.

Although the groundwater inflow to streams is often taken into account in water budgets, it is less commonly considered for surface water bodies, probably due to the greater difficulty of quantifying groundwater discharge in these settings. However, in recent years some studies have proven that groundwater flow into lakes can be reliably quantified. Interactions between lakes and groundwater depend on geology, soil, and sediment properties, and also on hydraulic gradient, which is strongly dependent on climatic conditions and recharge (Winter, 1999). Therefore, variation in groundwater fluxes may indicate a change in recharge in the lake catchment (Meinikmann et al., 2013).

In Quebec (Canada), more than 10 % of the surface is covered by freshwater, with more than one million lakes known to exist. In many cases, these are connected to underlying aquifers. However, lake–groundwater interactions are highly dynamic throughout the year, and, even if it is now possible to quantify groundwater inflow with a reasonable degree of confidence, it is difficult to determine how and to what extent lakes can be sensitive to changes in groundwater recharge. The lake water isotopic composition has been proven to be particularly useful for determining water balance parameter controls under changing conditions. For example, as shown in Turner et al. (2010), lake isotopic composition can highlight that (i) reduced winter precipitation could cause snowmelt-dominated lakes to become rainfall-dominated lakes, or that (ii) during longer ice-free seasons, lakes that are mainly rainfall-dominated (but also potentially snowmelt-dominated) may turn into evaporation-dominated lakes. Moreover, among all the methods used to quantify groundwater inflow to lakes, isotopic balances appear to be especially well adapted for quantifying groundwater flux variations on seasonal and yearly timescales (Arnoux et al., 2017a). Water stable isotopes are therefore expected to be very useful for monitoring seasonal and interannual variations in the water budget under changing recharge conditions.

The impact of climate change on groundwater recharge is not easy to determine, because of the complexity of interactions and processes involved, and can vary vastly depending on regions (Rivard et al., 2014; Crosbie et al., 2013). In addition, it is predicted to shift differentially under various climate scenarios and models (Jyrkama and Sykes, 2007; Levison et al., 2014). In Canada, highly variable recharge rates have been proposed in previous studies, for example, for the 2050 horizon (mainly the period 2041–2070) relative to modern (2000–2015) or past recharge rates (1950–2010), depending on study site, scenario, and model: +10 to +53 % in the Grand River watershed, Ontario (Jyrkama and Sykes, 2007), −41 to +15 % in the Chateauguay River watershed, Quebec (Croteau et al., 2010), −6 to +58 % in the Otter Brook watershed, New Brunswick (Kurylyk and MacQuarrie, 2013), −4 to +15 % at Covey Hill, Quebec (Levison et al., 2014), +14 to +45 % in the Annapolis Valley, Nova Scotia (Rivard et al., 2014), and −28 to +18 % for the Magdalen Islands, Quebec (Lemieux et al., 2015).

Recharge fluctuations can also impact lake water quality by changing groundwater fluxes, which are closely linked to phosphorous (P) loading to lakes. It is known that lake water quality is mainly driven by variations in P load, since it plays a critical role in limiting lake primary productivity and algal biomass, which in turn regulate lake trophic status. Increasing P concentration in the water column is the primary factor responsible for accelerated eutrophication and associated algae blooms (Schindler, 1977; Wang et al., 2008). At sites without urban drainage or point P sources, such as sewage treatment plants, domestic waste from septic systems may represent the largest anthropogenic source of P to lakes on the Canadian Shield (Dillon and Evans, 1993). Increases in shoreline development and population, combined with groundwater fluxes variations, can clearly impact lake quality, but still remain to be quantified.

For the present study, 10 lakes in southern Quebec were sampled to quantify their yearly groundwater inflows (see Arnoux et al., 2017a for more details), and one of these lakes was sampled over the course of a year to quantify its monthly groundwater inflows (see Arnoux et al., 2017b for more details). Small kettle lakes without surface inlets located in fluvioglacial deposits, and that are most likely well connected to shallow unconfined aquifers, were specifically targeted. The two main objectives of this study are (i) to determine how future groundwater recharge changes might affect lake water balance and geochemistry and (ii) to assess whether stable isotopes might be an effective tool for identifying lakes that are susceptible to change or are undergoing changes in water balance and water quality. To address these objectives, seasonal models of water and isotopic budgets were established for several lakes, and the models were then forced with future yearly and monthly timescale climate data from predictive global models to simulate anticipated conditions. Climate outputs of the Canadian Regional Climate Model were used, based on scenarios RCP 4.5 and RCP 8.5 (Moss et al., 2010; IPCC, 2014). It was assumed that recharge fluctuation was the main parameter influencing groundwater fluxes into lakes, and thus a percentage of recharge change leads to the same percentage of change of groundwater fluxes to lakes. Different recharge scenarios, which were translated into changes in groundwater inflow, were then tested to determine changes in water budget and isotopic evolution of the lakes. Predicted changes in recharge were then compared to predicted population growth in the study areas to discuss lake quality evolution. After the evolution of the lake geochemical signature is determined, the ways in which lakes that are connected to groundwater can be used to identify changes in groundwater recharge can also be determined, as can whether or not the isotopic composition of lakes can serve as an effective indicator of change or variability.

2 Method

2.1 Study sites

The 10 lakes chosen are located in four regions of southern Quebec characterized by contrasting climatic conditions: Laurentides (LAU), Outaouais (OUT), Abitibi-Témiscamingue (AT), and Saguenay–Lac-Saint-Jean (SAG). These kettle lakes, located in coarse-grained (sand and gravel) fluvioglacial deposits, are specifically targeted in this study, because they (i) are small enough to be sensitive to environmental changes on a short timescale, (ii) do not have permanent surface inflow streams and so are largely groundwater dependent, (iii) are generally characterized by predictable and uniform geomorphological features, and (iv) are likely connected to shallow, unconfined aquifers (Arnoux et al., 2017a; Isokangas et al., 2015). Kettle lakes originate as depressions in the landscape formed following the melting of ice blocks buried in the ground after glacial retreat of the Late Glacial to Holocene transition period (from −12 to −7 kyr). These kettle holes, becoming kettle lakes when they are filled with water, are mainly found in fluvioglacial deposits, such as outwash plains, deltas, eskers, and kame terraces (Benn and Evans, 2011). Figure 1 shows the locations of the 10 lakes analyzed here. Their main characteristics are described in Table 1.

2.2 Lake isotopic composition

2.2.1 Sampling

Water samples from each lake were retrieved during two field campaigns, in June–July and October–November 2014. When physicochemical parameters, measured in situ along the water column, revealed a well-mixed lake, the lake was considered to be homogeneous, and only one sample was collected, from close to the lake bottom, at its greatest depth. Otherwise, for stratified periods, two samples were collected: one from the top of the epilimnion and one from the base of the hypolimnion, in order to obtain the complete range of isotopic composition variation. Whenever possible, groundwater was sampled from private wells located in the vicinity of the studied lakes. Untreated groundwater samples were collected from residential wells from the tap after purging approximately 3 times the well volume.

Samples were transported in a cooler, and subsequently stored at 5 ‰ C until analyses were performed. Water stable isotopic compositions were measured with a Laser Water Isotope Analyser (OA ICOS DLT, Los Gatos Research, now ABB) at the GEOPS Laboratory (University of Paris-Sud–Paris-Saclay, France). The reproducibility of the sample measurement is ±1‰ versus VSMOW for δ^2H and ±0.2‰ versus VSMOW for δ^{18}O. Results are reported in δ values, representing deviations in per mill (‰) from the isotopic composition of the international standard (Vienna Standard Mean Ocean Water, VSMOW), such that δ^2H or $\delta^{18}\mathrm{O} = ((R_{\mathrm{sample}}/R_{\mathrm{VSMOW}}) - 1) \times 1000$, where R refers to ^{2}H/H or ^{18}O/^{16}O ratios.

One of the lakes, Lake Lacasse, was sampled in more detail throughout 2015–2016. Water samples were collected from the lake at 2-week to 1-month intervals, mainly from the deepest part of the lake, and at depth intervals of 1 to 2 m in order to monitor the vertical heterogeneity of the water column. Groundwater was sampled twice from eight private wells in the vicinity of the lake (see Arnoux et al., 2017b for more details).

2.2.2 Water mass balance

The lake water budget is defined as follows:

$$\frac{\mathrm{d}V}{\mathrm{d}t} = I - E - Q, \tag{1}$$

where V is the volume of the lake (m^3); t is time (days); E is evaporation (m^3 day^{-1}); I is the instantaneous inflow (m^3 day^{-1}), corresponding to the sum of upstream surface inflow (I_S; zero for the studied lakes because they do not have surface inlets), runoff (I_R; considered negligible because of

Table 1. Main lake characteristics.

Region	ID	Lake name	Lake surface area (10^3 m^2)	Lake volume (10^3 m^3)	Catchment Area (10^3 m^2)
AT	1	Clair	115	695	2646
AT	2	Paix	41	97	796
AT	3	Sauvage	44	142	89
OUT	4	Lachigan	33	142	336
OUT	5	Lacroix	236	1080	772
OUT	6	Lanthier	25	125	1134
LAU	7	Lacasse	27	67	148
SAG	8	Beau Portage	42	271	364
SAG	9	Girard	67	679	211
SAG	10	Ludovic	94	400	1829

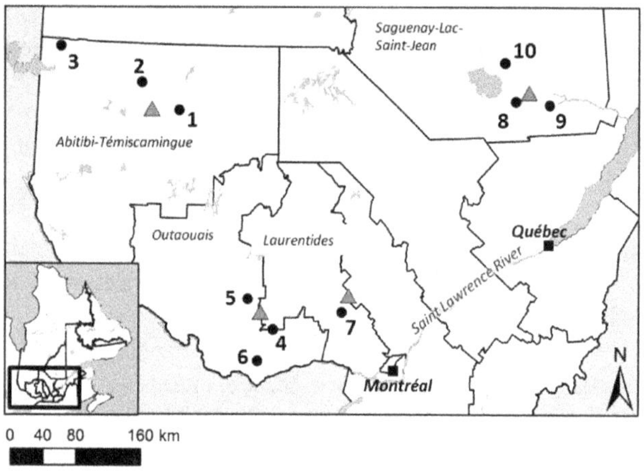

Figure 1. Locations of the study lakes (circles) and sources of climate data (triangles).

the permeable nature of the sandy soils of kettle lakes), groundwater inflow (I_G), and precipitation on the lake surface (P); and Q is the outflow (m^3 day^{-1}), which is the sum of surface (Q_S) and groundwater (Q_G) outflows. Under constant atmospheric and hydrologic conditions, a steady state is assumed (Gibson et al., 2016), implying that $dV/dt = 0$. Therefore $I_G = Q_S + Q_G + E - P$ for the entire lake.

2.2.3 Stable isotopic mass balance

Considering water stable isotopes, the lake isotopic mass balance is as follows:

$$V\frac{d\delta_L}{dt} + \delta_L\frac{dV}{dt} = I\delta_I - E\delta_E - Q\delta_Q, \tag{2}$$

where δ is isotopic composition of the lake (δ_L; equals to the mean if the lake is stratified – see Arnoux et al., 2017a and b, for more details about lake isotopic compositions used in the model); total inflow (δ_I), which includes runoff (δ_R), precipitation (δ_P), surface inflow (δ_S), and groundwater inflow (δ_G); and total outflow (δ_Q), which includes surface (δ_{QS}) and groundwater (δ_{QG}) outflows. The isotopic composition of evaporating water (δ_E) was estimated using the Craig and Gordon (1965) model, expressed by Gonfiantini (1986) as follows:

$$\delta_E = \frac{(\delta_L - \varepsilon^+)/\alpha^+ - h\delta_A - \varepsilon_K}{1 - h + 10^{-3}\varepsilon_K}, \tag{3}$$

where h is the relative humidity at the lake surface, δ_A is the local isotopic composition of the atmospheric moisture (‰), α^+ is the equilibrium isotopic fractionation, $\varepsilon^+ = (\alpha^+ - 1) \cdot 1000$ is the equilibrium isotopic separation (‰), and $\varepsilon_K = C_K(1 - h)$ is the kinetic isotopic separation (‰), with C_K being the ratio of molecular diffusivities between heavy and light molecules (Gibson et al., 2016).

In this study, C_K values were considered to be representative of fully turbulent wind conditions and a rough surface for both oxygen ($C_K = 14.2$‰) and hydrogen ($C_K = 12.5$‰), based on experimental data (Horita et al., 2008). For calculating equilibrium fractionation factors, experimental values of Horita and Wesolowski (1994) were used:

$$\alpha^+(^{18}O) = \exp(-7.685/10^3 + 6.7123/T - 1666.4/T^2 + 350410/T^3), \tag{4}$$

$$\alpha^+(^{2}H) = \exp(1158.8 \times T^3/10^{12} - 1620.1 \times T^2/10^9 + 794.84 \times T/10^6 - 161.04/10^3 + 2999200/T^3), \tag{5}$$

where T is temperature (K). The isotopic composition of atmospheric moisture (δ_A, ‰) was calculated assuming equilibrium isotopic exchange between precipitation and vapor:

$$\delta_A = \frac{\delta_P - \varepsilon^+}{1 + 10^{-3}\varepsilon^+}, \tag{6}$$

where δ_P (‰) is the mean annual isotopic composition of precipitation. Assuming well-mixed conditions in the lake, the combination of Eq. (3) and Eq. (2) yields the following:

$$V\frac{\mathrm{d}\delta_\mathrm{L}}{\mathrm{d}t}+\delta_\mathrm{L}\frac{\mathrm{d}V}{\mathrm{d}t}=P\delta_\mathrm{P}+I_\mathrm{G}\delta_\mathrm{G}-Q\delta_\mathrm{L} - \frac{E}{1-h+10^{-3}\varepsilon_\mathrm{K}}\left(\frac{\delta_\mathrm{L}-\varepsilon^+}{\alpha^+}-h\delta_\mathrm{A}-\varepsilon_\mathrm{K}\right). \quad (7)$$

A steady state was assumed, such that $\mathrm{d}V/\mathrm{d}t=0$. Equation (7) can therefore be simplified to the follows:

$$V\frac{\mathrm{d}\delta_\mathrm{L}}{\mathrm{d}t}=P\delta_\mathrm{P}+I_\mathrm{G}\delta_\mathrm{G}-(P+I_\mathrm{G}-E)\delta_\mathrm{L} - \frac{E}{1-h+10^{-3}\varepsilon_\mathrm{K}}\left(\frac{\delta_\mathrm{L}-\varepsilon^+}{\alpha^+}-h\delta_\mathrm{A}-\varepsilon_\mathrm{K}\right). \quad (8)$$

Resolving this calculation therefore allows isotopic composition of the lake water at time $t+\mathrm{d}t$ to be determined, expressed as a function of its value at the previous time step, t, and two established parameters, A (‰ $\mathrm{m^3\,yr^{-1}}$) and B ($\mathrm{m^3\,yr^{-1}}$):

$$\delta_\mathrm{L}^{t+\mathrm{d}t}=\frac{A}{B}+\left(\delta_\mathrm{L}^{t}-\frac{A}{B}\right)\exp\left(-\frac{B}{V}\mathrm{d}t\right) \quad (9)$$

with

$$A=P\delta_\mathrm{P}+I_\mathrm{G}\delta_\mathrm{G}-\frac{E}{1-h+10^{-3}\varepsilon_\mathrm{K}}\left(-h\delta_\mathrm{A}-\varepsilon_\mathrm{K}-\varepsilon^+/\alpha^+\right) \quad (10)$$

$$B=P+I_\mathrm{G}-E\left(1-\frac{1}{\alpha^+(1-h+10^{-3}\varepsilon_\mathrm{K})}\right). \quad (11)$$

The monthly-mean isotopic composition of precipitation (δ_P) was assessed in the four regions from the Global Network of Isotopes in Precipitation (GNIP) and Program for Groundwater Knowledge Acquisition (PACES) datasets. Future δ_P trends are uncertain; however, they have been shown to be mainly dependent on temperature evolution and local factors (Stumpp et al., 2014), and a recent study in Siberia showed that a long-term increase in precipitation $\delta^{18}O$ is close to the detection limit of the tracers (< 1‰ per 50 years) (Butzin et al., 2014). Monthly current means were therefore used in the current simulations. The mean value of groundwater isotopic composition (δ_Gi) was determined from the mean groundwater isotopic composition measured in wells, located in the same region and presenting no enrichment due to evaporation. The mean isotopic values used for groundwater are presented in Table 2.

The uncertainties associated with the Craig and Gordon (1965) model in the estimated isotopic composition of evaporating moisture (δ_E) can be substantial, especially if relative humidity is greater than 0.8 (Kumar and Nachiappan, 1999). Moreover, a sensitivity analysis of ^{18}O isotopic balance of a small lake in Austria (Yehdegho et al., 1997) indicates that for flow-though, groundwater-dominated systems with limited evaporation, the isotopic composition of the lake water and the inflow water are the parameters critical to the overall uncertainty. Horita et al. (2008) recommended using

Table 2. Mean isotopic composition of groundwater obtained for the four regions, in degrees versus VSMOW.

Region	$\delta^{18}O$	δ^2H
AT	−14.00	−101.3
OUT	−11.56	−81.6
LAU	−11.71	−80.9
SAG	−14.06	−103.1

time-averaged values of the parameters in the calculation of δ_E for the given period of interest. Moreover atmospheric parameters should be preferably evaporation-flux-weighted whereas liquid fluxes to a lake should be amount-weighted (Gibson, 2002; Gibson et al., 2016). Therefore, on an annual time step, δ_P is monthly precipitation-flux-weighted, except when it is used to estimate δ_A; in this case, δ_P is monthly evaporation-flux-weighted. On a monthly timescale, monthly values are used for each parameter of the model, and evaporation is considered to be null during the ice-covered period. Moreover, in winter, when monthly-mean temperature is below zero, precipitation is assumed to be zero in the model. Then, when monthly temperature becomes equal to or higher than zero, accumulated precipitation and amount-weighted δ_P are added to the calculation during the melt period. Moreover, sensitivity tests on this model have been performed by Arnoux et al. (2017a, b) and show that it is mostly sensitive to E, h, and δ_G.

2.3 Evolution scenarios

2.3.1 Climate models

Climatic parameters used in this study (evaporation, humidity, temperature, and precipitation) come from climate models. RCMs allow the downscaling of large-scale information from GCMs to scales closer to the watershed scale, leading to a better representation of surface forcings. In the present study, the fifth version of the Canadian RCM (CRCM5) was chosen, which has a 0.44‰ horizontal grid resolution (approx. 50 km; Sushama et al., 2010; Martynov et al., 2013; Šeparović et al., 2013). The CRCM5 is a gridpoint model, based on a two-time-level, semi-Lagrangian, (quasi) fully implicit time discretization scheme (Alexandru and Sushama, 2016). The model includes a terrain-following vertical coordinate based on hydrostatic pressure (Laprise, 1991; Alexandru and Sushama, 2015), and an horizontal discretization on a rotated latitude–longitude, Arakawa C grid (Arakawa and Lamb, 1977; Alexandru and Sushama, 2015). Following CRCM4, changes that have been introduced into CRCM5 include, for example, evolution in the planetary boundary layer parameterization to suppress both turbulent vertical fluxes under very stable conditions and the interactively coupled one-dimensional lake model (Flake; Mironov et al., 2010; Martynov et al., 2012; Šeparović et al., 2013).

CRCM5 uses the Canadian Land-Surface Scheme (CLASS, version 3.5; Verseghy, 1991; Alexandru and Sushama, 2015). This model is described in detail in Martynov et al. (2013) and Šeparović et al. (2013).

The CRCMs were driven by the second-generation Canadian Earth System Model (CanESM2, improved from CanESM1; Arora et al., 2011), developed by the Canadian Center for Climatic Modeling and Analysis (CCCma). As explained in Šeparović et al. (2013), it consists of a fourth-generation atmospheric general circulation model CanAM4, coupled with (i) the physical ocean component OGCM4 developed from the NCAR CSM Ocean Model (NCOM; Gent et al., 1998), (ii) the Canadian Model of Ocean Carbon (CMOC; Christian et al., 2010), and (iii) Canadian Terrestrial Ecosystem Model (CTEM; Arora and Boer, 2010). The CanAM4 is a spectral model employing T63 triangular truncation with physical tendencies calculated on a 2.81 linear grid and 35 vertical levels (Arora et al., 2011; Šeparović et al., 2013).

2.3.2 Climate data

Four greenhouse gas concentration scenarios (representative concentration pathways) have been adopted by the IPCC in its Fifth Assessment Report (AR5) in 2014: RCP 2.6, RCP 4.5, RCP 6.0, and RCP 8.5. The scenarios selected for the present study are RCP 4.5 and RCP 8.5, for which predicted climate data are available until 2100 for the study regions. The RCP 4.5 scenario considers that long-term global emissions of greenhouse gases, land use, and land cover stabilize radiative forcing at 4.5 W m^{-2} (approximately 650 ppm CO_2-equivalent) by the year 2100, without ever exceeding that value. The RCP 8.5 scenario corresponds to the highest greenhouse gas emissions pathway scenario, with gas emissions and CO_2 concentrations increasing considerably over time, thus leading to a radiative forcing of 8.5 W m^{-2} by the end of the century (approximately 1370 ppm CO_2 equivalent). The defining characteristics of these scenarios are enumerated in Moss et al. (2010).

In order to connect these RCP forecasts to our study and to visualize trends, yearly mean data are presented in Fig. 2. Based on previous literature on recharge changes (see Sect. 2.3.3), a reference period (2010–2040) is compared to a future period (2041–2071). It is noted that both evaporation and temperature display increases between the reference and future periods for both scenarios, although it is more pronounced for RCP 8.5. Moreover, precipitation and relative humidity do not show clear trends. However, it seems that precipitation variability will increase overall for both scenarios, although this is more pronounced for RCP 8.5. Moreover, the southern regions (i.e., OUT and LAU) have higher temperatures than the northern regions (i.e., AT and SAG), and precipitation is higher in LAU than in the other three regions. On a monthly timescale, surface temperatures in LAU show an increasing monthly trend, whereas evaporation increases mainly during summer and stays relatively constant the rest of the year (data not shown). Meanwhile, precipitation does not show any clear trend. However, as temperatures increases in winter, melt periods likely will shift more frequently occur earlier in the year.

2.3.3 Recharge evolution

The mean annual recharge for each lake basin was obtained by dividing the lake drainage area by the calculated mean annual groundwater inflow to the lake (Meinikmann et al., 2013). In this study, recharge evolution is thus expressed in terms of changes in groundwater inflow to the lakes.

In the first step, recharge is assumed to be constant for the 2006–2014 period. Over this period, recharge is adjusted to fit the calculated lake isotopic compositions to those measured. In the second step, the results of Rivard et al. (2014) was chosen for the simulation of recharge scenarios, since this study focusses on the Annapolis Valley (Nova Scotia, Canada), not far from southern Quebec and with a similar latitude, geology, and climate. Therefore, the future recharge dynamics determined for the Annapolis Valley are assumed to be similar to those of the present study sites. Rivard et al. (2014) found that all scenarios predict an annual recharge to the aquifer within the range of +14 to +45 % higher than at present by 2041–2071. They also predict, on a seasonal basis, that recharge will undergo (i) a marked decrease in summer (from 4 to 33 %) and (ii) a spectacular increase in winter (more than 200 %), due to an earlier melt period starting date.

The following section focused firstly on monthly lake isotopic composition evolution (Sect. 3.1) and secondly on yearly lake isotopic composition evolution (Sect. 3.2). Monthly and yearly values are compared for the two standard periods (i.e., for reference (2010–2040) and future (2041–2071) periods).

For the first part of the study, Lake Lacasse, located in the LAU region, has been chosen, since it was subject to continuous monitoring (Arnoux et al., 2017b). Its groundwater inflow and variability has therefore already been well constrained throughout the year 2015–2016 (Fig. 3b). For this lake, the model was run from 2006 to 2071, and four different recharge evolution scenarios were applied to the 2041–2071 period, following the predictions of Rivard et al. (2014) for scenarios S1 and S2, as described below.

- NC: no change in recharge (groundwater inflow follows the pattern described in Fig. 3, obtained from Arnoux et al., 2017b);
- S0: a recharge decrease of 33 % during the summer period (from June to October);
- S1: a 200 % increase in recharge during the melt period (from January to March) and a 4 % decrease in the summer period;

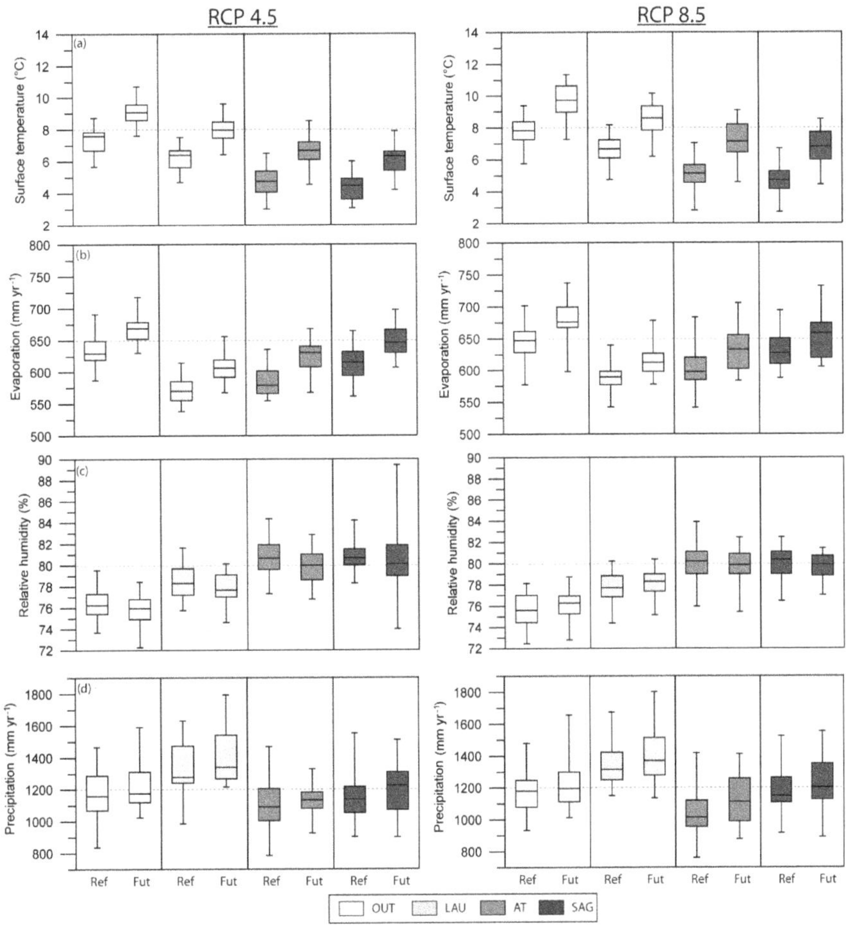

Figure 2. Climate data for the reference (Ref; 2010–2040) and future (Fut; 2041–2071) periods, obtained from CRCM5–CanESM2, with RCP 4.5 (left) and RCP 8.5 (right) scenarios for the four different study areas. The variables are **(a)** surface air temperature, **(b)** surface water evaporation (obtained from surface heat flux), **(c)** surface relative humidity (obtained from surface specific humidity), and **(d)** precipitation (Martynov et al., 2013; Šeparović et al., 2013); box whiskers describe median, first, and third quartiles as well as maximum and minimum values.

- S2: a 200 % increase in recharge during the melt period and a 33 % decrease during the summer period.

For the second part, three annual recharge evolution scenarios were tested, following the predictions of Rivard et al. (2014): no change (NC), a 14 % increase (Low), and a 45 % increase (High) in mean annual recharge.

2.4 Population growth

Variations in the quantity and/or quality of groundwater feeding lakes can obviously impact the geochemistry, and thus the water quality of lakes, especially for lakes displaying a high G index (the percentage of groundwater comprising the total lake inflow; Arnoux et al., 2017a). Moreover, in rural areas of Quebec, lake and groundwater quality is likely to be influenced by changes in population density. The population of Quebec is aging, and many seasonal residences (e.g., cottages) around lakes in rural areas are expected to become year-round residences. Furthermore, these residences are not connected to wastewater treatment plants; rather, owners have their own private wells for drinking water and private septic tanks with subsurface seepage beds for wastewater. The predicted population changes are summarized in Table 3. Population is mainly expected to increase in the southern regions (OUT and LAU), with a mean increase of 24 and 28 %, respectively (ISQ, 2014; Table 3). Scenarios of population growth are compared with scenarios of recharge evolution for each lake to assess their future quality evolution.

3 Results and discussion

3.1 Monthly evolution of lake isotopic composition

Figure 3 shows the measured (see Arnoux et al., 2017b, and the Supplement for more details about measured values) and

Table 3. Predicted population growth in the different study regions in 2036 relative to 2011 numbers, according to three different scenarios (ISQ, 2014).

Region	Scenarios		
	Reference (%)	Low (%)	High (%)
OUT	24	13	36
AT	5	0	10
LAU	28	21	34
SAG	0	−4	4

modeled isotopic compositions of Lake Lacasse. It can be observed that the modeled values are more variable than the measured ones, undoubtedly due to the higher evaporation rate in the climatic model (459 mm) than that measured during the field monitoring period (204 mm). It is also shown that the model attributes greater weight to the contribution of the depleted snow value than in reality. This is probably due to the snow column (which is close to 0 ‰ C during the snow melt) being less dense than the lake surface water (which has a mean temperature of close to 4 ‰ C) and therefore bypassing the lake, flowing rapidly out of the lake outlet. In such a case, the snow does not influence the lake isotopic composition as much as the model predicts. Since similar results are obtained for δ^2H values, only the δ^{18}O results from the model will be presented in the following sections.

Lake Lacasse has a mean G index (i.e. the percentage of groundwater that makes up total lake inflows) of 69 % during the reference period. Results for monthly simulations, with RCP 4.5 climate data, are illustrated in Fig. 4. Lake isotopic compositions are not significantly different between the reference and future periods if no change is applied to the recharge pattern (Fig. 4). Under scenarios S1 and S2, it can be observed that future δ^{18}O is nearly 100 % different from reference conditions during the first 2 months of the year (Fig. 4). It is at least 75 % different for the month of March, but this month shows important variation during the future period. Throughout the rest of the year, ranges of variation are not completely different, but increasing or decreasing trends can be observed, depending on the season.

Indeed, Fig. 5 shows the monthly differences between mean lake δ^{18}O in the reference period (which is the same for all scenarios) and mean lake δ^{18}O in the future period, for the four recharge evolution scenarios.

During the year:

- regarding the reference period, the highest variation is observed in March for S1 (−1‰), S2 (−1‰), and NC (−0.4‰), after the melt period. For S0, the greatest change regarding the reference period is observed in September and October (+0.4‰), after the evaporation period;

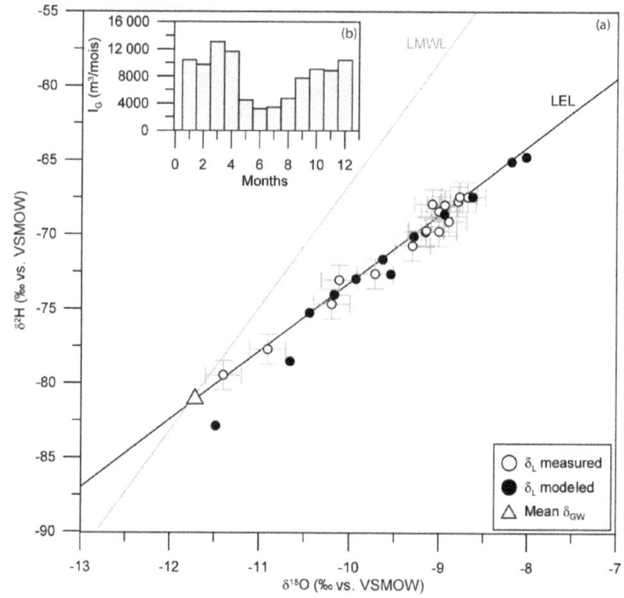

Figure 3. **(a)** Isotopic composition of Lake Lacasse between June 2015 and May 2016, measured and modeled from stable isotopic mass balance model using climate data from climate model CRCM5 – CanESM2 and scenario RCP 4.5; **(b)** the pattern of groundwater inflow (I_G) to Lake Lacasse (Arnoux et al., 2017b).

- regarding the NC future period, the greatest difference between winter recharge is in February (−0.6 and −0.5‰ for S1 and S2, respectively). This suggests that future changes in lake isotopic composition associated with recharge may be highest in February.

During the summer:

- regarding the reference period, the highest variation is in August for NC (+0.2‰), while it is in September and October for S0 (+0.4 % for both months) and S2 (+0.2 and +0.3‰ in September and October, respectively). S1 does not show any variation;

- regarding the NC future period, the greatest change occurs in October for S0 (+0.3‰) and for S2 (+0.2‰), and in September for S1 (−0.1‰).

Results of scenario S2, characterized by the greatest changes in recharge in both summer and winter, highlight that the impact of decreasing recharge during summer attenuates the substantial impact of increasing recharge during winter. Indeed, during winter, S1 shows more depleted values than S2 (−0.5 versus −0.4‰ in January, and −0.8 and −0.7‰ with respect to the reference period for S1 and S2, respectively). Therefore, the more recharge decreases in the summer, the more lake isotopic composition increases in the summer, due to increased evaporation. Meanwhile, the more recharge increases in the winter, the more lake isotopic composition is depleted in the winter. If both phenomena occur

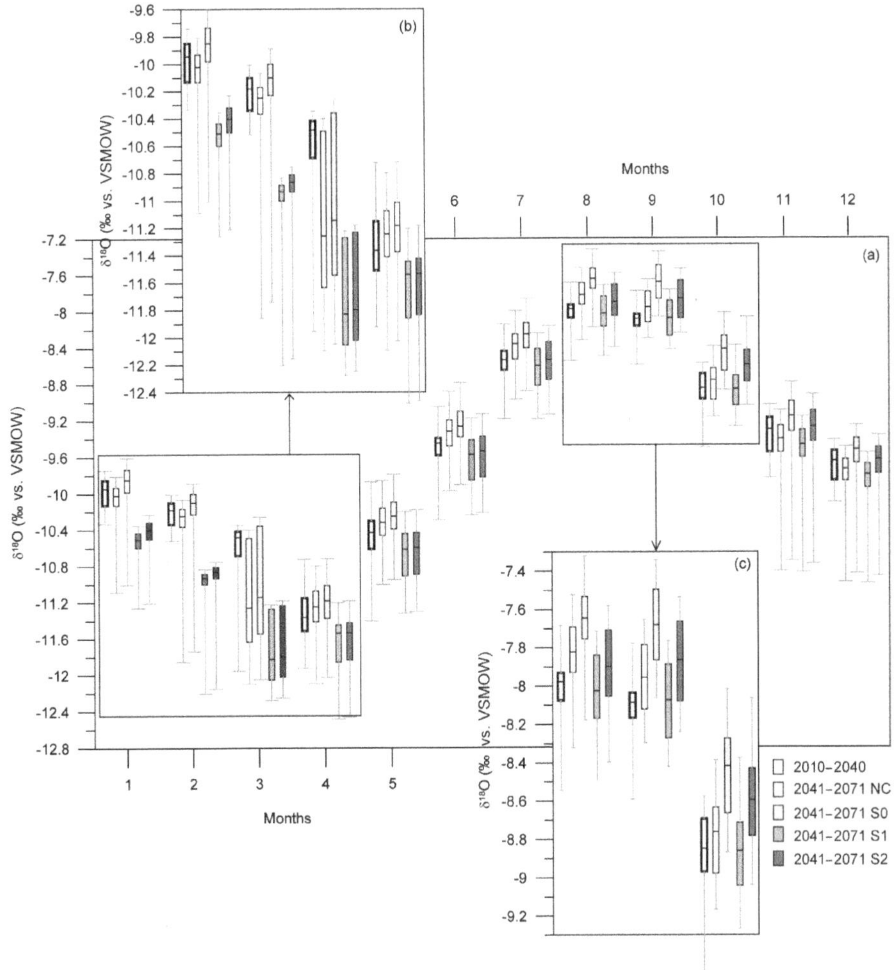

Figure 4. (a) Monthly Lake Lacasse isotopic composition, calculated using RCP 4.5 climatic data, for different periods and various recharge patterns: no change (NC), −33 % in the summer (from June to October; S0), +200 % during the melt period (from January to March) and −4 % in the summer (S1), and +200 % during the melt period and −33 % in the summer (S2); **(b)** close-up of the winter months; **(c)** close-up of the summer months; box whiskers describe median, first, and third quartiles as well as maximum and minimum values.

in a given year, the mean annual lake isotopic composition evolution will therefore not be expected to shift much, since their opposing impacts on lake isotopic composition will cancel each other out. As such, S1 is the scenario showing the highest variation in annual mean, of −3‰, compared with −2‰ for S2 and +2‰ for S0.

Based on these observations, it appears that isotopic signatures measured at the end of February and in September or October will provide information on the greatest changes during the winter and summer periods, respectively. The greatest changes in lake isotopic composition are likely to be at the end of the melt period.

Moreover, simulation results show that RCP 4.5 and 8.5 models provide similar results for Lake Lacasse isotopic composition evolution. Figure 6 shows the comparison of lake $\delta^{18}O$ composition for both RCP climate scenarios, from 2010 to 2071, assuming the NC recharge scenario. In Fig. 6, it can be observed that there is a small trend toward $\delta^{18}O$ enrichment due to a higher evaporation rate, which is more pronounced for the RCP 8.5 than for the RCP 4.5 scenario. However, on a yearly timescale, the impact of evaporation increase in the summer seems to be attenuated by a precipitation increase throughout the rest of the year, likely implying that these climate changes result in a nearly non-measurable impact on lake isotopic composition evolution.

Finally, all these results show that extreme caution is required when interpreting trends in lake isotopic composition, and that their interpretation requires (i) a minimum background knowledge – at least 1 year of data – of lake isotopic composition evolution in relation to its hydrological balance and (ii) an accurate evaluation of weather data variability in the year of monitoring, with respect to their annual means for the study lake. A long-term change in recharge will definitely impact lake isotopic composition, but the lake is also

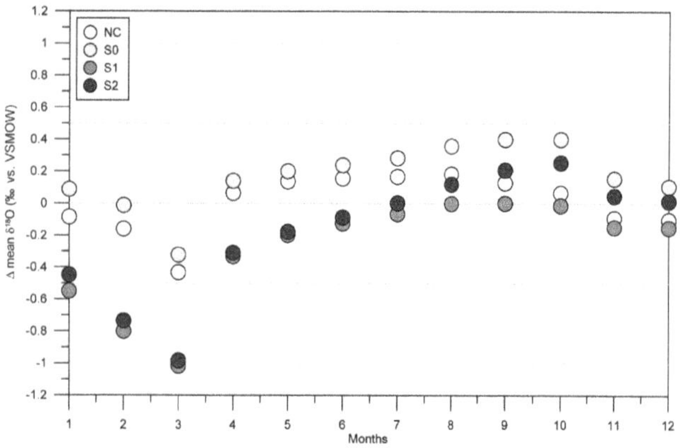

Figure 5. Differences between mean Lake Lacasse $\delta^{18}O$ in the reference period and mean Lake Lacasse $\delta^{18}O$ in the future period, for the RCP 4.5 climate scenario and four scenarios of recharge evolution: no change (NC), −33 % in the summer (from June to October; S0), +200 % during the melt period (from January to March) and −4 % in the summer (S1), and +200 % during the melt period and −33 % in the summer (S2).

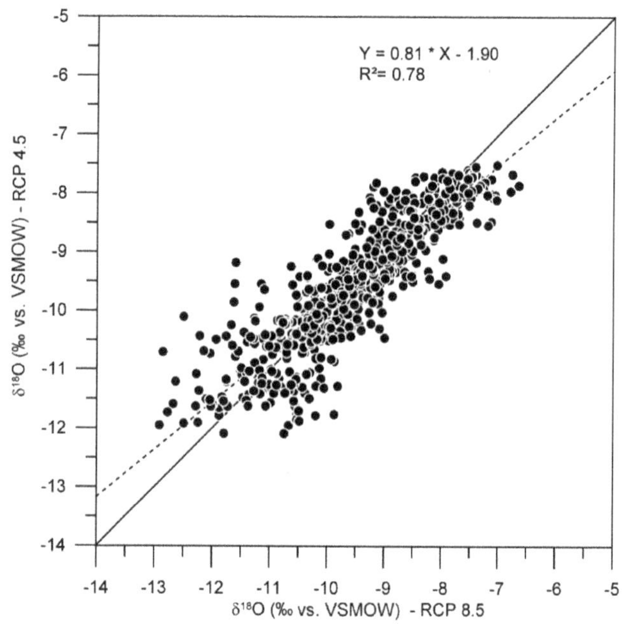

Figure 6. Comparison between monthly results in $\delta^{18}O$ for both scenarios RCP 4.5 and 8.5 for the 2010–2071 period.

sensitive to changes in other water budget parameters. It may therefore still be difficult to definitively isolate the effect of recharge over long time periods. As such, it is also important to consider evolution in the yearly mean lake isotopic composition.

3.2 Annual evolution of lake geochemistry

3.2.1 Isotopic signature evolution

The model was run for the 10 study lakes, including Lake Lacasse (Table 1 for main lake characteristics). Figure 7 illustrates differences in $\delta^{18}O$ in the reference period compared to the future period for lakes which have a range of G indices (see Arnoux et al., 2017a for more details about lakes measured values). It can be observed that, if the recharge is set as constant from 2010 to 2071 (NC recharge scenario), there is no significant difference between the reference and future period (Fig. 7), although evaporation shows a significant increase with time. The lack of a trend is probably mitigated by concurrent shifts in precipitation (Fig. 2). Without considering changes in groundwater inflow, it appears that lake isotopic composition will be impacted at least as much by changes in precipitation as by changes in evaporation.

Figure 7 illustrates that the range of lake isotopic composition variation depends significantly on climate conditions, lake volumes, and their associated G indices. It can be observed that lakes with a low G index and a small volume have higher potential variability in isotopic composition regarding climatic variations than those with a high G index and high volume. For example, for two lakes with a similar mean G index, such as Lake Ludovic (SAG; G index = 51 %) and Lake Lacroix (OUT; G index = 53 %), the former is expected to have a greater spread in isotopic compositions than the latter, even though the SAG region will likely undergo less evaporation increase compared with the OUT region (Fig. 2). This difference is due to the lower volume of Lake Ludovic ($V = 400\,000\ m^3$), compared with Lake Lacroix ($V = 1\,080\,000\ m^3$; Table 1). In addition, when lakes have a high G index, the groundwater flux tends to buffer lake isotopic variations, and so they tend to be less sensitive to changes in climate data. The dominant control on lake isotopic variability therefore appears to be the G index. Another example is Lake Lanthier, which has a smaller volume ($V = 125\,000\ m^3$) and a higher G index (G index = 94 %), and therefore shows a limited range of isotopic variation compared with Lake Lacroix, although both are located in the OUT region (Fig. 7).

If a changing recharge scenario is applied, a decreasing trend in lake isotopic composition is clearly observed (Fig. 7). However, it is also shown that lakes are sensitive to large changes in annual recharge (+45 %), but the differences are not significant if a smaller change (+14 %) occurs. Moreover, as the percentage of recharge change applied in the model is the same for all lakes, it can be observed that the trend intensity will depend on four main parameters: lake catchment size (which controls the intensity of the flux change), the region (which underlies climate condition), lake volume (which impacts the range of variation), and the G index. However, a relationship is only found with the latter.

Figure 8 illustrates variations in mean lake $\delta^{18}O$ versus G index in both reference and future periods. As shown, lake isotopic composition is more sensitive to changes in recharge for G indices ranging from 50 to 80 %, with a maximum of

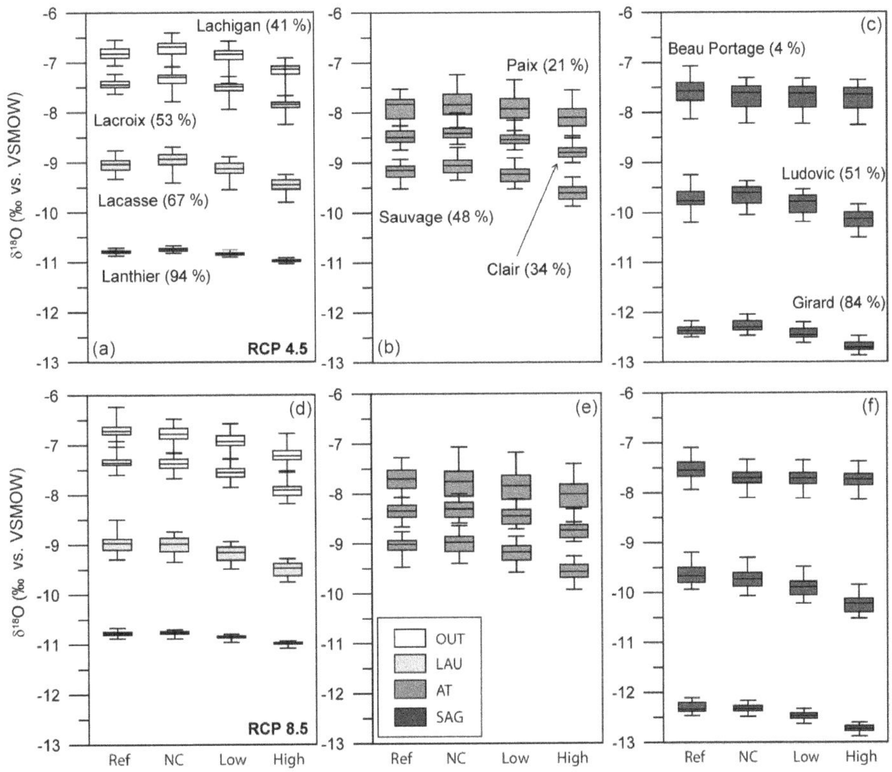

Figure 7. Reference period (Ref; 2010–2040) lake $\delta^{18}O$ composition and that corresponding to three different future period (2041–2071) recharge scenarios: no change (NC), +14 % (Low), and +45 % (High), for RCP 4.5 (top) and RCP 8.5 (bottom) scenarios. The values in brackets correspond to the mean G index (percentage of groundwater flow in the total inflow) for each lake calculated for the reference period; left panels show OUT and LAU regions, middle panels AT, and right panels SAG; box whiskers describe median, first, and third quartiles as well as maximum and minimum values.

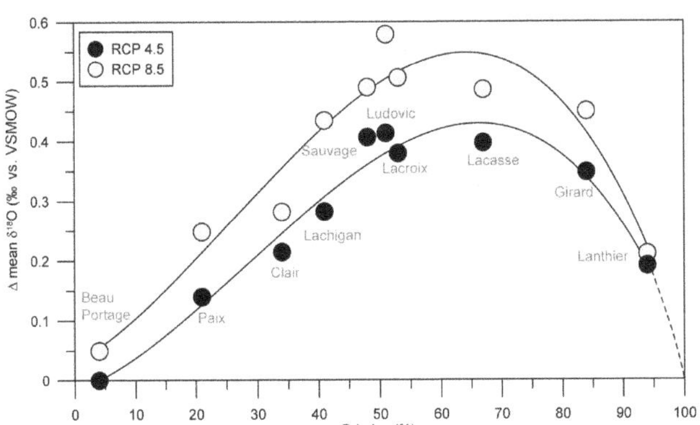

Figure 8. Differences between mean lake $\delta^{18}O$ in the reference period (2010–2040) and future period (2041–2071), for the higher recharge change scenario, versus lake G indices. RCP 4.5 (black dots) and 8.5 (grey dots) scenarios are represented.

sensitivity observed for a G index of around 65 %. It can also be observed that RCP 8.5 predicts a more depleted isotopic composition than does RCP 4.5. This implies that for the same recharge scenario, variations in precipitation and melt period (duration and time in the year) may impact the lake isotopic evolution more than precipitation. Finally, the polynomial relationship between the two variables in Fig. 8 highlights that the G index drives the response of lake isotopic composition to changes in recharge.

3.2.2 Lake quality evolution

As has been shown previously, isotopic composition can be sensitive to future recharge changes. This section is now a discussion about how lake water quality could be impacted by these future recharge changes depending on the location of the lake. In the study areas, one of the principal concerns about lake water quality, today and in the future, is to avoid blue–green algae blooms in limiting P loads to lakes. This study does not take into account several parameters that may impact blue–green algae blooms in lakes, such as the lake water biogeochemistry, chemical threshold processes, thermal and oxygen stratification, and warming of the water column. The purpose here is to show in which cases lakes could be under risk of too-high P loading, and therefore at risk of a decrease in water quality, depending on their catchment evolution as a function of recharge and population evolution. In such cases it will be important to prevent P loads and therefore changes on the lake catchment.

Turning to the predictions of population growth summarized in Table 3, population is predicted to increase mainly in the southern regions, OUT and LAU, with a mean increase of 24 and 28 % by 2036, respectively (ISQ, 2014). Assuming an identical per capita P load, total P load in groundwater originating from wastewater should increase by the same percentage.

Domestic sewage is the main contribution of anthropogenic sources to the total P load for most of Canadian lakes (Dillon and Evans, 1993; Paterson et al., 2006). The total P load from sewage systems is a function of (i) the population and (ii) the annual P consumption per capita (Paterson et al., 2006). Assuming an effluent concentration of $9\,mg\,L^{-1}$ (considering reductions in the phosphate content of detergents) and a daily water usage of $200\,L\,capita^{-1}\,day^{-1}$ (as was done by Paterson et al., 2006), the P contribution is estimated to be $0.66\,kg\,capita^{-1}\,yr^{-1}$. Investigated lakes in the OUT and LAU regions collect sewage from 4 (Lake Lachigan), 53 (Lake Lanthier), 117 (Lake Lacroix), and 17 houses (Lake Lacasse) within their catchments. If two habitants per house are assumed, P loading to groundwater will be increased from 1 to $39\,kg\,yr^{-1}$ in the studied lakes in these areas.

The impact of this P load increase on lakes can then roughly be estimated based on the ratio of change in annual P load versus change in annual recharge, as illustrated in Fig. 9. For an increase in recharge, if $\Delta_P/\Delta_R < 1$, the change in recharge over the catchment, and thus the evolution of the groundwater inflow to the lakes, will be greater than the P variation. In such a case, the lake water quality may not be impacted by this P variation. However, if $\Delta_P/\Delta_R > 1$, the lake water quality will be impacted, and precautions should be taken to minimize the risk of blue–green algae blooms and consequent eutrophication. For the study regions (Fig. 9), if recharge increases 14 % by 2036, as estimated by Rivard et al. (2014), lakes in the LAU and OUT areas will experience a decrease in their water quality. However, if the recharge change is closer to +45 % (Rivard et al., 2014), lake water quality should not be worse than today, providing all other things remain equal and assuming the population growth forecasts are accurate.

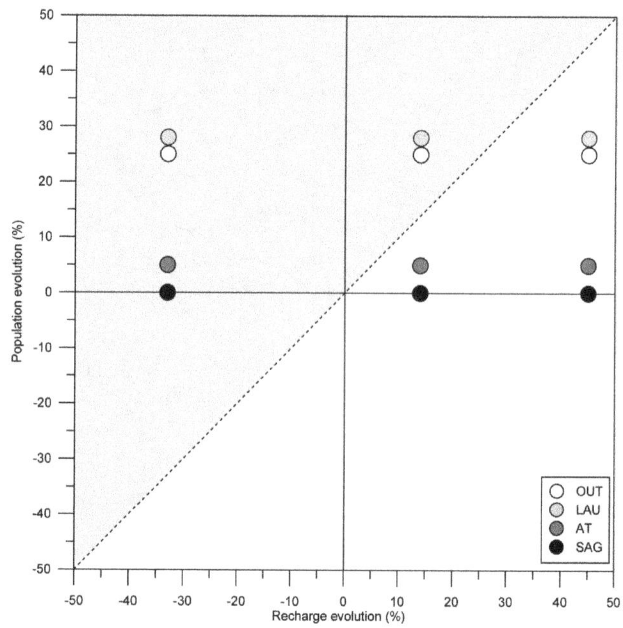

Figure 9. Population growth predictions versus changes in recharge. The shaded area represents the scenarios for which lakes may be under risk of too-high P loading, and therefore at risk of a decrease in water quality. Dots represent lakes in the four study areas for three recharge scenarios.

4 Conclusions

The main objectives of this study were to determine how future trends in groundwater recharge can affect lake geochemistry, and to assess whether stable isotopes might be an effective tool for identifying lakes that are susceptible to change, or are undergoing changes, in their water budget and quality.

Firstly, climate predictions from both RCP 4.5 and 8.5 scenarios and their impacts on future lake isotopic composition have been considered. By 2050, temperature and evaporation are expected to increase, and precipitation to exhibit a slightly increasing trend, all trends being more intense under the RCP 8.5 scenario. On a monthly time step, it has been highlighted that future lake isotopic signatures will be more depleted with respect to the reference period, mainly in March and February, because of an earlier melt period. In the summer, lake isotopic composition will be more enriched, mainly in August, due to the higher evaporation rate expected. However, future variations with respect to the reference period are smaller in the summer than in the winter. Scenario RCP 8.5 induces more intense monthly variations, but no significant difference in future lake isotopic signatures is observed on a yearly time step between the two scenarios. This means that enrichment caused by increased evaporation compensates for depletion induced by precipitation variation. It is therefore unclear whether lakes will be impacted more by increased evaporation or precipitation changes. Caution is therefore recommended in the interpretation of isotopic trends in lakes where background knowledge – for at least 1 year – of their isotopic composition evolution with respect to weather data and their hydrologic balance is lacking.

It has then been demonstrated that future lake isotopic composition will also depend on recharge fluctuations, in addition to climate conditions. On a monthly basis, the highest impact of recharge evolution on future lake isotopic composition will be in February. Moreover, if recharge decreases during the summer, the main difference will be observed at the end of the summer, after the evaporation period and before recharge stops decreasing, in September or October. Therefore, to clearly identify future changes in recharge through the lake isotopic signature evolution, sampling only at the end of February and in September or October will provide information on the greatest changes for the winter and summer periods, respectively.

On an annual time step, modeled evolutions of lake isotopic composition can clearly be sensitive to both +45 and +14 % changes in recharge – less so, nevertheless, to the latter. The intensity of the future trend of lake isotopic composition will depend on four main parameters: lake catchment (which controls the intensity of the flux change), the region (which drives climate conditions), lake volume (which impacts the range of variation), and the G index (which is the dominant control on water balance conditions). Based on these model simulations, stable isotopes appear to be especially useful for detecting changes in recharge to lakes with a G index of between 50 and 80 %.

It is important to keep in mind that if both a winter increase and summer decrease in recharge occur during the same year, the trend in mean annual lake isotopic composition will be nullified, because seasonal variation is impacted in opposing directions, canceling out the signal at the yearly time step. Consequently, if no clear annual trend is observed, it does not mean that recharge is not changing. Nevertheless, mean annual lake isotopic compositions will be observed to be impacted by recharge evolution only if it evolves in the same way throughout the year for the most part (i.e., consistently decreasing or increasing). In light of these results, it is a monthly time step is strongly suggested in such investigations, since seasonal recharge fluctuations can be canceled out in the yearly signal. Moreover, it is important to note that runoff has been considered negligible for the study lakes but can be important for other lakes and, in these cases, this model could underestimate the effect of spring melt on future lake isotopic composition.

It is also shown that changes in water quality in groundwater-connected lakes depend substantially on lake location and on the intensity of recharge change. For the studied lakes, in the case of a +14 % recharge increase by 2036, lakes in LAU and OUT regions may experience altered water quality (driven by phosphorous loading), but no change is expected in the case of a +45 % recharge intensification. If the percentage of recharge increase is at least equal to the percentage of population growth around the lake, lake quality should not become degraded, but if not, recharge evolution should be considered in lake management. Lake water quality in the SAG and AT areas may not decrease when considering population growth predictions. However, this study does not take into account several parameters that can impact blue–green algae blooms in lakes, such as the lake water residence time, chemical threshold processes, and the warming of the water column (Planas and Paquet, 2016).

Finally, even if small groundwater-fed lakes will be sensitive to climate, and especially to recharge and anthropogenic changes, it is still difficult to predict how their geochemistry will be impacted, as it is very reactive to each slight variation in water balance parameters. However, more indicators are now available to predict lake geochemistry evolution, mainly depending on their location and their G index. To go further, a recharge model adapted to lake catchments and coupled with melt dynamics, closely dependent on climate forecasts, could provide more details on lake geochemical evolution, for more sustainable lake management.

Competing interests. The authors declare that they have no conflict of interest.

Acknowledgements. We thank all participants of this project which is part of a PhD project funded by University Paris-Saclay/Paris-Sud and by grants to Florent Barbecot (CRSNG and FQRNT). We thank all participants of the field campaigns, Viorel Horoi and students for their help on the field, the GRIL team for their logistical support, and people living around the study lakes for their hospitality and support. We thank also the team of Laxmi Sushama for their help with climate data. Comments from the editor and two anonymous referees have helped to improve this paper and we gratefully acknowledge them.

Edited by: Bettina Schaefli

References

Alexandru, A. and Sushama, L.: Impact of land-use and land-cover changes on CRCM5 climate projections over North America for the twenty-first century, Clim. Dynam., 47, 1197, https://doi.org/10.1007/s00382-015-2896-3, 2016.

Arakawa, A. and Lamb, W. R.: Computational design of the basic dynamical processes of the UCLA general circulation model General circulation models of the atmosphere, Academic Press Inc, New York (A78-10662 01-47), 173–265, 1977.

Arnoux, M., Barbecot, F., Gibert-Brunet, E., Gibson, J., Rosa, E., and Noret, A., and Monvoisin, G.: Geochemical and isotopic mass balances of kettle lakes in southern Quebec (Canada) as tools to document variations in groundwater quantity and quality, Environ. Earth Sci., 76, 106, https://doi.org/10.1007/s12665-017-6410-6, 2017a.

Arnoux, M., Gibert-Brunet, E., Barbecot, F., Gillon, S., Gibson, J., and Noret, A.: Interactions between groundwater and seasonally ice-covered lakes: Using water stable isotopes and radon-222 multilayer mass balance models, Hydrol. Process., 31, 2566–2581, 2017b.

Arora, V. K. and Boer, G. J.: Uncertainties in the 20th century carbon budget associated with land use change, Glob. Change Biol., 16, 3327–3348, https://doi.org/10.1111/j.1365-2486.2010.02202.x, 2010.

Arora, V. K., Scinocca, J. F., Boer, G. J., Christian, J. R., Denman, K. L., Flato, G. M., Kharin, V. V., Lee, W. G., and Merryfield, W. J.: Carbon emission limits required to satisfy future representative concentration pathways of greenhouse gases, Geophys. Res. Lett., 38, https://doi.org/10.1029/2010GL046270, 2011.

Benn, D. I. and Evans, D. J. A.: Sediment-landform associations in: Glaciers and Glaciation, 2nd edn. Paperback, Hodder Arnold, London, 421–533, 2010.

ally ice-covered lakes: Using water stable isotopes and radon-222 multilayer mass balance models, Hydrol. Process., 31, 2566–2581, 2017b.

Arora, V. K. and Boer, G. J.: Uncertainties in the 20th century carbon budget associated with land use change, Glob. Change Biol., 16, 3327–3348, https://doi.org/10.1111/j.1365-2486.2010.02202.x, 2010.

Arora, V. K., Scinocca, J. F., Boer, G. J., Christian, J. R., Denman, K. L., Flato, G. M., Kharin, V. V., Lee, W. G., and Merryfield, W. J.: Carbon emission limits required to satisfy future representative concentration pathways of greenhouse gases, Geophys. Res. Lett., 38, https://doi.org/10.1029/2010GL046270, 2011.

Benn, D. I. and Evans, D. J. A.: Sediment-landform associations in: Glaciers and Glaciation, 2nd edn. Paperback, Hodder Arnold, London, 421–533, 2010.

Butzin, M., Werner, M., Masson-Delmotte, V., Risi, C., Frankenberg, C., Gribanov, K., Jouzel, J., and Zakharov, V. I.: Variations of oxygen-18 in West Siberian precipitation during the last 50 years, Atmos. Chem. Phys., 14, 5853–5869, https://doi.org/10.5194/acp-14-5853-2014, 2014.

Christian, J. R., Arora, V. K., Boer, G. J., Curry, C. L., Zahariev, K., Denman, K. L., Flato, G. M., Lee, W. G., Merryfield, W. J., Roulet, N. T., and Scinocca, J. F.: The global carbon cycle in the Canadian Earth system model (CanESM1): Preindustrial control simulation, J. Geophys. Res.-Biogeosci., 115, https://doi.org/10.1029/2008JG000920, 2010.

Craig, H. and Gordon, L. I.: Deuterium and oxygen-18 in the ocean and marine atmosphere, in: Stable Isotopes in Oceanographic Studies and Paleotemperatures, edited by: Tongiorgi, E., Spoleto, Italy, 9–130, 1965.

Crosbie, R. S., Scanlon, B. R., Mpelasoka, F. S., Reedy, R. C., Gates, J. B., and Zhang, L.: Potential climate change effects on groundwater recharge in the High Plains Aquifer, USA, Water Resour. Res., 49, 3936–3951, https://doi.org/10.1002/wrcr.20292, 2013.

Croteau, A., Nastev, M., and Lefebvre, R.: Groundwater Recharge Assessment in the Chateauguay River Watershed, Can. Water Resour. J., 35, 451–468, https://doi.org/10.4296/cwrj3504451, 2010.

Dillon, P. J. and Evans, H. E.: A comparison of phosphorus retention in lakes determined from mass balance and sediment core calculations, Water Res., 27, 659–668, https://doi.org/10.1016/0043-1354(93)90176-I, 1993.

Dragoni, W. and Sukhija, B. S.: Climate change and groundwater: A short review, Geological Society Special Publication, 1–12, 2008.

Gent, P. R., Bryan, F. O., Danabasoglu, G., Doney, S. C., Holland, W. R., Large, W. G., and McWilliams, J. C.: The NCAR climate system model global ocean component, J. Climate, 11, 1287–1306, 1998.

Gibson, J. J.: Short-term evaporation and water budget comparisons in shallow Arctic lakes using non-steady isotope mass balance, J. Hydrol., 264, 242–261, https://doi.org/10.1016/S0022-1694(02)00091-4, 2002.

Gibson, J. J., Birks, S. J., and Yi, Y.: Stable isotope mass balance of lakes: a contemporary perspective, Quaternary Sci. Rev., 131, 316–328, https://doi.org/10.1016/j.quascirev.2015.04.013, 2016.

Gonfiantini, R.: Environmental isotopes in lake studies. In: Fritz P, Fontes JCh (eds) Handbook of environmental isotope geochemistry, The terrestrial environment, B, vol 2, Elsevier, Amsterdam, 113–168, 1986.

Healy, R. W.: Estimating groundwater recharge, Estimating Groundwater Recharge, 1–245, 2011.

Horita, J. and Wesolowski, D.: Liquid-vapour fractionation of oxygen and hydrogen isotopes of water from the freezing to the critical temperature, Geochim. Cosmochim. Acta, 58, 3425–3437, 1994.

Horita, J., Rozanski, K., and Cohen, S.: Isotope effects in the evaporation of water: a status report of the Craig-Gordon model, Isotop. Environ. Health Stud., 44, 23–49, https://doi.org/10.1080/10256010801887174, 2008.

IPCC: the Fifth Assessment Report of the Intergovernmental Panel on Climate Change, available at: https://www.ipcc.ch/report/ar5/, 2014.

ISQ: Institut de la statistique du Québec, available at: http://www.stat.gouv.qc.ca (last access: 2016), 2014.

Isokangas, E., Rozanski, K., Rossi, P. M., Ronkanen, A.-K., and Kløve, B.: Quantifying groundwater dependence of a sub-polar lake cluster in Finland using an isotope mass balance approach, Hydrol. Earth Syst. Sci., 19, 1247–1262, https://doi.org/10.5194/hess-19-1247-2015, 2015.

Jyrkama, M. I. and Sykes, J. F.: The impact of climate change on spatially varying groundwater recharge in the grand river watershed (Ontario), J. Hydrol., 338, 237–250, https://doi.org/10.1016/j.jhydrol.2007.02.036, 2007.

Kumar, B. and Nachiappan, R. P.: On the sensitivity of Craig and Gordon model for the estimation of the isotopic composition of lake evaporates, Water Resour. Res., 35, 1689–1691, https://doi.org/10.1029/1999WR900011, 1999.

Kuniansky, E. L.: Geohydrology and simulation of groundwater flow in the "400-foot", "600-foot", and adjacent aquifers, Baton Rouge area, Louisiana, Louisiana Department of Transportation and Development Technical Report, 49, 1989.

Kurylyk, B. L. and MacQuarrie, K. T. B.: The uncertainty associated with estimating future groundwater recharge: A summary of recent research and an example from a small unconfined aquifer in a northern humid-continental climate, J. Hydrol., 492, 244–253, https://doi.org/10.1016/j.jhydrol.2013.03.043, 2013.

Laprise, R.: The Euler equations of motion with hydrostatic pressure as an independent variable, Mon. Weather Rev., 120, 197–207, 1991.

Lemieux, J.-M., Hassaoui, J., Molson, J., Therrien, R., Therrien, P., Chouteau, M., and Ouellet, M.: Simulating the impact of climate change on the groundwater resources of the Magdalen Islands, Québec, Canada, J. Hydrol.: Reg. Stud., 3, 400–423, https://doi.org/10.1016/j.ejrh.2015.02.011, 2015.

Levison, J., Larocque, M., Fournier, V., Gagné, S., Pellerin, S., and Ouellet, M. A.: Dynamics of a headwater system and peatland under current conditions and with climate change, Hydrol. Process., 28, 4808–4822, https://doi.org/10.1002/hyp.9978, 2014.

Martynov, A., Sushama, L., Laprise, R., Winger, K., and Dugas, B.: Interactive lakes in the Canadian Regional Climate Model, version 5: The role of lakes in the regional climate of North America, Tellus A, 64, 16226, https://doi.org/10.3402/tellusa.v64i0.16226, 2012.

Martynov, A., Laprise, R., Sushama, L., Winger, K., Šeparović, L., and Dugas, B.: Reanalysis-driven climate simulation over CORDEX North America domain using the Canadian Regional Climate Model, version 5: Model performance evaluation, Clim. Dynam., 41, 2973–3005, https://doi.org/10.1007/s00382-013-1778-9, 2013.

Meinikmann, K., Lewandowski, J., and Nützmann, G.: Lacustrine groundwater discharge: Combined determination of volumes and spatial patterns, J. Hydrol., 502, 202–211, https://doi.org/10.1016/j.jhydrol.2013.08.021, 2013.

Mironov, D., Heise, E., Kourzeneva, E., Ritter, B., Schneider, N., and Terzhevik, A.: Implementation of the lake parameterisation scheme FLake into the numerical weather prediction model COSMO, Boreal Environ. Res., 15, 218–230, 2010.

Moss, R. H., Edmonds, J. A., Hibbard, K. A., Manning, M. R., Rose, S. K., Van Vuuren, D. P., Carter, T. R., Emori, S., Kainuma, M., Kram, T., Meehl, G. A., Mitchell, J. F. B., Nakicenovic, N., Riahi, K., Smith, S. J., Stouffer, R. J., Thomson, A. M., Weyant, J. P., and Wilbanks, T. J.: The next generation of scenarios for climate change research and assessment, Nature, 463, 747–756, https://doi.org/10.1038/nature08823, 2010.

Paterson, A. M., Dillon, P. J., Hutchinson, N. J., Futter, M. N., Clark, B. J., Mills, R. B., Reid, R. A., and Scheider, W. A.: A review of the components, coefficients and technical assumptions of Ontario's Lakeshore Capacity Model, Lake and Reservoir Management, 22, 7–18, 2006.

Planas, D. and Paquet, S.: Importance of climate change-physical forcing on the increase of cyanobacterial blooms in a small, stratified lake, J. Limnol., 75, 201–214, https://doi.org/10.4081/jlimnol.2016.1371, 2016.

Rivard, C., Lefebvre, R., and Paradis, D.: Regional recharge estimation using multiple methods: an application in the Annapolis Valley, Nova Scotia (Canada), Environ. Earth Sci., 71, 1389–1408, https://doi.org/10.1007/s12665-013-2545-2, 2013.

Rivard, C., Paniconi, C., Vigneault, H., and Chaumont, D.: A watershed-scale study of climate change impacts on groundwater recharge (Annapolis Valley, Nova Scotia, Canada), Hydrol. Sci. J., 59, 1437–1456, https://doi.org/10.1080/02626667.2014.887203, 2014.

Rutledge, A. T.: Computer programs for describing the recession of ground-water discharge and for estimating mean ground-water recharge and discharge from streamflow data – update, US Geol Surv Water-Resour Invest Rep, 98–4148:43, 1998.

Rutledge, A. T.: Update on the use of the RORA program for recharge estimation, Ground Water, 45, 374–382, 2007.

Scanlon, B. R., Healy, R. W., and Cook, P. G.: Choosing appropriate techniques for quantifying groundwater recharge, Hydrogeol. J., 10, 18–39, https://doi.org/10.1007/s10040-001-0176-2, 2002.

Schindler, D. W.: Evolution of phosphorus limitation in lakes, Science, 195, 260–262, 1977.

Šeparović, L., Alexandru, A., Laprise, R., Martynov, A., Sushama, L., Winger, K., Tete, K., and Valin, M.: Present climate and climate change over North America as simulated by the fifth-generation Canadian regional climate model, Clim. Dynam., 41, 3167–3201, https://doi.org/10.1007/s00382-013-1737-5, 2013.

Stumpp, C., Klaus, J., and Stichler, W.: Analysis of long-term stable isotopic composition in German precipitation, J. Hydrol., 517, 351–361, https://doi.org/10.1016/j.jhydrol.2014.05.034, 2014.

Sushama, L., Khaliq, N., and Laprise, R.: Dry spell characteristics over Canada in a changing climate as simulated by the Canadian RCM, Global Planet. Change, 74, 1–14, https://doi.org/10.1016/j.gloplacha.2010.07.004, 2010.

Tiedeman, C. R, Goode, D. J., and Hsieh, P. A.: Numerical simulation of ground-water flow through glacial deposits and crystalline bedrock in the Mirror Lake area, Grafton County, New Hampshire, US Geological Survey Professional Paper, 1572, 1997.

Turner, K. W., Wolfe, B. B., and Edwards, T. W. D.: Characterizing the role of hydrological processes on lake water balances in the Old Crow Flats, Yukon Territory, Canada, using water isotope tracers, J. Hydrol., 386, 103–117, https://doi.org/10.1016/j.jhydrol.2010.03.012, 2010.

Verseghy, D. L.: Class – A Canadian land surface scheme for GCMS. I. Soil model, Int. J. Climatol., 11, 111–133, https://doi.org/10.1002/joc.3370110202, 1991.

Wang, H. J, Liang, X. M., Jiang, P. H., Wang, J., Wu, S. K., and Wang, H. Z.: TP ratio and planktivorous fish do not affect nutrient–chlorophyll relationships in shallow lakes, Freshw. Biol., 53, 935–944, 2008.

Winter, T. C.: Relation of streams, lakes, and wetlands to groundwater flow systems, Hydrogeol. J., 7, 28–45, 1999.

Yehdegho, B., Rozanski, K., Zojer, H., and Stichler, W.: Interaction of dredging lakes with the adjacentgroundwater field: An isotope study, J. Hydrol., 192, 247–270, https://doi.org/10.1016/S0022-1694(96)03102-2, 1997.

12

On the use of the GRACE normal equation of inter-satellite tracking data for estimation of soil moisture and groundwater in Australia

Natthachet Tangdamrongsub[1], Shin-Chan Han[1], Mark Decker[2], In-Young Yeo[1], and Hyungjun Kim[3]

[1]School of Engineering, University of Newcastle, Callaghan, New South Wales, Australia
[2]ARC Centre of Excellence for Climate System Science, University of New South Wales, Sydney, New South Wales, Australia
[3]Institute of Industrial Science, the University of Tokyo, Tokyo, Japan

Correspondence: Natthachet Tangdamrongsub (natthachet.tangdamrongsub@newcastle.edu.au)

Abstract. An accurate estimation of soil moisture and groundwater is essential for monitoring the availability of water supply in domestic and agricultural sectors. In order to improve the water storage estimates, previous studies assimilated terrestrial water storage variation (ΔTWS) derived from the Gravity Recovery and Climate Experiment (GRACE) into land surface models (LSMs). However, the GRACE-derived ΔTWS was generally computed from the high-level products (e.g. time-variable gravity fields, i.e. level 2, and land grid from the level 3 product). The gridded data products are subjected to several drawbacks such as signal attenuation and/or distortion caused by a posteriori filters and a lack of error covariance information. The post-processing of GRACE data might lead to the undesired alteration of the signal and its statistical property. This study uses the GRACE least-squares normal equation data to exploit the GRACE information rigorously and negate these limitations. Our approach combines GRACE's least-squares normal equation (obtained from ITSG-Grace2016 product) with the results from the Community Atmosphere Biosphere Land Exchange (CABLE) model to improve soil moisture and groundwater estimates. This study demonstrates, for the first time, an importance of using the GRACE raw data. The GRACE-combined (GC) approach is developed for optimal least-squares combination and the approach is applied to estimate the soil moisture and groundwater over 10 Australian river basins. The results are validated against the satellite soil moisture observation and the in situ groundwater data. Comparing to CABLE, we demonstrate the GC approach delivers evident improvement of water storage estimates, consistently from all basins, yielding better agreement on seasonal and inter-annual timescales. Significant improvement is found in groundwater storage while marginal improvement is observed in surface soil moisture estimates.

1 Introduction

The changes in terrestrial water storage (ΔTWS) derived from the Gravity Recovery And Climate Experiment (GRACE) data products have been used in the last decade to study global water resources, including groundwater depletion in India and the Middle East (Rodell et al., 2009; Voss et al., 2013), water storage accumulation in Canada (Lambert et al., 2013), and flood-influenced water storage fluctuation in Cambodia (Tangdamrongsub et al., 2016). The gravity data obtained from GRACE satellites are commonly processed and released in three different product levels (L) that increase in the amount of processing, L1B – satellite tracking data (e.g. Wu et al., 2006), L2 – global gravitational Stokes coefficients (e.g. Bettadpur, 2012), and L3 – global grids (e.g. Landerer and Swenson, 2012). The original (L1B) GRACE information is inevitably altered due to data processing and successive post-processing filterings, because the error covariance information is not propagated through each post-processing step.

The GRACE-derived ΔTWS has been computed widely from the higher-level products (e.g. L2 and L3) on which various ad hoc post-processing filters were applied (e.g. Gaussian smoothing filter, Jekeli, 1981; destripe filter, Swenson and Wahr, 2006). ΔTWS obtained from these filters lacks proper error covariance information and is attenuated and distorted. To overcome the signal attenuation in GRACE high-level products, empirical approaches have been developed, including the application of scale factors computed from land surface models (LSMs; Landerer and Swenson, 2012) to the GRACE L3 products. GRACE uncertainty in the high-level product is usually unknown or assumed. For example, Zaitchik et al. (2008) derived empirically a global average uncertainty that is variable depending on choices of post-processing filters (Sakumura et al., 2014). Furthermore, GRACE error and sensitivity is dependent on latitudes due to the orbit convergence toward poles (Wahr et al., 2006) and any post-processing filters will alter the GRACE data and their error information. Rigorous statistical error information is of equal importance to derivation of ΔTWS for data assimilation and model calibration (Tangdamrongsub et al., 2017; Schumacher et al., 2016, 2018). ΔTWS and its uncertainty estimates should be formulated directly from L1B data considering the complete statistical information.

The GRACE information is not fully exploited in many studies. For example, groundwater storage variation (ΔGWS) is often computed by subtracting the soil moisture variation (ΔSM) component simulated by the land surface model from GRACE-derived ΔTWS data (Rodell et al., 2009; Famiglietti et al., 2011), assuming the model ΔSM is error-free. This may result in the inaccurate ΔGWS and the associated error estimate as the uncertainties in observation and of the land surface model outputs are neglected in the combination (or regression) of two noisy data (e.g. Long et al., 2016). In data assimilation, the GRACE uncertainty is often derived empirically, not necessarily reflecting the actual GRACE error characteristics (e.g. Zaitchik et al., 2008; Tangdamrongsub et al., 2015; Tian et al., 2017). For example, Girotto et al. (2016) used the L3 product and showed that it was necessary to adjust GRACE observation and its uncertainty in order to make their water storage estimates more accurate. Similarly, Tian et al. (2017) reported the need for the application of a scale factor to GRACE uncertainty (from mascon, mass concentration, product) in their GRACE assimilation process. It is apparent that the use of post-processed GRACE products often requires data tuning, leading possibly to an integration of the altered gravity information into the data assimilation system. Some recent studies began to employ the full variance–covariance information in the data assimilation scheme to enhance the quality of the estimates (Eicker et al., 2014; Schumacher et al., 2016; Tangdamrongsub et al., 2017; Khaki et al., 2017a, b).

This study aims to use the GRACE information of ΔTWS measurement directly from the least-squares normal equation data. The approach optimally combines GRACE's normal equations with the model simulation results from the Community Atmosphere Biosphere Land Exchange (CABLE, Decker, 2015) to improve ΔSM and ΔGWS estimates. The proposed approach presents three main advantages. Firstly,

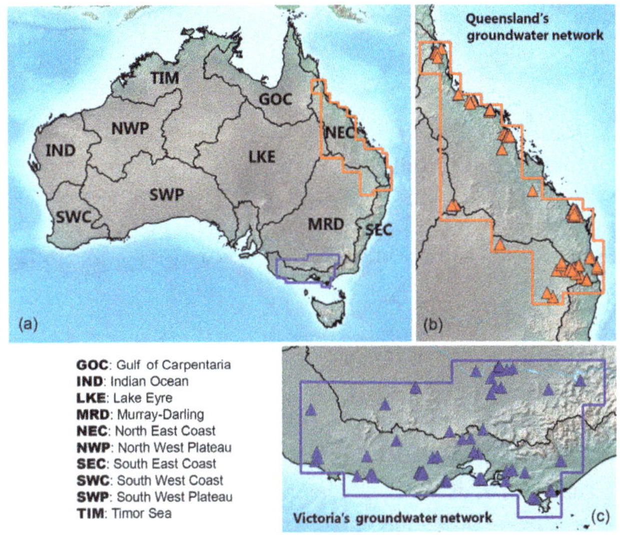

Figure 1. **(a)** Geographical location of 10 Australian river basins. Red and blue polygons indicate the boundaries of the groundwater networks in Queensland **(b)** and Victoria **(c)**, respectively. Triangles (in **b** and **c**) represent the selected bore locations used in this study.

one can exploit the full GRACE signal and error information by using the normal equation data sets. Secondly, the approach is developed for optimal least-squares combination (e.g. Ramillien et al., 2004), which maximizes the model and observation strength while simultaneously suppressing their weaknesses. Finally, the method bypasses empirical, multiple-step post-processing filters.

The main objective of this study is to present the GRACE-combined (GC) approach to improve the model-estimated ΔSM and ΔGWS on regional scales. We demonstrate our approach applied to 10 Australian river basins (Fig. 1a). One advantage of the study area is that the state vector can be simply defined by ΔSM and ΔGWS as other hydrological components (e.g. snow, glacier) are negligible. We validate the top layer of ΔSM estimates against the satellite soil moisture observation (the Advanced Microwave Scanning Radiometer aboard the Earth Observing System (EOS) – AMSR-E, Njoku et al., 2003) over all 10 basins and the ΔGWS estimates against the in situ groundwater data available over Queensland and Victoria (Fig. 1b and c).

This paper is outlined as follows: firstly, the derivation of the GC approach is presented in Sect. 2 while the description of GRACE data processing, including the use of the GRACE normal equation, is given in Sect. 3. Secondly, the CABLE modelling is outlined in Sect. 4. This includes the derivation of model uncertainty based on the quality of precipitation data and the model parameter inputs. The processing of validation data is also described in Sect. 4. Thirdly, Sect. 5 presents the result of ΔSM and ΔGWS estimates and comparison to in situ data. The long-term trends in the Australian mass variation over the last 13 years are also investigated in this section.

2 A method of combining GRACE L1B data with land surface model outputs

The statistical information of ΔTWS computed from a land surface model can be written as

$$\widetilde{\boldsymbol{h}} = \boldsymbol{h} + \boldsymbol{\epsilon}; \ \boldsymbol{\epsilon} \sim N(\mathbf{0}, \mathbf{C}), \tag{1}$$

where $\boldsymbol{h}$ is the "truth" (unknown) model state vector while $\widetilde{\boldsymbol{h}}$ is the calculated state vector characterized with the model error $\boldsymbol{\epsilon}$. The model error is assumed to have zero mean and covariance $\mathbf{C}$.

The term $\boldsymbol{h}$ is used to represent a vector including the global ΔTWS grid, and terms with a subscript "R" (e.g. $\boldsymbol{h}_{\mathrm{R}}$, $\mathbf{C}_{\mathrm{R}}$) are used to represent only a regional set of ΔTWS (for example, in Australia). As such, the observation equation over a region can be rewritten as

$$\widetilde{\boldsymbol{h}}_{\mathrm{R}} = \boldsymbol{h}_{\mathrm{R}} + \boldsymbol{\epsilon}; \ \boldsymbol{\epsilon} \sim N(\mathbf{0}, \mathbf{C}_{\mathrm{R}}). \tag{2}$$

As soil moisture and groundwater are the major components of ΔTWS in Australia (surface water storage being insignificant), the vector $\boldsymbol{h}_{\mathrm{R}}$ can be defined as

$$\boldsymbol{h}_{\mathrm{R}} = \begin{bmatrix} \Delta \boldsymbol{SM}_{\mathrm{top}} & \Delta \boldsymbol{SM}_{rz} & \Delta \boldsymbol{GWS} \end{bmatrix}^T, \tag{3}$$

where $\Delta \boldsymbol{SM}_{\mathrm{top}}$, $\Delta \boldsymbol{SM}_{rz}$, and $\Delta \boldsymbol{GWS}$ represent the vectors of top (surface) soil moisture, root zone soil moisture, and groundwater storage variations, respectively.

A least-squares normal equation of GRACE can be written as

$$\mathbf{N}\boldsymbol{x} = \boldsymbol{c}, \tag{4}$$

where $\mathbf{N}$ is a normal matrix, $\boldsymbol{x}$ contains the spherical harmonic coefficients (SHC) of the geopotential, and $\boldsymbol{c}$ is the normal vector. In this study, $\mathbf{N}$ and $\boldsymbol{c}$ can be obtained from the ITSG-Grace2016 products (Mayer-Gürr et al., 2016; https://www.tugraz.at/institute/ifg/downloads/gravity-field-models/itsg-grace2016, see more details in Sect. 3.1). Equation (4) can be written in terms of $\boldsymbol{h}$ as follows (see Appendix A for the derivation):

$$\left(\mathbf{H}^T\mathbf{Y}^T\mathbf{N}\mathbf{Y}\mathbf{H}\right)\hat{\boldsymbol{h}} = \mathbf{H}^T\mathbf{Y}^T\boldsymbol{c}, \tag{5}$$

where $\mathbf{Y}$ converts ΔTWS to geopotential coefficients considering the load Love numbers (e.g. Wahr et al., 1998) and $\mathbf{H}$ is the operational matrix converting $\Delta \boldsymbol{SM}_{\mathrm{top}}$, $\Delta \boldsymbol{SM}_{rz}$, and $\Delta \boldsymbol{GWS}$ to ΔTWS. Equation (5) is based on the assumption that the GRACE orbital perturbation is a result of ΔTWS variation on the surface. If M is the number of model grid cells, N_{max} is the maximum degree of the geopotential coefficients, and $L = (N_{\mathrm{max}} + 1)^2 - 4$ is the number of geopotential coefficients from GRACE; the dimension of $\mathbf{Y}$, $\mathbf{H}$, and $\boldsymbol{h}$ are $L \times M$, $M \times 3M$, and $3M \times 1$, respectively. Note that Eq. (5) is defined with the global grid of $\boldsymbol{h}$. For a regional application, Eq. (5) can be modified as

$$\left[\mathbf{H}_{\mathrm{R}}^T\mathbf{Y}_{\mathrm{R}}^T | \mathbf{H}_{\mathrm{o}}^T\mathbf{Y}_{\mathrm{o}}^T\right]\mathbf{N}\begin{bmatrix} \mathbf{Y}_{\mathrm{R}}\mathbf{H}_{\mathrm{R}} \\ \mathbf{Y}_{\mathrm{o}}\mathbf{H}_{\mathrm{o}} \end{bmatrix}\begin{bmatrix} \hat{\boldsymbol{h}}_{\mathrm{R}} \\ \hat{\boldsymbol{h}}_{\mathrm{o}} \end{bmatrix} = \left[\mathbf{H}_{\mathrm{R}}^T\mathbf{Y}_{\mathrm{R}}^T | \mathbf{H}_{\mathrm{o}}^T\mathbf{Y}_{\mathrm{o}}^T\right]\boldsymbol{c}, \tag{6}$$

where the subscript "R" indicates the grid ΔTWS only in a region of interest and "o" for the rest of the globe. The number of the model grid cells associated with R is J, and the number of grid cells outside cells is $M - J$. As such, the dimensions of $\mathbf{Y}_{\mathrm{R}}$, $\mathbf{H}_{\mathrm{R}}$, $\hat{\boldsymbol{h}}_{\mathrm{R}}$, $\mathbf{Y}_{\mathrm{o}}$, $\mathbf{H}_{\mathrm{o}}$, $\hat{\boldsymbol{h}}_{\mathrm{o}}$ are $L \times J$, $J \times 3J$, $3J \times 1$, $L \times (M - J)$, $(M - J) \times 3(M - J)$, $3(M - J) \times 1$, respectively. The dimension of $\mathbf{N}$ and $\boldsymbol{c}$ remain unchanged, since they are essentially from the normal equations of the original GRACE L1B data (to be discussed in the following section).

From Eq. (6), the normal equations associated with ΔTWS in the region of interest can then be written as

$$\mathbf{H}_{\mathrm{R}}^T\mathbf{Y}_{\mathrm{R}}^T\mathbf{N}\mathbf{Y}_{\mathrm{R}}\mathbf{H}_{\mathrm{R}}\hat{\boldsymbol{h}}_{\mathrm{R}} = \mathbf{H}_{\mathrm{R}}^T\mathbf{Y}_{\mathrm{R}}^T\boldsymbol{c} - \mathbf{H}_{\mathrm{R}}^T\mathbf{Y}_{\mathrm{R}}^T\mathbf{N}\mathbf{Y}_{\mathrm{o}}\mathbf{H}_{\mathrm{o}}\hat{\boldsymbol{h}}_{\mathrm{o}} \tag{7}$$

or

$$\mathbf{N}_{\mathrm{R}}\hat{\boldsymbol{h}}_{\mathrm{R}} = \boldsymbol{c}_{\mathrm{R}}, \tag{8}$$

where $\mathbf{N}_{\mathrm{R}} = \mathbf{H}_{\mathrm{R}}^T\mathbf{Y}_{\mathrm{R}}^T\mathbf{N}\mathbf{Y}_{\mathrm{R}}\mathbf{H}_{\mathrm{R}}$ and $\boldsymbol{c}_{\mathrm{R}} = \mathbf{H}_{\mathrm{R}}^T\mathbf{Y}_{\mathrm{R}}^T\boldsymbol{c} - \mathbf{H}_{\mathrm{R}}^T\mathbf{Y}_{\mathrm{R}}^T\mathbf{N}\mathbf{Y}_{\mathrm{o}}\mathbf{H}_{\mathrm{o}}\hat{\boldsymbol{h}}_{\mathrm{o}}$. As seen, Eq. (7) is the regional representation of Eq. (5) where only the grid cells inside the study region are used, while the contribution from the grid cells outside the region needs to be removed or corrected. Combining the normal equation of Eqs. (2) and (8), the optimal combined solution of $\hat{h}_{\mathrm{R}}$ can be resolved as follows:

$$\hat{\boldsymbol{h}}_{\mathrm{R}} = \left(\mathbf{C}_{\mathrm{R}}^{-1} + \mathbf{N}_{\mathrm{R}}\right)^{-1}\left(\mathbf{C}_{\mathrm{R}}^{-1}\widetilde{\boldsymbol{h}}_{\mathrm{R}} + \boldsymbol{c}_{\mathrm{R}}.\right). \tag{9}$$

The computation of model covariance matrix $\mathbf{C}_{\mathrm{R}}$ will be discussed in Sect. 4.2. The a posteriori covariance of $\hat{\boldsymbol{h}}_{\mathrm{R}}$ can be estimated as follows:

$$\hat{\mathbf{\Sigma}} = \left(\mathbf{C}_{\mathrm{R}}^{-1} + \mathbf{N}_{\mathrm{R}}\right)^{-1}, \tag{10}$$

and the uncertainty estimate of $\hat{h}_{\mathrm{R}}$ is simply calculated as

$$\sigma_{\hat{h}} = \sqrt{\mathrm{diag}(\hat{\mathbf{\Sigma}})}, \tag{11}$$

where diag( ) represents the diagonal element of the given matrix.

3 GRACE data

3.1 GRACE least-squares normal equations

In this study, the least-squares normal equations are obtained from the ITSG-Grace2016 products between January 2003 and March 2016. All L1B data including K-band range (KBR) inter-satellite tracking data, attitude, accelerometer, GPS-based kinematic orbit data, and AOD1B corrections are reduced in terms of the normal equations. These data products are usually used to compute the Earth's geopotential field to the maximum harmonic degree and order of 90, or at a spatial resolution of ~ 220 km. The products contain the information of the normal matrix $\mathbf{N}$ and the vector $\boldsymbol{c}$ (as shown in Eq. 4) as well as the a priori time-varying gravity field coefficients predicted with the GOCO05s solution (Mayer-Gürr et al., 2015). Note that the solution of the ITSG-Grace2016 normal equation is the anomalous geopotential coefficient vector ($\Delta \boldsymbol{x}$), which is referenced to the a priori time-varying gravity field ($\boldsymbol{x}_0$), through:

$$\mathbf{N}\Delta \boldsymbol{x} = \boldsymbol{d}, \tag{12}$$

where $\boldsymbol{d}$ and $\boldsymbol{x}_0$ are given. To obtain a complete gravity field variation between the study period ($\boldsymbol{x}$ term in Eq. 4), the a priori time-varying gravity field, $\boldsymbol{x}_0$, is firstly restored to Eq. (12), and the mean gravity field ($\overline{\boldsymbol{x}}_0$) computed from all $\boldsymbol{x}_0$ between January 2003 and March 2016 is then removed as follows:

$$\mathbf{N}(\Delta \boldsymbol{x} + \boldsymbol{x}_0 - \overline{\boldsymbol{x}}_0) = \boldsymbol{d} + \mathbf{N}(\boldsymbol{x}_0 - \overline{\boldsymbol{x}}_0), \tag{13}$$

$$\mathbf{N}\boldsymbol{x} = \boldsymbol{d} + \mathbf{N}(\boldsymbol{x}_0 - \overline{\boldsymbol{x}}_0). \tag{14}$$

Therefore, in Sect. 2 (e.g. Eq. 5), the matrix $\mathbf{N}$ remains unchanged while the vector $\boldsymbol{c}$ can be simply replaced by $\boldsymbol{c} = \boldsymbol{d} + \mathbf{N}(\boldsymbol{x}_0 - \overline{\boldsymbol{x}}_0)$.

3.2 GRACE-derived ΔTWS products

Three monthly GRACE-derived ΔTWS products are also used: the ITSG-Grace2016 DDK5 solution (ITSG-DDK5 for short, http://icgem.gfz-potsdam.de/series/99_non-iso/ITSG-Grace2016), the CNES/GRGS Release 3 (RL3) (GRGS for short, Lemoine et al., 2015; http://grgs.obs-mip.fr/grace/variable-models-grace-lageos/grace-solutions-release-03), and the JPL RL05M mascon-CRI version 2 product (mascon for short, Watkins et al., 2015; Wiese et al., 2016; http://grace.jpl.nasa.gov/data/get-data/jpl_global_mascons). The ITSG-DDK5 product is the post-processed version of the ITSG L2 solution where the non-isotropic filter DDK5 (Kusche et al., 2009) is applied. The DDK5 solution is empirically selected here to be a good balance between the over-smoothed (e.g. DDK1) and noisy (e.g. DDK8) solutions. The GRGS solution provides ΔTWS at $1° \times 1°$ globally, derived from the Earth's geopotential coefficients up to the maximum degree and order 80, and no filter nor scale factor is applied (L2 data product). Mascon provides ΔTWS at equal-area 3° spherical cap grid globally. In contrast to the ITSG-DDK5 and GRGS solutions, the mascon uses a gain factor derived from the land surface model to restore mitigated signals and reduce leakage errors (L3 data products) (Watkins et al., 2015; Wiese et al., 2016). Additionally, mascon provides the ΔTWS uncertainty together with the solution. The uncertainty is computed based on several geophysical models (see Watkins et al., 2015, and Wiese et al., 2016, for more details). The uncertainty information is not available in the ITSG-DDK5 or GRGS product.

The GRACE data are obtained between January 2003 and March 2016. After retrieval, the long-term mean value between January 2003 and March 2016 is computed and subtracted from the monthly products. To be consistent with CABLE grid spacing (see Sect. 4), the ΔTWS is computed using 0.5° spatial resolution. The coarse-scale datasets (e.g. mascon, GRGS) are resampled to $0.5° \times 0.5°$ using the nearest grid values.

In this study, the independent GRACE solutions are used for two main reasons:

1. To obtain the ΔTWS values outside Australia. As shown in Eq. (7), the $\hat{\boldsymbol{h}}_o$ vector needs to be known, which can be from the GRACE-derived ΔTWS solution. We use the GRGS solutions as the GRGS solution is not subject to the filter choice and it provides ΔTWS at a spatial resolution comparable to the normal equation data.
2. To compare with the ΔTWS estimates from our approaches. All solutions are used to compare and validate our ΔTWS estimates.

4 Hydrology model and validation data

4.1 Model setup

The extensive description of the CABLE model is given in Decker (2015) and Ukkola et al. (2016). This section describes the model setup and specific changes applied to this study. CABLE is a public available land surface model and can be used to estimate soil moisture and groundwater in terms of volumetric water content every 3 h at a $0.5° \times 0.5°$ spatial resolution. The soil moisture and groundwater storage can be simply computed by multiplying the estimates with thicknesses of various layers. For soil moisture, the thickness of six soil layers is 0.022, 0.058, 0.154, 0.409, 1.085, and 2.872 m, from top to bottom, respectively. The thickness of the groundwater layer is modelled to be 20 m uniformly. Recalling Eq. (3), $\Delta \boldsymbol{SM}_{\mathrm{top}}$ is defined as the soil moisture storage variation at the top 0.022 m thick layer, while $\Delta \boldsymbol{SM}_{rz}$ is the variation accumulated over the second to the bottom soil layers (depth between 0.022 and 4.6 m).

CABLE is initially forced with the data from the Global Soil Wetness Project Phase 3 (GSWP3), which is currently available until December 2010 (http://hydro.iis.u-tokyo.ac.jp/GSWP3, https://doi.org/10.20783/dias.501). We replace GSWP3 forcing data with GLDAS data (Rodell et al., 2004)

Table 1. Precipitation data from seven different products used in this study, the Global Soil Wetness Project Phase 3 (GSWP3), the Global Land Data Assimilation System (GLDAS), the Tropical Rainfall Measuring Mission (TRMM), the Modern-Era Retrospective Analysis for Research and Applications (MERRA), the European Centre for Medium-Range Weather Forecasts (ECMWF), Princeton's Global Meteorological Forcing Dataset (Princeton), and the Precipitation Estimation from Remotely Sensed Information using Artificial Neural Networks (PERSIANN). The temporal resolution of all products is 3 h. Most products are available to present while GSWP3, MERRA, and Princeton terminate earlier.

Product	Availability	Spatial resolution	References
GSWP3	Jan 1901–Dec 2010	$0.5° \times 0.5°$	http://dx.doi.org/10.20783/DIAS.501
GLDAS (NOAH025SUBP 3H)	Mar 2000–present	$0.25° \times 0.25°$	Rodell et al. (2004)
TRMM (3B42)	Jan 1998–present	$0.25° \times 0.25°$	Huffman et al. (2007)
MERRA (MSTMNXMLD.5.2.0)	Jan 1980–Feb 2016	$0.5° \times 0.67°$	Rienecker et al. (2011)
ECMWF (ERA-Interim)	Jan 1979–present	$0.75° \times 0.75°$	Dee et al. (2011)
Princeton (V2 0.5°)	Jan 1987–Dec 2012	$0.5° \times 0.5°$	Sheffield et al. (2005)
PERSIANN (3 h)	Mar 2002–present	$0.25° \times 0.25°$	Sorooshian et al. (2000)

Table 2. Model parameters that are sensitive to SM and GWS estimates. The following parameters were perturbed using the additive noise with the boundary conditions given in the last column. The further parameter description can be found in Decker (2015) and Ukkola et al. (2016).

Parameter	Name	Spatial variability	Perturbed range
f_{clay}, f_{sand}, f_{silt}	Fraction of clay, sand, and silt	Yes	0–1
f_{sat}	Fraction of grid cell that is saturated	No	810–990
q_{sub}	Maximum rate of subsurface drainage assuming a fully saturated soil column	No	0.009–0.01
f_p	Tuneable parameter controlling drainage speed	No	1.9–2.2

to compute the water storage changes to 2016. The forcing data used in CABLE are precipitation, air temperature, snowfall rate, wind speed, humidity, surface pressure, and shortwave and long-wave downward radiations. To investigate the impact of different forcing data, the offline sensitivity study is conducted by comparing the water storage estimates computed using

1. all eight forcing data components of GSWP3 and
2. GSWP3 data with the replacement of one component obtained from GLDAS forcing data.

It is found that the water storage estimate is most sensitive to the replacement of precipitation data, as expected, and relatively less sensitive to the change in other forcing components. We use the GLDAS forcing data in this study and also further test seven different precipitation data products (see more details in Sect. 4.2). The forcing data are up- or down-sampled to a $0.5° \times 0.5°$ spatial grid to reconcile with the CABLE spatial resolution.

4.2 Model uncertainty

In this study, the CABLE uncertainty is derived from 210 ensemble estimates associated with different forcing data and model parameters. The seven different precipitation products (see Table 1) are used to run the model independently. Most products are available to present day while GSWP3, Princeton, and MERRA are only available until December 2010, December 2012, and February 2016, respectively. For each precipitation forcing, 30 ensembles are generated by perturbing the model parameters within ±10 % of the nominal values. The perturbed size of 10 % is similar to Dumedah and Walker (2014). Based on the CABLE structure, the $\Delta \boldsymbol{SM}$ and $\Delta \boldsymbol{GWS}$ estimates are most sensitive to the model parameters listed in Table 2. For example, the fractions of clay, sand, and silt (f_{clay}, f_{sand}, f_{silt}) are used to compute soil parameters including field capacity, hydraulic conductivity, and soil saturation which mainly affect soil moisture storage. Similarly, the drainage parameters (e.g. q_{sub}, f_p) control the amount of subsurface runoff, which has a direct impact on root zone soil moisture and groundwater storages.

From ensemble generations, total $K = 210$ sets of the ensemble water storage estimates ($\boldsymbol{h}_e$) are obtained:

$$\mathcal{H}_R = \left[\ \boldsymbol{h}_e|_{k=1} \quad \boldsymbol{h}_e|_{k=2} \quad \boldsymbol{h}_e|_{k=3} \quad \cdots \quad \boldsymbol{h}_e|_{k=K} \right], \tag{15}$$

and the mean value of $\mathcal{H}_R$ is computed as follows:

$$\widetilde{\boldsymbol{h}}_R = \frac{1}{K}\sum_{k=1}^{K} \boldsymbol{h}_e|_k. \tag{16}$$

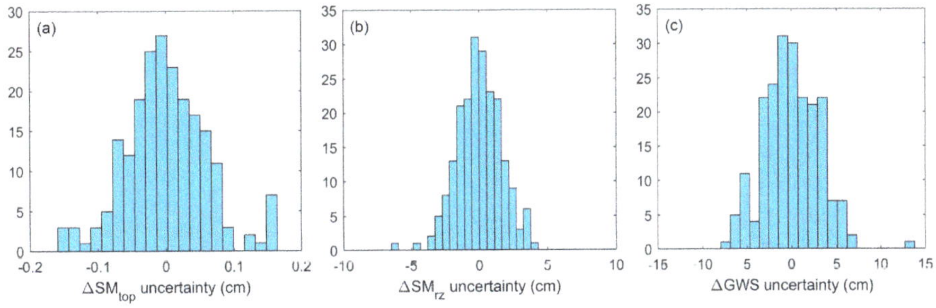

Figure 2. Histograms of the model errors computed from 210 ensemble members ($\mathcal{H}'_{\mathrm{R}}$) without the mean. The basin averaged values (from all 10 Australian basins) of January 2003, for example, are shown.

Note that due to the absence of GSWP3, Princeton, and MERRA data, the number of ensembles reduces to $K = 180$ after December 2010, $K = 150$ after December 2012, and $K = 120$ after February 2016, respectively. The GC approach assumes that model errors are normally distributed with zero mean. Any violation of this assumption will yield a bias in the combined solutions. Therefore, the mean value is removed from each ensemble member, $\mathcal{H}'_{\mathrm{R}} = \mathcal{H}_{\mathrm{R}} - \widetilde{\boldsymbol{h}}_{\mathrm{R}}$, and the error covariance matrix of the model is empirically computed as

$$\mathbf{C}_{\mathrm{R}} = \mathcal{H}'_{\mathrm{R}}(\mathcal{H}'_{\mathrm{R}})^T/(K-1). \tag{17}$$

The $\widetilde{\boldsymbol{h}}_{\mathrm{R}}$ (Eq. 16) and $\mathbf{C}_{\mathrm{R}}$ (Eq. 17) terms can be directly used in Eq. (9). The distribution of model errors is demonstrated in Fig. 2. The figure illustrates the histogram of model errors ($\mathcal{H}'_{\mathrm{R}}$) computed using 210 ensemble members of the model-estimated $\Delta \boldsymbol{SM}$ and $\Delta \boldsymbol{GWS}$ in January 2003. The histogram indicates that the model error may be approximately described by a normal distribution as introduced in Eq. (1).

Furthermore, in practice, the sampling error caused by finite sample size might lead to spurious correlations in the model covariance matrix (Hamill et al., 2001). The effect can be reduced by applying an exponential decay with a particular spatial correlation length to $\mathbf{C}_{\mathrm{R}}$. In this study, the correlation length is determined based on the empirical covariance of model-estimated ΔTWS. The covariance function of ΔTWS is firstly assumed isotropic, and it is computed empirically based on the method given in Tscherning and Rapp (1974). The distance where the maximum value of the variance decreases to half is defined as the correlation length. The obtained values vary month to month, and the mean value of 250 km is used in this study.

It is emphasized that the model omission error caused by imperfect modelling of hydrological processes within the LSM is not taken into account in the above description. The omission error may increase the model covariance and introduce a bias as well. We account for the omission error by increasing 20 % of the model covariance. (i.e. multiplying $\mathbf{C}_{\mathrm{R}}$ by 1.2). We determine such omission error based on trial and error such that it increases the model error (due to the omission error) but does not exceed the model error value reported by Dumedah and Walker (2014). We acknowledge that this is only a simple practical way of accounting for the omission error into the total model error.

4.3 Validation data

4.3.1 Satellite soil moisture observation

The satellite-observed surface soil moisture data are obtained from the Advanced Microwave Scanning Radiometer aboard the Earth Observing System using the Land Parameter Retrieval Model (Njoku et al., 2003). The observation is used to validate our estimates of top soil moisture changes ($\Delta\mathrm{SM}_{\mathrm{top}}$). The AMSR-E product provides volumetric water content in the top layer derived from a passive microwave data (from NASA EOS Aqua satellite) and forward radiative transfer model. In this study, the level 3 product, available daily between June 2002 and June 2011 at $0.25^\circ \times 0.25^\circ$ spatial resolution, is used (Owe et al., 2008). The measurements from ascending and descending overpasses are averaged for each frequency band (C and X). Then, the monthly mean value is computed by averaging the daily data within a month. To obtain the variation in the surface soil moisture, the long-term mean between June 2002 and June 2011 is removed from the monthly data. Regarding the different depth measured in CABLE and AMSR-E, the cumulative density function (CDF)-matching technique (Reichle and Koster, 2004) is used to reduce the bias between the top soil moisture model and the observation. The CDF is built using the 2003–2004 data, and it is used for the entire period. There are no satellite-observed or ground-measured root zone soil moisture data for meaningful comparison with our results, particularly on a continental scale. Validation of $\Delta\mathrm{SM}_{rz}$ on regional and continental scales is currently unachievable due to a complete lack of observations on this spatial scale.

4.3.2 In situ groundwater

The in situ groundwater levels from bore measurements are obtained from two different ground observation networks (see Fig. 1). The data in Queensland are obtained from the Department of Natural Resources and Mines (DNRM) while the data in Victoria are from the Department of Environment and Primary Industries (DWPI). More than 10 000 measurements are available from each network, but the data gap and outliers are present. Therefore, the bore measurement is firstly filtered by removing the sites that present no data or data gap longer than 30 months during the study period.

To obtain the monthly mean value, the hourly or daily data are averaged in a particular month. The outliers are detected and fixed using the Hampel filter (Pearson, 2005) where the remaining data gaps are filled using the cubic spline interpolation. To obtain the groundwater level variation, the long-term mean groundwater level computed between the study period is removed from the monthly values. The groundwater level variation (ΔL) is then converted to ΔGWS using $\Delta\mathrm{GWS} = S_y \cdot \Delta L$, where S_y is specific yield. Based on Chen et al. (2016), $S_y = 0.1$ is used for the Victoria network. Specific yields of Queensland's network have been found ranging from 0.045 (Rassam et al., 2013) to 0.06 (Welsh, 2008), and an averaged $S_y = 0.05$ is used in this study. Finally, the mean value computed from all data (in each network) is used to represent the in situ data of the network.

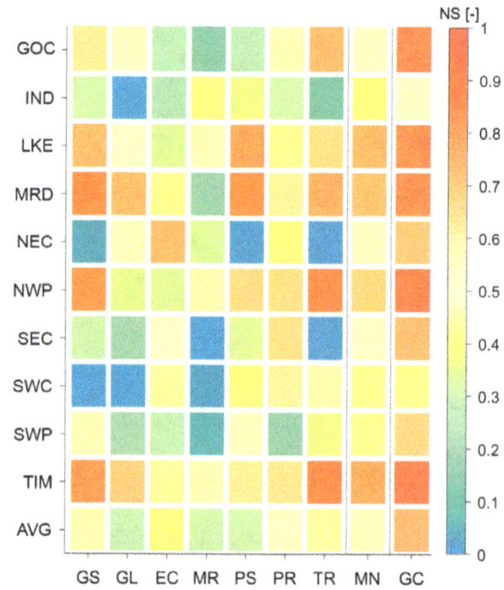

Figure 3. NS coefficients between the model and GRACE mascon ΔTWS over 10 Australian basins (in ordinate). The NS values were computed based on CABLE ΔTWS computed with seven different precipitation data (in abscissa), including GSWP3 (GS), GLDAS (GL), ECMWF (EC), MERRA (MR), PERSIANN (PR), TRMM (TR). The NS value of the mean ΔTWS estimates (the average of seven variants) is also shown (MN). The area-weighted average NS value over all basins is also shown (AVG). The NS value of ΔTWS from the GRACE-combined (GC) approach is shown in the last column. The full name of the basins can be found in Fig. 1.

5 Results

5.1 Model-only performance

We study the model ΔTWS changes under different meteorological forcing and land parameterization. A total of 210 estimates of monthly TWS (sum of $\mathrm{SM_{top}}$, SM_{rz}, and GWS) are obtained between January 2003 and March 2016 from the ensemble run based on seven different precipitation inputs. Then, the averaged values of the TWS estimates are computed from the 30 precipitation-associated ensemble members. This results in seven sets of monthly mean TWS estimates from seven different precipitation data. For each set, the monthly ΔTWS is computed by removing the long-term mean computed between January 2003 and March 2016.

The precipitation-based ΔTWS values are then compared with the GRACE mascon solution (see Sect. 3.2) over 10 different Australian basins. The comparison is carried out between January 2003 and March 2016. Due to the availability of the data, the periods used are shorter in cases of GSWP3, Princeton, and MERRA precipitation (see Table 1). The metric used to evaluate a goodness of fit between the CABLE run and GRACE mascon estimates is the Nash–Sutcliffe (NS) coefficient (see Eq. B1) (Fig. 3).

Figure 3 demonstrates CABLE ΔTWS varies noticeably by precipitation as well as locations. The area-weighted average values (see Eq. B2) computed from Princeton, GSWP3, and TRMM yield the model ΔTWS, reasonably agreeing with GRACE by giving an NS coefficient greater than 0.45, while MERRA, PERSIANN, and GLDAS show NS $= \sim 0.3$. The lower agreement is mainly due to the quality of rainfall estimates over Australia. The NS of ECMWF is around 0.4.

All model ensembles are consistent with the GRACE data over the Timor Sea and inner parts of Australia (e.g. Lake Eyre, LKE; Murray–Darling, MRD; North West Plateau, NWP) where the NS value can reach as high as 0.9 (see e.g. TRMM over TIM, Timor Sea). On the contrary, the lower agreement is found mostly over the coastal basins. Very small or even negative NS values indicate the misfit between the CABLE and GRACE mascon solutions, and they are observed over the Indian Ocean (see GLDAS), North East Coast (NEC; see GSWP3, PERSIANN, TRMM), South East Coast (see MERRA, TRMM), South West Coast (SWC; see GSWP3, GLDAS, MERRA), and South West Plateau (SWP; see MERRA).

By averaging all ΔTWS estimates from seven different precipitation datasets, the mean-ensemble estimate (MN) delivers the best agreement with GRACE as seen by the highest average NS value (MN of AVG $= 0.55$) among all ensembles. Particularly, NS values are greater than 0.4 in all basins and no negative NS values are presented in MN. On average, it can be clearly seen that using the mean value (MN) is a viable option to increase the overall performance of the ΔTWS estimates. Therefore, only the CABLE MN result will be used in further analyses. The comparison with the GRGS GRACE solution was also evaluated (not shown here) and the overall results are similar to Fig. 3.

5.2 Impact of GRACE on storage estimates

5.2.1 Contribution of GRACE

This section investigates the impact of the GC approach on the estimates of various water storage components. The ΔTWS estimate obtained from the GC approach is demonstrated in Sect. 5.1, by comparing with the independent GRACE mascon solution. Figure 3 shows the GC result yields the highest NS values in all basins, outperforming all other CABLE runs. In the average (AVG), the NS value increases by $\sim 35\,\%$ (0.55 to 0.74) from the MN case. Similar behaviour is also seen when compared with the GRGS GRACE solution (not shown); the average NS value increases from 0.50 to 0.74. This is not surprising as the GC approach uses the fundamental GRACE tracking data as GRACE mascon and GRGS solutions do. Improvement of the NS coefficient indicates merely the successfulness of integrating GRACE data and the model estimates.

Figures 4 and 5 show the GC results of ΔTWS as well as ΔSM_{top}, ΔSM_{rz}, and ΔGWS in different basins. The monthly time series and the deseasonalized time series are shown. In general, GRACE tends to increase ΔTWS when the model ΔTWS (MN) is predicted to be underestimated (see e.g. LKE, MRD, NWP, SWP, TIM between 2011 and 2012) and decrease ΔTWS when it is determined to be overestimated (see all basins between 2008 and 2010). A clear example is seen over the Gulf of Carpentaria (GOC, Fig. 4d), where CABLE overestimates ΔTWS and produces phase delay between 2008 and 2010. The overestimated amplitude and phase delay seen in CABLE ΔGWS during this above period (Fig. 4c) is caused by an overestimation of soil and groundwater storage. The positively biased soil and groundwater storage causes a phase delay by increasing the amount of time required for the subsurface drainage (baseflow) to reduce to soil and groundwater stores. The overestimation of water storage is the result of overestimated precipitation or underestimated evapotranspiration. The amplitude and phase of the water storage estimate are adjusted toward GRACE observation in the GC approach.

The impact of GRACE varies across the individual storage as well as across the geographical location (climate regime). In general, the major contributors to ΔTWS are ΔSM_{rz} and ΔGWS. Due to a small store size (only ~ 2 cm thick), ΔSM_{top} contributes only $\sim 2\,\%$ to ΔTWS. As such, ΔSM_{rz} and ΔGWS have greater variations, which commonly lead to greater uncertainty compared to ΔSM_{top}, and, therefore, the stores anticipate greater shares from the GRACE update. This behaviour is seen over all basins where the differences between the CABLE-simulated and GC ΔSM_{rz}, and ΔGWS estimates are greater (compared to ΔSM_{top}).

Furthermore, the impact of GRACE on ΔSM_{rz} and ΔGWS is different across the continent. For example, over central and southern Australia (see e.g. LKE, MRD, NWP, SWP), the dry climate is responsible for a small amount of groundwater recharge and most of the infiltration is stored in soil compartments. In this climate condition, ΔSM_{rz} amplitude is significantly larger than ΔGWS and it plays a greater role in ΔTWS, and, consequently, the GRACE contribution is mostly seen in the ΔSM_{rz} component. Different behaviour is seen over northern Australia (GOC, NEC, TIM) where ΔGWS amplitudes are greater ($\sim 40\,\%$ of ΔTWS) compared to other basins (only $\sim 17\,\%$ of ΔTWS). This is due to the sufficient amount of rainfall over the wet climate region, replenishing groundwater recharges and resulting in greater variability in ΔGWS. Therefore, compared to the dry climate basin, GRACE contributes to ΔGWS over these basins by a larger amount.

5.2.2 Impact on long-term trend estimates

The spatial patterns of the long-term trends of water storage changes over January 2003 and March 2016 are analysed before and after applying the GC approach (Fig. 6). For comparison, the long-term trends of ΔTWS derived from the ITSG-DDK5, mascon, and GRGS solutions are shown in Fig. 7. From Fig. 6b, GRACE effectively changes the long-term trend estimates in most basins in a way the spatial pattern of the ΔTWS trend of the GC solution consistent to the GRACE solutions while satisfying the model processes and keeping the spatial resolution. The trend of ΔSM_{top} is insignificant (Fig. 6c) and the GC approach does not change (Fig. 6d). The largest adjustment is seen in ΔSM_{rz} and ΔGWS components, to be consistent with the GRACE data in most basins (Fig. 6 and h).

GRACE shows significant changes in the ΔTWS trend estimates particularly over the northern and western parts of the continent (Fig. 7). The model estimates around the Gulf of Carpentaria basin show a strong negative trend that is inconsistent from the GRACE data. It is found that underestimated precipitation after 2012 is likely the cause of such an incompatible negative trend (see Fig. 4d). Applying the GC approach clearly improves the trend (Fig. 6a vs. Fig. 6b). The other example is seen over the western part of the continent (see rectangular area in Fig. 6a and b) where the averaged long-term trend of ΔTWS was predicted to be $-0.4\,\mathrm{cm\,yr^{-1}}$ but changed to be $-1.2\,\mathrm{cm\,yr^{-1}}$ (see also Sect. 5.4) by the GC approach. The precipitation over Western Australia is understood to be overestimated after 2012, evidently seen by the fact that the model ΔTWS is always greater than the GC solution (see e.g. Figs. 4h, 5d and p). The GC approach reveals that the water loss over Western Australia is at least 2 times greater than what has been predicted by the CABLE model.

In addition, the shortage of water storage in the south-eastern part of the continent from the Millennium Drought

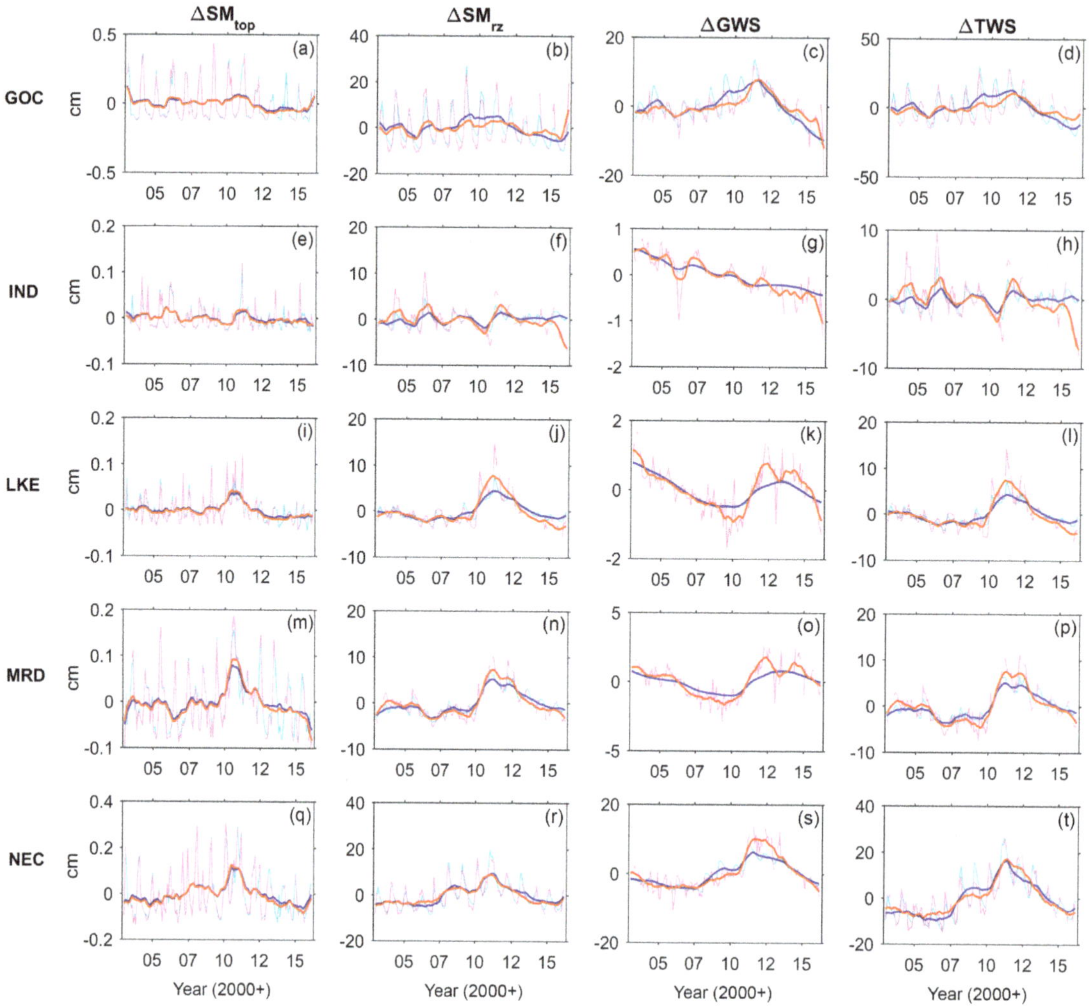

Figure 4. The monthly time series of ΔSM_{top}, ΔSM_{rz}, ΔGWS, and ΔTWS estimated from model (blue) and GC (red) solutions over the Gulf of Carpentaria (GOC), Indian Ocean (IND), Lake Eyre (LKE), Murray–Darling (MRD), and North East Coast (NEC). The deseasonalized time series is also shown.

(McGrath et al., 2012) has been recovered (seen as a positive water storage trend in Fig. 6) after the rainfall between 2009 and 2012, while the western part is still drying out (seen as negative trends). The trend estimates in terms of mass change are discussed in more detail in Sect. 5.4.

5.2.3 Reduction of uncertainty

Influenced by climate pattern, the uncertainty in water storage estimates significantly varies across Australia. The uncertainty in the model estimate is computed from the variability induced by different precipitation and model parameters while the uncertainty in the GC solution is computed using Eq. (11). As expected, larger uncertainties are observed in ΔSM_{rz} and ΔGWS than in ΔSM_{top} (an order of magnitude smaller) since ΔSM_{top} is smaller than others (Fig. 8). Over the wet basins, larger amplitude of the water storage leads to larger uncertainty, seen over the Gulf of Carpentaria, North East Coast, South East Coast, and Timor Sea where the CABLE-simulated ΔTWS uncertainty is approximately 28 % larger than other basins. The smaller uncertainty is found over the dry regions (e.g. LKE, SWP). In most basins, the uncertainty in ΔSM_{rz} is larger than the ΔGWS, except the wet basins (e.g. GOC, NEC, TIM) where the greater groundwater recharge leads to a larger uncertainty in ΔGWS.

Figure 8 demonstrates how much the formal error of each of the storage components is reduced by the GC approach. Overall, the estimated CABLE uncertainties averaged over all basins (AVG) are 0.2, 4.0, 4.0, and 5.7 cm for ΔSM_{top}, ΔSM_{rz}, ΔGWS, and ΔTWS, respectively. With the GC approach, the uncertainties in ΔSM_{top}, ΔSM_{rz}, ΔGWS, and ΔTWS decrease by approximately 26, 35, 39, and 37 %, respectively.

It is worth mentioning that the model uncertainty is mainly influenced by the meteorological forcing data. The uncertainty in precipitation derived from seven different precipi-

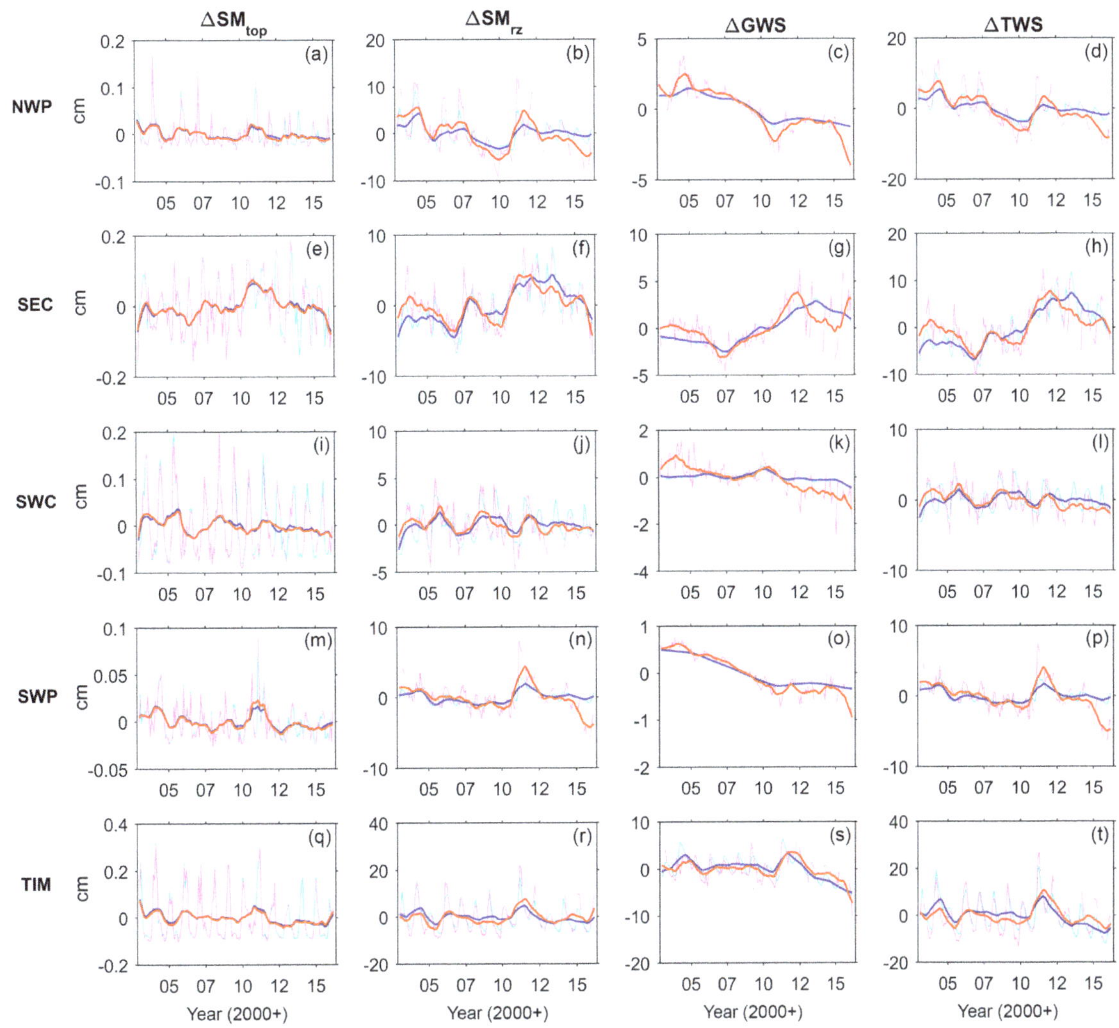

Figure 5. Similar to Fig. 3, but estimated over the North West Plateau (NWP), South East Coast (SEC), South West Coast (SWC), South West Plateau (SWP), and Timor Sea (TIM).

tation products is shown in Fig. 8e. The spatial pattern of the precipitation uncertainty is correlated with the uncertainty in water storage estimates. The larger water storage uncertainty is deduced from the larger precipitation uncertainty. The quality of precipitation forcing data is found to be an important factor to determine the accuracy of water storage computation.

5.3 Comparison with independent data

5.3.1 Soil moisture

The ΔSM_{top} estimates are compared with the AMSR-E-derived soil moisture. The processing of AMSR-E data is described in Sect. 4.3.1. The performance is assessed using Nash–Sutcliffe coefficients, given in Table 3. In general, CABLE (MN) shows a good performance in the top soil moisture simulation showing an NS value of > 0.4 for most of the basins. The top soil moisture estimate shows slightly better agreement with the C-band measurement of the AMSR-E product. This is likely caused by the greater emitting depth of the C-band measurement (~ 1 cm), which is closer to the depth of the top soil layer (~ 2 cm) used in this study (Njoku et al., 2003).

The GC approach leads to a small bit of improvement of the top soil estimate consistently from C- and X-band measurements and from all basins. No degradation of the NS value is observed in the GC solutions. The largest improvement is seen over LKE and NEC, where NS increases by 10–15 %. For other regions, the change in the NS coefficient may be incremental.

5.3.2 Groundwater

The ΔGWS estimates from the model and the GC method are compared with the in situ data obtained from two different ground networks in Queensland and Victoria. For each network, all ΔGWS data inside the groundwater network

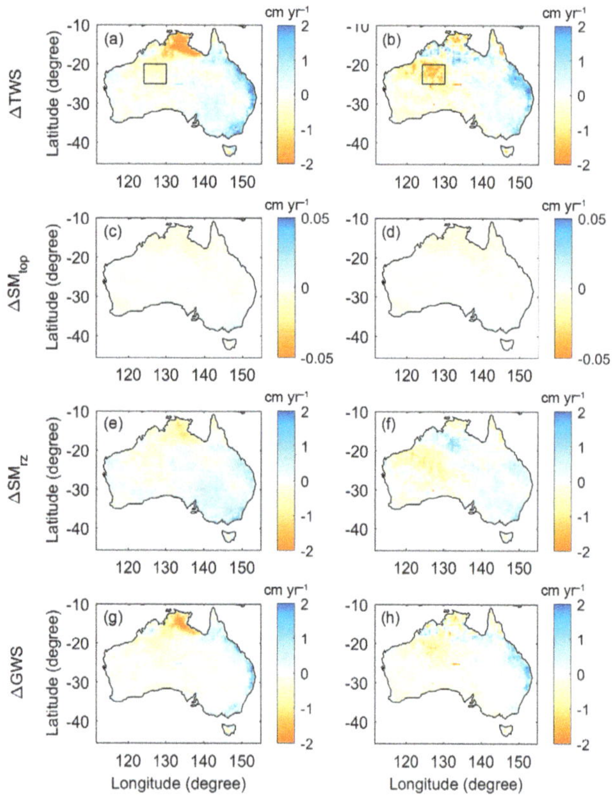

Figure 6. Long-term trends of ΔTWS **(a, b)**, ΔSM_{top} **(c, d)**, ΔSM_{rz} **(e, f)**, and ΔGWS **(g, h)** estimated from the model-only (left) and the GC solutions (right). The eastern part of North West Plateau basin is shown as a rectangle polygon in **(a)** and **(b)**.

boundary (see polygons in Fig. 1) are used to compute the average ΔGWS time series. From the comparison given in Fig. 9, it is found that the GC solutions of ΔGWS follow the overall inter-annual pattern of CABLE but with a considerably larger amplitude. This results in a better agreement with the in situ ΔGWS data seen from both networks. The NS coefficient of ΔGWS between the estimates and the in situ data are given in Table 4. The CABLE ΔGWS performs significantly better in Queensland (NS = ∼ 0.5) than Victoria (NS = ∼ 0.3). Significant improvement is found from the GC solutions in both networks, where the NS value increases from 0.5 to 0.6 (∼ 22 %) in Queensland and from 0.3 to 0.6 (∼ 85 %) in Victoria. Even greater improvement is seen when the inter-annual patterns are compared. The NS value increases from 0.5 to 0.7 (∼ 32 %) and 0.4 to 0.8 (∼ 93 %) in Queensland and Victoria, respectively.

The comparison of the long-term trend of ΔGWS is also evaluated. The estimated trends in Queensland and Victoria are given in Table 4. Beneficially from the GC approach, the ΔGWS trend is improved by approximately 20 % (from 0.4 to 0.6, compared to 1.6 cm yr^{-1}) in Queensland. The increase in ΔGWS is mainly influenced by the large amount of rainfall during the 2009–2012 La Niña episodes (see Fig. 9a). In

Table 3. NS coefficients between top soil moisture estimates and the satellite soil moisture observations from AMSR-E products over 10 different Australian basins. The area-weighted average value (AVG) is also shown.

	C band		X band	
	CABLE	GC	CABLE	GC
GOC	0.67	0.68	0.58	0.60
IND	0.53	0.54	0.41	0.41
LKE	0.48	0.53	0.36	0.42
MRD	0.77	0.80	0.75	0.78
NEC	0.34	0.39	0.14	0.19
NWP	0.33	0.36	0.38	0.42
SEC	0.68	0.68	0.69	0.71
SWC	0.85	0.85	0.89	0.89
SWP	0.55	0.56	0.46	0.48
TIM	0.44	0.45	0.16	0.16
AVG	0.53	0.56	0.47	0.50

Victoria, a significant improvement of the ΔGWS trend by about 76 % (from 0.1 to −0.2, compared to −0.3 cm yr^{-1}) is observed. Similar improvement of long-term trend estimates is seen in deseasonalized time series (improves by ∼ 15 % in Queensland and by ∼ 74 % in Victoria). The decrease in ΔGWS in Victoria is mainly due to the highly demanded groundwater consumption by agriculture and domestic activities (Van Dijk et al., 2007; Chen et al., 2016). As the groundwater consumption is not parameterized in CABLE, the decrease in the ΔGWS estimate cannot be properly captured in the model simulation. Applying GC approach effectively reduces the model deficiency and improves the quality of the groundwater estimations.

5.4 Assessment of mass variation in the past 13 years

Australia experiences significant climate variability; for example, the Millennium Drought starting from late 1990 (Van Dijk et al., 2013) and extremely wet conditions during several La Niña episodes (Trenberth, 2012; Han, 2017). These periods are referred as "Big Dry" and "Big Wet" (Ummenhofer et al., 2011; Xie et al., 2016). To understand the total water storage (mass) variation influenced by these two distinct climate variabilities, the water storage change obtained from the GC approach during Big Dry and Big Wet is separately investigated over 10 basins. The time window between January 2003 and December 2009 is defined as the Big Dry period while between January 2010 and December 2012 it is defined as the Big Wet period following Xie et al. (2016). In each period, the long-term trends of GC estimates of ΔTWS, ΔSM_{top}, ΔSM_{rz}, and ΔGWS are firstly calculated. Then, the total water storage variation (in m) is simply obtained by multiplying the long-term trend (in m yr^{-1}) with the number of years in the specific period – 7 years for Big Dry and 3 years for Big Wet. To obtain the mass variation,

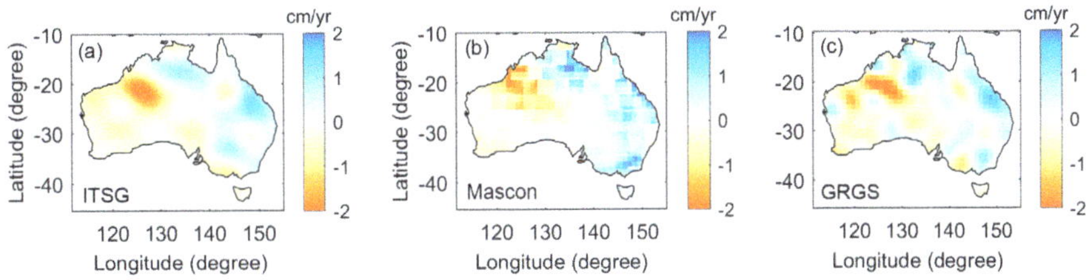

Figure 7. Long-term trends of GRACE-derived ΔTWS from ITSG-DDK5 **(a)**, mascon **(b)**, and GRGS solution **(c)**.

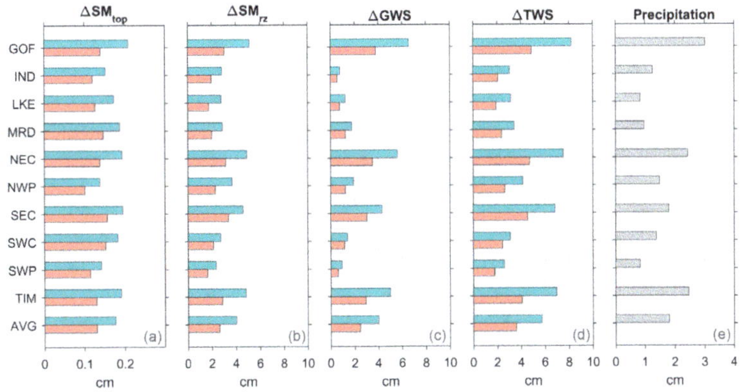

Figure 8. Uncertainties in ΔSM_{top}, ΔSM_{rz}, ΔGWS, and ΔTWS estimated from the model (blue) and the GC solutions (red) in 10 different Australian basins. The uncertainty in the precipitation is shown in **(e)**. The area-weighted average value (AVG) is also shown.

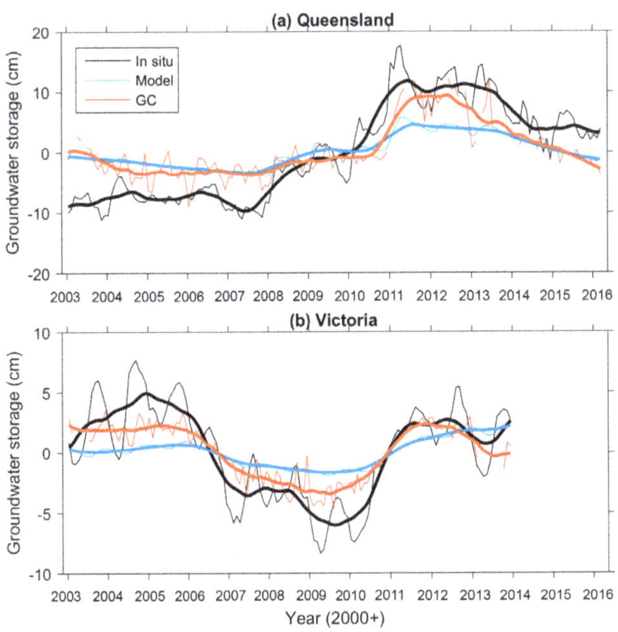

Figure 9. The monthly time series of ΔGWS estimated from the model, GC solutions, and measured from the in situ groundwater network in Queensland **(a)** and Victoria **(b)**. Deseasonalized time series are shown in thick lines.

the water storage variation is multiplied by the area of the basin and the density of water ($1000\,kg\,m^{-3}$). The estimated mass variations during Big Dry and Big Wet are displayed in Fig. 10. The long-term mass variation in the entire period (January 2003–March 2016) is also shown.

During Big Dry (2003–2009), a significant loss of total storage (40–60 Gt over 7 years) is observed over the LKE, MRD, NWP, and SWP basins. The largest groundwater loss of > 20 Gt is found from LKE and MRD. No significant change is observed over the tropical climate regions (e.g. GOC, NEC). The mass loss mostly occurs in the root zone and groundwater compartments where the sum of ΔSM_{rz} and ΔGWS explains more than 90 % of the ΔTWS value. The mass loss is also observed in ΔSM_{top} but is > 10 times smaller than ΔSM_{rz} and ΔGWS.

During Big Wet (2010–2012), the basins such as LKE, MRD, and TIM exhibit the significant total storage gain of > 100 Gt. The gain is particularly larger in ΔSM_{rz} over the basins that experienced the significant loss during Big Dry. For example, over LKE and MRD, the gain of ΔSM_{rz} is approximately 2–3 times greater than ΔGWS. It implies that most of the infiltration (from the 2009–2012 La Niña rainfall) is stored as soil moisture through the long drought period and that the groundwater recharge is secondary to the ΔSM_{rz} increase.

The opposite behaviour is observed over the basins (such as NEC and GOC) that experienced mass gain during Big

Table 4. NS coefficient and long-term trend of ΔGWS estimated from the model-only and GC solutions in Queensland and Victoria groundwater network. The long-term trend of the in situ data is also shown.

	Queensland			Victoria		
	In situ	CABLE	GC	In situ	CABLE	GC
Original time series						
NS (–)	–	0.49	0.60	–	0.34	0.63
Trend (cm yr^{-1})	1.60 ± 0.05	0.39 ± 0.02	0.63 ± 0.05	-0.27 ± 0.05	0.10 ± 0.02	-0.18 ± 0.03
Deseasonalized time series						
NS (–)	–	0.50	0.66	–	0.43	0.83
Trend (cm yr^{-1})	1.60 ± 0.05	0.39 ± 0.02	0.57 ± 0.04	-0.25 ± 0.05	0.10 ± 0.02	-0.16 ± 0.03

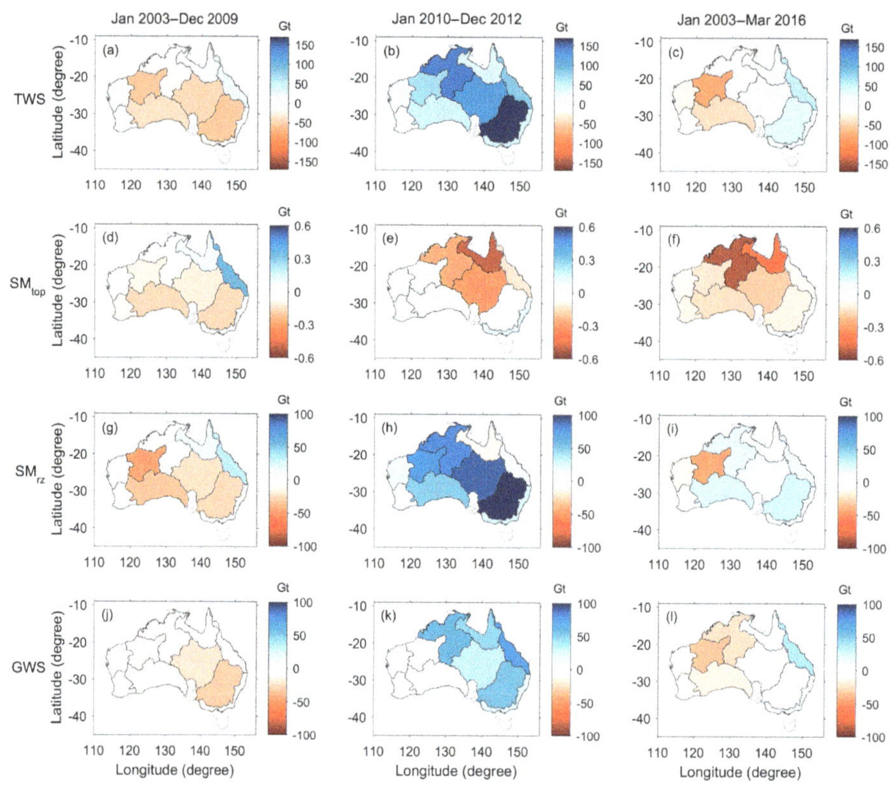

Figure 10. Mass changes (Gt, Gigatonne) of ΔTWS, ΔSM_{top}, ΔSM_{rz}, and ΔGWS estimated from GC solutions over 10 Australian basins in three different periods: Big Dry (January 2003–December 2009), Big Wet (January 2010–December 2012), and the entire period (January 2003–March 2016).

Dry. The water storage gain is greater in ΔGWS compared to ΔSM_{rz}. In NEC, ΔGWS gain is ~ 8 times larger than ΔSM_{rz} during Big Wet. The soil compartment may be saturated during Big Dry and additional infiltration from the Big Wet precipitation leads to an increased groundwater recharge. The ΔSM_{rz} loss observed over GOC is simply caused by the timing selection of the Big Wet period, which ends earlier (~ 2011) in GOC than in other basins. The ΔSM_{rz} gain becomes ~ 26 Gt if the Big Wet period is defined as 2008–2011. During the post-Big-Wet period (2012 and afterwards), the decreasing trend of water storage is observed from all basins (see Figs. 4 and 5). This is mainly caused by the decrease in precipitation after 2012 and by gradual water loss through evapotranspiration (Fasullo et al., 2013).

The overall water storage change in the last 13 years demonstrates that the severe water loss from most basins during Big Dry (the Millennium Drought) is balanced with the gain during Big Wet (the La Niña). The negative ΔTWS estimated during Big Dry becomes positive in LKE, MRD, and

SEC and less negative in TIM, and the greatest gain is observed from NEC by ∼ 50 Gt during the 13-year period (see Fig. 10c). However, the water mass loss is still detected over the western basins (e.g. IND, NWP, SWP, SWC), and their magnitudes are even larger than the mass loss during Big Dry. For example, the greatest ΔTWS loss of ∼ 79 Gt is observed over NWP, which is ∼ 25 Gt greater than the loss during Big Dry (see Fig. 10a and c). The basin is less affected by the La Niña, and the rainfall during Big Wet is clearly inadequate to support the water storage recovery in the basin. Rainfall deficiency also reduces the groundwater recharge, resulting in even more decreasing of ΔGWS, compared to the Millennium Drought period (see Fig. 10j and l). The continual decrease in water storage over western basins is likely caused by the interaction of complex climate patterns such as the El Niño–Southern Oscillation, Indian Ocean Dipole, and Southern Annular Mode cycles (Australian Bureau of Meteorology, 2012; Xie et al., 2016).

5.5 Comparison of the GC approach with alternatives

The simplest approach to estimate ΔGWS is to subtract the model soil moisture component from GRACE ΔTWS data, without considering uncertainty in the model output, as used in Rodell et al. (2009) and Famiglietti et al. (2011). This method is called Approach 1 (App 1). In Approach 2 (App 2) as in Tangdamrongsub et al. (2017), by accounting for the uncertainty in model outputs and GRACE data, the water storage states are updated through a Kalman filter:

$$\hat{\boldsymbol{h}}_{\mathrm{R}} = \widetilde{\boldsymbol{h}}_{\mathrm{R}} + \mathbf{H}\mathbf{C}_{\mathrm{R}}^{T}\left(\mathbf{H}\mathbf{R}\mathbf{H}^{T} + \mathbf{C}_{\mathrm{R}}\right)^{-1}\left(\boldsymbol{b} - \mathbf{H}\widetilde{\boldsymbol{h}}_{\mathrm{R}}\right), \tag{18}$$

where $\widetilde{\boldsymbol{h}}_{\mathrm{R}}$, $\mathbf{H}$, $\mathbf{C}_{\mathrm{R}}$ are described in Sect. 2; $\boldsymbol{b}$ is an observation vector containing GRACE-derived ΔTWS; and $\mathbf{R}$ is an error variance–covariance matrix of the observation. The GRACE-derived ΔTWS and its error information is obtained from the mascon solution. The matrix $\mathbf{R}$ is a (diagonal) error variance matrix since no covariance information is given in the mascon product. Note that the model uncertainty remains the same as in GC approach (Sect. 4.2). The different results from App 1 and App 2 are mainly attributed to the different estimates of the uncertainty.

The ΔGWS estimates from App 1, App 2, and GC in Queensland and Victoria are shown in Fig. 11. It is clearly seen that ΔGWS values from App 1 are overestimated while the one from App 2 fits the ground data significantly better. This behaviour was also seen in Tangdamrongsub et al. (2017), where the water storage estimates tend to be overestimated when error components such as spatial correlation error were neglected as in App 1. ΔGWS from App 2 shows clear improvements in terms of NS coefficients in both networks. Considering the deseasonalized ΔGWS estimates, in Queensland, the trend increases from 0.39 ± 0.03 to $0.42 \pm 0.03\,\mathrm{cm\,yr^{-1}}$ (improves by 1.5 %) and the NS value increases from 0.46 to 0.53. In Victoria, the trend decreases

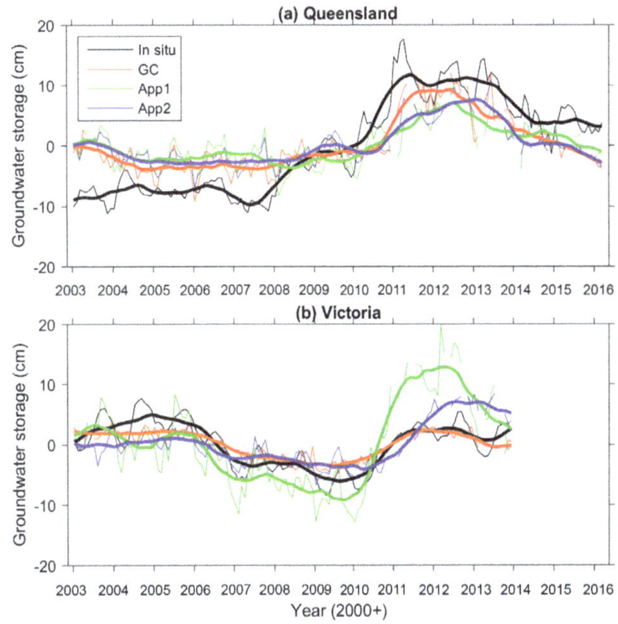

Figure 11. ΔGWS estimated from Approach 1 (App 1) and Approach 2 (App 2) in Queensland **(a)** and Victoria **(b)**. The in situ groundwater network data and the GC solutions are also shown. Deseasonalized time series are shown in thick lines.

from 0.73 ± 0.10 to $0.46 \pm 0.05\,\mathrm{cm\,yr^{-1}}$ (improves by 27 %) and the NS value increases from −0.89 to 0.30. Although App 2 is not yet as good as the GC solution based on the most comprehensive error propagation, this simple test demonstrates the importance of considering the uncertainty. The reason for App 2 being less accurate than GC is likely the too-simplified error information implemented in App 2.

6 Conclusion

This study presents an approach to combine the raw GRACE observation with model simulation to improve water storage estimates over Australia. Distinct from other methods, we exploit the fundamental GRACE satellite tracking data and the full data error variance–covariance information to avoid alteration of signal and measurement error information present in higher-level data products.

We compare groundwater storage estimates from the GC approach and two other approaches, subject to inclusion of GRACE uncertainty in the ΔGWS calculation. Validating three results of ΔGWS against the in situ groundwater data, we find that the GC approach delivers the most accurate groundwater estimate, followed by the approach based on incomplete information of GRACE's data error. The poorest estimate of groundwater storage is seen when the GRACE uncertainty is completely ignored. This confirms the critical value of using the complete GRACE signal and error information at the raw data level.

The analysis of water storage change between 2003 and 2016 reveals that half of the continent (5 out of 10 basins) has still not fully recovered from the Millennium Drought. The TWS decrease in Western Australia has been most characteristic, and the GC approach finds that the water loss mainly occurs in the groundwater layer. Rainfall inadequacy is attributed to the continual dry condition, leading to a greater decrease in groundwater recharge and storage over Western Australia.

The land surface model we used is deficient in anthropogenic groundwater consumption. The model calibration will never help, and the groundwater consumption must be brought in by external sources. On the contrary, statistical approaches similar to our GC approach may be useful to fill in the missing component and lead to a more comprehensive water storage inventory.

However, it is difficult to constrain different water storage components by only using total storage observation such as GRACE. In addition, it is challenging to improve surface soil moisture varying rapidly in time, using a monthly mean GRACE observation. Tian et al. (2017) utilized the satellite soil moisture observation from Soil Moisture and Ocean Salinity (SMOS, Kerr et al., 2001) in addition to GRACE data for their data assimilation and showed a clear improvement in the top soil moisture estimate. The GC approach with complementary observations at higher temporal resolution should be considered, particularly to enhance the surface soil moisture computation.

Furthermore, the GC approach can be simply extended for GRACE data assimilation. Assimilating the raw GRACE data into land surface models like CABLE enables the model state and parameter to be adjusted with the realistic error information, allowing more reliable storage computation.

Appendix A: Least-squares normal equation of GRACE

A linearized GRACE satellite-tracking observation equation is formulated as

$$\boldsymbol{y} = \mathbf{A}\boldsymbol{x} + \boldsymbol{e};\ \boldsymbol{e} \sim N(\mathbf{0}, \boldsymbol{\Sigma}), \tag{A1}$$

where $\boldsymbol{y}$ is the observation vector containing various kinds of L1B data including the inter-satellite ranging data; $\mathbf{A}$ is the design (partial derivative) matrix relating the data and the Earth gravity field variations; $\boldsymbol{x}$ contains the Stokes coefficients of time-varying geopotential fields (e.g. Wahr et al., 1998); and $\boldsymbol{e}$ is the L1B data noise, which has zero mean and covariance $\boldsymbol{\Sigma}$. Equation (A1) can be modified explicitly in terms of soil moisture and groundwater storage variations as

$$\boldsymbol{y} = \mathbf{A}\mathbf{S}\overline{\mathbf{Y}}\mathbf{H}\boldsymbol{h} + \boldsymbol{e};\ \boldsymbol{e} \sim N(\mathbf{0}, \boldsymbol{\Sigma}), \tag{A2}$$

where $\mathbf{S}$ contains a factor used to convert ΔTWS to geopotential coefficients considering the load Love numbers (e.g. Wahr et al., 1998) and $\overline{\mathbf{Y}}$ converts the gridded data into the corresponding spherical harmonic coefficients. For convenience, the term $\mathbf{Y} = \mathbf{S}\overline{\mathbf{Y}}$ is used in the further derivation. A least-squares solution of Eq. (A2) is given as

$$\left(\mathbf{H}^T\mathbf{Y}^T\mathbf{A}^T\boldsymbol{\Sigma}^{-1}\mathbf{A}\mathbf{Y}\mathbf{H}\right)\hat{\boldsymbol{h}} = \mathbf{H}^T\mathbf{Y}^T\mathbf{A}^T\boldsymbol{\Sigma}^{-1}\boldsymbol{y}. \tag{A3}$$

It can be simplified as

$$\mathbf{H}^T\mathbf{Y}^T\mathbf{N}\mathbf{Y}\mathbf{H}\hat{\boldsymbol{h}} = \mathbf{H}^T\mathbf{Y}^T\boldsymbol{c}, \tag{A4}$$

where $\mathbf{N} = \mathbf{A}^T\boldsymbol{\Sigma}^{-1}\mathbf{A}$ and $\boldsymbol{c} = \mathbf{A}^T\boldsymbol{\Sigma}^{-1}\boldsymbol{y}$. Equation (A4) is identical to Eq. (5).

Appendix B: Nash–Sutcliffe coefficient and area-weighted average

The Nash–Sutcliffe coefficient (NS) is computed as follows:

$$\mathrm{NS} = 1 - \frac{\sum_{i=1}^{N}\left(y_i - \hat{x}_i\right)^2}{\sum_{i=1}^{N}\left(y_i - \overline{y}\right)^2}, \tag{B1}$$

where $\boldsymbol{y}$ is an observation vector, $\overline{\boldsymbol{y}}$ is the mean of the observation, $\hat{\boldsymbol{x}}$ is a vector containing the simulated result, i is the index of observation, and N is the number of observations.

Area-weighted average ($\overline{Z}$) is compute as follows:

$$\overline{Z} = \frac{\sum_{j=1}^{M} w_j \overline{z}_j}{\sum_{j=1}^{M} w_j}, \tag{B2}$$

where w is the area size, $\overline{z}$ is the mean value inside the considered area, j is the area index, and M is the number of considered areas.

Competing interests. The authors declare that they have no conflict of interest.

Acknowledgements. This work is funded by The University of Newcastle to support NASA's GRACE and GRACE Follow-On projects as an international science team member to the missions. Mark Decker was supported by the ARC Centre of Excellence for Climate Systems Science. Hyungjun Kim was supported by Japan Society for the Promotion of Science KAKENHI (16H06291). We thank Torsten Mayer-Gürr for GRACE data products in the form of the least-squares normal equations. We also thank three anonymous reviewers for helping us improve the manuscript.

Edited by: Louise Slater

References

Australian Bureau of Meteorology: Record-breaking La Niña events: An analysis of the La Niña life cycle and the impacts and significance of the 2010–11 and 2011–12 La Niña events in Australia, National Climate Centre, Bureau of Meteorology, http://www.bom.gov.au/climate/enso/history/La-Nina-2010-12.pdf (last access: 5 January 2017), 2012.

Bettadpur, S.: CSR Level-2 Processing Standards Document for Product Release 05, GRACE 327-742, Center for Space Research, The University of Texas, Austin, 2012.

Chen, J. L., Wilson, C. R., Tapley, B. D., Scanlon, B., and Güntner, A.: Long-term groundwater storage change in Victoria, Australia from satellite gravity and in situ observations, Global Planet. Change, 139, 56–65, https://doi.org/10.1016/j.gloplacha.2016.01.002, 2016.

Decker, M.: Development and evaluation of a new soil moisture and runoff parameterization for the CABLE LSM including subgrid-scale processes, J. Adv. Model. Earth Syst., 7, 1788–1809, https://doi.org/10.1002/2015MS000507, 2015.

Dee, D. P., Uppala, S. M., Simmons, A. J., Berrisford, P., Poli, P., Kobayashi, S., Andrae, U., Balmaseda, M. A., Balsamo, G., Bauer, P., Bechtold, P., Beljaars, A. C. M., van de Berg, L., Bidlot, J., Bormann, N., Delsol, C., Dragani, R., Fuentes, M., Geer, A. J., Haiberger, L., Healy, S. B., Hersbach, H., Hólm, E. V., Isaksen, L., Kållberg, P., Köhler, M., Matricardi, M., McNally, A. P., Monge-Sanz, B. M., Morcrette, J. J., Park, B. K., Peubey, C., de Rosnay, P., Tavolato, C., Thépaut, J. N., and Vitart, F.: The ERA-Interim reanalysis: configuration and performance of the data assimilation system, Q. J. Roy. Meteorol. Soc., 137, 553–597, https://doi.org/10.1002/qj.828, 2011.

Dumedah, G. and Walker, J. P.: Intercomparison of the JULES and CALBE land surface models through assimilation of remote sensed soil moisture in southeast Australia, Adv. Water Resour., 74, 231–244, https://doi.org/10.1016/j.advwatres.2014.09.011, 2014.

Eicker, A., Schumacher, M., Kusche, J., Döll, P., and Müller Schmied, H.: Calibration data assimilation approach for integrating GRACE data into the WaterGAP Global Hydrology Model (WGHM) using an Ensemble Kalman Filter: First Results, Surv. Geophys., 35, 1285–1309, https://doi.org/10.1007/s10712-014-9309-8, 2014.

Famiglietti, J. S., Lo, M., Ho, S. L., Bethune, J., Anderson, K. J., Syed, T. H., Swenson, S. C., de Linage, C. R., and Rodell, M.: Satellites measure recent rates of groundwater depletion in California's Central Valley, Geophys. Res. Lett., 38, L03403, https://doi.org/10.1029/2010GL046442, 2011.

Fasullo, J. T., Boening, C., Landerer, F. W., and Nerem, R. S.: Australia's unique influence on global sea level in 2010–2011, Geophys. Res. Lett., 40, 4368–4373, https://doi.org/10.1002/grl.50834, 2013.

Girotto, M., De Lannoy, G. J. M., Reichle, R. H., and Rodell, M.: Assimilation of gridded terrestrial water storage observations from GRACE into a land surface model, Water Resour. Res., 52, 4164–4183, https://doi.org/10.1002/2015WR018417, 2016.

Hamill, T. M., Whitaker, J. S., and Snyder, C.: Distance-Dependent Filtering of Background Error Covariance Estimates in an Ensemble Kalman Filter, Mon. Weather Rev., 129, 2776–2790, 2001.

Han, S.-C.: Elastic deformation of the Australian continent induced by seasonal water cycles and the 2010–11 La Niña determined using GPS and GRACE, Geophys. Res. Lett., 44, 2763–2772, https://doi.org/10.1002/2017GL072999, 2017.

Huffman, G. J., Adler, R. F., Bolvin, D. T., Gu, G., Nelkin, E. J., Bowman, K. P., Hong, Y., Stocker, E. F., and Wolf, D. B.: The TRMM multisatellite precipitation analysis (TMPA): Quasi-global, multiyear, combined-sensor precipitation estimates at fine scales, J. Hydrometeorol., 8, 38–55, https://doi.org/10.1175/JHM560.1, 2007.

Jekeli, C.: Alternative methods to smooth the Earth's gravity field, Rep. 327, Dept. of Geod. Sci. and Surv., Ohio State Univ., Columbus, 1981.

Kerr, Y. H., Waldteufel, P., Wigneron, J.-P., Martinuzzi, J.-M., Font, J., and Berger, M.: Soil moisture retrieval from space: The soil moisture and ocean salinity (SMOS) mission, IEEE T. Geosci. Remote, 39, 1729–1735, 2001.

Khaki, M., Hoteit, I., Kuhn, M., Awange, J., Forootan, E., van Dijk, A., Schumacher, M., and Pattiaratchi, C.: Assessing sequential data assimilation techniques for integrating GRACE data into a hydrological model, Adv. Water Resour., 107, 301–316, https://doi.org/10.1016/j.advwatres.2017.07.001, 2017a.

Khaki, M., Schumacher, M., Forootan, E., Kuhn, M., Awange, J., and van Dijk, A.: Accounting for spatial correlation errors in the assimilation of GRACE into hydrological models through localization. Adv. Water Resour., 108, 99–112, https://doi.org/10.1016/j.advwatres.2017.07.024, 2017b.

Kusche, J., Schmidt, R., Petrovic, S., and Rietbroek, R.: Decorrelated GRACE time-variable gravity solutions by GFZ, and their validation using a hydrological model, J. Geodesy, 83, 903–913, https://doi.org/10.1007/s00190-009-0308-3, 2009.

Lambert, A., Huang, J., van der Kamp, G., Henton, J., Mazzotti, S., James, T. S., Courtier, N., and Barr, A. G.: Measuring water accumulation rates using GRACE data in areas experiencing glacial isostatic adjustment: The Nelson River basin, Geophys. Res. Lett., 40, 6118–6122, https://doi.org/10.1002/2013GL057973, 2013.

Landerer, F. W. and Swenson, S. C.: Accuracy of scaled GRACE terrestrial water storage estimates, Water Resour. Res., 48, W04531, https://doi.org/10.1029/2011WR011453, 2012.

Lemoine, J. M., Bourgogne, S., Bruinsma, S., Gégout, P., Reinquin, F., and Biancale, R.: GRACE RL03-v2 monthly time series of solutions from CNES/GRGS, EGU2015-14461, EGU General Assembly 2015, Vienna, Austria, 2015.

Long, D., Chen, X., Scanlon, B. R., Wada, Y., Hong, Y., Singh, V. P., Chen, Y., Wang, C., Han, Z., and Yang, W.: Have GRACE satellites overestimated groundwater depletion in the Northwest India Aquifer?, Sci. Rep., 6, 24398, https://doi.org/10.1038/srep24398, 2016.

Mayer-Gürr, T., Behzadpour, S., Ellmer, M., Kvas, A., Klinger, B., and Zehentner, N.: ITSG-Grace2016 – Monthly and Daily Gravity Field Solutions from GRACE, GFZ Data Services, https://doi.org/10.5880/icgem.2016.007, 2016.

Mayer-Gürr, T., Pail, R., Gruber, T., Fecher, T., Rexer, M., Schuh, W.-D., Kusche, J., Brockmann, J.-M., Rieser, D., Zehentner, N., Kvas, A., Klinger, B., Baur, O., Höck, E., Krauss, S., and Jäggi, A.: The combined satellite gravity field model GOCO05s, EGU 2015, Vienna, 2015.

McGrath, G. S., Sadler, R.,Fleming, K., Tregoning, P., Hinz, C., and Veneklaas, E. J.: Tropical cyclones and the ecohydrology of Australia's recent continental-scale drought, Geophys. Res. Lett., 39, L03404, https://doi.org/10.1029/2011GL050263, 2012.

Njoku, E. G., Jackson, T. L., Lakshmi, V., Chan, T., and Nghiem, S. V.: Soil Moisture Retrieval from AMSR-E, IEEE T. Geosci. Remote, 41, 215–229, 2003.

Owe, M., de Jeu, R., and Holmes, T.: Multisensor historical climatology of satellite-derived global land surface moisture, J. Geophys. Res., 113, F01002, https://doi.org/10.1029/2007JF000769, 2008.

Pearson, E. K.: Mining imperfect data: Dealing with contamination and incomplete records, ProSanos Corporation, Harrisburg, Pennsylvania, https://doi.org/10.1137/1.9780898717884, 2005.

Ramillien, G., Cazenave, A., and Brunau, O., Global time variations of hydrological signals from GRACE satellite gravimetry, Geophys. J. Int., 158, 813–826, 2004.

Rassam, D. W., Peeters, L., Pickett, T., Jolly, I., and Holz, L.: Accounting for surfaceegroundwater interactions and their uncertainty in river and groundwater models: A case study in the Namoi River, Australia, Environ. Model. Softw., 50, 108–119, https://doi.org/10.1016/j.envsoft.2013.09.004, 2013.

Reichle, R. H. and Koster, R. D.: Bias reduction in short records of satellite soil moisture, Geophys. Res. Lett., 31, L19501, https://doi.org/10.1029/2004GL020938, 2004.

Rienecker, M. M., Suarez, M. J., Gelaro, R., Todling, R., Bacmeister, J., Liu, E., Bosilovich, M. G., Schubert, S. D., Takacs, L., Kim, G.-K., Bloom, S., Chen, J., Collins, D., Conaty, A., da Silva, A., Gu, W., Joiner, J., Koster, R. D., Lucchesi, R., Molod, A., Owens, T., Pawson, S., Pegion, P., Redder, C. R., Reichle, R., Robertson, F. R., Ruddick, A. G., Sienkiewicz, M., and Woollen, J.: MERRA – NASA's Modern-Era Retrospective Analysis for Research and Applications, J. Climate, 24, 3624–3648, https://doi.org/10.1175/JCLI-D-11-00015.1, 2011.

Rodell, M., Houser, P. R., Jambor, U., Gottschalck, J., Mitchell, K., Meng, C. J., Arsenault, K., Cosgrove, B., Radakovich, J., Bosilovich, M., Entin, J. K., Walker, J. P., Lohmann, D., and Toll, D.: The global land data assimilation system, B. Am. Meteorol. Soc., 85, 381–394, 2004.

Rodell, M., Velicogna, I., and Famiglietti, J. S.: Satellite-based estimates of groundwater depletion in India, Nature, 460, 999–1002, https://doi.org/10.1038/nature08238, 2009.

Sakumura, C., Bettadpur, S., and Bruinsma, S.: Ensemble prediction and intercomparison analysis of GRACE time-variable gravity field models, Geophys. Res. Lett., 41, 1389–1397, https://doi.org/10.1002/2013GL058632, 2014.

Schumacher, M., Kusche, J., and Döll, P.: A Systematic Impact Assessment of GRACE Error Correlation on Data Assimilation in Hydrological Models, J. Geodesy, 90, 537–559, https://doi.org/10.1007/s00190-016-0892-y, 2016.

Schumacher, M., Forootan, E., van Dijk, A., Schmied, H. M., Crosbie, R., Kusche, J., and Döll, P.: Improving drought simulations within the Murray-Darling Basin by combined calibration/assimilation of GRACE data into the WaterGAP Global Hydrology Model, Remote Sens. Environ., 204, 212–228, https://doi.org/10.1016/j.rse.2017.10.029, 2018.

Sheffield, J., Goteti, G., and Wood, E. F.: Development of a 50-yr high-resolution global dataset of meteorological forcings for land surface modeling, J. Climate, 19, 3088–3111, 2005.

Sorooshian, S., Hsu, K., Gao, X., Gupta, H. V., Imam, B., and Braithwaite, D.: Evaluation of PERSIANN System Satellite-Based Estimates of Tropical Rainfall, B. Am. Meteorol. Soc., 81, 2035–2046, 2000.

Swenson, S. C. and Wahr, J.: Post-processing removal of correlated errors in GRACE data, Geophys. Res. Lett., 33, L08402, https://doi.org/10.1029/2005GL025285, 2006.

Tangdamrongsub, N., Ditmar, P. G., Steele-Dunne, S. C., Gunter, B. C., and Sutanudjaja, E. H.: Assessing total water storage and identifying flood events over Tonlé Sap basin in Cambodia using GRACE and MODIS satellite observations combined with hydrological models, Remote Sens. Environ., 181, 162–173, https://doi.org/10.1016/j.rse.2016.03.030, 2016.

Tangdamrongsub, N., Steele-Dunne, S. C., Gunter, B. C., Ditmar, P. G., and Weerts, A. H.: Data assimilation of GRACE terrestrial water storage estimates into a regional hydrological model of the Rhine River basin, Hydrol. Earth Syst. Sci., 19, 2079–2100, https://doi.org/10.5194/hess-19-2079-2015, 2015.

Tangdamrongsub, N., Steele-Dunne, S. C., Gunter, B. C., Ditmar, P. G., Sutanudjaja, E. H., Xie, T, Wang, Z.: Improving estimates of water resources in a semi-arid region by assimilating GRACE data into the PCR-GLOBWB hydrological model, Hydrol. Earth Syst. Sci., 21, 2053–2074, https://doi.org/10.5194/hess-21-2053-2017, 2017.

Tian, S., Tregoning, P., Renzullo, L. J., van Dijk, A. I. J. M., Walker, J. P., Pauwels, V. R. N., and Allgeyer, S.: Improved water balance component estimates through joint assimilation of GRACE water storage and SMOS soil moisture retrievals, Water Resour. Res., 53, 1–21, https://doi.org/10.1002/2016WR019641, 2017.

Trenberth, K. E.: Framing the way to relate climate extremes to climate change, Climatic Change, 115, 283–290, https://doi.org/10.1007/s10584-012-0441-5, 2012.

Tscherning, C. C. and Rapp, R. H.: Closed covariance expressions for gravity anomalies, geoid undulations, and deflections of the vertical implied by anomaly degree variance models, Rep. 208, Dep. of Geod. Sci. and Surv., Ohio State Univ., Columbus, 1974.

Ukkola, A. M., Pitman, A. J., Decker, M., De Kauwe, M. G., Abramowitz, G., Kala, J., and Wang, Y.-P.: Modelling evapotranspiration during precipitation deficits: identifying critical processes in a land surface model, Hydrol. Earth Syst. Sci., 20, 2403–2419, https://doi.org/10.5194/hess-20-2403-2016, 2016.

Ummenhofer, C. C., Gupta, A., Briggs, P. R., England, M. H., McIntosh, P. C., Meyers, G. A., Pook, M. J., Raupach, M. R., and Risbey, J. S.: Indian and Pacific Ocean influences on Southeast Australian drought and soil moisture, J. Climate, 24, 1313–1336, https://doi.org/10.1175/2010JCLI3475.1, 2011.

Van Dijk, A., Podger, G., and Kirby, M.: Integrated water resources management in the Murray-Darling Basin, in: Increasing demands on decreasing supplies, Reducing the Vulnerability of Societies to Water Related Risks at the Basin Scale, edited by: Schumann, A. and Pahlow, M., IAHS Publ., Bochum, Germany, 24–30, 2007.

Van Dijk, A., Beck, H. E., Crosbie, R. S., De Jeu, E. A. M., Liu, Y. Y., Podger, G. M., Timbal, B., and Viney, N. R.: The Millennium Drought in southeast Australia (2001–2009): Natural and human causes and implications for water resources, ecosystems, economy, and society, Water Resour. Res., 49, 1040–1057, https://doi.org/10.1002/wrcr.20123, 2013.

Voss, K. A., Famiglietti, J. S., Lo, M., de Linage, C., Rodell, M., and Swenson, S. C.: Groundwater depletion in the Middle East from GRACE with implications for transboundary water management in the Tigris-Euphrates-Western Iran region, Water Resour. Res., 49, 904–914, https://doi.org/10.1002/wrcr.20078, 2013.

Wahr, J., Molenaar, M., and Bryan, F.: Time variability of the Earth's gravity field: Hydrological and oceanic effects and their possible detection using GRACE, J. Geophys. Res., 103, 30205–30229, 1998.

Wahr, J., Swenson, S., and Velicogna, I.: Accuracy of GRACE mass estimates, Geophys. Res. Lett., 33, L06401, https://doi.org/10.1029/2005GL025305, 2006.

Watkins, M. M., Wiese, D. N., Yuan, D.-N., Boening, C., and Landerer, F. W.: Improved methods for observing Earth's time variable mass distribution with GRACE using spherical cap mascons, J. Geophys. Res.-Solid, 120, 2648–2671, https://doi.org/10.1002/2014JB011547, 2015.

Welsh, W. D.: Water balance modelling in Bowen, Queensland, and the ten iterative steps in model development and evaluation, Environ. Model. Softw., 23, 195–205, 2008.

Wiese, D. N., Landerer, F. W., and Watkins, M. M.: Quantifying and reducing leakage errors in the JPL RL05M GRACE mascon solution, Water Resour. Res., 52, 7490–7502, https://doi.org/10.1002/2016WR019344, 2016.

Wu, S. C., Kruizinga, G., and Bertiger, W.: Algorithm theoretical basis document for GRACE Level-1B data processing V1.2, JPL D-27672, Jet Propul. Lab., Pasadena, California, 2006.

Xie, Z., Huete, A., Restrepo-Coupea, N., Maa, X., Devadasa, R., and Caprarellib, G.: Spatial partitioning and temporal evolution of Australia's total water storage under extreme hydroclimatic impacts, Remote Sens. Environ., 183, 43–52, https://doi.org/10.1016/j.rse.2016.05.017, 2016.

Zaitchik, B. F., Rodell, M., and Reichle, E. H.: Assimilation of GRACE terrestrial water storage data into a land surface model: Results for the Mississippi basin, J. Hydrometeorol., 9, 535–548, 2008.

13

Aquifer configuration and geostructural links control the groundwater quality in thin-bedded carbonate–siliciclastic alternations of the Hainich CZE, central Germany

Bernd Kohlhepp[1], Robert Lehmann[1], Paul Seeber[1], Kirsten Küsel[2,3], Susan E. Trumbore[4], and Kai U. Totsche[1]

[1]Friedrich Schiller University Jena, Institute of Geosciences, Chair of Hydrogeology, Burgweg 11, 07749 Jena, Germany

[2]Friedrich Schiller University Jena, Institute of Ecology, Chair of Aquatic Geomicrobiology, Dornburger Strasse 159, 07743 Jena, Germany

[3]German Centre for Integrative Biodiversity Research (iDiv), Halle-Jena-Leipzig, Deutscher Platz 5d, 04103 Leipzig, Germany

[4]Max Planck Institute for Biogeochemistry Jena, Department of Biogeochemical Processes, Hans-Knöll-Str. 10, 07745 Jena, Germany

Correspondence to: Kai U. Totsche (kai.totsche@uni-jena.de)

Abstract. The quality of near-surface groundwater reservoirs is controlled, but also threatened, by manifold surface–subsurface interactions. Vulnerability studies typically evaluate the variable interplay of surface factors (land management, infiltration patterns) and subsurface factors (hydrostratigraphy, flow properties) in a thorough way, but disregard the resulting groundwater quality. Conversely, hydrogeochemical case studies that address the chemical evolution of groundwater often lack a comprehensive analysis of the structural buildup. In this study, we aim to reconstruct the actual spatial groundwater quality pattern from a synoptic analysis of the hydrostratigraphy, lithostratigraphy, pedology and land use in the Hainich Critical Zone Exploratory (Hainich CZE). This CZE represents a widely distributed yet scarcely described setting of thin-bedded mixed carbonate–siliciclastic strata in hillslope terrains. At the eastern Hainich low-mountain hillslope, bedrock is mainly formed by alternated marine sedimentary rocks of the Upper Muschelkalk (Middle Triassic) that partly host productive groundwater resources. Spatial patterns of the groundwater quality of a 5.4 km long well transect are derived by principal component analysis and hierarchical cluster analysis. Aquifer stratigraphy and geostructural links were deduced from lithological drill core analysis, mineralogical analysis, geophysical borehole logs and mapping data. Maps of preferential recharge zones and recharge potential were deduced from digital (soil) mapping, soil survey data and field measurements of soil hydraulic conductivities (K_S). By attributing spatially variable surface and subsurface conditions, we were able to reconstruct groundwater quality clusters that reflect the type of land management in their preferential recharge areas, aquifer hydraulic conditions and cross-formational exchange via caprock sinkholes or ascending flow. Generally, the aquifer configuration (spatial arrangement of strata, valley incision/outcrops) and related geostructural links (enhanced recharge areas, karst phenomena) control the role of surface factors (input quality and locations) vs. subsurface factors (water–rock interaction, cross-formational flow) for groundwater quality in the multi-layered aquifer system. Our investigation reveals general properties of alternating sequences in hillslope terrains that are prone to forming multi-layered aquifer systems. This synoptic analysis is fundamental and indispensable for a mechanistic understanding of ecological functioning, sustainable resource management and protection.

1 Introduction

Near-surface groundwater reservoirs are increasingly threatened by anthropogenic impact, like the intensification of agricultural land use and water production, while their ecological functioning is still rather unexplored. Generally, the hydrogeochemical and (micro)biological compositions of shallow (< 200 m) groundwater systems are a result of biogeochemical processes and the fluid–rock/soil interactions while travelling through the overburden of the aquifers (Urich, 2002; Canora et al., 2008). Infiltrating precipitation, as the main component of recharge of the surface near groundwater systems, first passes the soils. Thereby, the seepage collects among others dissolved organic carbon (Guggenberger and Kaiser, 2003), colloidal, mineral, organic and organo-mineral suspended particles (Münch et al., 2002; Totsche et al., 2006, 2007; Schrumpf et al., 2013; Gleixner, 2013) and even biota (Dibbern et al., 2014). The residence time of the percolating water in the soils as well as in the unsaturated and saturated zones regulates the chemical and biological quality by the extent to which the partial equilibrium of the rate-limited dissolution, retardation and release processes is achieved (Weigand and Totsche, 1998; Münch et al., 2002). Residence times and thus groundwater quality depend on various surface factors (duration and intensity of precipitation, land management/vegetation, presence of macropores) and subsurface factors (i.e. soil thickness, hydraulic parameters and preferential flow paths).

However, subsurface factors are spatially very variable in fractured rock and carbonate-rock terrains (Cook, 2003), and their relative importance for recharge and groundwater quality is still poorly understood (Gleeson et al., 2009). Sustainable management and resource protection therefore call for a holistic investigation of surface–subsurface architecture and interactions for the reconstruction of the hydrogeological functioning and groundwater quality. Concepts for vulnerability assessment (Hölting et al., 1995; Goldscheider et al., 2000; Witkowski et al., 2003; Vías et al., 2006), for instance, already consider the variable interplay of surface–subsurface factors and compartments in a thorough way, but disregard the evolution of pristine groundwater quality. Conversely, hydrogeochemical case studies (i.e. Suk and Lee, 1999; Moore et al., 2009) often lack a comprehensive analysis of the structural buildup and composition of the entire compartments, whereas the important "biogeoreactor" of the overburden soils is vastly neglected. Studies that include land use and soil groups (Pronk et al., 2009; Steward et al., 2011; Allocca et al., 2014) typically leave groundwater chemistry out of their focus. In this study, we follow a holistic approach for reconstructing the pristine groundwater quality in the aquifers of the Hainich Critical Zone Exploratory (Hainich CZE) in central Germany. Here, the bedrock of fractured, mixed carbonate–siliciclastic alternations represents a widely distributed yet scarcely described geological setting. We synoptically characterize all surface and subsurface compartments involved in the flow of water and transport of matter in the subsurface of the Hainich CZE. The goals of our study are (1) to understand the hydrogeological and biogeochemical functioning of the subsurface in a setting provided by the Hainich CZE that so far has not been adequately addressed, (2) to explore the links and feedbacks between the surface and subsurface, and (3) to demonstrate that a holistic multi-method approach that considers the surface and subsurface factors is indispensable for the mechanistic understanding of the coupling of surface and subsurface compartments, fluid dynamics, biogeochemical element cycling and ecology in the critical zone.

To reach these goals, the following research questions are answered. (i) How is the critical zone comprised and connected in the hillslope setting of a thin-bedded, mixed carbonate–siliciclastic succession? (ii) How do spatial arrangement, particularly outcrop patterns and geostructural links (karst features like caprock sinkhole lineaments), impact compartment connection (intrinsic vulnerability) and groundwater quality? Does groundwater quality in contrasting summit/midslope and footslope wells reflect surface influences? And finally: (iii) what are the main control parameters for groundwater quality? What are the reasons for particular hydrogeochemical conditions within the multi-storey/hillslope aquifer system?

2 Site description

The Hainich Critical Zone Exploratory covers 430 km^2 of a hillslope subcatchment of the Unstrut River in north-western Thuringia, central Germany (Fig. 1). It is bounded by the recipient Unstrut River and the distribution area of the Upper Muschelkalk formations. The Hainich represents a NW–SE orientated geological anticline that is developed topographically as a low-mountain range with a steep western flank and a moderately inclined eastern flank (Jordan and Weder, 1995). The study area is located at the eastern flank of the Hainich hillslope that shows a geostructural buildup with NE dipping strata in the direction of the syncline of Mühlhausen–Bad Langensalza (Kaiser, 1905; König, 1930; Patzelt, 1998; Wätzel, 2007). Tectonic uplift, faulting and tilting of strata are assumed for the Late Cretaceous in analogy to the surrounding horst structures Thüringer Wald (in the south) and Harz (in the north; compare to Voigt et al., 2004; Kley and Voigt, 2008). The outcropping strata in the study area comprise sedimentary rocks (Middle/Upper Muschelkalk and Lower Keuper – subgroups of the Middle Triassic). According to the German stratigraphy (Deutsche Stratigraphische Kommission, 2002), the Diemel formation (Middle Muschelkalk), Trochitenkalk, Meissner, and Warburg formation (Upper Muschelkalk) and the Erfurt formation (Lower Keuper) outcrop in the study area. The Upper Muschelkalk subgroup, which hosts the target aquifers of the Hainich CZE, is organized by bio- and litho-stratigraphic marker beds (Ockert

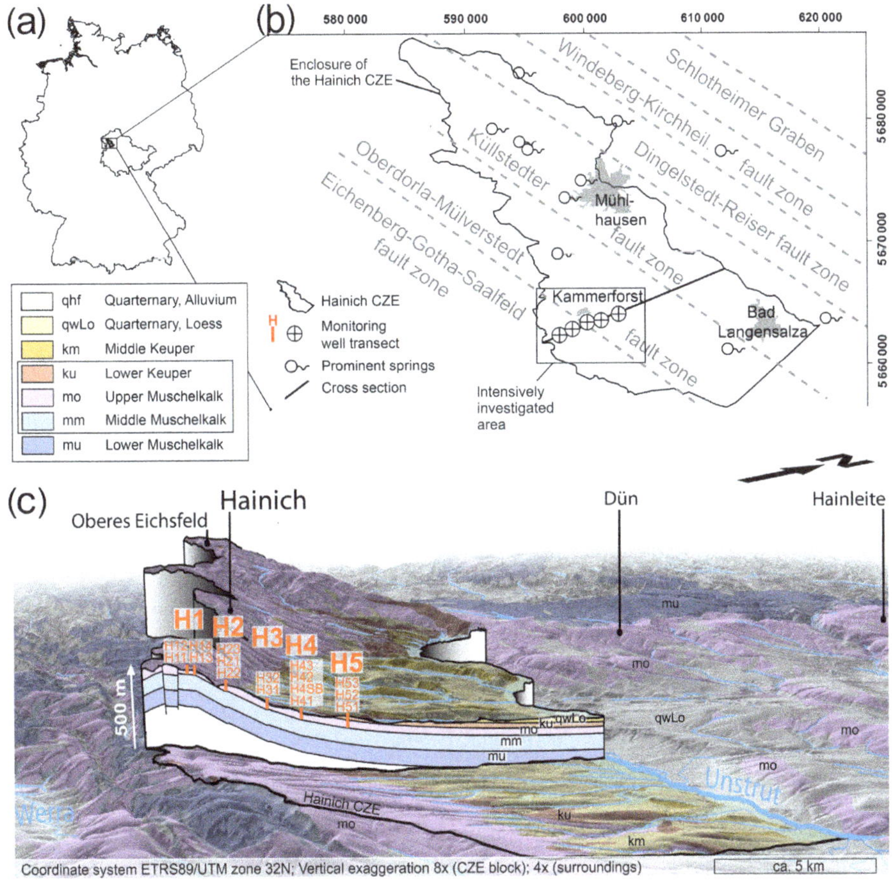

Figure 1. Location of the Hainich CZE **(a, b)**: prominent karst springs in the Hainich CZE are coupled to NE–SW orientated fault zones (**b**; modified from Mempel, 1939, and Jordan and Weder, 1995). **(c)**: geological setting of the eastern Hainich hillslope with monitoring wells of the research transect, accessing Upper Muschelkalk target formations (mo). Data sources: DEM© GeoBasisDE/TLVermGeo, gen. no.: 7/2016.

and Rein, 2000; Kostic and Aigner, 2004). Previous studies hydrostratigraphically organize the Upper Muschelkalk subgroups into a Hainich Transect Lower Aquifer Assemblage (HTL) and a Hainich Transect Upper Aquifer Assemblage (HTU; Küsel et al., 2016). The area belongs to the Cfb climate region (C: warm temperate, f: fully humid, b: warm summer) according to the Köppen–Geiger classification (Kottek et al., 2006) and exhibits a leeward decline in areal precipitation and increasing mean air temperature from the Hainich ridge ($>900\,\mathrm{mm\,year^{-1}}$; 7.5–8 °C) to the Unstrut Valley ($<600\,\mathrm{mm\,year^{-1}}$; 9–9.5 °C; long-term average 1970–2010, TLUG, 2016). The intensively investigated study area is limited to a 29 $\mathrm{km^2}$ subarea of the Hainich CZE, which surrounds the soil and groundwater monitoring transect (Küsel et al., 2016).

The Hainich CZE is a multifarious environmental laboratory for multiscale geo- and bio-science research, as

1. both the geological setting so scarcely described (alternations of marine thin-bedded limestones and marlstones) and the hillslope relief/sloping aquifer configuration are common and widely distributed;

2. the monitoring plot and well transect (see Küsel et al., 2016) provide unique access to the multi-layered aquifer system of the common setting;

3. it represents a rare anthropogenically low-impacted (non-contaminated) cultural region in central Europe with a very extensive type of land management during the last centuries, allowing the investigation of natural (surface) signal transformation in the pristine aquifer systems and of ecosystem functioning;

4. the geostructural and lithological properties were found to be predictable, and thus enable the tracking of biogeochemical cycling and quality development within single aquifer storeys; and

5. the Hainich is a regionally important groundwater recharge area in Thuringia (Fig. 1) and an example of peripheral groundwater supply for water-deficient sed-

imentary basins (Rau and Unger, 1997; Hiekel et al., 2004).

3 Material and methods

For the detailed and synoptic analysis of the critical zone and the aimed reconstruction of the groundwater quality control factors, we apply a multi-method approach that comprises drill core and geophysical-log analysis, geological surveys, soil and groundwater samplings, soil hydraulic measurements and a statistical analysis of groundwater hydrogeochemistry.

3.1 Groundwater well transect

The Upper Muschelkalk strata are accessed by a hillslope transect that consists of multiple wells at five sites for groundwater monitoring, termed H1 to H5. Each site contains one to four drilled wells. Sites H1 (wells H11 to 14) and H2 (H21 to H23) are located in the summit/shoulder region of the catchment, which is covered by forest. Sites H3 (H31/32), H4 (H41 to H43 and additional sweep well H4SB) and H5 (H51 to H53) are situated in the agriculturally used, midslope and footslope regions of the Hainich low-mountain range.

3.2 Hydrogeochemistry and hydraulics

Groundwater levels were recorded by permanently installed data loggers (Orpheus Mini, Ecolog 500/800, OTT Hydromet GmbH, Germany). Groundwater was sampled at least every 4 weeks over 3 hydrological years (November 2013 to October 2015) in 15 permanent monitoring wells (5.1 to 88.5 m final depth below the surface) of five sites along the 5.4 km long well transect (Fig. 1) between 417 and 244 m a.s.l. (metres above mean sea level). The groundwater wells were sampled with submersible motor pumps (MP1, Grundfos, Denmark) or, in the case of very low water levels, with bladder pumps/bailers. The physicochemical parameters of groundwater samples were measured on-site using a flow-through cell equipped with probes for temperature (T), pH (SenTix 980, WTW GmbH, Germany), electrical conductivity, temperature corrected to 25 °C (EC_25) (TetraCon 925), redox potential (Eh) (SenTix ORP 900), and dissolved oxygen (O_2) (FDO 925). Hydrochemical analyses (duplicates) comprised major and minor ions (< 0.45 µm, PES filter) by ICP-OES (725 ES, Varian/Agilent, USA) and ICP-MS (Thermo Fisher Scientific, Germany; Thermo Electron, UK), acid and base neutralizing capacity by acid–base titration, major anions (SO_4^{2-}, Cl^-, NO_3^-, NO_2^-, PO_4^{3-}; PES filter < 0.45 µm) by ion chromatography (DX-120, DIONEX, USA), redox-sensitive parameters (Fe^{2+}, NO_2^-, NH_4^+, HS^-) by colorimetry (DR/890, Hach, USA) and determination of carbon sum parameters (TOC, TIC, DIC, DOC; PES filter < 0.45 µm) by high temperature catalytic oxidation (multi N/C 2100S, AnalytikJena, Germany). Groundwater was classified using a Piper plot (Piper, 1944) following Furtak and Langguth (1967) and Kralik et al. (2005).

3.3 Statistical analysis

SPSS 22 (IBM Corp., USA) and Origin Pro 2015 (OriginLab Corp., USA) were used for descriptive and multivariate statistics of the hydrochemical data. This includes hierarchical cluster analysis (HC) for distinguishing hydrogeochemical groups independently from lithostratigraphy-based hydrostratigraphy. Two different parameter sets were statistically examined at the same number of groundwater analysis from three hydrological years. A complete parameter set (a), including all measured parameters, was chosen to identify the parameters that control the hydrochemical compositions of the groundwater domains by means of principal component analysis (PCA). As redox processes seem to be very distinct in the studied groundwaters (masking other processes), a limited parameter set (b) was defined that does not include ions which are strongly affected by redox processes. The complete parameter set (a) contains EC_25, pH, O_2, Eh, TOC, anions (NO_3^-, Cl^-, SO_4^{2-}), cations (Ca^{2+}, Mg^{2+}, $Mn^{2\text{ and }4+}$, K^+, Na^+, $Fe^{2\text{ and }3+}$, Zn^{2+}, Sr^{2+}, Ba^{2+}) and silica (Si^{4+}). The limited parameter set (b) includes EC_25, pH, Cl^-, Ca^{2+}, Mg^{2+}, K^+, Na^+, Zn^{2+}, Sr^{2+}, Ba^{2+} and Si^{4+}. Data preparation for HC includes a z-score normalization of concentrations. Strongly correlating variables (> 0.9) and constant values were excluded, outliers were eliminated (next neighbour method) and list-wise deletion was carried out (following Backhaus et al., 2016). Then, a Euclidian distance measure was applied and five clusters were chosen based on visual inspection of the scree plot elbow criterion (Cattell, 1966). The phenon line was chosen at a linkage distance of about 67. Thus, samples with a linkage distance lower than 67 were grouped into the same cluster. Finally, Ward's method for clustering was applied (Ward, 1963).

3.4 Field survey

A total of 482 bedrock outcrops were investigated for lithology, flow paths, karst features and the thickness of Quaternary cover sediment. Strike and dip directions were constructed based on the mapped boundary between two prominent formations (Trochitenkalk/Meissner formation). Soil profiles were described for 117 ramcore drillings (GSH 27, Robert Bosch GmbH, Germany; 50 mm and 60 mm diameters, drilled to the bedrock; up to 5 m depths) and 154 supplementary soundings (Pürckhauer, 22 mm). Soil description was carried out according to the German soil survey instruction (Bodenkundliche Kartieranleitung KA5, AG Boden, 2005) and the world reference base (WRB) scheme (IUSS Working Group WRB, 2006). Soil colours were determined using a Munsell soil colour chart.

Soil hydraulic conductivities (K_S) were measured at two depths (10 and 30 cm) at 16 locations (duplicates) within the potential groundwater catchment area by using a Guelph Permeameter (2800K1, Soilmoisture, USA). Soil hydraulic conductivities were calculated according to the equations in Elrick and Reynolds (1992, p. 320ff.). Hydraulic conductivities for each site are calculated as the harmonic average of the topsoil and subsoil (or parent rock in Ah/C soils). Mean values of hydraulic conductivities of soil groups are given as spatially weighted arithmetical means.

3.5 Analysis of drill cores and borehole logs

For the determination of the stratigraphic succession, aquifer properties and mineralogical indicators for groundwater flow, 395 m of drill cores from 12 drillings were investigated. Lithology, fractures/pores and rock weathering were described according to DIN EN ISO 14689-1. We extended this geotechnical rock classification to sedimentary structures, limestone classification (Dunham, 1962), pore classification (Lucia, 1983), degrees of karstification, aquifer types, fracture angles and fracture colours, as well as secondary mineralization on fractures. Additional information about stratigraphy, clay content, fracturing and groundwater inflow was revealed by analysing geophysical open-borehole logs of 10 drillings (sites H1/2/3/4/5, Fig. 1). This included data on calipers, passive gamma-ray radiation, sonic velocity (delay time of sound waves), specific electrical resistivity of rocks, as well as the temperature and specific electrical resistivity of the well water. Gamma-ray curves are interpreted based on the graphical interpretation of high gamma-ray peaks (marlstones) and intersections of gamma-ray curves with an empirically defined "shale line", here at 90 API (separating limestones from marlstones).

Rock-forming minerals were analysed by X-ray diffraction (Bruker D8 Advance, Cu–Kα, 40 kV, 40 mA, Bruker AXS Inc., USA) and Fourier transform infrared spectroscopy (FTIR; Nicolet iS10; Thermo Fisher Scientific; USA).

3.6 Grain size analysis of soils

Grain size analysis of all major soil groups (144 duplicates) was carried out on filtered (< 125 µm), decarbonized (HCl) and organic-free (H_2O_2) samples (Laser Particle Sizer, Analysette 22, Fritsch, Germany).

3.7 Geospatial analysis and construction of preferential groundwater recharge areas

Maps of soil groups and geology were created and jointly analysed with aerial imagery and a digital elevation model (DEM, 2 m resolution) using ArcGIS 10.3 (ESRI Inc., USA). The primary data sets were interpreted in terms of land use types, surface morphology and drainage patterns, with a special focus on karst phenomena like caprock sinkholes, which are considered preferential input structures for funnelled infiltration (Nennstiel, 1933; Mempel, 1939; Hecht, 2003). The two-dimensional correlation of geophysical and geological well logs was carried out with the GeoDIN V.8 well management software (Fugro Consult GmbH, Germany) by graphical correlation of prominent high gamma-ray peaks and stratigraphic marker beds (grainstones/rudstones). The spatial correlation of marlstones presumes that basin-centre marlstones are laterally more continuous than shallow water limestones (Aigner, 1985). In accordance with published examples for multi-storey subsurface architecture (Haag and Kaupenjohann, 2001; Heinz and Aigner, 2003; Klimchouk, 2005; Sharp, 2007), we use the term "aquifer storey" to emphasize our conceptualization of the fine-stratified setting by arbitrary definition of intervals that are dominated by fractured limestone beds and confined at the top and base by unfractured or low permeable beds, the latter with an effective minimum thickness of 80 cm. The degree of aggregation is a compromise between increased detail to account for different hydrogeochemical patterns and the necessary well yield for recovering water samples.

3.8 Soil group map

The soil group map was constructed from different geospatial data sources, including (1) a land management map (interpreted from satellite images and field observations), (2) outcrop areas of surface geology (mapping data) and (3) DEM2. After conversion into raster data, a (4) slope gradient map and a (5) topographical position index (TPI; Weiss, 2001) map with an empirically chosen radius parameter of 75 m were calculated using the "Slope" function of the ArcMap toolbox. The slope position/relief position (i.e. shoulder, midslope) was calculated using the "Relief Analysis" tool of ArcGIS (Deumlich et al., 2010). It considers slope gradient, relative elevation and profile curvature (concave/convex) by using the slope gradient (4) and the TPI raster (5). The spatial distribution of soil groups was calculated with a query function, which considers land management, slope position and slope gradient for the following substrata: "carbonate soil series", "mixed carbonate–siliceous soil series", "claystone series" and "loess loam series".

3.9 Groundwater recharge potential map

For the determination of preferential zones for infiltration/recharge, we calculated qualitative maps of the recharge potential (compare to Muir and Johnson, 1979; Shaban et al., 2006; Deepa et al., 2016) in ArcGIS 10.3 by a weighted linear combination of surface–subsurface properties that influence water infiltration and percolation. The input raster data sets and the respective weight factors were chosen based on expert knowledge and the best fit to measured soil hydraulic conductivities: aquifer storey overburden thickness (40 %), type of bedrock (limestone-dominated vs. marlstone-dominated strata; 15 %), soil classes (15 %), fracture/karst zones (10 %), vegetation and type of land management (10 %), soil thickness (5 %) and slope angle classes (5 %).

4 Results

4.1 Surface characteristics of the Hainich CZE

4.1.1 Relief and surface water network in the groundwater catchment

The study area ranges from the Hainich ridge (topographic water divide) towards the Unstrut Valley and covers altitude regions between 170 and 494 m a.s.l. with an average slope of 35 m km^{-1}. Two orders of relief types are distinguished here: the first order (regional) relief grades from the culmination of the Hainich low-mountain range towards the low-angle midslope (< 5°) and footslope. The second order (local) relief is formed by a bundle of 10 straight, parallel and roughly equidistant, SW–NE orientated transverse valleys with slopes > 5°. Second order relief elements are also the two NW–SE orientated lineaments of more than 80 caprock sinkholes (which are passive karst phenomena) with up to 80 m in diameter. Caprock sinkholes are mostly exhibited on local ridges. A second and parallel lineament of three shallow elongated (uvala-like) karst depressions with a horizontal extent of up to 400 m crosses the lower Hainich hillslope (Fig. 2a). A few small contact springs occur in the study area. Larger springs occur at the lower eastern slopes of the Hainich ridge, tracing regional fault/fracture zones. Headwater areas of creeks, small rivers, transverse valleys and even agricultural drainage ditches are mostly dry. During our monitoring period (November 2013 to May 2016) stream runoff typically occurred from December to March.

4.1.2 Land management and history of land management

Major types of land management are agriculture (crop, pasture) and forest (unmanaged and managed deciduous forest). Forests are to a large extent within the unmanaged Hainich National Park that occupies summit to midslope positions. Extensive forest management has been the dominant type of land management during all time periods, even with no deforestation during the Middle Ages (Otto, 2000). Mode of forest operation shifted from random selection of wood via coppice use (since the 15th century) and planter forestry (since the 20th century) to unmanaged woodland with the foundation of the Hainich National Park in December 1997 (Otto, 2000; Röhling and Safar, 2004). Parts of the study area (12 %), which are actually unmanaged grassland/scrubland areas within the Hainich National Park, had been previously used as a military training area since 1964, particularly for tank training from 1980 until 1990 (Otto, 2000; Poser, 2004). Cultivated grasslands outside the Hainich National Park, which are used both as meadows and pastures, cover parts of the midslope and locally some cleared glades at the shoulder as well as riparian areas of small rivers. Cropland (locally used for wheat, corn and canola production) covers mainly the midslopes and footslopes (Fig. 2a). Agricultural plots are typically large due to organization by the GDR's agricultural production cooperatives (from 1945 to 1990). These plots are now (since 1990) managed by privately owned agricultural cooperatives.

4.2 Soils in the groundwater catchment

4.2.1 Soil distribution and soil development

Soils cover the entire landscape with major soil series developed from carbonate rocks ("carbonate soil series": Rendzic Leptosols to Chromic Cambisols), siliciclastic rocks ("siliceous soil series": Luvisols, Stagnosols) or alluvial sediments (WRB and German soil groups: Table 1). Culmination areas and adjacent shoulder positions are covered with Cambisols. Small plateaus and spurs between transverse valleys in shoulder positions exhibit Chromic Cambisols which grade into Cambic Regosols on the shoulder and Calcaric Regosols on the midslope. Chromic Cambisols are also found in local depressions and old caprock sinkholes. Rendzic Leptosols occur in form of narrow patches in western crestal areas (close to well H11). Luvisols are coupled to the spatial distribution of loess loam in the central and eastern midslope/footslope areas. In case of a very thin loess loam cover, soils are developed as Pelosol-Cambisol and Cambisol. Fluvial soils cover the central parts of headwater areas and the complete valley floor in the lower parts (in the north-east) of the study area. Colluvisols occur at the margins of local valley flanks in the shoulder and midslope area (Fig. 2b). Typical sequences of two superimposed soils comprise (i) Chromic Cambisols (paleosols) developed from marlstones and (ii) Luvisols developed from loess loam with windblown loess sedimentation after the formation of soil (i) (Fier, 2012). Average soil thicknesses are 21 cm (for Rendzic Leptosols), 32 cm (Rendzic Leptosol-Cambisol transitions), 66 cm (Cambisols), 54 cm (Chromic Cambisols), 71 cm (Luvisols), 132 cm (Fluvisols), 91 cm (Stagnosols) and 78 cm (Colluvisols). Subsoils show higher average clay and fine silt content compared to the topsoils of the same soil group (Fig. 3).

4.2.2 Soil hydraulic properties

Average (median) soil hydraulic conductivities (K_s) of the five major soil groups infer that Cambisols and Luvisols (1.3 to 1.5×10^{-4} m s^{-1}) form the most conductive soil cover in the study area, followed by Rendzic Leptsols and Chromic Cambisols (2.5 to 5×10^{-5} m s^{-1}) and Stagnosols (about 6×10^{-7} m s^{-1}; Table 1, Figs. 3 and 8). Topsoils of Chromic Cambisols are considerably more conductive than those in Rendzic Leptosols and Luvisols. Generally subsoils are less

Table 1. Soil groups (WRB, 2003, and AG Boden, 2005), soil texture and soil hydraulic properties.

WRB soil group	Rendzic Leptosols, Calcaric Regosols	Chromic Cambisols	Cambisols	Luvisols	Stagnosols	Fluvic Cambisol
German soil group	Rendzina, Pararendzina	Terra fusca	Braunerde	Parabraunerde	Pseudogley	Vega
Substratum	limestone, marlstone	limestone	marlstone, loess loam	loess loam, marlstone	loess loam, marlstone	alluvial silt + clay
Thickness	5–35 cm	85–100 cm	35–80 cm	50–100 cm	80–190 cm	up to 330 cm
Soil category	medium silty clay	slightly silty clay	medium silty clay	slightly clayey silt	medium silty clay	strongly silty clay
Median grain size (µm) % clay + fine silt	12.6 ± 3.6 34.0 ± 7.8	13.3 ± 4.9 33.9 ± 9.7	13.2 ± 3.6 32.3 ± 7.0	17.3 ± 4.2 27.0 ± 9.4	not measured	17.8 ± 5.0 25.6 ± 9.1
Water storage capacity	low	high	high	medium	very high	very high
Roots	< 35 cm	30–60 cm	20–80 cm	20–180 cm	30–80 cm	15–40 cm
Cracks and borrows	earthworm borrows down to the host rock	earthworm borrows down to the host rock	frequent	deep borrows (voles, worms)	limited to shallow soil horizons	limited to shallow soil horizons
Decarbonatization	no	complete	almost complete	complete	incomplete	no
Typical colour (subsoil)	yellowish grey	dark yellowish grey	brownish grey	yellowish grey	greenish grey	yellowish grey
	10YR3/2	7.5YR4/3	10YR5/6	10YR4/6	2.5Y4/2	10YR4/3
Hydromorphic attributes	no	oxidative	oxidative	oxidative	oxidative + reductive motteling	oxidative + reductive motteling
Soil water	> 200 cm	rarely and > 80 cm	rarely and > 100 cm	rarely and > 100 cm	commonly in 1–2 m	very commonly > 40–60 cm
Morphologic position	midslope + all steep local slopes	culmination + local plateaus	culmination to midslope	footslope, (and midslope)	footslope, valley	valley centre
Land use	forest + military training area	forest + some grassland	cropland, grassland, forest	cropland, grassland, uncommonly forest	grassland, uncommonly forrest	grassland
Anthropogenic changes	compacted (tanks), drained	uncommon	ploughed, compacted	ploughed, compacted	drained	drained
Hydraulic conductivity K_s ($m\,s^{-1}$) (median, ± SD)	5.2×10^{-5} ± 4.9×10^{-5}	2.5×10^{-5} ± 1.2×10^{-5}	1.5×10^{-4} ± 5.2×10^{-5}	1.2×10^{-4} ± 1.2×10^{-4}	6.5×10^{-7}	not measured

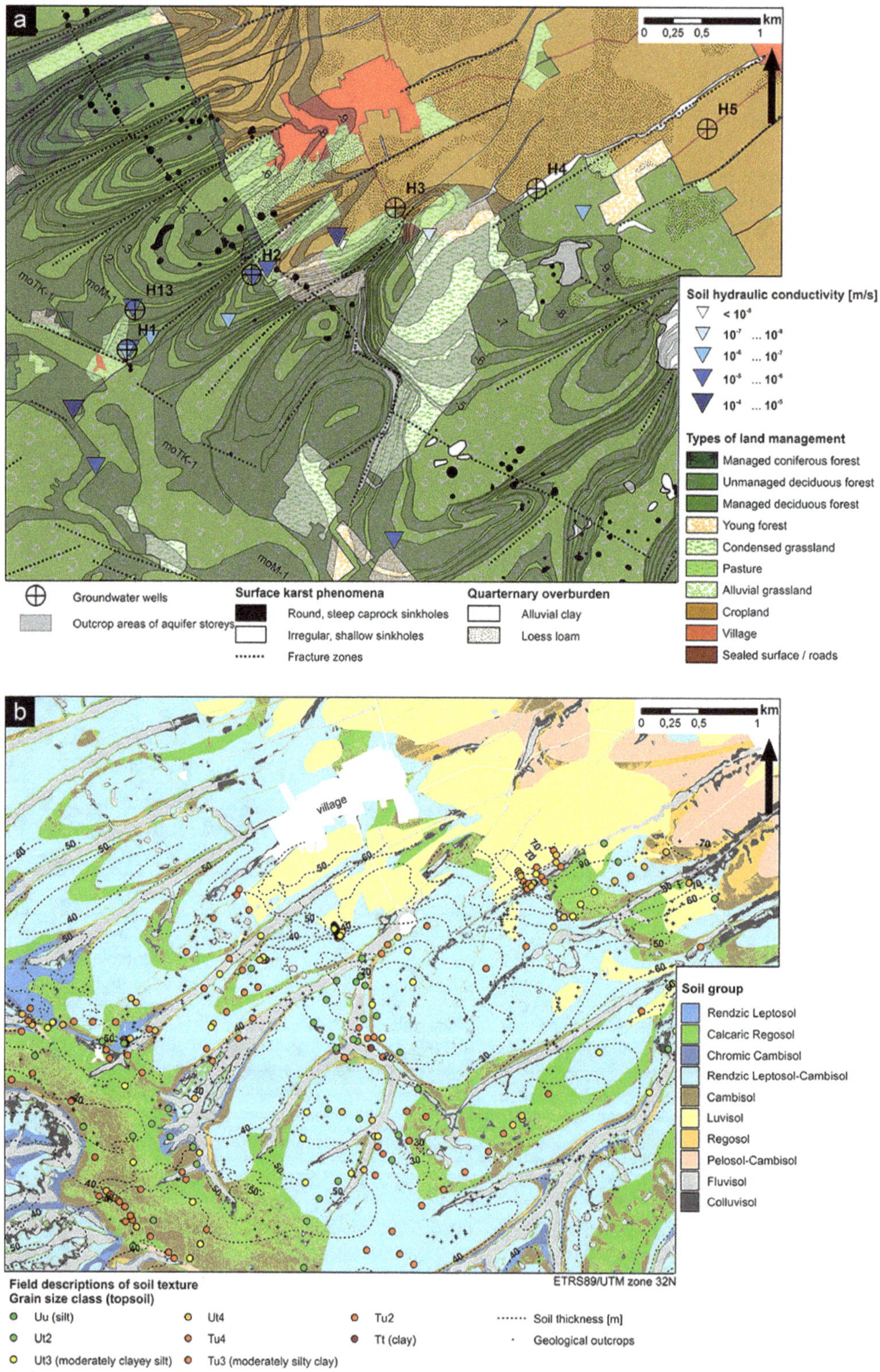

Figure 2. **(a)** Types of land management, outcrop zones of aquifer storeys, sinkhole lineaments, potential fracture zones, and measurement points for soil hydraulic conductivities. Aquifer storeys which are lower in stratigraphy outcrop in higher positions on the Hainich hillslope. The AquaDiva well transect H1 to H5 (also shown) covers hillslope regions from the summit (H1) to the footslope (H5). **(b)** Soil map showing calculated soil groups and mapped calibration data points. Signatures of the mapped soil profiles represent the grain size class (soil category) of the topsoil. Interpolated isolines of mapped soil thickness show increasing thickness towards the north-east and towards the transverse valleys.

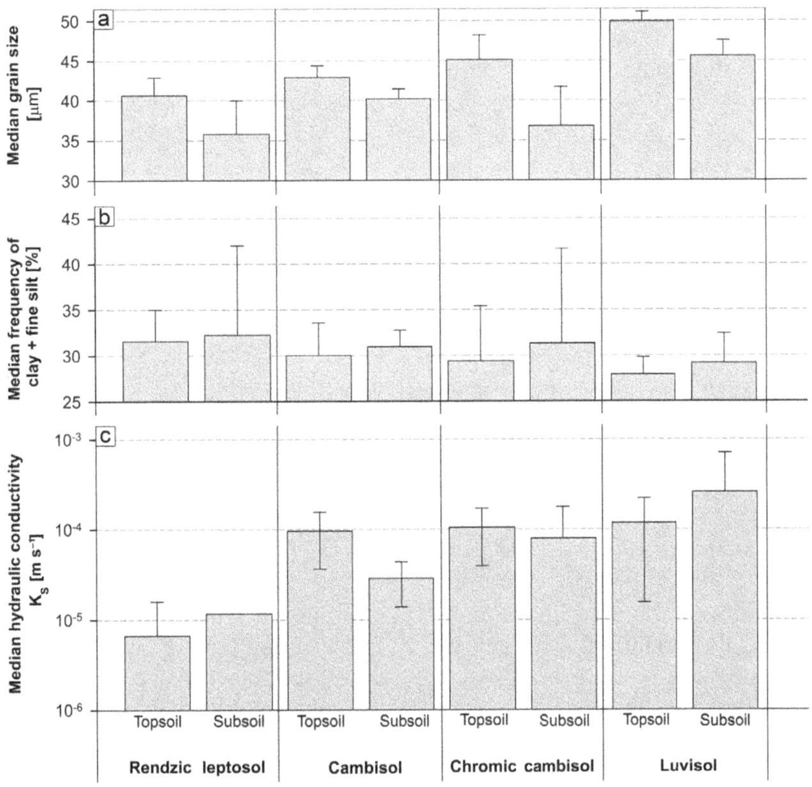

Figure 3. Physical properties of topsoils/subsoils and parent rocks of the four major soil groups. Median grain size decrease **(a)** and the frequencies of the fine grain size fractions (clay + fine silt) increase towards the subsoil **(b)**, respectively. **(c)**: median soil hydraulic conductivities of soil groups showing higher median soil hydraulic conductivity in Chromic Cambisols and Luvisols compared to the Rendzic Leptosols and Cambisols and general decreases in hydraulic conductivity towards the subsoil. The error bars describe the root of the estimation variance of the average.

conductive than topsoils of the same location. Soil hydraulic conductivities (K_S) are essentially uncorrelated with the soil texture (i.e. correlation K_S vs. median grain size: Spearman $r^2 = +0.17$). Soil thickness is uncorrelated with the slope gradient ($r^2 = -0.24$ and barely better if slope positions are correlated individually).

4.2.3 Geological outcrop zones

The landscape of the Hainich low-mountain range has been developed within the Triassic formations of the Upper Muschelkalk and Lower Keuper. All formations outcrop in distinct zones along the eastern slope with dip angles steeper than the slope: from 5° (footslope) to 8° (shoulder position); and lower/older strata outcrop in higher relief positions (Fig. 2a). A NW–SE orientated normal fault is located close to the Hainich summit with offsets of about 10 m (Fig. 2a). Fault-bound troughs (up to 700 m long and 150 m wide) are also found parallel to this fault. The Triassic rocks are covered by young loess loam in sheltered, concave depressions of the east-exposed slopes. Alluvial soils and colluvia fill the valley bottoms with increasing thickness (maximum 3.5 m) and areal cover from the midslope to the footslope (Fig. 2a).

4.3 Subsurface properties of the Hainich CZE

4.3.1 Lithostratigraphy and bedrock mineralogy

The stratigraphic succession is characterized by thin-bedded carbonate–siliciclastic alternations of which only the limestone beds are fractured and therefore represent potential aquifers. The Diemel formation (mmDO) comprises thin (1–3 cm) yellowish dolomitic marlstones and fine crystalline dolomite, and rarely cavernous dolomites without any fossils, gypsum or salt. The 7 m thick Trochitenkalk formation (moTK) with thick (5–30 cm), grey, coarse bioclastic limestones (mainly rudstones with the rock-forming fossil *Encrinus liliformis*) forms a carbonate-rock-fracture aquifer with minor karstification (intrastratal karst according to Ford and Williams, 2007). Large-scale (metre-scale) and consistent fractures, a high fracture index and dissolution-enlarged fractures and vugs (Lucia, 1983), as well as karst breccia at the base of the formation are common features. The fractures found in rock cores are mainly developed as strata-bound (Odling et al., 1999) fractures that are restricted to the limestone beds, as is common in limestone–marlstone alternations (Meier et al., 2015). All fractures and pores are stained with red brownish (10R6/6) coatings. Karst features in the Trochitenkalk formation occur from the near surface down to 90 m in depth. Rock mineralogy is dominated by calcite and very low contents of dolomite with trace amounts of quartz, muscovite, chlorite and Na-/K-feldspar. Based on a consistent (moTK) formation thickness, bed thickness and limestone types (cf. Dunham, 1962) in all wells, lateral facies changes within the Trochitenkalk formation are negligible.

The 34.6 m thick Meissner formation (moM) consists of 5 m thick basal marlstones with three discrete bioclastic limestone beds (Kalkbank $\alpha/\beta/\gamma$), overlain by 29 m of alternating thin (2–5 cm) limestones (mudstones to grainstones) and marlstones. These are covered by a thick (> 60 cm) bioclastic, regional biostratigraphic marker horizon, the Cycloidesbank, which is a known regional aquifer (Grumbt et al., 1997; Hecht, 2003) of minor importance. Limestones in the Meissner formation are fracture aquifers and their rock matrices are predominantly composed of calcite and trace amounts of dolomite, quartz, illite and feldspar. The overlaying 16 m thick Warburg formation (moW) with predominantly marlstones (mineralogically: calcite, dolomite, quartz) and the 35 m thick Erfurt formation (kuE) with dolomite rocks (dolomite, calcite and quartz), siltstones, silty sandstones and claystones (illite, muscovite, chlorite and gypsum) represent both unfractured aquitards (Hoppe, 1952; Hecht, 2003) and form the low permeable top-seal strata of the Upper Muschelkalk aquifer storeys at the footslope positions (Table 2). The Triassic strata are bounded by an erosional unconformity on top, which is overlain by aeolian loess loam developed from loess deposits of the last glacial period (Weichsel glacial in Germany; Greitzke and Fiedler, 1996) and alluvial/colluvial sediments of Holocene age (Rau and Unger, 1997).

4.3.2 Hydrostratigraphy

The correlation of sedimentological core logs and geophysical borehole logs infers that a multi-layer hydrostratigraphy is applicable and that all aquifers and aquitards are presumably continuous (and with a similar thickness and limestone type) on the scale of the research transect (Fig. 4). In the Upper Muschelkalk strata Trochitenkalk formation (moTK) and Meissner formation (moM), the alternated bedding of limestones, marlstones and claystones results in permeable fracture aquifers (dense matrix, varying fracturing) and low permeable marlstones (dense matrix, unfractured). Ten aquifer storeys are newly defined for the Hainich CZE. Of these, the Trochitenkalk formation contains one aquifer storey (moTK-1) and the Meissner formation contains nine aquifer storeys (moM-1 to moM-9; Table 2). All footslope aquifer storeys are overlain by impermeable cap rocks of the Warburg formation (Upper Muschelkalk) and the Erfurt formation (Lower Keuper). Flow paths are predominantly fractures and matrix porosity is lower than 5 % in all stratigraphic intervals. Although it is of a karst-fracture type with partially solution-enlarged fractures, the karstification and the development of conduits are limited and concentrated at the formations' very base. Below moTK-1, the impermeable dolomitic marlstones of the Diemel formation (mmDO) form a hydraulic basis seal. We observed an intact and unweathered base seal in all wells that were drilled to the base of moTK-1. The Meissner formation (moM) contains limestone-fracture aquifers which are interbedded marlstone aquitards on the decimetre to metre scale. Limestones of this formation are almost exclusively fracture aquifers with very little matrix porosity, concentrated at certain thickly bedded limestone marker beds.

4.3.3 Land use types and soils within the outcrop zones of aquifer storeys

The outcrop zones of the two lowermost aquifer storeys (moTK-1, moM-1) are predominantly covered by forest, whereas the outcrop zones of stratigraphically higher storeys show mixed types of land management, including forest, cropland and pasture (Fig. 5a). Within the aquifer outcrop zones, the four major soil groups (Rendzic Leptosol, Leptic Cambisol, Cambisol and Chromic Cambisol) are present in different spatial proportions (Fig. 5b). Significant proportions of Luvisol cover the outcrop areas of moM-7, moM-9 and moTK-1. Stagnosol and Colluvisol are restricted to the outcrop area of mm, moTK-1 and moM1-4. The grain size classes of subsoils show a slight trend of increasing silt content towards the outcrop zones of stratigraphically higher aquifer storeys (moTK-1 to moM-9). Argillaceous soils occur frequently in moM-4 to moM-6. Pure clay as a subsoil category is restricted to moM-2 and moM-7 (Fig. 5c).

4.4 Groundwater chemistry

4.4.1 Groundwater classification

The groundwater in the carbonate rock landscape is classified as weakly to strongly mineralized earthalkaline, bicarbonatic type (site H1/2/3/4), and earthalkaline, bicarbonatic-sulfatic type (cf. TLUG, 1996) for groundwater of deep well H51 (moTK-1). The order of abundance for dissolved ions is $Ca^{2+} > Mg^{2+} > Na^{+} > K^{+}$ and $HCO_3^{-} > SO_4^{2-} > Cl^{-}$. According to the groundwater subtypes of Kralik et al. (2005), the moM waters (except for well H14) plot in the "dolomite" field, and moTK-1 waters (wells H13/21) and water from H14 (moM) plot in the "calcite" field. All other moTK-1 wells (H31/41/51) are classified as the mixed subtype ("calcite+dolomite"). Two contact springs in the recharge area (Grauröder Quelle, Ihlefeldquelle) and two karst springs in the discharge area (Kainspring, Melchiorbrunnen, coupled to NW–SE orientated fault zones) are classified as carbonate-earth alkaline type predominantly in contact with calcite in the recharge area and a calcite–dolomite mixing type in the discharge area. Groundwater chemistry in aquifer storeys in

Table 2. Hydrostratigraphic standard section: relative positions of aquifer storeys (this study) in comparison with a published classification (Küsel et al., 2016) in the context of the German stratigraphy and the timescale.

Timescale (DSK, 2002; Gradstein et al., 2004) (Ma)	Level above base of moTK (m)	German stratigraphy (DSK, 2002)	Regional stratigraphy (central Germany, Seidel, 2003)	Aquifer assemblages (Küsel et al., 2016)		Aquifer storeys (this study) X aquifer –aquitard	Level above base of moTK (m)	Groundwater well (screen section)
0.01...0			soils alluvial sediments (qhf)					
0.1...0.01			loess (qwLo)					
232.5...0.1					erosional unconformity			
235.0...232.5	61.3...>90.0	Erfurt formation (kuE)	Graue Mergel (kuGM)		–			
			Sandsteinkomplex 1 (kuS1)		X	kuS1-2	74.5...76.5	
					X	kuS1-1	68.0...70.0	
			Grenzschichten (kuGR)		–			
	43.4...61.3	Warburg formation (moW)	Glasplatten (moCGP)		–			
			Glaukonitbank (moCG)		X	moW-1	53.0...54.0	
			Zinkblendebank		–			
			Fischschuppenschichten (moCFU)		–	moM-9	42.0...45.0	
					X			
			Cycloidesbank (moCC)		X			
238.5...235.0	8.7...43.4	Meissner formation (moM)	Discitesschichten (moCD)	HTU (Hainich transect upper aquifer assemblage)	X	moM-8	38.0...40.5	H42, H43, H53
			Gervilleienschichten (moCGV)		–			
					X	moM-7	34.0...36.0	H4SB
					–			
					X	moM-6	31.0...33.7	H32, H4SB
					–			
					X	moM-5	27.0...29.0	H32
					–			
					X	moM-4	22.9...24.5	H52
					–			
					X	moM-3	18.9...21.8	H52
					–			
					X	moM-2	15.2...18.3	H23
					–			
					X	moM-1	8.0...13.2	H14
					–			
>238.5	0.0...8.7	Trochitenkalk formation (moTK)	Trochitenkalk (moT)	HTL (lower aquifer assemblage)	X	moTK-1	0.0...6.6	H11, H12, H13, H21, H22, H31, H41, H51

deeper stratigraphic positions exhibits high O_2, SO_4^{2-}, and Sr^{2+} and low Mg^{2+}, HCO_3^-, Si^{4+}, K^+, Na^+, and NO_3^- concentrations. Shallow aquifer storeys are partly anoxic in mid-slope/footslope positions and deeper aquifer storeys are oxic in all slope positions (Fig. 6). Aquifers in footslope wells show higher concentrations of K^+, Mg^{2+}, Si^{4+}, Sr^{2+}, SO_4^{2-} and Cl^- compared to the same aquifer in summit positions.

4.4.2 Statistics of hydrogeochemistry and chemical composition of groundwater groups

Hierarchical cluster analysis revealed five clusters. Cluster 1 comprises the groundwater samples from upper slope moTK-1 wells (H13/21) and moM-1 well H14. Cluster 2 encompasses the groundwater samples from moTK-1 wells H31/41

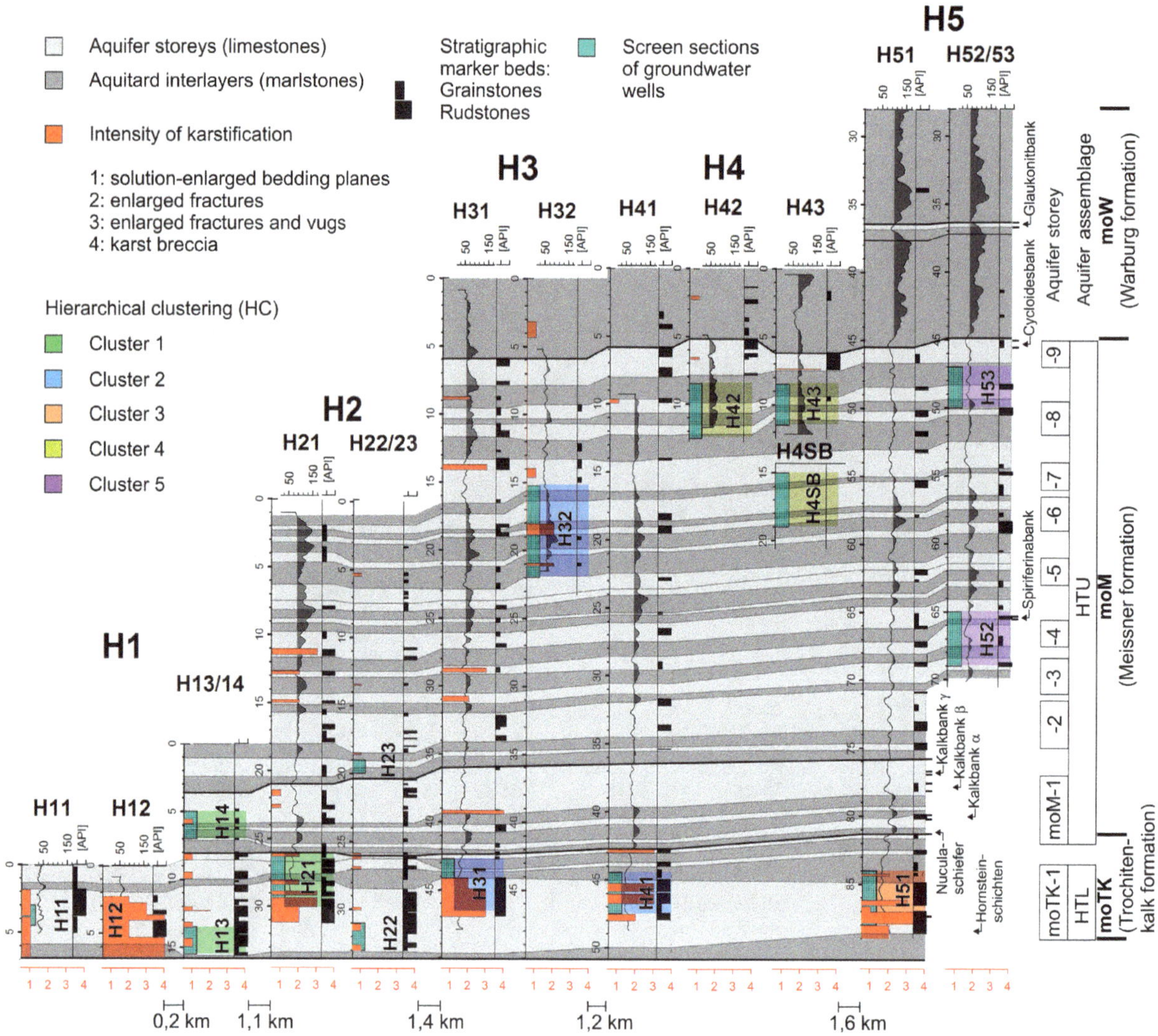

Figure 4. Graphical correlation of marlstone–claystone intervals in gamma-ray logs, biostratigraphic limestone marker beds (grainstones/rudstones) and the intensity of karstification (red bars and red scale bar below: karstification increases in the direction: solution-enlarged bedding planes to karst breccia). The geological aquifer correlation is cross-checked with the hierarchical clustering of hydrochemical parameters.

and also from moM-5/6 well H32. Groundwater samples from moTK-1 well H51 are plotted in cluster 3. Cluster 4 includes moM-6/7/8 wells H42/43/4SB, while cluster 5 consists of two moM-3/4/8 wells of the same site (H52/53; Fig. 7). Multi-case clustering does not occur. As a result of the PCA using the complete parameter set (Fig. 7a) and the parameter set without redox-sensitive parameters (Fig. 7b), data points fall into very similar groupings compared to the HC. According to the PCA of the complete parameter set, the first two components (PC1 plus PC2) explain 54.9 % (variances of PC1: 34.6 %; and PC2: 20.4 %), and according to the parameter set without redox-sensitive parameters, PC1 plus PC2 explain 62.1 % (variances of PC1: 35.4 %; and PC2: 26.8 %) of the total variability, respectively. For both PCAs, the parameters with the highest factor loads are Si^{4+}, K^+, Na^+, HCO_3^-, and Mg^{2+} for PC1 and Sr^{2+}, SO_4^{2-}, EC_25, and Cl^- for PC2.

The average chemical composition of the five groundwater clusters is significantly different with respect to their physico-chemistry and major ion composition. Cluster 1 shows high pH and Eh, as well as high concentrations in dissolved O_2, TOC and Ca^{2+}. High concentrations in HCO_3^-, NO_3^{2-}, TOC, Cl^- and Mg^{2+} are distinctive for cluster 2, whereas cluster 3 is characterized by very high SO_4^{2-}, Ca^{2+}, Cl^-, and Sr^{2+} concentrations in combination with high redox potential and moderate K^+ and Na^+ content. Cluster 4 shows very high concentrations in $Fe^{2+/3+}$, $Mn^{2+/4+}$, TOC, and Cl^- in combination with very low redox potential and very low dissolved oxygen. The latter is also applied to clus-

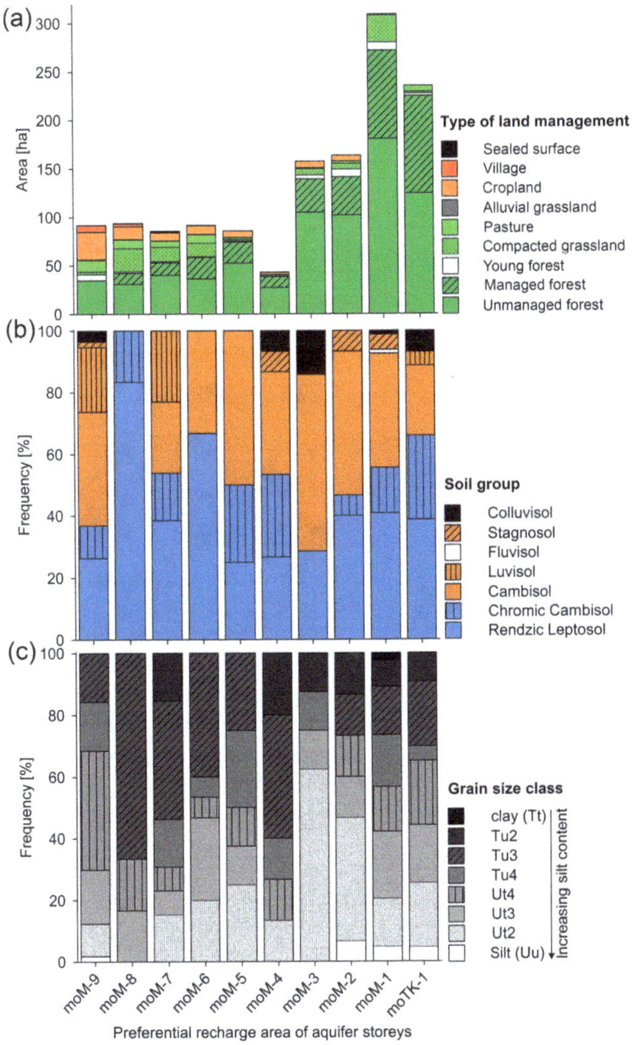

Figure 5. Surface and subsurface properties of the aquifer recharge zones (moTK-1 to moM-9) on the Hainich hillslope. Area sizes are related to the 29 km^2 area of this study. **(a)** Absolute abundances of land management types in the preferential recharge areas of the aquifer storeys. The two basal aquifer storeys are characterized by the largest aquifer outcrop areas and the highest amounts of forest within these areas. Agricultural land management increases towards the higher aquifer storeys (moM-2 to moM-9). **(b)** Relative abundances of soil groups within the aquifer outcrop zones showing high diversity in all aquifer outcrop zones depending on slope positions and Quaternary loess loam/alluvial clay coverage. **(c)** Grain size classes of the topsoils and subsoils in the aquifer outcrop zones.

ter 5, which additionally contains high concentrations in K^+, Na^+ and Mg^{2+} (Fig. 9).

4.4.3 Fluctuations of groundwater levels and quality

Groundwater levels are confined in footslope wells. In the wells of site H5, groundwater rises to 30 m (H53), 50 m (H52) and 70 m (H51) higher than the base of the screen section. Fluctuations of groundwater levels in our monitoring wells range from 1 to 3 m in the hilltop recharge area (well site H13) to more than 25 m in the groundwater transit area (well site H5). The groundwater level fluctuations show a strong seasonality, with highest levels in March to April and lowest levels in October to December. Monitoring wells of different sites and screen depths differ in average concentrations of the major solutes. These spatial differences are generally higher than the seasonal fluctuations in the wells. Exceptions to these conditions are marked seasonal fluctuations of Ca^{2+} (monitoring well H31/41), Cl^- (H52/53), K^+ (H41), Mg^{2+} (H41), Na^+ (H31/41), Si^{4+} (H41) and SO_4^{2-} (H31/41).

4.4.4 Recharge potential map

The recharge potential is defined here as a qualitative measure of the probability of infiltration, percolation and groundwater recharge. The maps (Fig. 8) visualize potential spatial variation in infiltration recharge, waiving spatiotemporal variable conditions like precipitation characteristics and antecedent soil moisture. The recharge area maps of two selected aquifer storeys (Fig. 8a and b) show the largest recharge potential for their outcrop areas on the Hainich hillslope. Increasing overburden thickness is accompanied by a drastic decrease in recharge potential towards the NE. With respect to the relief position, the greatest recharge potential is assumed for local valleys and in the extension of sinkhole lineaments, although these areas show increased soil thicknesses. In areas with greater thicknesses of the overburden strata, valleys are considered areas of preferential recharge. A slightly higher recharge potential is assumed for pasture and cropland areas compared to forests. The same holds for the flat summit/culmination and upper slope areas, while the slopes of transverse valleys show greater surface water runoff.

5 Discussion

5.1 Hydrostratigraphy

5.1.1 Aquifer structure evaluated by lithology

The lithologically defined aquifer–aquitard succession in combination with the increase in karstification towards the base, and the preferential presence of groundwater at the base of limestone packages, point to confined aquifers with stratigraphic flow control (Klimchouk and Ford, 2000; Goldscheider and Drew, 2007). As hydraulic heads and chemical groundwater compositions of different aquifers at the same sites (H13/14 (cluster 1); H41/42 (clusters 2 and 4); H51/53 (clusters 3 and 5)) are significantly distinct, vertical communication between aquifer storeys is highly reduced. Thus we interpret the thin aquifer beds as confined, "sandwich flow type" aquifers (term: White, 1969; Klimchouk, 2005) which are interbedded with confining aquitards. The assumed lat-

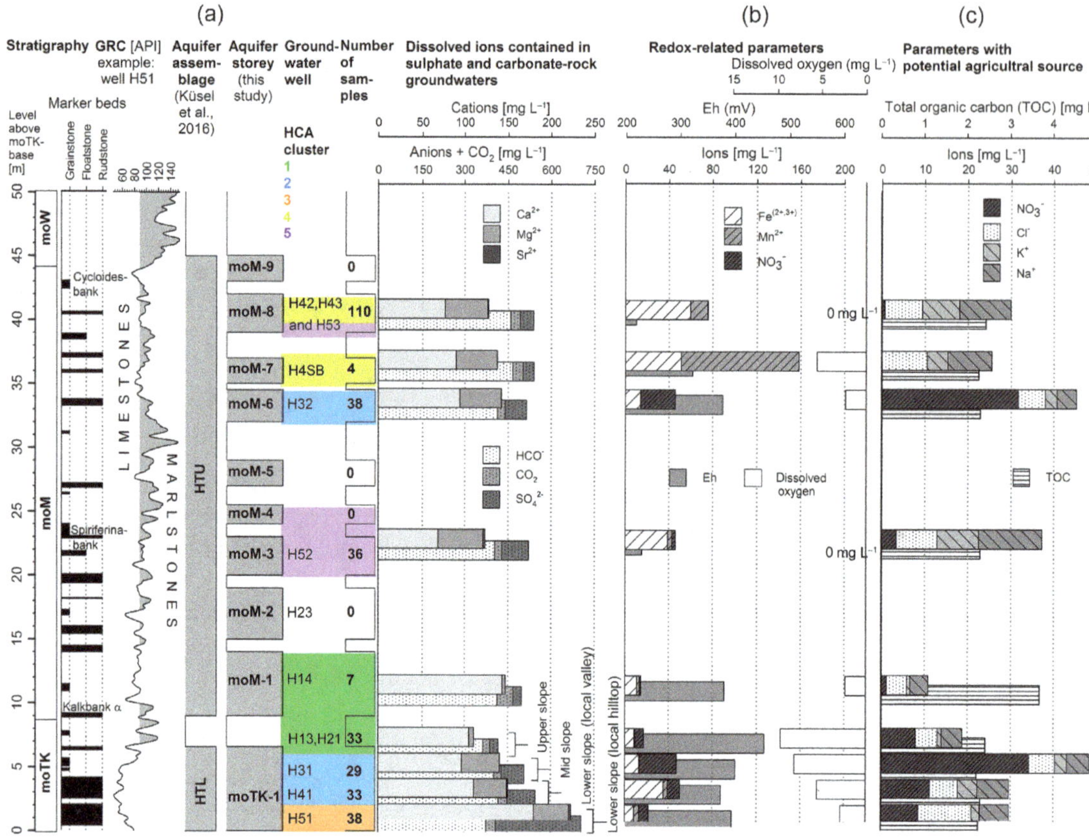

Figure 6. Stratigraphic succession of the Upper Muschelkalk with stratigraphic marker beds **(a)**, a gamma-ray log, aquifer assemblages HTL/HTU, aquifer storeys moTK-1 to moM-9 and the average chemical compositions of monitored groundwater (**a**: ions related to carbonate/sulfate minerals; **b**: redox-sensitive ions; **c**: ions which are potentially related to the type of land management). The colour code of the hierarchical clusters is identical to Fig. 4.

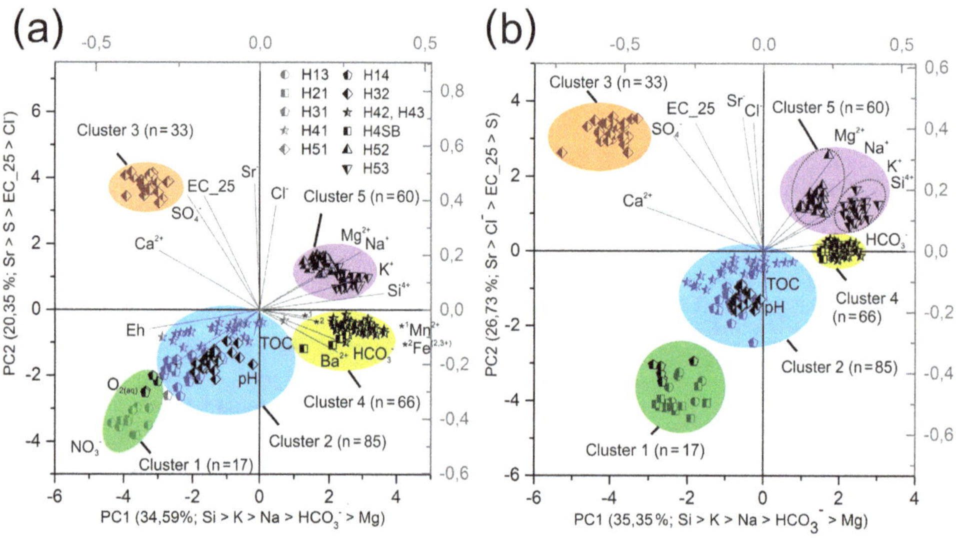

Figure 7. Principal component analysis (PCA) biplot for the complete parameter set **(a)** and for the limited parameter set **(b)** without redox-related parameters. Five clusters can be distinguished in both parameter sets. Samples within the clusters are identical for both PCA plots. Factor loads for PC1 and PC2 are displayed as labels of the x/y-axis as well as the sequence of factor loads for individual components. The colour code of the hierarchical clusters is identical to Fig. 4.

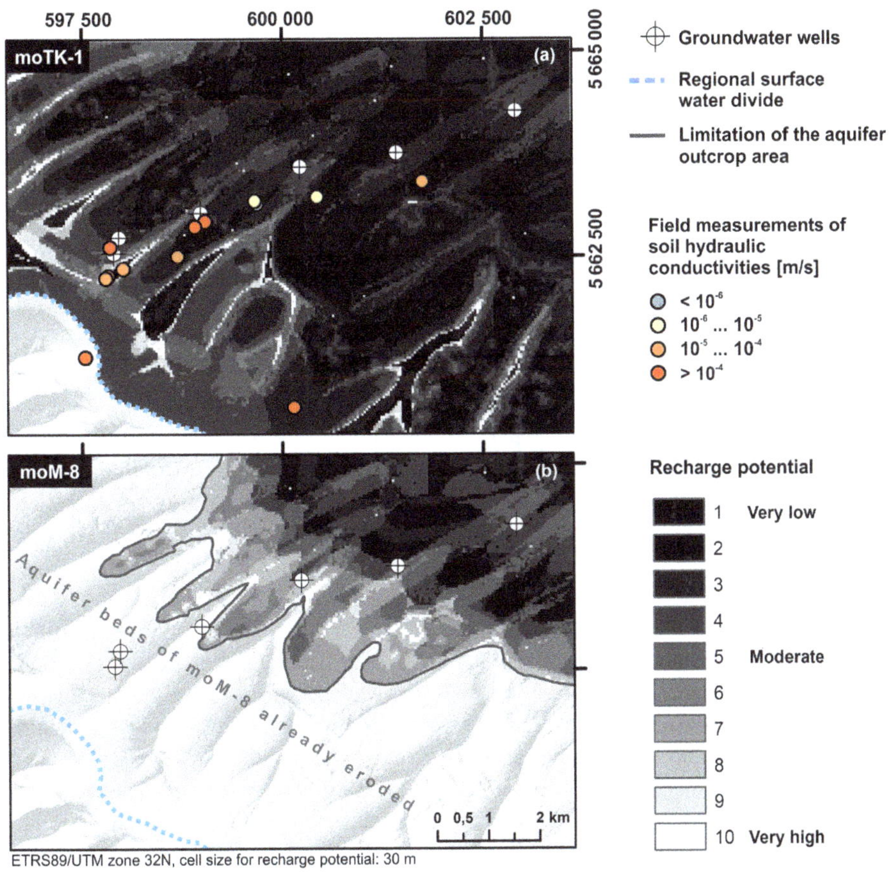

Figure 8. Recharge potential maps for the two contrasting aquifer storeys moTK-1 **(a)** and moM-8 **(b)**. The recharge potential is a qualitative indicator for infiltration and percolation towards selected aquifer storeys: the brighter the output colour, the higher the recharge potential. Aquifer outcrop zones show the highest recharge potentials followed by sinkhole lineaments and (NE–SW orientated) transverse valleys. The recharge potential decreases drastically with increasing thickness of overburden strata towards the NE. Data points of field measurements are displayed in the upper map (moTK-1 recharge potential) and are colour-coded with respect to their soil hydraulic conductivity. Data sources: DEM© GeoBasisDE/TLVermGeo, gen. no.: 7/2016.

eral continuity of aquifers and aquitards results in a "layer-cake" aquifer architecture which has been demonstrated for the Upper Muschelkalk of central Europe (Aigner, 1982, 1985; Merz, 1987; Simon, 1997; Borkhataria et al., 2005). Secondary mineralizations, which indicate groundwater flow (for instance Liesegang banding), occur on limestone fracture walls. Fracturing of aquifer rocks increases close to the (NE–SW striking) fault zones (Hoppe, 1962) and karstification follows the network of faults (Goldscheider and Drew, 2007). Red and brown iron and manganese oxides on upper to midslope fractures (H1/2/3; clusters 1 and 2) indicate temporarily unsaturated conditions, whereas green/grey fracture minerals in the footslope domains (H4/H5; clusters 4 and 5) point to permanently water-saturated conditions due to the lack of oxidized Fe/Mn minerals. Corrosion occurs in the form of intrastratal karstification, i.e. one or more layers of soluble strata are covered or sandwiched between insoluble beds (Palmer, 1995). Karstification is consistently high in the thick-bedded aquifer storey (moTK-1) and less pronounced in thin-bedded aquifer storeys (moM-1/4/5/6; Fig. 4) as the bed thickness controls the continuity, spacing and width of joints (Goldscheider and Drew, 2007). In general, the lithological definition of hydrostratigraphy allows a structuring with a high spatial resolution.

5.1.2 Evaluation of groundwater chemistry

Stratigraphically deep aquifers in summit positions are grouped in cluster 1 (aquifer storeys moTK-1 and moM-1). Cluster 2 (moTK-1 and moM-5+6) encompasses deep and stratigraphically intermediate aquifer storeys in midslope positions and cluster 3 (moTK-1) a deep aquifer storey in the footslope position. Shallow aquifer storeys (moM-6/7/8) in the midslope position within a local valley are grouped in cluster 4 and intermediately deep aquifer storeys (moM-3/4/8) are grouped in cluster 5. According to this grouping, the slope position and the depth below the surface is important, likewise to the depth level within the hydrostratigraphy (the aquifer storey).

PCA using the complete parameter set (a) infers that each aquifer storey contains more than one type of groundwater chemistry (Fig. 7). By contrast, the PCA (b) carried out with the limited parameter set (without redox-related parameters) shows a clear separation of moTK-1 and moM-1-9 aquifer storeys (except for cluster 2). According to the PCA, cluster 1 encompasses high factor charges in pH, Eh, O_2, and NO_3^- and low factor charges in Mg^{2+}, Na^+, K^+, and Si^{4+}. The same is less markedly applied for cluster 2, whereas factor charges in cluster 5 point in opposite directions. We explain this linear trend of cluster grouping as a consequence of a trend of the chemical evolution from summit wells (cluster 1) with small recharge areas and shallow aquifers towards midslope (cluster 2) and footslope (cluster 5) passages with thicker overburden strata. The latter are affected by minor surface influences and longer groundwater travel times/residence times within the bedrock. Data points of different aquifer storeys are grouped (but within the clusters), and slope position and aquifer depths seem to be more important for groundwater chemistry than the position in hydrostratigraphy, here. Apart from this trend, cluster 3, which is limited to the deepest well (in footslope position), exhibits high factor charges in Ca^{2+}, SO_4^{2-}, EC_25, and Sr^{2+} and low charges in Ba^{2+}, HCO_3^-, $Fe^{3/4+}$, and $Mn^{2/4+}$, whereas cluster 4 (limited to shallow wells in midslope position) shows opposite factor charges for these parameters. In general, clusters are grouped in a line of chemical development and by oxic/anoxic conditions and different mineralizations, depending on the position in aquifer stratigraphy.

5.1.3 Modes of groundwater flow

Within the aquifer–aquitard sandwich, fast conduit groundwater flow, which is typical for karstified carbonate rocks (Wong et al., 2012), likely takes place in the moTK-1 and partially in the moM-1 aquifer storeys, whereas slow diffusion in slightly fractured, thin aquifers beds is anticipated in the moM-2 to moM-9 aquifer storeys. Confined diffuse flow is considered laminar and takes place in interparticle pores and fractures of dense limestones with low primary porosity (Smart and Hobbs, 1986). This results in the well oxygenated moTK-1 and moM-1 groundwaters and a significant oxygen consumption/deficiency (coupled to the mobility of Fe^{2+} and Mn^{2+} ions and the low mobility of NO_3^- and SO_4^- ions) in the moM-2 to moM-9 groundwater, resulting in completely different milieu conditions for the biogeochemical processes and for the life in the subsurface as well.

5.2 Surface–subsurface connectivity

Potential zones for interactions between surface and subsurface waters are the outcrop areas of the aquifers, incised valleys (cutting into the stratigraphic succession), springs and surface karst phenomena. Nearby faults and valleys, an increased number of additional non-strata-bound, connected fractures that provide pathways for pronounced and preferential fluid flow (Meier et al., 2015) are likely.

5.2.1 Preferential recharge areas

As a typical feature of aquifer–aquitard alternations in a tilted hillslope setting, preferential recharge takes place at the aquifer outcrop zones (Andreu et al., 2011; Fig. 2a). Based on the tectonically tilted and subsequently exhumed hillslope, the main aquifer recharge area is situated in the upper hillslope/summit area, which is covered by forest with very low anthropogenic impact. A predictable geological structure allows a general tracking of flow paths from the recharge to discharge areas. The second route for preferential recharge is related to the lines of caprock sinkhole lineaments, which can be tracked over more than 4 km in the midslope between transect locations H2/3 and H4. The origin of these sinkholes does not lie in the karstification of the Upper Muschelkalk strata itself, as the combination of aquifer fracture networks (tight conduits) and stabilizing, insoluble aquitard beds does not cause sufficient mass deficits that would allow hanging wall collapses. Here, caprock sinkholes are related to mass deficits by subrosion in the underlying evaporite rocks with gypsum and halite (Mempel, 1939; Malcher, 2014). As caprock sinkholes are arranged in lineaments (Fig. 2a) parallel to regionally known fault orientations, it is very likely that they are coupled to the penetration of surface–subsurface water at fracture zones (compare to Smart and Hobbs, 1986; Worthington, 1999; Klimchouk, 2005) that promote preferential recharge (Smart and Hobbs, 1986; Suschka, 2007). Although dissolution of soluble rocks is limited within the target aquifers of this study, it is reasonable that collapse structures are accompanied by enhanced rock fracturing and permeability.

5.2.2 Influences of soils in the recharge area

The spatial distribution of the soil groups reflects the outcrop zones of limestones/marlstones (Greitzke and Fiedler, 1996; Brandtner, 1997; Rau and Unger, 1997) and the succession of aquifer storeys in the summit/shoulder area of the hillslope which exhibits little Quaternary rock coverage and thin soils (Rendzic Leptosol, Chromic Cambisol) with favourable infiltration properties in the aquifer outcrop areas of moTK-1 and moM-1/6/8. Thicker relictic Chromic Cambisols which are formed by intensive decarbonatization (Rau and Unger, 1997; AG Boden, 2005) are restricted to accumulations in former depressions (i.e. caprock sinkholes) in shoulder regions of the hillslope (moTK-1 and moM-1 outcrops). Besides these depressions, Chromic Cambisols with low hydraulic conductivities are typically the subsoil layer of sequences with two layer superimposed soils, resulting in lateral soil interflow (Ali et al., 2011) as is observed in our shoulder and midslope regions during the measurement of soil hydraulic conductivities. Soils on unfractured marlstone/claystone aquitards are typically Calcaric Regosols and Stagnosols with low hydraulic conductivities of both soils and parent rocks. According to our data set, soil thickness is primarily controlled by the slope gradient of transverse valleys with increasing soil thickness and water storage options towards the centre of valleys. As a result of the increasing coverage of parent rocks by loess loam from the regional midslope to the footslope (related to solifluction of wind-blown dust; Kleber, 1991; Bullmann, 2010), loess loam becomes the dominating soil substratum in the outcrop areas of aquifer storeys moM 7/8/9. In this area, Luvisols cover local ridges and Luvisol–Stagnosol transitions (related to the continuous vertical clay relocation, Rau and Unger, 1997) cover local valleys, whereas older caprock sinkholes are either filled by Colluvisols or with loess loam, leading to either poor or very good infiltration/recharge properties. In the same context, Bachmair et al. (2009) consider the role of micro-depressions in soils as one of the major factors in preferential recharge.

Although median grain size, grain size sorting and the standard deviation of grain size fractions are comparable in all soils, "carbonate series soils" (Rendzic Leptosol and Chromic Cambisol) offer lower hydraulic conductivities than "siliceous series soils" (Cambisol, Luvisol, Stagnosol) by 1 order of magnitude. Soil hydraulic conductivities (K_S) are uncorrelated with soil texture properties. For this reason a linear transfer function (K_S vs. median grain size or grain size category) for extending the 16 K_S measurements to the 271 mapped soil profiles was not possible with the actual data set. A low correlation between K_S and texture is related to strong influences of soil structure and aggregation (Totsche et al., 2017) rather than texture on K_S. Structural parameters are for instance the content and proportion of macro- and micro-aggregates, a hierarchical aggregate system, the presence and frequency of secondary pores and anthropogenic changes like former plough traces or traces of forestry machines. Bachmair et al. (2009) for instance identify tillage-related macropores as a main factor in deep infiltration into cropland soils, whereas litter layers inhibit infiltration into forest soils. Moreover, preferential infiltration through biomacropores (i.e. earthworm borrows, root channels) that bypass the soil matrix are of great significance in aggregated and argillaceous soils with low matrix conductivity for the infiltration during heavy rain events (Deurer et al., 2003; Weiler and Naef, 2003; Blouin et al., 2013). Klaus et al. (2013) identify vertical macropores of anecic earthworms as a major flow control in a hillslope tile drain system. Wienhöfer and Zehe (2014) additionally consider loose, litter-rich topsoils, lateral preferential pathways, preferential pathways at the soil–bedrock surface and bedrock topography.

5.2.3 Influences of land management in the recharge area on groundwater quality

The forest areas as a source of groundwater from near-catchment summit/shoulder wells are confirmed by their aquifer outcrop areas (moTK-1, moM-1) within the managed (and partly unmanaged) forest. In contrast, we detected potentially agriculture-related substances (NO_3^-, K^+, Cl^-) in the wells drilled to all aquifer storeys in midslope location wells (H31/41), although these groundwater types are recharged mainly within the forest (up to 96.5 % forest). These surface signals are interpreted as being related to the cropland and village areas within the preferential recharge zones with sinkhole lineaments. Anoxic groundwaters in aquifers (moM8/9) partially recharged from outcrop areas with agricultural and village land use are likely due to microbial oxygen depletion resulting from the degradation of organic carbon or oxidation of inorganic electron donors. For shallow wells in valleys (H42/43/4SB), lateral soil water inflow with high organic/fertilizer load towards the valley provides an alternative way for the enhancement of oxygen depletion. Much lower NO_3^-, K^+, and Cl^- concentrations in the footslope wells (H52/53) are likely a consequence of the argillaceous caprocks that increase in thickness towards the footslope.

5.2.4 Utilization of the recharge potential for interpreting groundwater quality

Due to the recharge potential map (Fig. 8), thick, limestone-dominated aquifer storeys in the deeper sections of the hydrostratigraphy (i.e. moTK-1) are mainly recharged in their outcrop areas and subordinarily in sinkhole lineaments and transverse valleys. Within these areas, Ah-Cv soils (i.e. Rendzic Leptosol) and fractured limestone bedrocks offer the highest recharge potential. A drastic decrease in recharge potential with increasing overburden strata is confirmed by low concentrations in surface-related substances (i.e. NO_3^-.Cl^-, K^+) in well sites H1/2/4/5 for the aquifer storey moTK-1. Increasing concentrations of ions, which are related to the dissolution of the carbonate (aquifer) bedrock (Ca^{2+}, Mg^{2+}, HCO_3^- and Sr^{2+}) point to a chemical evolution of groundwater in the direction of discharge (H1 to H5). A moderate to high recharge potential combined with cropland/pasture areas) in the recharge direction of site H3 is reflected by increased concentrations in TOC, NO_3^-, Cl^-, K^+, Na^+ and O_2 (Figs. 6 and 9). Aquifer storeys with thin-bedded limestone–marlstone alternations (i.e. moM-8, Fig. 9) generally show a lower recharge potential in their aquifer outcrop zones compared to thick, limestone-dominated aquifers (moTK-1), which is related to thicker soils and a high marlstone–limestone ratio. This is reflected in the groundwater quality of moM-8 wells (H42/43/53) with low Eh and low concentrations in O_2, pointing to slow flow velocities and long residence time within the argillaceous soils and marlstone-dominated and thinly fractured bedrocks.

5.2.5 Flow directions

The four modes of presumed groundwater flow are (1) vertical percolation, (2) bedding-parallel flow, (3) descending, cross-formational flow and (4) ascending cross-formational flow. (1) Vertical percolation through the soils and the unsaturated zone takes place in the whole Hainich CZE catchment area, following diffuse/non-point infiltration. As the aquitard interbeds highly reduce the vertical flow connections, outcrop zones of the aquifer storeys as well as caprock sinkholes act as important preferential infiltration pathways which are typical for hillslope recharge zones. Since caprock sinkholes remain dry, even directly after precipitation events, high infiltration rates can be assumed for these structures. (2) An undisturbed sequence of aquifers and aquitards with constant bed characteristics, comparable groundwater chemistry within the same aquifer storey and the high pressure heads (well H51/52/53: 70/50/20 m rise within the well pipe) are interpreted to indicate bedding-parallel flow (stratigraphic flow control; Goldscheider, 2005) in the groundwater transit zone. This is supported by the spatial variation of groundwater chemistry within the same aquifer storey (i.e. an increase in K^+, Na^+, Mg^{2+} and Si^{4+} within moM-8 from well sites H4 to H5), given a cut-off from surface influences there. Also, the combination of saturated conditions within the shallow storey and temporarily unsaturated conditions within the lower aquifer storey in well site H3 points to highly reduced vertical flow. Bedding-parallel flow is also supposed for fractures with small fault displacements that do not exceed the thickness of the interbedded aquitard (compare to Goldscheider and Drew, 2007). (3) Zones with enhanced fracture indices (fractures / m drill core) in well site H3 as well as consistent Fe/Mn-oxide fracture walls in all aquifer storeys of this site point to a quick and cross-formational descending flow of oxygenated groundwater (thus with low Fe/Mn mobility; Hem, 1985) via vertical master joints (term: Dreybrodt, 1988; Ford and Williams, 2007) close to potential fracture zones. This is supported by higher concentrations in Na^+, K^+, NO_3^-, Cl^- and TOC in the deeper site of H3 compared to the shallower well of this site, related to agriculture and fertilizing (Matthess, 1994; Kunkel et al., 2004). Cross-formational descending flow in shattered or fractured rocks is assumed to take place in fracture zones (Worthington, 1999; Goldscheider and Drew, 2007), tracked by lineaments of caprock sinkholes (Mempel, 1939; Hoppe, 1962; Smart and Hobbs, 1986; Jordan and Weder, 1995). (4) Groundwater chemistry of footslope well H51 (screen base at the base of the moTK-1 aquifer) bears a combination of high concentrations in Ca^{2+}, SO_4^{2-}, Sr^{2+}, and Cl^- (and SO_4^{2-} / HCO_3^- ratios) and low concentrations in K^+, Na^+, and NO_3^- that points to a non-surface import of sulfate (chloride) groundwater due to dissolution of evaporite rocks (Edmunds and Smedley, 2000). The location of this lowermost aquifer storey directly above dolomite, gypsum and halite beds (Middle Muschelkalk; Jordan and Weder, 1995) points to ascending cross-formational groundwater flow (Garleb, 2002; Völker and Völker, 2002). The ascent of groundwater, which is well known for artesian settings (Klimchouk, 2005), could be related to hydraulically confined conditions as measured for site H5 and the assessed groundwater flow crossing a sinkhole lineament between well sites H4 and H5 (Fig. 10). Alternatively, the ascent of Middle Muschelkalk groundwaters is probably related to a decrease in permeability within the Middle Muschelkalk aquifers, as subrosion of sulfate rocks (succeeding from the summit to the footslope region) has not reached the groundwater transit zone yet, resulting in the presence of low permeable sulfate rocks and an ascent of groundwaters discharging the Middle Muschelkalk (Hoppe and Seidel, 1974; Garleb, 2002).

5.3 Controls on groundwater quality and groundwater provenance

Cluster 1 comprises the shallow to intermediate deep moTK-1 (H13/21) and moM-1 (H14) wells with base depths of screen sections between 7.0 and 30.5 m. Groundwater chemistry with high HCO_3^- and Ca^{2+} as well as low Mg^{2+}, SO_4^{2-}, Na^+, K^+, and Si^{4+} concentrations (Fig. 9) correspond to the pure calcite limestone ($CaCO_3$) aquifer lithology. Low concentrations in Cl^-, K^+, Na^+, and Mg^{2+} are related to parts of the catchment area that are exclusively used as forest and to soils (mainly Rendzic Leptosols) with high hydraulic conductivities and with little clay content, resulting in short residence times of water in soils. Relatively high concentrations in TOC are linked to little organic degradation due to short travel times. Fast infiltration also arises from aquifer outcrop zones in culmination areas, allowing a high ratio of infiltration vs. lateral soil interflow. According to the PCA, cluster 1 wells could be interpreted as one end-member at the beginning of a "groundwater development series". Generally, groundwater compositions in summit regions of hillslopes are comparable to the composition of the bedrock. Surface signals (i.e. soil or vegetation-derived organic carbon) can be traced more clearly due to little organic degradation (and low filtering) along the flow path by the soils and bedrock pores/fractures.

Cluster 2 encompasses moderately deep moTK-1 wells (H31/41) and one moM-5/6 well (H32) with base depths of screen sections between 22.0 and 47.5 m. The hydrochemistry with moderate concentrations in Ca^{2+}, Mg^{2+}, and SO_4^{2-} and low K^+ and Na^+ as well as high Mg / Ca ratios is not linked to the pure calcite limestones (moTK-1) or the calcite/illite/kaolinite/chlorite mineral composition (moM-5/6) and requires an external source of Mg^{2+}. Nitrate and TOC contents are higher than expected by the forest (moTK-1)/forest-dominated type of land management (moM-5/6) in the aquifer outcrop zones and are likely ex-

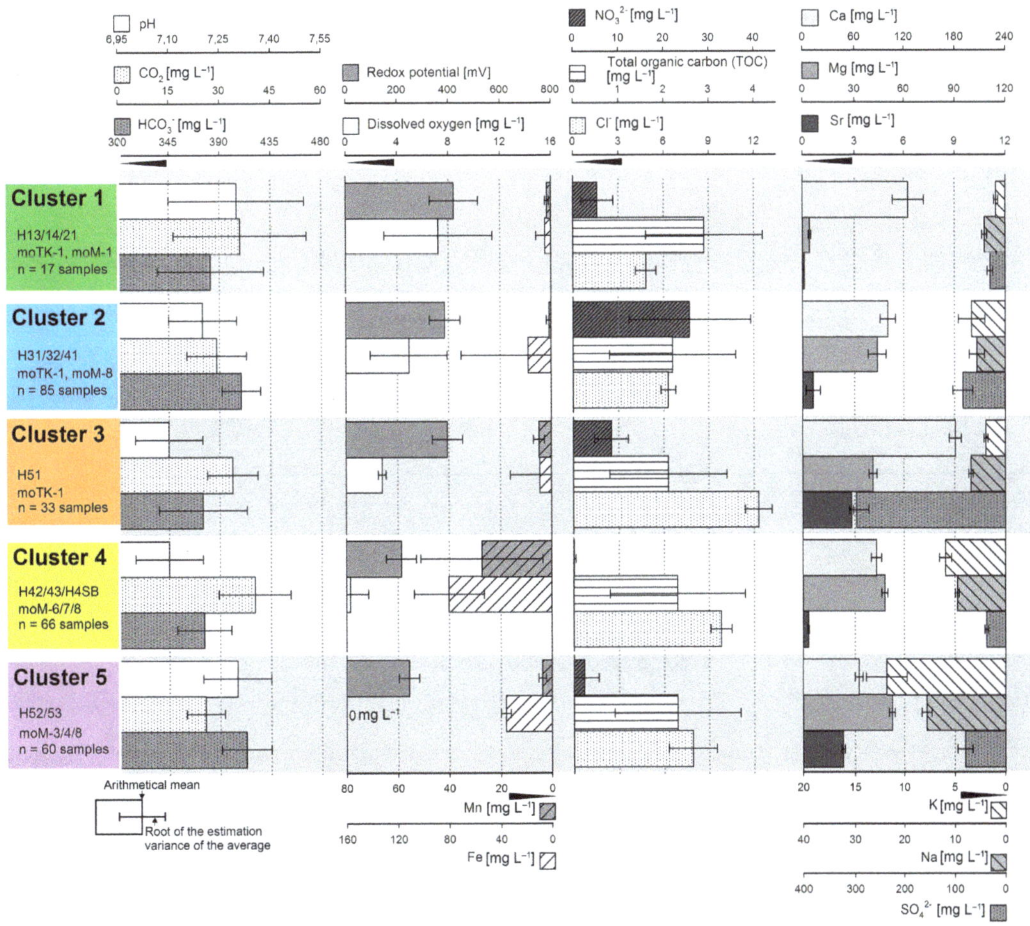

Figure 9. Average groundwater chemistry of hierarchical clusters 1–5 (left: groundwater wells and aquifer storeys within the clusters). First column: parameters related to the carbonate–CO_2 equilibrum; second column: dissolved O_2, redox potential and redox-sensitive ions; third column: ions which are potentially related to land management; fourth column: ions related to the dissolution of either carbonate/sulfate or clay minerals. The colour code of hierarchical clusters is identical to Fig. 4.

plained by cropland/pasture in a sinkhole lineament in the recharge zones of sites H3/4. Oxygenated groundwater in both aquifer storeys (thus with low Fe/Mn mobility; Hem, 1985) and a high degree of rock fracturing and fracture mineralization with Fe-/Mn-oxide minerals (aquifer storeys moM-9/8/7 and 6), point to a vertical penetration with near-surface groundwater via fracture zones through all aquifer storeys. Enhanced concentrations in Na^+, K^+, NO_3^-, Cl^- and TOC (Fig. 9) are likely related to agriculture and fertilizing (Matthess, 1994; Kunkel et al., 2004) around the sinkhole lineaments. As nitrate, which is generally derived from agricultural fertilizers (Agrawal et al., 1999; Jeong, 2001), is still present in the deep aquifer waters of site H3, vertical bypassing through master joints (term: Dreybrodt, 1988; Ford and Williams, 2007) must be faster than the denitrification process. Quick infiltration is here mostly related to the preferential sinkhole recharge, as the soils (Luvisol/Cambisol, both with low-conductive subsoils and a high degree of lateral soil interflow) in the outcrop zones bear only moderate to poor soil hydraulic conductivities. In general, groundwater in the discharge of exceptionally highly conductive preferential recharge spots can reflect surface signals, even in stratified aquifer–aquitard successions. These zones are therefore suggested to be of uppermost importance for groundwater protection.

Cluster 3 contains groundwater of the deepest moTK-1 well (screen base 88 m) drilled to a cropland plot in the foot-slope area. As is discussed in Sect. 5.2.5, ascending sulfatic groundwater is likely responsible for the SO_4^{2-}–HCO_3^- type of groundwater. High chloride concentration is interpreted with a long groundwater travel distance. Very weak surface signals (in the form of low nitrate and TOC concentrations) attest to the inhibition of vertical infiltration by aquitards and a likely result from long groundwater travel distances (high residence time) from the forest recharge zone of the moTK-1 aquifer. However, high groundwater oxygen concentration in more than 5 km from the capture zones requires a very quick oxygen supply by more rapid flow events. In general, deeper

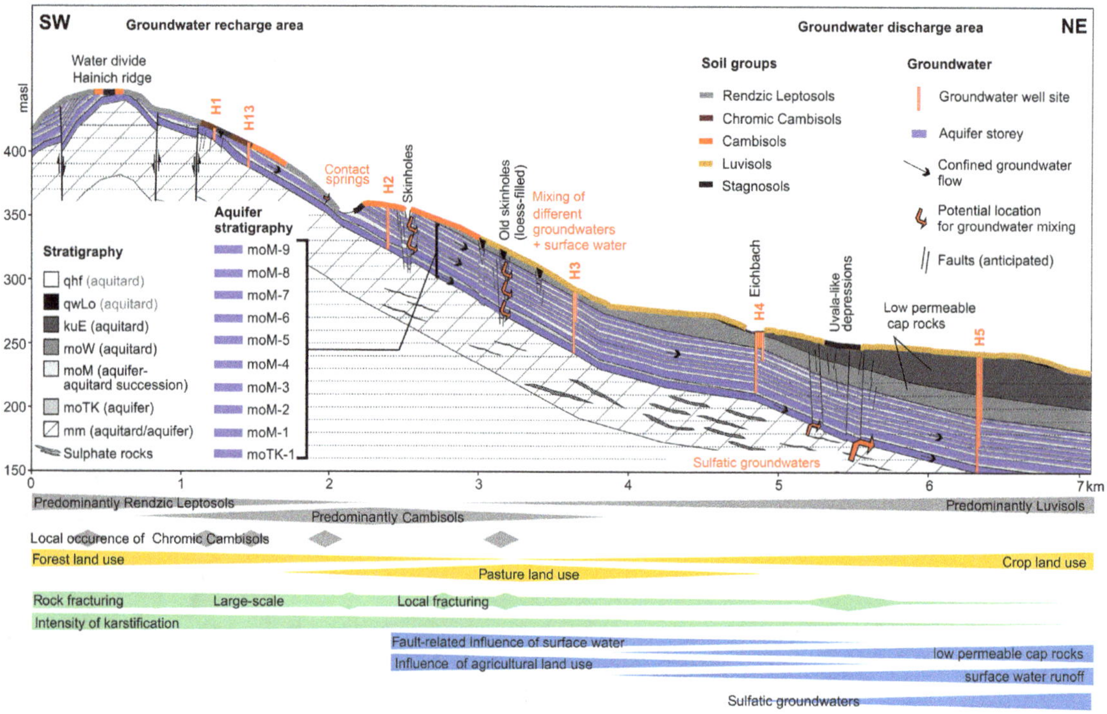

Figure 10. Conceptual cross section of the Hainich hillslope showing the overall geological structure with the 10 aquifer storeys illustrating preferential aquifer recharge zones in the summit to midslope region as well as the discharge direction. Lineaments of caprock sinkholes crossing the potential groundwater flowpath represent potential zones for descending or ascending cross-formational flow and also the mixing of different types of groundwater.

groundwater in the transit zone may bear surface signals, if flow velocities are great enough for transporting young groundwater into the subsurface. This is applied for the more karstified aquifer storeys in aquifer–aquitard successions.

Cluster 4 includes groundwater of the shallow moM-6/7/8 wells H42/43/H4SB with screen depths in 11.5 to 19.0 m. Low-conductive alluvial cover sediments reduce vertical infiltration drastically. High HCO_3^- and low Na^+, K^+, Si^{4+}, and Sr^{2+} concentrations as well as the moderately high Mg / Ca ratio in the groundwater reflect the calcite and dolomite aquifer mineralogy. High $Fe^{(2+ \text{ and } 3+)}$, $Mn^{(2+ \text{ and } 4+)}$ and low SO_4^{2-} concentrations correspond to a low redox potential and the absence of dissolved oxygen, controlling the mobility of these ions (Hem, 1985; Hsu et al., 2010). Although the aquifer catchment of moM-6/7/8 is of a mixed forest/agriculture/village type, TOC and NO_3^- concentrations are low due to the consumption of organic matter and the denitrification under anoxic conditions (Agrawal et al., 1999). Relatively high Cl^-, K^+ concentrations are probably remnants of agriculture-/village-related surface signals. The discrepancy of a close distance to the recharge area and the low oxygen content could be explained by low hydraulic subsoil conductivities of the moM-7 outcrop area (Luvisols, Stagnosols, Cambisols) and very narrow fractures in the aquifer rocks of moM-8. Generally, aquifer confinements in valleys of hillslope aquifers could result in low aquifer quality, directly depending on the fracture sizes and soil hydraulic conductivities.

Cluster 5 encompasses groundwater from deep (50.0–69.0 m screen depths) aquifer storeys moM-3/4/8 (wells H52/53) which are covered by very thick (> 40 m), low permeable cover strata (Erfurt formation, Warburg formation). High K^+, Na^+,Mg^{2+} and Si^{4+} concentrations correspond to long residence times of groundwater (Bakalowicz, 1994; Khan and Umar, 2010) in marlstone–limestone successions due to the distant (> 3 km) recharge zones. If the grouping of groundwater types in the PCA represents a "chemical evolution series", then cluster 5 stands for the most developed/oldest groundwater. In general, caprock-covered footslope aquifers with narrow fractures likely contain more developed groundwater that mostly reflects subsurface influences and fewer surface influences (i.e. in comparison with more conductive aquifers of the same well site – H5).

5.4 Assessment of vulnerability factors

The configuration of the aquifer system in the hillslope setting also controls the groundwater resource vulnerability. With our multi-method investigation of the different subsurface compartments, we also revealed factors of the areas' intrinsic vulnerability. Characteristics of intrinsic vulnerability are solely controlled by hydrogeological properties of the aquifer and overburden (Vrba and Zoporozec, 1994), and integrate the inaccessibility (i.e. by low-permeable cover strata) of the saturated zone and the attenuation capacity (retention, turnover) of the overburden (Adams and Foster, 1992).

The stacking of aquifers/aquitards, the coverage with caprocks and the lateral continuity of strata reduce the overall dominating disperse infiltration and, thus, the intrinsic vulnerability. Threats to single aquifer storeys or assemblages are predominantly related to fractures/faults (valleys) or karst phenomena bypassing the protective cover of soils and unsaturated zones (Fig. 8). A moderate degree of physical filtering by the narrow fractures, which are the predominant flow paths, is assumed. Also, the claystone and marlstone interlayers bear a certain filtering of contaminants by retention. The preferential recharge/outcrop zones of the aquifer storeys are characterized by generally the highest vulnerability. However, these zones, located in the summit to upper midslope of low-mountain ranges, are mostly covered by forest. Generally, the summit position of outcrop zones of the main aquifer storey (moTK-1) lowers the risk of contamination, paradoxically due to the thin soils that prevented lasting agricultural use and settlements. Further zones of higher vulnerability are located in the discharge of sinkhole lineaments or fracture zones that bypass surface water or drains directly into the aquifers. A mapping and structural investigation of these geostructural links will be recommended for proper dimensioning of drinking water protection zones. In comparison to classical karst sites (i.e. massive carbonates) with pronounced karst phenomena (Doerfliger et al., 1999; Witowski et al., 2002), our portrayed setting shows an overall low to moderate vulnerability that is locally elevated by inherent factors (outcrop zones) and inherited features (i.e. karstification of the underlying Middle Muschelkalk subgroup).

6 Conclusions

We applied a multi-method approach for a detailed investigation of Critical Zone functioning and reconstruction of groundwater quality in the hitherto scarcely described setting of thin-bedded alternating carbonate-/siliciclastic rocks in a hillslope terrain. For the Hainich Critical Zone Exploratory that offers unique access to a multi-storey groundwater system in the common and widely distributed geological setting, we found the following factor manifestations of local geology, relief, soils and land use for recharge and chemical evolution of groundwater:

- Low-permeable marlstone beds within a marine succession of high lateral continuity represent a number of aquitards that cause a multi-storey hydrostratigraphy of the Upper Muschelkalk formations.
- As a multi-storey hydrostratigraphy exhibits limited vertical percolation, the outcrop zones of dipping aquifer storeys become very important as preferential surface-recharge areas for inputs of matter and energy.
- Diffuse fracture flow dominates over karst/conduit flow in the mixed/multi-layered lithology. Subsurface water flow predominantly takes place in bedding-plane parallel mode/in stratabound fractures of the limestone beds, and it is trackable from the recharge areas along the storeys.
- From summit to footslope positions, travel distances and presumably groundwater ages generally increase. For the individual storeys, however, travel distances to monitoring wells decrease in the downslope direction, whereas their groundwater ages very likely increase due to lower fracturing and higher retention. In the same direction, surface controls (i.e. nutrient input) decrease and subsurface controls (water–rock interaction) increase.
- Compared to more vulnerable settings (i.e. massive carbonate karst, open karst), the mixed carbonate-siliciclastic alternations exhibit moderate intrinsic vulnerability. This is due to lateral continuity of low permeable interbeds, soil covers and caprocks, of which the latter successively increase in thickness towards the footslope. Areas downstream the caprock sinkhole lineaments (and likely transverse valleys) are likely more threatened by anthropogenic (mostly agricultural) inputs.
- The quality of groundwater resources with peripheral hillslope recharge benefits from extensive land management or, ideally (managed/unmanaged) forest coverage, and reveals the importance of recharge area protection.

In general, for mixed carbonate–siliciclastic rock alternations that are prone to developing multi-layered aquifer systems, both the aquifer configuration (spatial arrangement of strata, hillside cutting, outcrop positions) and the related geostructural links (preferential recharge areas, karst phenomena) are major controls of impacting surface and subsurface factors. For the studied type of thin-bedded carbonate aquifer setting, we could demonstrate that a comprehensive investigation of aquifer connectivity in the transit–discharge area as well as soil cover and land use in the recharge area is mandatory and must be rated as indispensable for a thorough understanding of the state and evolution of groundwater quality. Linking groundwater hydrochemistry mostly to surface factors such as land use would result in a contradiction, as groundwater chemistry does not reflect the type of land use in immediate proximity to the wells. Furthermore, footslope wells' hydrochemistry is strongly impacted by aquifer stratigraphy, karst phenomena input and the cross-formational ascent of sulfatic groundwater that could not be evaluated without spatial lithostratigraphic data. Geostructural investigation and mapping are essential for the localization of aquifer outcrop areas and the assumption of stratigraphy-controlled flow directions. In case the soil group and the type of land use had been neglected, discrimination between influences by natural and anthropogenic controls would have been ambiguous. If the data set included hydrochemistry and multivariate statistics only, the interpretation of this data set would have been ambiguous also, as different chemical surface–subsurface sources result in similar hydrogeochemical compositions.

Recent studies of the CRC AquaDiva focus on signal transit and transformations of surface signals. This is for instance applied to surface-sourced organic matter and microorganisms (Küsel et al., 2016; Schwab et al., 2017; Lazar et al., 2017) to further investigate subsurface connectivity, surface–subsurface interactions and functions of microbial life in groundwater environments. Further CZ exploration should also aim at the investigation of deeper strata connection by regional groundwater flow, hydraulic properties and proportions of unsaturated zones and matter processing within.

Author contributions. BK and RL contributed equally to this article. RL and BK organized and BK conducted the geological mapping, soil mapping, as well as the analysis and correlation of drill cores and geophysical logs. RL organized and partly conducted the groundwater monitoring and obtained authorization from authorities and landowners/companies. PS carried out parts of the soil mapping/description and hydrochemical analyses and carried out the descriptive and multivariate statistics. KUT, SET and KK provided editorial comments on the manuscript. KUT designed and coordinated the project and supervised the field work, data analysis, data presentation and manuscript preparation. KUT, KK and SET are speakers of the CRC AquaDiva.

Competing interests. The authors declare that they have no conflict of interest.

Acknowledgements. The work has been (partly) funded by the Deutsche Forschungsgemeinschaft (DFG) CRC 1076 "AquaDiva" and the state of Thuringia "ProExzellenz" initiative AquaDiv@Jena (107-1). Field work permits were issued by the responsible state environmental offices of Thuringia, local authorities and landowners. We thank Christine Hess and Maria Fabisch for scientific coordination. Heiko Minkmar supported the field surveys and conducted water and ramcore samplings. Tanja Reiff conducted the Guelph Permeameter measurements. Sören Drabesch and Marie Mollenkopf constructed the conceptual soil map. Matthias Händel carried out the FT-IR and XRD measurements on bulk rock samples.

Edited by: Alberto Guadagnini

References

Adams, B. and Foster, S. S. D.: Land-Surface Zoning for Groundwater Protection, Water Environ. J., 6, 312–319, 1992.

AG Boden: Bodenkundliche Kartieranleitung, 5. Edition, Ad-hoc-Arbeitsgruppe Boden der Geologischen Landesämter und der Bundesanstalt für Geowissenschaften und Rohstoffe der Bundesregierung Deutschland, Schweizerbart, Stuttgart, Hannover, 438 pp., 2005.

Agrawal, G. D., Lunkad, S. K., and Malkhed, T.: Diffuse agricultural nitrate pollution of groundwaters in India, Water Sci. Technol., 39, 67–75, 1999.

Aigner, T.: Calcareous tempestites: storm-dominated stratification in Upper Muschelkalk limestones (Middle Triassic, SW-Germany), in: Cyclic event stratification, edited by: Einsele, G. and Seilacher, A., Springer, Berlin, Heidelberg, New York, 190–198, 1982.

Aigner, T.: Storm depositional systems: Dynamic stratigraphy in modern and ancient shallow-marine sequences, Lect. Notes Earth Sci., 3, Springer, Berlin, Heidelberg, New York, 174 pp., 1985.

Ali, G. A., L'Heureux, C., Roy, A. G., Turmel, M.-C., and Courchesne, F.: Linking spatial patterns of perched groundwater storage and stormflow generation processes in a headwater forested catchment, Hydrol. Process, 25, 3843–3857, https://doi.org/10.1002/hyp.8238, 2011.

Allocca, V., Manna, F., and De Vita, P.: Estimating annual groundwater recharge coefficient for karst aquifers of the southern Apennines (Italy), Hydrol. Earth Syst. Sci., 18, 803–817, https://doi.org/10.5194/hess-18-803-2014, 2014.

Andreu, J. M., Alcalá, F. J., Vallejos, Á., and Pulido-Bosch, A.: Recharge to mountainous carbonated aquifers in SE Spain: Different approaches and new challenges, J. Arid Environ., 75, 1262–1270, https://doi.org/10.1016/j.jaridenv.2011.01.011, 2011.

Bachmair, S., Weiler, M., and Nützmann, G.: Controls of land use and soil structure on water movement: Lessons for pollutant transfer through the unsaturated zone, J. Hydrol., 369, 241–252, https://doi.org/10.1016/j.jhydrol.2009.02.031, 2009.

Backhaus, K., Erichson, B., Plinke, W., and Weiber, R.: Multivariate Analysemethoden: eine anwendungsorientierte Einführung, 14, Springer, Berlin, Heidelberg, 647 pp., 2016.

Bakalowicz, M.: Water geochemistry: water quality and dynamics, Groundw. Ecol., 1, 97–127, 1994.

Blouin, M., Hodson, M. E., Delgado, E. A., Baker, G., Brussaard, L., Butt, K. R., Dai, J., Dendooven, L., Peres, G., Tondoh, J. E., Cluzeau, D., and Brun, J.-J.: A review of earthworm impact on soil function and ecosystem services, Eur. J. Soil. Sci., 64, 161–182, https://doi.org/10.1111/ejss.12025, 2013.

Borkhataria, R., Aigner, T., Poeppelreiter, M. C., and Pipping, J. C. P.: Characterization of epeiric "layer-cake" carbonate reservoirs: Upper Muschelkalk (Middle Triassic), the Netherlands, J. Petrol. Geol., 28, 15–42, https://doi.org/10.1111/j.1747-5457.2005.tb00076.x, 2005.

Brandtner, W.: Böden, in: Erläuterungen zur Geologischen Karte 1:25 000 von Thüringen, Blatt 4729 (Schlotheim) und 4730 (Ebeleben) edited by: Grumbt, E., Kästner, H., and Lützner, H., 210–216, Thüringer Landesanstalt für Umwelt und Geologie, Weimar, 1997.

Bullmann, H.: Eigenschaften und Genese periglazialer Deckschichten des Muschelkalks in einem Teilgebiet der Ostthüringer Triaslandschaft, PhD-thesis University of Leipzig, 210 pp., 2010.

Canora, F., Fidelibus, M. D., Sciortino, A., and Spilotro, G.: Variation of infiltration rate through karstic surfaces due to land use changes, Eng. Geol., 99, 210–227, https://doi.org/10.1016/j.enggeo.2007.11.018, 2008.

Cattell, R. B.: The scree test for the number of factors, Multivar. Behav. Res., 1, 245–276, https://doi.org/10.1207/s15327906mbr0102_10, 1966.

Cook, P. G.: A guide to regional groundwater flow in fractured rock aquifers, 151 pp., Australia, CSIRO, 2003.

Deepa, S., Venkateswaran, S., Ayyandurai, R., Kannan, R., and Prabhu, M. V.: Groundwater recharge potential zones mapping in upper Manimuktha Sub basin Vellar river Tamil Nadu India using GIS and remote sensing techniques, Modeling Earth Systems and Environment, 2, 1–13, 2016.

Deumlich, D., Schmidt, R., and Sommer, M.: A multiscale soil-landform relationship in the glacial-drift area based on digital terrain analysis and soil attributes, J. Plant. Nutr., 173, 843–851, https://doi.org/10.1002/jpln.200900094, 2010.

Deurer, M., Green, S. R., Clothier, B. E., Böttcher, J., and Duijnisveld, W. H. M.: Drainage networks in soils: A concept to describe bypass-flow pathways, J. Hydrol., 272, 148–162, 2003.

Deutsche Stratigraphische Kommission: Stratigraphische Tabelle von Deutschland (STD 2002), 1. Edition, Stein, Potsdam, 2002.

Dibbern, D., Schmalwasser, A., Lüders, T., and Totsche, K. U.: Selective transport of plant root-associated bacterial populations in agricultural soils upon snowmelt, Soil Biol. Biochem., 69, 187–196, 2014.

DIN EN ISO 14689-1: Geotechnical investigation and testing – Identification and classification of rock – Part 1: Identification and description (ISO 14689-1:2003), 26, 2013.

Doerfliger, N., Jeannin, P. Y., and Zwahlen, F.: Water vulnerability assessment in karst environments: a new method of defining protection areas using a multi-attribute approach and GIS tools (EPIK method), Environ. Geol., 39, 165–176, 1999.

Dreybroth, W.: Processes in karst systems – physics, chemistry and geology, Springer, Heidelberg, New York, 288 pp., 1988.

Dunham, R. J.: Classification of carbonate rocks according to depositional texture, in: Classification of carbonate rocks, edited by: Ham, W. E., AAPG Memoir, 1, 108–121, 1962.

Edmunds, W. M. and Smedley, P. L.: Residence time indicators in groundwater: the East Midlands Triassic sandstone aquifer, Appl. Geochem., 15, 737–752, 2000.

Elrick, D. E. and Reynolds, W. D.: Methods for Analyzing Constant-Head Well Permeameter Data, Soil Sci. Soc. Am. J., 56, 320–323, 1992.

Fier, A.: Entstehung, Eigenschaften und Klassifikation tonreicher Unterbodenhorizonte auf Carbonatgesteinen in norddeutschen Berg- und Hügelländern, PhD-thesis, University of Halle, 190 pp., 2012.

Ford, D. C. and Williams, P. W.: Karst Hydrogeology and Geomorphology, Wiley, Chichester, 561 pp., 2007.

Furtak, H. and Langguth, H. R.: Zur hydrochemischen Kennzeichnung von Grundwässern und Grundwassertypen mittels Kennzahlen, Mem. IAH-Congress, Hannover, 86–96, 1967.

Garleb, H.: Zu einigen hydrogeochemischen Problemen bei der Erkundung von Grundwasser in der Trias des Thüringer Beckens, Mitteilungsblatt des Thüringischen Geologischen Vereins, 15, 41–45, 2002.

Gleeson, T., Novakowski, K., and Kyser, T. K.: Extremely rapid and localized recharge to a fractured rock aquifer, J. Hydrol., 376, 496–509, 2009.

Gleixner, G.: Soil organic matter dynamics: a biological perspective derived from the use of compound-specific isotopes studies, Ecol. Res., 28, 683–695, https://doi.org/10.1007/s11284-012-1022-9, 2013.

Goldscheider, N.: Fold structure and underground drainage pattern in the alpine karst system Hochifen-Gottesacker, Eclogae Geol. Helv., 98, 1–17, 2005.

Goldscheider, N., Klute, M., Sturm, S., and Hötzl, H.: The PI method – a GIS-based approach to mapping groundwater vulnerability with special consideration of karst aquifers, Z. Angew. Geol., 46, 157–166, 2000.

Goldscheider, N. and Drew, D.: Methods in Karst Hydrogeology, International Contributions to Hydrogeology, 26, Taylor & Francis, Lodon, Leiden, New York, Philadelphia, Singapore, 264 pp., 2007.

Gradstein, F. M., Ogg, J. G., and Smith, A. G.: A Geologic Time Scale, 1, Cambridge University Press, Cambridge, 2004.

Greitzke, A. and Fiedler, H. J.: Schuttdecken und Bodentypen entlang einer Catena auf Muschelkalk in Buchenbeständen des Hainich (Nordwest-Thüringen), Archiv für Naturschutz und Landschaftforschung, 34, 257–268, 1996.

Grumbt, E., Kästner, H., and Lützner, H.: Erläuterungen zur Geologischen Karte 1 : 25 000 von Thüringen, Blatt 4729 (Schlotheim) und 4730 (Ebeleben), Thüringer Landesanstalt für Umwelt und Geologie, Weimar, 262 pp., 1997.

Guggenberger, G. and Kaiser, K.: Dissolved organic matter in soil: challenging the paradigm of sorptive preservation, Geoderma, 113, 293–310, https://doi.org/10.1016/s0016-7061(02)00366-x, 2003.

Haag, D. and Kaupenjohann, M.: Landscape fate of nitrate fluxes and emissions in Central Europe: A critical review of concepts, data, and models for transport and retention, Agr. Ecosyst. Environ., 86, 1–21, 2001.

Hecht, G.: Grundwässer, in: Geologie von Thüringen, edited by: Seidel, G., Schweizerbart, Stuttgart, 529–586, 2003.

Heinz, J. and Aigner, T.: Hierarchical dynamic stratigraphy in various Quaternary gravel deposits, Rhine glacier area (SW Germany): implications for hydrostratigraphy, Int. J. Earth Sci., 92, 923–938, 2003.

Hem, J. D.: Study and Interpretation of the chemical characteristics of natural water, third edition, U.S. Geological Survey, Water Supply Paper 2254, U.S. Geological Survey, Alexandria, 263 pp., 1985.

Hiekel, W., Fritzlar, F., Nöllert, A., and Westhus, W.: Die Naturräume Thüringens, Naturschutzreport 21, Thüringer Landesantstalt für Umwelt und Geologie, Jena, 364 pp., 2004.

Hölting, B., Haertlé, T., Hohberger, K. H., Nachtigall, K. H., Villinger, E., Weinzierl, W., and Wrobel, J. P.: Konzept zur Ermittlung der Schutzfunktion der Grundwasserüberdeckung, Geol. Jb., C63, 5–24, Hannover, 1995.

Hoppe, W.: Die hydrogeologischen Grundlagen der Wasserversorgung in Thüringen, Fischer, Jena, 109 pp., 1952.

Hoppe, W.: Die Bedeutung der herzynischen Störungszonen für die Grundwasserführung des Thüringer Beckens, Geologie, 11, 679–699, 1962.

Hoppe, W. and Seidel, G.: Geologie von Thüringen, Haack, Gotha, 1000 pp., 1974.

Hsu, C.-H., Han, S.-T., Kao, Y.-H., and Liu, C.-W.: Redox characteristics and zonation of arsenic-affected multi-layers aquifers in the Choushui River alluvial fan, Taiwan, J. Hydrol., 391, 351–366, 2010.

IUSS Working Group WRB: World reference base for soil resources (WRB), World soil resources reports, FAO, Rom, 103 pp., 2006.

Jeong, C. H.: Effect of land use and urbanization on hydrochemistry and contamination of groundwater from Taejon area, Korea, J. Hydrol., 253, 194–210, 2001.

Jordan, H. and Weder, H. J.: Hydrogeologie und Methoden, Regionale Hydrogeologie: Mecklenburg-Vorpommern, Brandenburg und Berlin, Sachsen-Anhalt, Sachsen, Thüringen, Enke, Stuttgart, 601 pp., 1995.

Kaiser, E.: Erläuterungen zur geologischen Karte von Preussen und benachbarten Bundesstaaten, Blatt 4828 (Langula), Königlich Preussische Geologische Landesanstalt und Bergakademie, Berlin, 58 pp., 1905.

Khan, M. M. A. and Umar, R.: Significance of silica analysis in groundwater in parts of Central Ganga Plain, Uttar Pradesh, India, Curr. Sci., 98, 1237–1240, 2010.

Klaus, J., Zehe, E., Elsner, M., Külls, C., and McDonnell, J. J.: Macropore flow of old water revisited: experimental insights from a tile-drained hillslope, Hydrol. Earth Syst. Sci., 17, 103–118, https://doi.org/10.5194/hess-17-103-2013, 2013.

Kleber, A.: Gliederung und Eigenschaften der Hang-Schuttdecken und ihre Bedeutung für die Bodengenese, Mitteilungen der deutschen bodenkundlichen Gesellschaft, 66, 807–809, 1991.

Kley, J. and Voigt, T.: Late Cretaceous intraplate thrusting in central Europe: Effect of Africa-Iberia-Europe convergence, not Alpine collision, Geology, 36, 839–842, https://doi.org/10.1130/G24930A.1, 2008.

Klimchouk, A. B.: Conceptualisation of speleogenesis in multistorey artesian systems: a model of transverse speleogenesis, Int. J. Speleol., 34, 45–64, https://doi.org/10.5038/1827-806X.34.1.4, 2005.

Klimchouk, A. B. and Ford, D. C.: Types of Karst and Evolution of Hydrogeologic Setting, in: Speleogenesis – Evolution of karst aquifers, edited by: Klimchouk, A. B., Ford, D. C., Palmer, A. N. and Dreybrodt, W., and Huntsville, A., National Speleological Society of America, 45–53, 2000.

König, E.: Das Hainichgebiet und seine Karsterscheinungen, Thüringer Höhlen, 4, 92–101, 1930.

Kostic, B. and Aigner, T.: Sedimentary and poroperm anatomy of shoal-water carbonates (Muschelkalk, South-German Basin): an outcrop-analogue study of inter-well spacing scale, Facies, 50, 113–131, https://doi.org/10.1007/s10347-004-0003-z, 2004.

Kottek, M. J., Grieser, C., Beck, B., Rudolf, F., and Rubel, F.: World map of the Koeppen-Geiger climate classification updated, Meteorol. Z., 15, 259–263, https://doi.org/10.1127/0941-2948/2006/0130, 2006.

Kralik, M., Zieritz, I., Grath, J., Vincze, G., Philippitsch, R., and Pavlik, H.: Hydrochemische Karte Österreichs: Oberflächennaher Grundwasserkörper und Fliessgewässer, Mittelwerte von Wassererhebungsdaten (WGEV-Daten) 1991–2001, Berichte BE-269, 2nd revised edition, Vienna, 14 pp., 2005.

Kunkel, R., Voigt, H.-J., Wendland, F., and Hannappel, S.: Die natürliche, ubiquitär überprägte Grundwasserbeschaffenheit in Deutschland, Schriften des Forschungszentrums Jülich, Reihe Umwelt/Environment, 47, Forschungszentrum Jülich, Jülich, 204 pp., 2004.

Küsel, K., Totsche, K. U., Trumbore, S. E., Lehmann, R., Steinhäuser, C., and Herrmann, M.: How Deep Can Surface Signals Be Traced in the Critical Zone? Merging Biodiversity with Biogeochemistry Research in a Central German Muschelkalk Landscape, Front. Earth Sci., 32, 1–18, https://doi.org/10.3389/feart.2016.00032, 2016.

Lazar, C. S., Stoll, W., Lehmann, R., Herrmann, M., Schwab, V. F., Akob, D. M., Nawaz, A., Wubet, T., Buscot, F., Totsche, K. U., and Küsel, K.: Archaeal Diversity and CO_2 Fixers in Carbonate-/Siliciclastic-Rock Groundwater Ecosystems, Archaea, 2017, 2136287, https://doi.org/10.1155/2017/2136287, 2017.

Lucia, F. J.: Petrophysical parameters estimated from visual descriptions of carbonate rocks: A field classification of carbonate pore space, J. Petrol. Technol., 216, 221–224, https://doi.org/10.2118/10073-PA, 1983.

Malcher, G.: Karstland Thüringen – Morphologie Hydrologie in Thüringen, in: Thüringen – Karst und Höhle 2011–2014, edited by: Verband der Deutschen Höhlen und Karstforscher, München, 9–24, 2014.

Matthess, G.: Lehrbuch der Hydrogeologie, Band 2, Die Beschaffenheit des Grundwassers, Bornträger, Berlin, Stuttgart, 483 pp., 1994.

Meier, S., Bauer, J. F., and Philipp, S. L.: Fault zone characteristics, fracture systems and permeability implications of Middle Triassic Muschelkalk in Southwest Germany, J. Struct. Geol., 70, 170–189, 2015.

Mempel, G.: Die hydrogeologischen Verhältnisse von Nordwest-Thüringen, Jahrbuch der Preußischen Geologischen Landesanstalt, 59, Preußische Geologische Landesanstalt, Berlin, 587–646, 1939.

Merz, G.: Zur Petrographie, Stratigraphie, Paläogeographie und Hydrogeologie des Muschelkalks (Trias) im Thüringer Becken, Zeitschrift der geologischen Wissenschaften, 15, 457–473, 1987.

Moore, P. J., Martin, J. B., and Screaton, E. J.: Geochemical and statistical evidence of recharge, mixing, and controls on spring discharge in an eogenetic karst aquifer, J. Hydrol., 376, 443–455, https://doi.org/10.1016/j.jhydrol.2009.07.052, 2009.

Münch, J. M., Totsche, K. U., and Kaiser, K.: Physicochemical factors controlling the release of dissolved organic dissolved organic carbon release in forest subsoils – a column study, Eur. J. Soil Sci., 53, 311–320, https://doi.org/10.1046/j.1365-2389.2002.00439.x, 2002.

Muir, K. S. and Johnson, M. J.: Classification of ground-water recharge potential in three parts of Santa Cruz County, California, U.S. Geological Survey, Open-File Report 79-1065, 1979.

Nennstiel, K.: Springquellen und andere starke Quellen Thüringens, Beiträge zur Geologie von Thüringen, 3, 33–66, 1933.

Ockert, W. and Rein, S.: Biostratigraphische Gliederung des Oberen Muschelkalkes von Thüringen, Beiträge zur Geologie von Thüringen, 7, 195–228, 2000.

Odling, N. E., Gillespie, P., Bourgine, B., Castaing, C., Chiles, J. P., Christensen, N. P., Fillion, E., Genter, A., Olsen, C., Thrane, L., Trice, R., Aarseth, E., Walsh, J. J., and Watterson, J.: Variations in fracture system geometry and their implications for fluid flow in fractured hydrocarbon reservoirs, Petrol. Geosci., 5, 373–384, 1999.

Otto, C.: Historische Landschaftsanalyse im Nationalpark Hainich und deren Moeglichkeiten der Anwendung, diploma thesis, Fachhochschule Eberswalde, 87 pp., 2000.

Palmer, A. N.: Geochemical models for the origin of macroscopic solution porosity in carbonate rocks, in: Unconformities and porosity in carbonate strata, edited by: Budd A. D., Saller A. H., and Harris, P. M., AAPG Memoir, 63, Tulsa (Oklahoma), 77–101, 1995.

Patzelt, G.: Der Hainich, Cordier, Heiligenstadt, 48 pp., 1998.

Piper, A. M.: A graphic procedure in the geochemical interpretation of water analysis, Eos T. Am. Geophys. Un., 25, 914–928, https://doi.org/10.1029/TR025i006p00914, 1944.

Poser, S.: Rekonstruktion der Nutzungsgeschichte für einen Teilbereich des NLP Hainich mit Anlage vegetationskundlicher Dauerbeobachtungsflächen und Vorschlägen zur Besucherinformation, Diploma thesis, Fachhochschule Anhalt, 108 pp., 2004.

Pronk, M., Goldscheider, N., Zopfi, J., and Zwahlen, F.: Percolation and Particle Transport in the Unsaturated Zone of a Karst Aquifer, Ground Water, 47, 361–369, https://doi.org/10.1111/j.1745-6584.2008.00509.x, 2009.

Rau, D. and Unger, K. P.: Erläuterungen zur Bodenkarte von Thüringen, Blatt 4732 (Kindelbrück), Thüringer Landesanstalt für Geologie und Umwelt, Weimar, 108 pp., 1997.

Röhling, S. and Safar, D.: Forstgeschichte und waldbauliche Nutzung des Weberstedter Holzes ab dem 18. Jahrhundert im heutigen Nationalpark Hainich, Belegarbeit Lehrgebiet Forstgeschichte, University of Dresden, 24 pp., 2004.

Schrumpf, M., Kaiser, K., Guggenberger, G., Persson, T., Kögel-Knabner, I., and Schulze, E.-D.: Storage and stability of organic carbon in soils as related to depth, occlusion within aggregates, and attachment to minerals, Biogeosciences, 10, 1675–1691, https://doi.org/10.5194/bg-10-1675-2013, 2013.

Schwab, V. F., Herrmann, M., Roth, V.-N., Gleixner, G., Lehmann, R., Pohnert, G., Trumbore, S., Küsel, K., and Totsche, K. U.: Functional diversity of microbial communities in pristine aquifers inferred by PLFA- and sequencing-based approaches, Biogeosciences, 14, 2697–2714, https://doi.org/10.5194/bg-14-2697-2017, 2017.

Seidel, G.: Geologie von Thüringen, Schweizerbart, Stuttgart, 601 pp., 2003

Shaban, A., Khawlie, M., and Abdallah, C.: Use of remote sensing and GIS to determine recharge potential zones: the case of Occidental Lebanon, J. Hydrol., 14, 433–443, 2006.

Sharp Jr., J. M.: A Glossary of Hydrogeological Terms, Department of Geological Sciences, The University of Texas: Austin (Texas), 63 pp., 2007.

Simon, T.: The Muschelkalk karst in southwest Germany, in: Tracer hydrology 97 – proceedings of the international symposium on water tracing, edited by: Kranjc, A., Portoroz (Slovenia), 26–31 May 1997, 297–286, 1997.

Smart, P. L. and Hobbs, S. L.: Characterization of carbonate aquifers – A conceptual basc, in: Environmental problems in karst terranes and their solutions: Bowling Green, KY, 1–14, 1986.

Steward, D. R., Yang, X., Lauwo, S. Y., Staggenborg, S. A., Macpherson, G. L., and Welch, S. M.: From precipitation to groundwater baseflow in a native prairie ecosystem: a regional study of the Konza LTER in the Flint Hills of Kansas, USA, Hydrol. Earth Syst. Sci., 15, 3181–3194, https://doi.org/10.5194/hess-15-3181-2011, 2011.

Suk, H. and Lee, K. K.: Characterization of a Ground Water Hydrochemical System Through Multivariate Analysis: Clustering Into Ground Water Zones, Ground Water, 37, 348–366, https://doi.org/10.1111/j.1745-6584.1999.tb01112.x, 1999.

Suschka, A.: Erdfälle im Nationalpark Hainich (Thüringen): Natürliche und anthropogen bedingte Veränderungen von Erdfallstandorten in die Umweltbildung, diploma thesis, University of Greifswald, 96 pp., 2007.

TLUG (Thüringer Landesanstalt für Umwelt: Grundwasser in Thüringen): Bericht zu Menge und Beschaffenheit, Thüringer Landesanstalt für Umwelt und Thüringer Landesanstalt für Geologie, Erfurt, 163 pp., 1996.

TLUG (Thüringer Landesanstalt für Umwelt und Geologie): Umwelt regional, available at: http://www.tlug--jena.de/uw_raum/umweltregional/uh/uh09.html, last access: 18 January 2016.

Totsche, K. U., Jann, S., and Kögel-Knabner, I.: Release of polycyclic aromatic hydrocarbons, dissolved organic carbon, and suspended matter from disturbed NAPL-contaminated gravelly soil material, Vadose Zone J., 5, 469–479, https://doi.org/10.2136/vzj2005.0057, 2006.

Totsche, K. U., Jann, S., and Kögel-Knabner, I.: Single event driven release of PAH, colloids and suspended matter under natural conditions, Vadose Zone J., 6, 233–243, https://doi.org/10.2136/vzj2006.0083, 2007.

Totsche, K. U., Amelung, W., Gerzabek, M. H., Guggenberger, G., Klumpp, E., Knief, C., Lehndorff, E., Mikutta, R., Peth, S., Prechtel, A., Ray, N., and Kögel-Knabner, I.: Microaggregates in soils, J. Plant. Nutr. Soil. Sc., 1–33, https://doi.org/10.1002/jpln.201600451, 2017.

Urich, P. B.: Land use in karst terrain: review of impacts of primary activities on temperate karst ecosystems, Department of Conservation, Wellington, 58 pp., 2002.

Vías, J. M., Andreo, B., Perles, M. J., Carrasco, F., Vadillo, I., and Jiménez, P.: Proposed method for groundwater vulnerability mapping in carbonate (karstic) aquifers: the COP method, Hydrogeol. J., 14, 912–925, 2006.

Voigt, T., von Eynatten, H., and Franzke, H. J.: Late Cretaceous unconformities in the Subhercynian Cretaceous Basin (Germany), Acta Geol. Pol., 54, 673–694, 2004.

Völker, C. and Völker, R.: Bemerkenswertes zu Karstquellen im Bereich Mühlhausen und des Eichsfeldes, in: Die Geologie von NW Thüringen, edited by: Thüringer Geologischer Verein (TGV), Exkursionsführer, 10–17, 2002.

Vrba, J. and Zaporozec, A.: Guidebook on mapping groundwater vulnerability, International Association of Hydrogeologists, International Contributions to Hydrogeology, 16, 131 pp., Heise, Hannover, 1994.

Ward, J. H.: Hierarchical Grouping to Optimize an Objective Function, J. Am. Statist. Assoc., 48, 236–244, 1963.

Wätzel, A.: Geologische Heimatkunde des Unstrut-Hainich-Kreises, Mühlhäuser Beiträge, 30, 20–41, 2007.

Weigand, H. and Totsche, K. U.: Flow and reactivity effects on dissolved organic matter transport in soil columns, Soil Sci. Soc. Am. J., 62, 1268–1274, https://doi.org/10.2136/sssaj1998.03615995006200050017x, 1998.

Weiler, M. and Naef, F.: An experimental tracer study of the role of macropores in infiltration in grassland soils, Hydrol. Process., 17, 477–493, https://doi.org/10.1002/hyp.1136, 2003.

Weiss, A. D.: Topographic Position and Landforms Analysis, Conference Poster, ESRI International User Conference, San Diego, CA, USA, 2001.

White, W. B.: Conceptual Models for Carbonate Aquifers (reprint 2012), Ground Water, 7, 180–186, 1969.

Wienhöfer, J. and Zehe, E.: Predicting subsurface stormflow response of a forested hillslope – the role of connected flow paths, Hydrol. Earth Syst. Sci., 18, 121–138, https://doi.org/10.5194/hess-18-121-2014, 2014.

Witkowski, A. J., Rubin, K., Kowalczyk, A., Różkowski, A., and Wrobel, J.: Groundwater vulnerability map of the Chrzanów karst-fissured Triassic aquifer (Poland), Env. Geol., 44, 59–67, https://doi.org/10.1007/s00254-002-0735-4, 2003.

Wong, C. I., Mahler, B. J., Musgrove, M., and Banner, J. L.: Changes in sources and storage in a karst aquifer during a transition from drought to wet conditions, J. Hydrol., 468, 159–172, 2012.

Worthington, S. R. H.: A Comprehensive strategy for understanding flow in carbonate aquifer, in: Karst Modeling, Proceedings of the symposium held 24–27 February, 1999 Charlottesville (Virginia), edited by: Palmer, A. N., Palmer, M. V., and Sasowsky, I. D., Karst Waters Institute Special Publication 5, Karst Waters Institute, Charles Town (West Virginia), 30–37, 1999.

Permissions

All chapters in this book were first published in HESS, by Copernicus Publications; hereby published with permission under the Creative Commons Attribution License or equivalent. Every chapter published in this book has been scrutinized by our experts. Their significance has been extensively debated. The topics covered herein carry significant findings which will fuel the growth of the discipline. They may even be implemented as practical applications or may be referred to as a beginning point for another development.

The contributors of this book come from diverse backgrounds, making this book a truly international effort. This book will bring forth new frontiers with its revolutionizing research information and detailed analysis of the nascent developments around the world.

We would like to thank all the contributing authors for lending their expertise to make the book truly unique. They have played a crucial role in the development of this book. Without their invaluable contributions this book wouldn't have been possible. They have made vital efforts to compile up to date information on the varied aspects of this subject to make this book a valuable addition to the collection of many professionals and students.

This book was conceptualized with the vision of imparting up-to-date information and advanced data in this field. To ensure the same, a matchless editorial board was set up. Every individual on the board went through rigorous rounds of assessment to prove their worth. After which they invested a large part of their time researching and compiling the most relevant data for our readers.

The editorial board has been involved in producing this book since its inception. They have spent rigorous hours researching and exploring the diverse topics which have resulted in the successful publishing of this book. They have passed on their knowledge of decades through this book. To expedite this challenging task, the publisher supported the team at every step. A small team of assistant editors was also appointed to further simplify the editing procedure and attain best results for the readers.

Apart from the editorial board, the designing team has also invested a significant amount of their time in understanding the subject and creating the most relevant covers. They scrutinized every image to scout for the most suitable representation of the subject and create an appropriate cover for the book.

The publishing team has been an ardent support to the editorial, designing and production team. Their endless efforts to recruit the best for this project, has resulted in the accomplishment of this book. They are a veteran in the field of academics and their pool of knowledge is as vast as their experience in printing. Their expertise and guidance has proved useful at every step. Their uncompromising quality standards have made this book an exceptional effort. Their encouragement from time to time has been an inspiration for everyone.

The publisher and the editorial board hope that this book will prove to be a valuable piece of knowledge for researchers, students, practitioners and scholars across the globe.

List of Contributors

Z. Zhang and H. Hu
State Key Laboratory of Hydroscience and Engineering, Department of Hydraulic Engineering, Tsinghua University, Beijing 100084, China

F. Tian
State Key Laboratory of Hydroscience and Engineering, Department of Hydraulic Engineering, Tsinghua University, Beijing 100084, China
State Key Laboratory of Hydroscience and Engineering, Department of Hydraulic Engineering, Tsinghua University, Beijing, China

X. Yao
Xinjiang Tarim River Basin Management Bureau, Bayangol Prefecture, Xinjiang 841000, China

M. Sivapalan
Department of Civil and Environmental Engineering, Department of Geography and Geographic Information Science, University of Illinois at Urbana-Champaign, Urbana, IL 61801, USA

Christian Reszler and Hermann Stadler
JR-AquaConSol GmbH, Steyrergasse 21, 8010 Graz, Austria

Jürgen Komma and Günter Blöschl
Vienna University of Technology, Institute for Hydraulic and Water Resources Engineering, and Centre for Water Resource Systems, Karlsplatz 13, 1040 Vienna, Austria

Elmar Strobl
Technisches Büro – Ingenieurbüro Mag. Dr. Elmar Strobl, Grubbergweg 4, 8062 Kumberg, Austria

Chinchu Mohan, Andrew W. Western and Margarita Saft
Department of Infrastructure Engineering, University of Melbourne, Melbourne, Victoria, Australia

Yongping Wei
School of Geography, Planning and Environmental Management, The University of Queensland, Brisbane, Australia

Paula Rodríguez-Escales and Arnau Canelles
Dept. of Civil and Environmental Engineering. Universitat Politècnica de Catalunya, Barcelona, Spain Hydrogeology Group (UPC-CSIC)

Xavier Sanchez-Vila and Albert Folch
Dept. of Civil and Environmental Engineering. Universitat Politècnica de Catalunya, Barcelona, Spain Hydrogeology Group (UPC-CSIC)

Daniel Kurtzman
Institute of Soil, Water and Environmental Sciences, The Volcani Center, Agricultural Research Organization, Rishon LeZion, Israel

Rudy Rossetto
Institute of Life Sciences, Scuola Superiore Sant'Anna, Pisa, Italy

Enrique Fernández-Escalante and Jon San-Sebastián
Empresa de Transformación Agraria (TRAGSA), R&D department, Madrid, Spain

João-Paulo Lobo-Ferreira
Laboratorio Nacional de Engenharia Civil, Lisbon, Portugal

Manuel Sapiano
Energy and Water Agency, Luqa, Malta

Christoph Schüth
Institute of Applied Geosciences, Technische Universität Darmstadt, Darmstadt, Germany

Yoram Rubin, Ching-Fu Chang, Jiancong Chen, Karina Cucchi, Bradley Harken and Heather Savoy
Civil and Environmental Engineering, University of California, Berkeley, 94720, USA

Falk Heße
Computational Hydrosystems, Helmholtz Centre for Environmental Research (UFZ), 04318 Leipzig, Germany

Etienne Bresciani
National Centre for Groundwater Research and Training, School of the Environment, Flinders University, Adelaide, SA 5001, Australia
Korea Institute of Science and Technology, Seoul, 02792, Republic of Korea

Roger H. Cranswick
National Centre for Groundwater Research and Training, School of the Environment, Flinders University, Adelaide, SA 5001, Australia
Department of Environment, Water and Natural Resources, Government of South Australia, Adelaide, SA 5000, Australia

Eddie W. Banks, Peter G. Cook and Okke Batelaan
National Centre for Groundwater Research and Training, School of the Environment, Flinders University, Adelaide, SA 5001, Australia

Jordi Batlle-Aguilar
National Centre for Groundwater Research and Training, School of the Environment, Flinders University, Adelaide, SA 5001, Australia
Kansas Geological Survey, University of Kansas, Lawrence, KS 66047, USA

Jost Hellwig and Kerstin Stahl
Faculty of Environment and Natural Resources, University of Freiburg, 79098 Freiburg, Germany

Adrian A. S. Barfod
Department of Groundwater and Quaternary Geological Mapping, Geological Survey of Denmark & Greenland (GEUS), C.F. Møllers Allé 8, 8000 Aarhus C, Denmark
Hydrogeophysics Group, Department of Geoscience, Aarhus University, C.F. Møllers Allé 4, 8000 Aarhus C, Denmark

Troels N. Vilhelmsen and Anders V. Christiansen
Hydrogeophysics Group, Department of Geoscience, Aarhus University, C.F. Møllers Allé 4, 8000 Aarhus C, Denmark

Flemming Jørgensen, Anne-Sophie Høyer and Ingelise Møller
Department of Groundwater and Quaternary Geological Mapping, Geological Survey of Denmark & Greenland (GEUS), C.F. Møllers Allé 8, 8000 Aarhus C, Denmark

Julien Straubhaar
Centre d'Hydrogéologie et de Géothermie (CHYN), Université de Neuchâtel, Neuchâtel, Switzerland

Hubert Savenije and Markus Hrachowitz
Water Resources Section, Faculty of Civil Engineering and Geosciences, Delft University of Technology, 2600 GA Delft, the Netherlands

Laurène Bouaziz
Water Resources Section, Faculty of Civil Engineering and Geosciences, Delft University of Technology, 2600 GA Delft, the Netherlands
Department Catchment and Urban Hydrology, Deltares, Boussinesqweg 1, 2629 HV Delft, the Netherlands

Albrecht Weerts
Department Catchment and Urban Hydrology, Deltares, Boussinesqweg 1, 2629 HV Delft, the Netherlands
Hydrology and Quantitative Water Management Group, Department of Environmental Sciences, Wageningen University, Wageningen, the Netherlands

Jaap Schellekens
Vander Sat, Wilhelminastraat 43A, 2011 VK Haarlem, the Netherlands

Eric Sprokkereef and Jasper Stam
Ministry of Infrastructure and Water Management, Zuiderwagenplein 2, 8224 AD Lelystad, the Netherlands

Antonio-Juan Collados-Lara and Leticia Baena-Ruiz
Instituto Geológico y Minero de España, Granada, Spain

Javier Senent-Aparicio
Universidad Católica de Murcia, Murcia, Spain

David Pulido-Velazquez
Instituto Geológico y Minero de España, Granada, Spain
Universidad Católica de Murcia, Murcia, Spain

Arianna Renau-Pruñonosa and Ignacio Morell
Jaume I University, Castellón, Spain

Carlos Llopis-Albert
Universitat Politècnica de València, Valencia, Spain

Marie Arnoux
GEOTOP, Université du Québec à Montréal, Montréal, Québec, H3C 3P8, Canada
GEOPS, UMR 8148, CRNS-Université Paris Saclay/ Paris-Sud, Orsay, France

Florent Barbecot
GEOTOP, Université du Québec à Montréal, Montréal, Québec, H3C 3P8, Canada

Elisabeth Gibert-Brunet and Aurélie Noret
GEOPS, UMR 8148, CRNS-Université Paris Saclay/ Paris-Sud, Orsay, France

John Gibson
Alberta Innovates Technology Futures, 3-4476 Markham Street, Victoria, BC V8Z 7X8, Canada

Index

www.ingramcontent.com/pod-product-compliance
Lightning Source LLC
LaVergne TN
LVHW080950210926
848485LV00004B/814